AF333655

Statics and Strength of Materials

IRVING GRANET, P.E.

Queensborough Community College
New York Institute of Technology

HOLT, RINEHART AND WINSTON

New York   Chicago   San Francisco   Philadelphia
Montreal   Toronto   London   Sydney
Tokyo   Mexico City   Rio de Janeiro   Madrid

This book is dedicated to my lovely and devoted wife Arlene.
Her unselfish love, devotion, patience, and forbearance
made it possible for me to complete it.

LIBRARY OF CONGRESS CATALOGING IN PUBLICATION DATA

Granet, Irving.
  Statics and strength of materials.

  Includes bibliographical references and index.
  1. Statics.   2. Strength of materials.   I. Title.
TA351.G723   1982   620.1'053   82-11878
ISBN   0-03-060309-9

CBS COLLEGE PUBLISHING
Holt, Rinehart and Winston
The Dryden Press
Saunders College Publishing

Contents

Contents

9 Torsion   221

10 Riveted and Welded Structures   239

14 Columns ... 379

15 Combined Stresses ... 399

Contents

Contents

Preface

The needs of many technology students can be met by a single textbook combining both statics and strength of materials into a single unified one semester course. This procedure is already used in technical institutes, community colleges, and senior colleges. This book has therefore been prepared with the needs of these students in mind. In addition, it is also intended that this book serve the technically minded individual who will use it for self-study or as a refresher source while pursuing other courses such as machine design, structural design, architectural technology, and so on.

The first five chapters of the book are devoted to statics, with an introductory section on the SI system of units. The adoption of this system of units throughout the world has made its ultimate adoption in the United States inevitable. However, industry in the United States has been slow to adopt the SI system of units, which has caused some concern in educational circles. In order to resolve this problem for the present text, I have adopted a dual set of units, both the conventional English units and the SI units. Illustrative problems and exercises for students to solve are given almost equally in both systems. At some time in the future, the SI system will be fully accepted by industry, and such organizations as the American Institute of Steel Construction will make available tables conforming to SI usage. Until then, the procedure that I have adopted seems to be the best course.

The second part of the book, strength of materials, has been taken, for the most part, from my earlier text, *Strength of Materials for Engineering Technology*. Portions of the material have been rearranged, rewritten, and made compatible with the needs of the user of this textbook. The proven features of all of my books have been retained in this book; it has 630 problems of which 138 are fully solved illustrative problems that are integrated with the text material; calculus is completely avoided; emphasis is placed on developing the material to fulfill the needs of the student and then to applying it to practical problems of the type that will be encountered in practice; extensive useful tables from the 8th Edition of the AISC Steel Construction Manual are given in the appendix; answers to even-numbered problems are also given in the appendix.

It is almost impossible for me to thank individually all the people who have at some time helped me to carry this work to fruition. My colleagues and friends at Queensborough Community College and New York Institute of Technology have been most supportive of this endeavor from the time of its inception to its completion. The reviewers of this book in its early version made constructive suggestions that were incorporated into the final version. Countless students also contributed to the improvement of the book as a tool for learning.

My family—my wife Arlene, and our children, Ellen, Kenny, David, and Jeffrey, has supplied the love, patience, and devotion that permitted me to complete this book. I am fortunate and blessed for their understanding support.

IRVING GRANET

Preface

Statics and Strength of Materials

Introduction to the SI System of Units

At the time of the French Revolution, the systems of weights and measures used throughout the world were an incoherent and almost hopeless jumble. International trade as well as the interchange of scientific information suffered greatly due to this condition. French scientists and scholars of this era developed a rational system of weights and measures, which was called the metric system, and which was adopted by most countries of the world. In 1960 the General Conference of Weights and Measures extensively revised and simplified the older metric system and gave it the French title, Système International d'Unités (International System of Units), commonly abbreviated SI. The latest revisions and additions were made in an international conference in 1971, and work still continues on these standards.

The SI system consists of three classes of units, namely,

1. Base units
2. Supplementary units
3. Derived units
   a. With special names
   b. Without special names

Table I.1 gives the seven base units of the SI system. Several observations concerning this table should be made. The unit of length is the metre (not meter), and the kilogram is a unit of mass, not weight. Also, symbols are never pluralized, never written with a period, and the use of upper- and lowercase symbols *must* be followed, as shown, *without exception*.

Quantity	Name of Base SI Unit	Symbol
Length	metre	m
Mass	kilogram	kg
Time	second	s
Electric current	ampere	A
Thermodynamic temperature	kelvin	K
Amount of substance	mole	mol
Luminous intensity	candela	cd

Table I.2 gives the supplementary units of the SI system. These units can be regarded as either base units or derived units.

Table I.3 gives the derived units (with and without symbols) often used in engineering mechanics. These derived units are formed by the algebraic combination of base and supplementary units. It is noted that where the unit is named for a person, the first letter of the symbol appears as a capital letter, such as N for newton. Otherwise the convention is to make the symbol lowercase.

For the engineer the greatest confusion concerned the units for mass and weight. The literature abounds with units such as slugs, pounds mass, pound force, poundal, kilogram force, kilogram mass, dyne, and so on. In the SI system, the base unit for *mass* (not weight or force) is the kilogram, which is equal to the mass of the international standard kilogram located at the International Bureau of Weights and Measures. It is used to specify the quantity of matter in a body. The mass of a body never varies, and it is independent of gravitational force.

The SI *derived* unit for force is the newton (N). The unit of force is defined from Newton's law of motion, namely, force is equal to mass times acceleration ($F = ma$). Thus by this definition, 1 newton applied to a mass of 1 kilogram gives the mass an acceleration of 1 metre per second squared ($N = kg{\cdot}m/s^2$). The newton is used in all combinations of units that include force, that is, pressure or stress (N/m^2), energy ($N{\cdot}m$), or power ($N{\cdot}m/s = W$). By this procedure, the unit of force is not related to gravity as was the older kilogram-force.

Weight is defined as a measure of gravitational force acting on a material object at a specified location. Thus weight is a force that has both a mass component and an acceleration component (gravity). Gravitational forces vary by about 0.5 percent over the earth's surface. For nonprecision measurements, these variations can normally be ignored. Thus a constant mass has an approximate constant weight on the

TABLE I.2   Supplementary SI Units

Quantity	Supplementary SI Unit	Symbol
Plane angle	radian	rad
Solid angle	steradian	sr

Introduction to the SI System of Units

TABLE I.3   Derived SI Units

Quantity	Name	Symbol	Formula	Expressed in Terms of Base Units
Acceleration	acceleration	—	m/s^2	m/s^2
Area	square metre	m^2	m^2	m^2
Density	kilogram per cubic metre	—	kg/m^3	$kg{\cdot}m^{-3}$
Energy or work	joule	J	$N{\cdot}m$	$m^2{\cdot}kg{\cdot}s^{-2}$
Force	newton	N	$m{\cdot}kg/s^2$	$m{\cdot}kg{\cdot}s^{-2}$
Length	metre	m	m	m
Mass	kilogram	kg	kg	kg
Moment	newton metre	—	$N{\cdot}m$	$m^2{\cdot}kg{\cdot}s^{-2}$
Moment of inertia of area	—	m^4	m^4	m^4
Plane angle	radian	rad	rad	rad
Power	watt	W	J/s	$m^2{\cdot}kg{\cdot}s^{-3}$
Pressure or stress	pascal	Pa	N/m^2	$N{\cdot}m^{-2}$
Rotational frequency	revolutions per second	rev/s	s^{-1}	s^{-1}
Temperature	degree celsius	°C	°C	$1°C = 1 K$
Time	second	s	s	s
Torque (see moment)	newton metre	—	$N{\cdot}m$	$m^2{\cdot}kg{\cdot}s^{-2}$
Velocity (speed)	metre per second	—	m/s	$m{\cdot}s^{-1}$
Volume	cubic metre	—	m^3	m^3

surface of the earth. The agreed standard value (standard acceleration) of gravity is 9.806 650 m/s^2. Figure I.1 illustrates the difference between mass (kilogram) and force (newton).

The term *mass*, or *unit mass*, should only be used to indicate the quantity of matter in an object, and the old practice of using weight in such cases should be avoided in engineering and scientific practice. However, since the determination of an object's mass will be accomplished by the use of a weighing process, the common usage of the term *weight* instead of mass is expected to continue, but should be avoided.

ILLUSTRATIVE PROBLEM I.1

A body has a mass of 5 kg. How much will it weigh on earth?

SOLUTION

The weight of the body will be its mass multiplied by the local acceleration of gravity, that is,

$$w = mg$$

Thus

$$w = 5 \text{ kg} \times 9.81 \, \frac{m}{s^2} = 49.05 \, \frac{kg{\cdot}m}{s^2}$$

Introduction to the SI System of Units

and since 1 N = 1 kg·m/s^2,

$$w = 49.05 \text{ N}$$

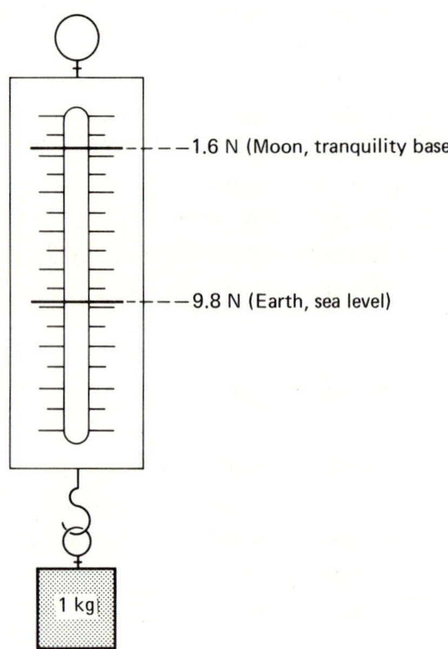

FIGURE I.1   Mass and force.

ILLUSTRATIVE PROBLEM I.2

The body of Illustrative Problem I.1 is taken to the moon, whose local acceleration of gravity is 1/6 that of the earth. What is the weight of the body on the moon?

SOLUTION

The mass of the body remains constant. The weight will depend only on the local acceleration of gravity. Since

$$w = mg$$

$$w = 5 \text{ kg} \times \frac{9.81}{6} \frac{\text{m}}{\text{s}^2} = 8.175 \frac{\text{kg·m}}{\text{s}^2}$$

or

$$w = 8.175 \text{ N}$$

which is 1/6 of the earth weight of this body.

In order for the SI system to be universally understood, it is most important that the symbols for the SI units and the conventions governing their use be strictly adhered to. Care should be taken to use the correct case for symbols, units, and their multiples,

                          Introduction to the SI System of Units

for example, K for kelvin, k for kilo, m for milli, M for mega. As noted earlier, unit *names* are never capitalized, except at the beginning of a sentence. SI unit *symbols* derived from proper names are written with the first letter in uppercase; all other symbols are written in lowercase, for example, m (metre), s (second), K (kelvin), Wb (weber). Also, unit names form their plurals in the usual manner. Unit symbols are always written in singular form, for example, 350 megapascals, or 350 MPa; 50 milligrams, or 50 mg. Since the unit symbols are standardized, the symbols should always be used in preference to the unit names. An exception is made when a number written out in words precedes the unit, such as seven metres, *not* seven m. Unit symbols are not followed by a period, unless they occur at the end of a sentence, and the numerical value associated with a symbol should be separated from that symbol by a space, for example, 1.81 mm, *not* 1.81mm. The period is only to be used as a decimal marker. Since the comma is used by some countries as a decimal marker, the SI system does not use the comma. A space is used to separate large numbers in groups of threes starting from the decimal in either direction. Thus 3 807 747.0 and 0.030 704 254 indicate this type of grouping. Notice that for numerical values less than 1, the decimal point is preceded by a zero. For a number of four digits, the space can be omitted.

In addition, certain style rules should also be adhered to.

1. When a product is to be indicated, use a space between unit names, for example, newton metre.
2. When a quotient is indicated, use the word "per," for example, metre per second.
3. When a product is indicated, use the word "square," "cubic," and so on, for example, square metre.
4. In designating the product of units, use a centered dot, for example, $N \cdot s$, $kg \cdot m$.
5. For quotients, use a solidus (/) or a negative exponent, for example, m/s or $m \cdot s^{-1}$. The solidus (/) should not be repeated in the same expression unless ambiguity is avoided by using parentheses. Thus one should use m/s^2 or $m \cdot s^{-2}$, but *not* m/s/s. Also, use $m \cdot kg/(s^3 \cdot A)$ or $m \cdot kg \cdot s^{-3} \cdot A^{-1}$, but *not* $m \cdot kg/s^3/A$.

One of the features of the older metric system and the current SI system that is most useful is the fact that multiples and submultiples of the units are in terms of factors of 10. Thus the prefixes given in Table I.4 are used in conjunction with SI units to form names and symbols of multiples of SI units. Certain general rules apply to the use of these prefixes.

1. The prefix becomes part of the name or symbol without separation, for example, kilometre, megagram.
2. Compound prefixes should not be used. Use GPa, *not* kMPa.
3. In calculations, use powers of 10 in place of prefixes.
4. Try to select a prefix where the numerical value will fall between 0.1 and 1000. This rule may be disregarded when it is better to use the same multiple

TABLE I.4   Factors of 10 for SI Units

Prefix	Symbol		Factor
tera	T	10^{12}	1 000 000 000 000
giga	G	10^{9}	1 000 000 000
mega	M	10^{6}	1 000 000
kilo	k	10^{3}	1 000
hecto	h	10^{2}	100
deka	da	10^{1}	10
deci	d	10^{-1}	0.1
centi	c	10^{-2}	0.01
milli	m	10^{-3}	0.001
micro	μ	10^{-6}	0.000 001
nano	n	10^{-9}	0.000 000 001
pico	p	10^{-12}	0.000 000 000 001
femto	f	10^{-15}	0.000 000 000 000 001
atto	a	10^{-18}	0.000 000 000 000 000 001

for all items. It is also recommended that prefixes representing 10 raised to a power that is a multiple of 3 be used, for example, 100 mg, *not* 10 cg.

5. The prefix is combined with the unit to form a new unit, which can be provided with a positive or negative exponent. Therefore mm^3 is $(10^{-3}\text{m})^3$ or 10^{-9}m^3.

6. Where possible, avoid the use of prefixes in the denominator of compound units. The exception to this rule is the prefix k in the base unit kg (kilogram).

There are certain units outside the SI that may be used together with the SI units and their multiples. These are recognized by the International Committee for Weights and Measures as having to be retained because of their practical importance. These are listed in Table I.5.

TABLE I.5   Retained Common Units

Quantity	Name of Unit	Unit Symbol	Definition
Time	minute	min	1 min = 60 s
	hour	h	1 h = 60 min = 3600 s
	day	d	1 d = 24 h = 86 400 s
Plane angle	degree	°	$1° = 1/(\pi/180)$ rad
	minute	′	$1' = (1/60)° = 2.909 \times 10^{-4}$ rad
	second	″	$1'' = (1/60)' = 4.848 \times 10^{-6}$ rad
Volume	litre	l	$1\,l = 1\ dm^3 = 10^{-3}m^3$
Mass	tonne	t	$1\ t = 1\ Mg = 10^3$ kg

Introduction to the SI System of Units

It is almost universally agreed that when a new language is to be learned the student should be completely immersed and made to "think" in the new language. This technique has been proven most effective by the Berlitz language schools and the Ulpan method of language teaching. A classical joke about this is of the American traveling in Europe who was amazed that two-year-old children were able to speak "foreign" languages. In dealing with the SI system, the student should not "think" in terms of customary units and then perform a mental conversion. It is better to learn to "think" in terms of the SI system, which will then become a second language. However, there will be times when it may be necessary to convert from customary U.S. units to SI units. In order to facilitate such conversions, Table I.6 gives some commonly used conversion factors.

TABLE I.6   Conversion Factors

Multiply	By	To Obtain
Atmospheres	2.992×10^1	Inches mercury (32°F)
Atmospheres	1.033×10^4	Kilograms per square metre
Atmospheres (760 torr)	1.013×10^2	Kilopascals
Bars	9.869×10^{-1}	Atmospheres
Bars	1.000×10^2	Kilopascals
British thermal units (Btu)	3.927×10^{-4}	Horsepower hours
British thermal units (Btu)	1.056	Kilojoules
British thermal units (Btu)	2.928×10^{-4}	Kilowatt hours
British thermal units (Btu)	1.221×10^{-8}	Megawatt days
Btu/h·ft²	3.153×10^{-4}	Watts per square centimetre
Btu/h·ft²·°F	5.676×10^{-4}	Watts per square centimetre degree Celsius
Btu/min	2.356×10^{-2}	Horsepower
Btu/min	1.757×10^1	Watts
Calories	4.190	Joules
Cubic feet	2.832×10^{-2}	Cubic metres
Cubic feet	2.832×10^1	Litres
Cubic feet per minute	4.720×10^{-4}	Cubic metres per second
Cubic metres	8.107×10^{-4}	Acre feet
Cubic metres	3.531×10^1	Cubic feet
Cubic metres	2.642×10^2	Gallons (U.S.A.)
Cubic metres per second	2.119×10^3	Cubic feet per minute
Cubic metres per second	1.585×10^4	Gallons per minute
Degrees Celsius	(9/5)°C + 32	Degrees Fahrenheit
Degrees Fahrenheit	5/9 (°F − 32)	Degrees Celsius
Feet	3.048×10^{-1}	Metres
Feet of H₂O (39.2°F)	3.048×10^2	Kilograms per square metre
Feet of H₂O (39.2°F)	4.335×10^{-1}	Pounds per square inch
Feet per second	3.048×10^{-1}	Metres per second

Introduction to the SI System of Units

Multiply	By	To Obtain
Foot pounds (force)	1.356	Joules
Foot pounds (force) per minute	2.260×10^{-2}	Watts
Gallons	3.785×10^{-3}	Cubic metres
Gallons per minute	6.309×10^{-5}	Cubic metres per second
Horsepower	4.244×10^{1}	British thermal units per minute
Horsepower	7.457×10^{-1}	Kilowatts
Horsepower hours	2.547×10^{3}	British thermal units
Horsepower hours	7.457×10^{-1}	Kilowatt hours
Inches of H_2O (39.2°F)	2.491×10^{-1}	Kilopascals
Inches mercury (32°F)	3.342×10^{-2}	Atmospheres
Inches mercury (32°F)	3.453×10^{2}	Kilograms per square metre
Inches mercury (32°F)	3.386	Kilopascals
Inches mercury (32°F)	4.912×10^{-1}	Pounds per square inch
Joules	7.376×10^{-1}	Foot pounds (force)
Joules	1.000	Watt seconds
Joules	2.387×10^{-1}	Calories
Kilograms	2.205	Pounds
Kilograms	1.102×10^{-3}	Tons (short)
Kilograms per cubic metre	6.243×10^{-2}	Pounds per cubic foot
Kilograms per square metre	9.678×10^{-5}	Atmospheres
Kilograms per square metre	3.281×10^{-3}	Feet of H_2O (39.2°F)
Kilograms per square metre	2.896×10^{-3}	Inches mercury (32°F)
Kilograms per square metre	1.422×10^{-3}	Pounds per square inch
Kilojoules	9.471×10^{-1}	British thermal units
Kilopascals	4.015	Inches of H_2O (39.2°F)
Kilopascals	1.450×10^{-1}	Pounds (force) per square inch
Kilopascals	2.953×10^{-1}	Inches mercury (32°F)
Kilopascals	1.000×10^{-2}	Bars
Kilopascals	9.869×10^{-3}	Atmospheres (760 torr)
Kilowatts	1.341	Horsepower
Kilowatt hours	3.413×10^{3}	British thermal units
Kilowatt hours	1.341	Horsepower hours
Kilowatt hours	4.167×10^{-5}	Megawatt days
Litre	3.531×10^{-2}	Cubic feet
Megawatt days	8.189×10^{7}	British thermal units
Megawatt days	2.400×10^{4}	Kilowatt hours
Metres	3.281	Feet
Newtons	2.248×10^{-1}	Pounds (force)
Pounds	4.536×10^{-1}	Kilograms

Introduction to the SI System of Units

TABLE I.6 (*continued*)

Multiply	By	To Obtain
Pounds (force)	4.448	Newtons
Pounds per cubic feet	1.602×10^1	Kilograms per cubic metre
Pounds per square inch	2.307	Feet of H_2O (39.2°F)
Pounds per square inch	2.036	Inches mercury (32°F)
Pounds per square inch	7.031×10^2	Kilograms per square metre
Pounds per square inch	6.895	Kilopascals
Square feet	9.290×10^{-2}	Square metres
Square metres	2.471×10^{-4}	Acres
Square metres	1.076×10^1	Square feet
Tonnes	2.205×10^3	Pounds
Tons (short)	9.072×10^2	Kilograms
Watts	5.688×10^{-2}	British thermal units per minute
Watts	4.427×10^1	Foot pounds (force) per minute
Watt seconds	1.000	Joules
Watts per square centimetre	3.171×10^3	Btu/h·ft^2
Watts per square centimetre degree Celsius	1.762×10^3	Btu/h·ft^2·°F

ILLUSTRATIVE PROBLEM I.3

Table I.6 lists the conversion factor for obtaining square metres from square feet as 9.290×10^{-2}. Starting with the definition that 1 in. equals 2.54 cm (0.0254 m), derive this conversion factor.

SOLUTION

In solving this type of conversion problem, it must be kept in mind that units must be consistent and that we can use the familiar rules of algebra to manipulate units.

Proceeding as noted, we have

$$1 \text{ in.} = 0.0254 \text{ m}$$

or

$$1 = 0.0254 \frac{\text{m}}{\text{in.}} \tag{a}$$

Since 1 ft = 12 in.

$$1 = 12 \frac{\text{in.}}{\text{ft}} \tag{b}$$

If we now multiply Eq. (a) by Eq. (b), we obtain

$$1 = 0.0254 \frac{\text{m}}{\text{in.}} \times 12 \frac{\text{in.}}{\text{ft}}$$

Introduction to the SI System of Units

Since inches cancel, we have

$$1 = 0.0254 \times 12 \, \frac{m}{ft}$$

We can now square both sides to obtain

$$(1)^2 = (0.0254 \times 12)^2 \, \frac{m^2}{ft^2} = 9.290 \times 10^{-2} \, \frac{m^2}{ft^2}$$

or the desired result,

$$1 \, ft^2 = 9.290 \times 10^{-2} \, m^2$$

which states that $ft^2 \times 9.290 \times 10^{-2} = m^2$.

The foregoing can also be obtained in a single chain-type calculation as

$$\left(0.0254 \, \frac{m}{in.} \times \frac{12 \, in.}{ft} \right)^2 = 9.290 \times 10^{-2} \, \frac{m^2}{ft^2}$$

REFERENCES

American Iron and Steel Institute, AISI METRIC PRACTICE GUIDE—SI UNITS AND CONVERSION FACTORS FOR THE STEEL INDUSTRY. Washington, DC: American Iron and Steel Institute, 1975.

American Society of Mechanical Engineers, ASME ORIENTATION AND GUIDE FOR USE OF SI (METRIC) UNITS, 5th ed. 1974.

American Society of Mechanical Engineers, ASME TEXT BOOKLET—SI UNITS IN STRENGTH OF MATERIALS. 1975.

Meriam, J. L., STATICS, 2nd ed., SI version. New York: John Wiley, 1975.

Walker, K. M., APPLIED MECHANICS FOR ENGINEERING TECHNOLOGY, 2nd ed. Reston, VA: Reston Publishing, 1978.

Basic Concepts

1.1 INTRODUCTION

"Statics" and "strength of materials" are two subdivisions of the general science known as "mechanics." Mechanics is defined as the study of the effects of forces on bodies. Statics is the study of bodies that are at rest or are moving with constant velocity while subjected to force systems. In engineering mechanics we are concerned with the external effects of forces acting on rigid bodies. When the changes of shape of the body and the internal state of the body due to the effects of external force systems become important, the study is then known as strength of materials. Historically the study of mechanics dates back to the time of the building of the pyramids and to the writings of Aristotle (384–322 B.C.) and Archimedes (287–212 B.C.). During the Middle Ages progress was made by Stevinus (1548–1620) who formulated the parallelogram law of the combination of vectors. Rapid advancement was made subsequently by Galileo (1564–1642), and Sir Isaac Newton (1642–1727) completed the basic formulation of mechanics by his famous laws of motion and the law of universal gravitation.

This chapter is devoted to defining certain basic concepts and to the review of the basic mathematics needed for the study of statics.

1.2 SCALARS AND VECTORS

The concepts of a quantity (scalar) and a directed quantity (vector) are generally well known and accepted by most people. The statement, a dozen eggs, is the description of the quantity of eggs in a given box, which we denote as a scalar. We can generalize this concept to define a scalar quantity as one that has magnitude only. A vector quantity, however, has both direction as well as magnitude. As an example we completely describe the motion of a train by specifying its speed (a scalar) and its direction by stating that it is going at a speed of 40 miles per hour in an easterly direction. Thus velocity, the description of the motion of a body, includes magnitude and direction and is a vector. A vector can also be described as a directed line segment of a given length. In Fig. 1.1 the line segment shown has an arrowhead placed at one end. The length of the line segment is proportional to the magnitude of the quantity that the vector describes (for example, 1 in. = 40 miles per hour), and the arrowhead denotes the direction of the motion. The direction with regard to a reference axis is given by the angle with respect to that axis, that is, 45° to the horizontal.

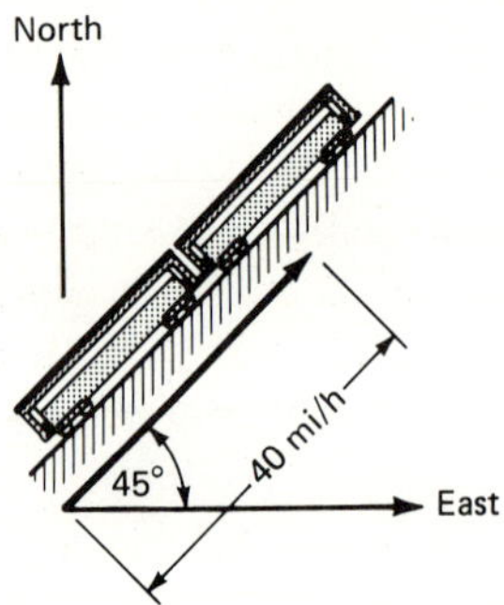

FIGURE 1.1   Vector representation of a train proceeding at 40 miles per hour in a northeasterly manner.

At this point it is necessary to distinguish between a free vector and a fixed (or localized) vector. A free vector has a specified magnitude and direction but is not defined to be acting through any specified point in space. Figure 1.2 shows a number of equivalent free vectors. A fixed or localized vector has either a specified line of action or a definite point through which it acts. Therefore the complete definition of a fixed vector must specify its magnitude, direction, and point (or line) of application.

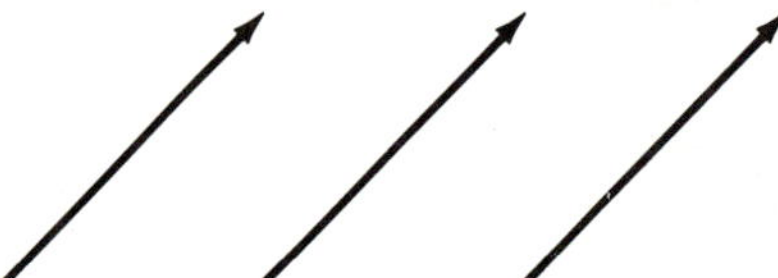

FIGURE 1.2   Free vectors.

An example of a free vector is the vector used to represent the displacement of a body that moves without rotation. The movement (or displacement) of *any* point in the body can be represented by a vector which will describe the direction and magnitude of the movement of *every* point in the body. However, if a force is applied to a body which can deform, the forces and movements internal to the body depend also upon the point of application of the force, giving rise to a fixed vector.

The concept of a force is fundamental to the study of mechanics. From everyday observation it is commonplace to define a force to be a push or pull exerted on a body. While this definition is not rigorous enough for the study of mechanics, it does embody a very fundamental idea, that is, that a force is the action of one body on another body

                                                                                 Basic Concepts

which tends to alter the state of the body being acted upon. Forces must always act in pairs since the body acted upon resists the action of the other body. It is also noted that this effect of a force on a body is dependent upon the magnitude of the force, the direction along its line of action, and the position of the line of action of the force on the body upon which it acts. Therefore a force is a fixed vector.

The external effect of a force on a rigid body does not depend upon where along its line of action the force acts on the body. A car pushed by a given force will attain the same velocity as when it is pulled by the same force acting in the same direction. However, the internal effects caused by a force acting on a body will be influenced by the location of the point of application of the force in the body. This concept as to the effect of a force on a rigid body is formalized and called the *principle of transmissibility,* which states that *the external effect of a force on a rigid body is independent of the point of application of the force along its line of action.* We can therefore take a force to be acting at any point along its line of action and find the same external effect. Figure 1.3 shows this concept applied to a car being pushed or pulled by a force having the same direction and magnitude and acting on the line of action of the force.

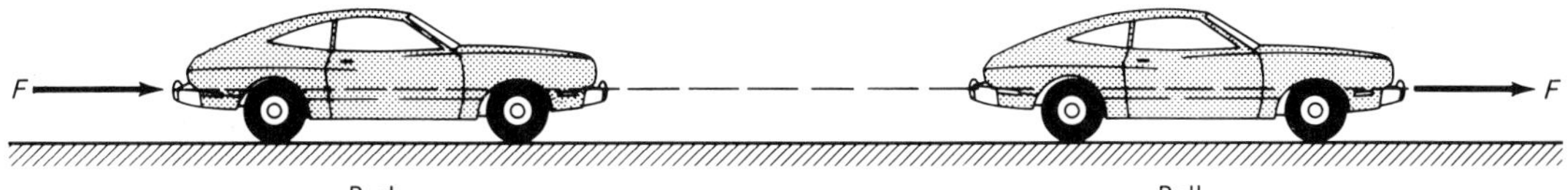

FIGURE 1.3   Principle of transmissibility.

When several forces act on a rigid body we call this situation a force system. For the purposes of our study we will classify force systems using the arrangement of the lines of force of such systems to describe them. The two general categories of force systems are coplanar systems, in which the force will lie in one plane, and noncoplanar systems, in which the forces do not lie in one plane.

Coplanar Force Systems

COLLINEAR
All forces of the system have the same line of action. Figure 1.4 shows a collinear force system acting on a body.

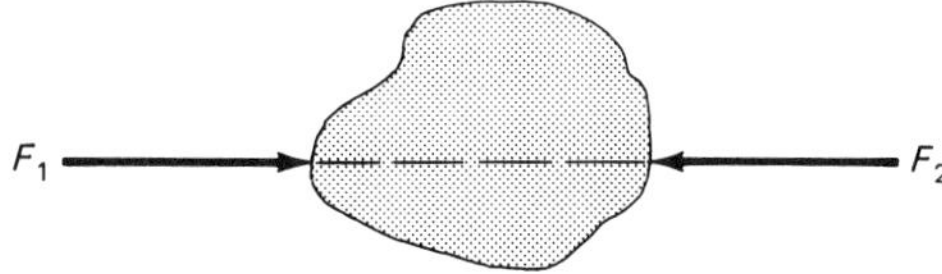

FIGURE 1.4   A collinear force system.

CONCURRENT, COPLANAR
The lines of action of all forces lie in one plane and intersect at a common point. Figure 1.5 shows this type of force system.

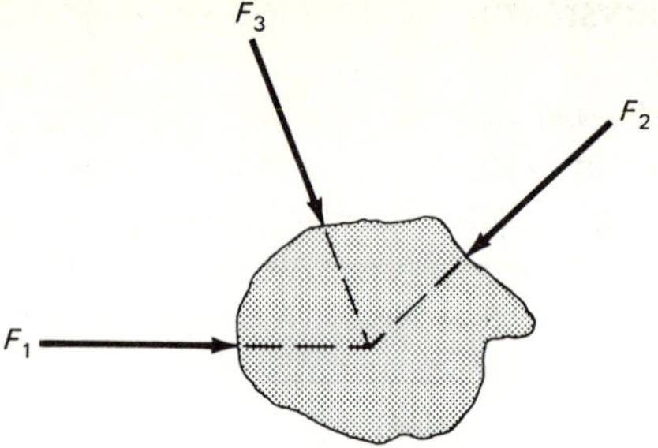

FIGURE 1.5   A concurrent, coplanar force system.

PARALLEL, COPLANAR

As the name indicates, all of the forces in the system lie in one plane and are parallel to each other, as shown in Fig. 1.6.

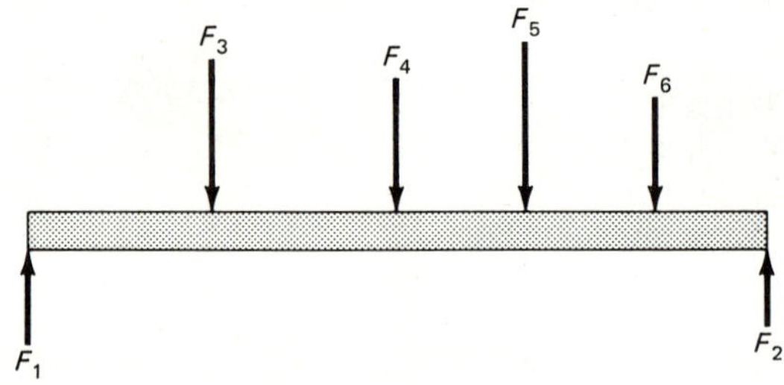

FIGURE 1.6   A parallel, coplanar force system.

NONCONCURRENT, NONPARALLEL, COPLANAR

All of the lines of action of the forces of this system lie in the same plane, but they do not intersect at a common point and are not parallel, as shown in Fig. 1.7.

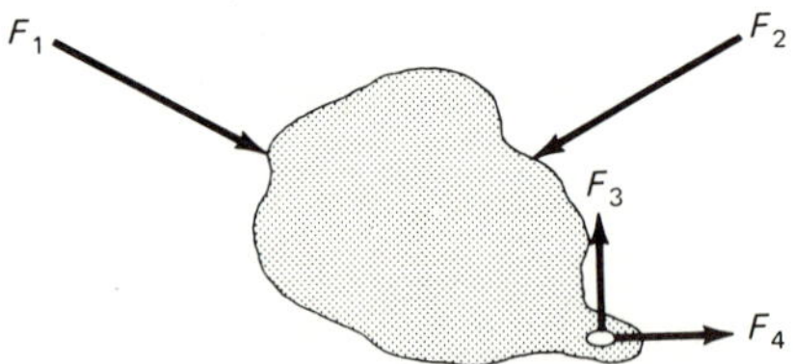

FIGURE 1.7   A nonconcurrent, nonparallel, coplanar force system.

                                    Basic Concepts

Noncoplanar Force Systems

CONCURRENT, NONCOPLANAR

All of the lines of action of this force system intersect at a common point, but the lines of action do not lie in a common plane, as shown in Fig. 1.8.

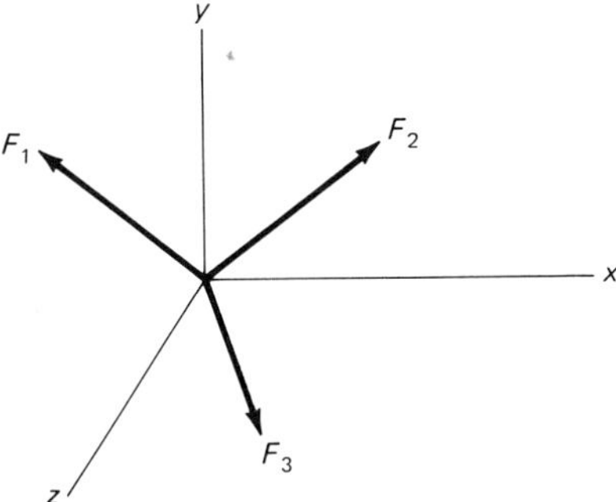

FIGURE 1.8   A concurrent, noncoplanar force system.

PARALLEL, NONCOPLANAR

While the lines of action of the forces in this system are parallel to each other they do not lie in the same plane, as shown in Fig. 1.9.

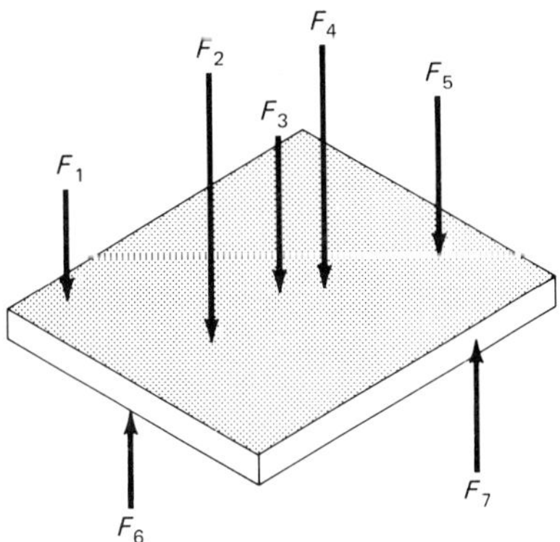

FIGURE 1.9   A parallel, noncoplanar force system.

NONPARALLEL, NONCURRENT, NONCOPLANAR

This is the most general system in which the lines of force are not parallel, do not intersect at a common point, and do not lie in one plane, as illustrated in Fig. 1.10.

Scalars and Vectors

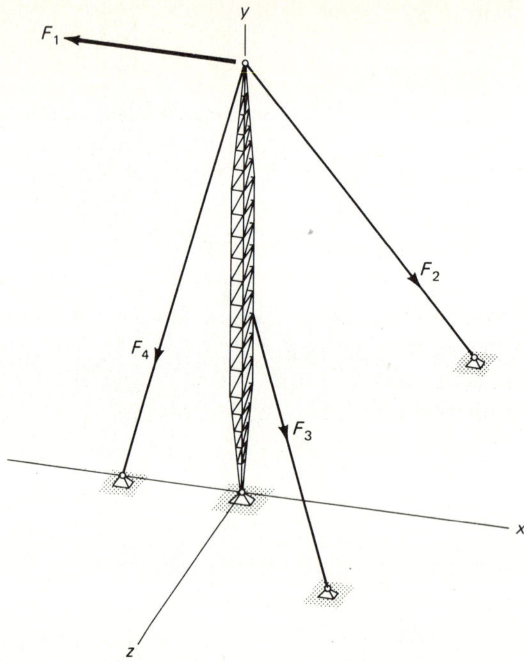

FIGURE 1.10   A nonparallel, noncurrent, noncoplanar force system.

1.3   MATHEMATICS OF STATICS

Since statics is a quantitative study, it will be necessary to utilize certain branches of mathematics to express the concepts to be developed in the succeeding chapters of this book in a quantitative manner. For the most part we will use algebra, geometry, and trigonometry as the analytic tools of our study. While most students readily recall and use algebra and geometry, there often is not the same facility with trigonometry. For this reason we will briefly review those trigonometric concepts that we will need later on.

The right triangle is the basis of the formulation of the three most common trigonometric functions, namely, sine, cosine, and tangent, and the inverse of these functions, the cosecant, secant, and cotangent. The right triangle shown in Fig. 1.11 has its angles labeled by the capital letters A, B, and C and the sides of the triangle opposite each of these angles are labeled a, b, and c, respectively. Angle C is a right angle, and the side c opposite the right angle is known as the hypotenuse of the triangle. By definition, the three functions of the angles A and B can be written as

$$\text{sine } A \text{ (or } B) = \frac{\text{side opposite}}{\text{hypotenuse}}$$

$$\text{cosine } A \text{ (or } B) = \frac{\text{side adjacent}}{\text{hypotenuse}} \tag{1.1}$$

$$\text{tangent } A \text{ (or } B) = \frac{\text{side adjacent}}{\text{side opposite}}$$

                                                Basic Concepts

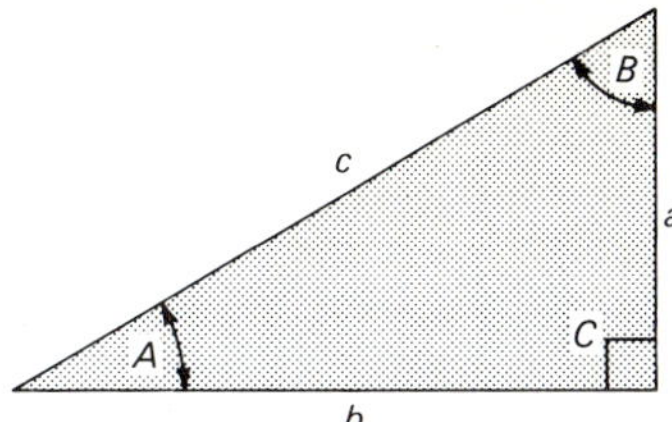

FIGURE 1.11   The right triangle.

A very common associative memory device to remember these functions is to remember SOH-CAH-TOA (sometimes said as "soak your toe"), where sine is opposite over hypotenuse, cosine is adjacent over hypotenuse, and tangent is opposite over adjacent. Using the abbreviations sin, cos, and tan we have from Eq. (1.1),

$$\sin A = \frac{a}{c} \qquad \sin B = \frac{b}{c}$$

$$\cos A = \frac{b}{c} \qquad \cos B = \frac{a}{c} \tag{1.2}$$

$$\tan A = \frac{a}{b} \qquad \tan B = \frac{b}{a}$$

Note that

$$\tan A = \frac{\sin A}{\cos A} \qquad \tan B = \frac{\sin B}{\cos B}$$

In addition, the three sides of a right triangle are related by the pythagoras relation

$$a^2 + b^2 = c^2 \tag{1.3}$$

and the angles of any triangle are related by

$$A + B + C = 180° \tag{1.4}$$

Table 1.1 gives the functions of various angles up to 90°.

Most students will have an electronic calculator available to them which will have trigonometric functions. These pocket machines will give all the required trigonometric information to an even greater accuracy than Table 1.1. It is suggested, however, that the student review the subsequent material in this chapter in order to understand it better and also not to become totally dependent on the machine.

ILLUSTRATIVE PROBLEM 1.1

A right triangle has a hypotenuse of 10 in. and one of the angles is 21°. Determine the length of each of the sides of the triangle.

Mathematics of Statics

TABLE 1.1 Trigonometric Functions

Angle	Sine	Cosine	Tangent	Angle	Sine	Cosine	Tangent
0°	0.000	1.000	0.000				
1°	0.018	1.000	0.018	46°	0.719	0.695	1.036
2°	0.035	0.999	0.035	47°	0.731	0.682	1.072
3°	0.052	0.999	0.052	48°	0.743	0.669	1.111
4°	0.070	0.998	0.070	49°	0.755	0.656	1.150
5°	0.087	0.996	0.088	50°	0.766	0.643	1.192
6°	0.105	0.995	0.105	51°	0.777	0.629	1.235
7°	0.122	0.993	0.123	52°	0.788	0.616	1.280
8°	0.139	0.990	0.141	53°	0.799	0.602	1.327
9°	0.156	0.988	0.158	54°	0.809	0.588	1.376
10°	0.174	0.985	0.176	55°	0.819	0.574	1.428
11°	0.191	0.982	0.194	56°	0.829	0.559	1.483
12°	0.208	0.978	0.213	57°	0.839	0.545	1.540
13°	0.225	0.974	0.231	58°	0.848	0.530	1.600
14°	0.242	0.970	0.249	59°	0.857	0.515	1.664
15°	0.259	0.966	0.268	60°	0.866	0.500	1.732
16°	0.276	0.961	0.287	61°	0.875	0.485	1.804
17°	0.292	0.956	0.306	62°	0.883	0.470	1.881
18°	0.309	0.951	0.325	63°	0.891	0.454	1.963
19°	0.326	0.946	0.344	64°	0.899	0.438	2.050
20°	0.342	0.940	0.364	65°	0.906	0.423	2.145
21°	0.358	0.934	0.384	66°	0.914	0.407	2.246
22°	0.375	0.927	0.404	67°	0.921	0.391	2.356
23°	0.391	0.921	0.425	68°	0.927	0.375	2.475
24°	0.407	0.914	0.445	69°	0.934	0.358	2.605
25°	0.423	0.906	0.466	70°	0.940	0.342	2.747
26°	0.438	0.899	0.488	71°	0.946	0.326	2.904
27°	0.454	0.891	0.510	72°	0.951	0.309	3.078
28°	0.470	0.883	0.532	73°	0.956	0.292	3.271
29°	0.485	0.875	0.554	74°	0.961	0.276	3.487
30°	0.500	0.866	0.577	75°	0.966	0.259	3.732
31°	0.515	0.857	0.601	76°	0.970	0.242	4.011
32°	0.530	0.848	0.625	77°	0.974	0.225	4.331
33°	0.545	0.839	0.649	78°	0.978	0.208	4.705
34°	0.559	0.829	0.675	79°	0.982	0.191	5.145
35°	0.574	0.819	0.700	80°	0.985	0.174	5.671
36°	0.588	0.809	0.727	81°	0.988	0.156	6.314
37°	0.602	0.799	0.754	82°	0.990	0.139	7.115
38°	0.616	0.788	0.781	83°	0.993	0.122	8.144
39°	0.629	0.777	0.810	84°	0.995	0.105	9.514
40°	0.643	0.766	0.839	85°	0.996	0.087	11.43
41°	0.656	0.755	0.869	86°	0.998	0.070	14.30
42°	0.669	0.743	0.900	87°	0.999	0.052	19.08
43°	0.682	0.731	0.933	88°	0.999	0.035	28.64
44°	0.695	0.719	0.966	89°	1.000	0.018	57.29
45°	0.707	0.707	1.000	90°	1.000	0.000	∞

SOLUTION

From Table 1.1, sin 21° is 0.358 and cos is 0.934. The side labeled y in Fig. 1.12 can be found using the definition of the sine.

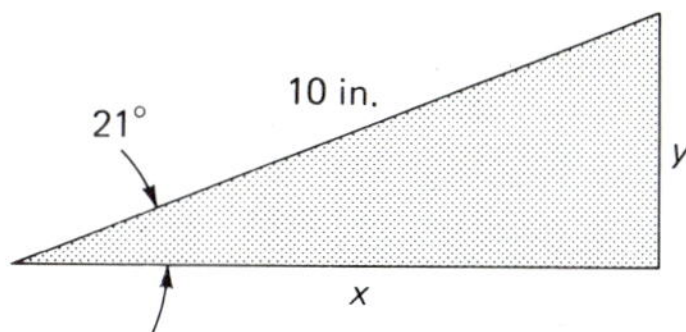

FIGURE 1.12   Illustrative Problem 1.1.

Thus, $y/10 = \sin 21°$ or $y = 10 \sin 21° = 10 \times 0.358 = 3.58$ in., and similarly using the cosine we have $x/10 = \cos 21°$ or $x = 10 \cos 21° = 10 \times 0.934 = 9.34$ in. As a check, $x^2 + y^2 = 10^2$, $3.58^2 + 9.34^2 = 87.24 + 12.82 = 100.06$. This is sufficient accuracy for this type of problem. If greater accuracy is required, it would necessitate the use of a table with more significant figures than given in Table 1.1, or a calculator.

ILLUSTRATIVE PROBLEM 1.2

If the sine of an angle is 0.510, determine the angle.

SOLUTION

From Table 1.1 we find the sine of 30° = 0.500 and the sine of 31° = 0.515. The desired angle thus lies between 30° and 31°, and it becomes necessary to interpolate between 30° and 31° to obtain it. A simple method of interpolation is shown in Fig. 1.13. The desired angle is

$$30° + \frac{0.010}{0.015} \times 1 = 30.7°$$

	Sine	Angle
0.010	0.50	30
0.015	0.510	
	0.515	31

FIGURE 1.13   Illustrative Problem 1.2.

The angles 30° and 45° will be found to recur very often in our study. These angles are shown in Fig. 1.14 as parts of right triangles. It will be noted in the 30–60–90 triangle that the side opposite the 30° is half the hypotenuse, and using Eq. (1.3) we obtain the third side as $\sqrt{3}$. In the 45–45–90 triangle we first note that each of the sides must be equal since the triangle is isosceles (since two angles are

Mathematics of Statics

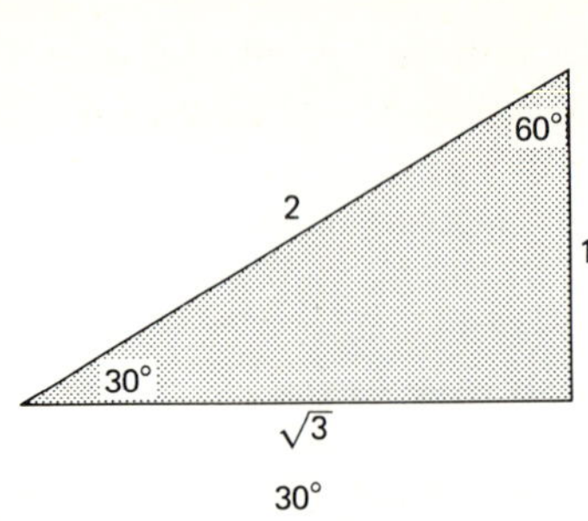

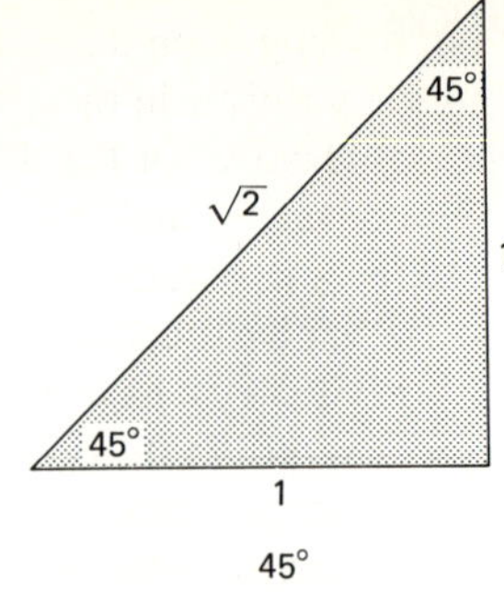

FIGURE 1.14 Useful triangle relations.

equal). If each side is denoted to be unity, then the hypotenuse is $\sqrt{2}$. From these simple figures we can readily obtain the trigonometric functions of 30°, 45°, and 60°. For reference the $\sqrt{3} = 1.732$ and $\sqrt{2} = 1.414$.

ILLUSTRATIVE PROBLEM 1.3

Determine the sin, cos, and tan of 30°, 45°, and 60°.

SOLUTION

From Fig. 1.14 we have the following directly,

$$\sin 30° = \frac{1}{2} = 0.500 \qquad\qquad \sin 60° = \frac{\sqrt{3}}{2} = \frac{1.732}{2} = 0.866$$

$$\cos 30° = \frac{\sqrt{3}}{2} = \frac{1.732}{2} = 0.866 \qquad \cos 60° = \frac{1}{2} = 0.500$$

$$\tan 30° = \frac{1}{\sqrt{3}} = \frac{1}{1.732} = 0.577 \qquad \tan 60° = \frac{\sqrt{3}}{1} = 1.732$$

$$\sin 45° = \frac{1}{\sqrt{2}} = \frac{1}{1.414} = 0.707$$

$$\cos 45° = \frac{1}{\sqrt{2}} = \frac{1}{1.414} = 0.707$$

$$\tan 45° = \frac{1}{1} = 1.0$$

The trigonometric functions studied thus far have been applied only to angles up to 90°. However, they do extend to angles larger than 90°. If we divide the 360° in a circle into four quadrants denoting quadrant I as 0°–90°, quadrant II as 90°–180°, quadrant III as 180°–270°, and quadrant IV as 270°–360°, we obtain Fig. 1.15(a). The angle θ is shown in Fig. 1.15(b), (c), (d), and (e) in each of the four quadrants. Since it is conventional to take distances to the right and up from 0 as positive, while those to the left and down are taken to be negative, we find that the functions will be positive

   Basic Concepts

or negative depending upon the quadrant under discussion. In the first quadrant all of the functions are positive; in the second quadrant the sine is positive while the cosine and tangent are negative; in the third quadrant the tangent is positive while the sine and cosine are negative; and in the fourth quadrant the cosine is positive while the sine and tangent are negative. A simple method of remembering the sign of the functions in the various quadrants is obtained by use of the word CAST [see Figure 1.15(a)]. Using this aid we have the cosine positive in the fourth quadrant, all functions positive in the first quadrant, the sine positive in the second quadrant, and the tangent positive in the third quadrant.

In order to determine the values of the functions for angles greater than 90°, it is necessary to first obtain the angle ϕ by finding the difference between the angle θ and 180° or 360°, as shown in Figure 1.15. Values of sin ϕ, cos ϕ, and tan ϕ are obtained from Table 1.1, and the appropriate algebraic signs are then assigned to the functions.

SOLUTION

Referring to Figure 1.15(c) we find ϕ to be 180° − 135° = 45°. From Table 1.1, sin 45° = 0.707, cos 45° = 0.707, and tan 45° = 1.0. Therefore sin 135° = 0.707, cos 135° = −0.707, and tan 135° = −1.0.

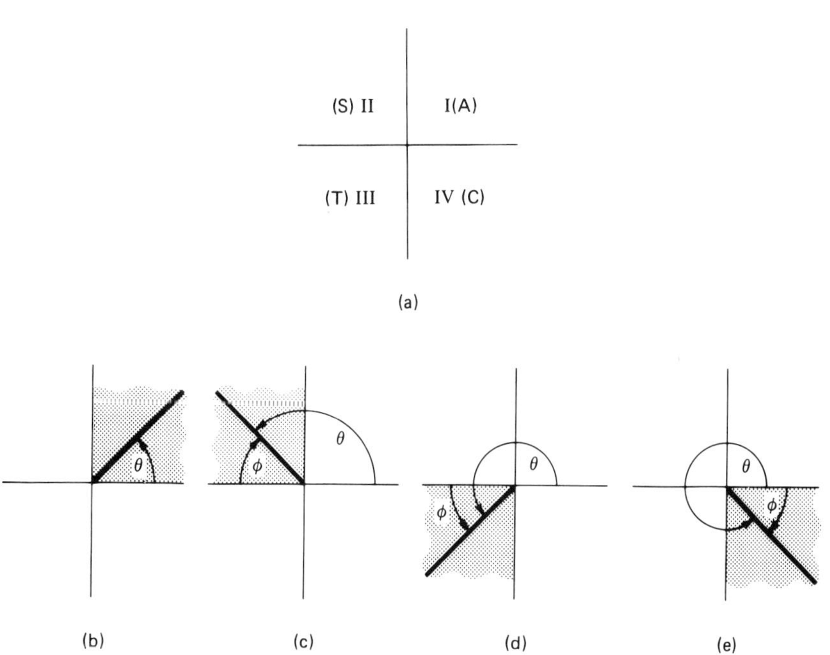

FIGURE 1.15 Trigonometric functions in all quadrants.

Before leaving the right triangle it will be noted that any triangle has five quantities that can be varied, namely three sides and two angles. In order to fully determine all of these quantities, at least three must be known (angle–side–angle, side–angle–side or side–side–side). However, in the right triangle one angle is specified to be 90°. Therefore if any combination of two variables (side and angle) are specified, the entire triangle is specified.

While the solution of a right triangle will be quite often required in our study, we will also find occasion to have to solve oblique triangles. For the case where two angles and the included side of an oblique triangle are known, the other sides can be calculated from the law of sines. Referring to Fig. 1.16, we express the law of sines as

$$\frac{a}{\sin A} = \frac{b}{\sin B} = \frac{c}{\sin C} \tag{1.5}$$

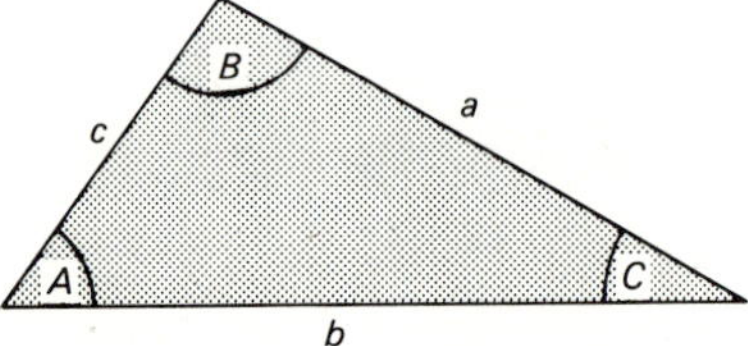

FIGURE 1.16   The oblique triangle.

ILLUSTRATIVE PROBLEM 1.5

If angle A in Fig. 1.16 is 45° while angle $B = 80°$ and $c = 5$ in., determine sides a and b.

SOLUTION

From the given angles we have angle $C = 180° - 80° - 45° = 55°$. From the law of sines,

$$\frac{a}{\sin A} = \frac{c}{\sin C} \qquad \frac{a}{\sin 45°} = \frac{5}{\sin 55°}$$

Thus

$$a = \frac{5 \sin 45°}{\sin 55°} = \frac{5\,(0.707)}{(0.819)} = 4.32 \text{ in.}$$

Side b is found in a similar manner,

$$\frac{b}{\sin B} = \frac{c}{\sin C} \qquad \frac{b}{\sin 80°} = \frac{5}{\sin 55°}$$

Thus

$$b = \frac{5 \sin 80°}{\sin 55°} = \frac{5\,(0.985)}{(0.819)} = 6.01 \text{ in.}$$

                                                        Basic Concepts

When two sides and the included angle between the sides are known (side–angle–side), the law of cosines can be used to determine the third side. Once the third side is known, the law of sines can be used to determine the rest of the angles. The law of cosines can be stated in words as: the square of any side of a triangle is equal to the sum of the squares of the other two sides minus twice their product multiplied by the cosine of the included angle. Once again we can refer to Figure 1.16 and obtain the mathematical expression for the law of cosines as

$$a^2 = b^2 + c^2 - 2bc\cos A \tag{1.6}$$

ILLUSTRATIVE PROBLEM 1.6

Two sides of a triangle are 6 in. and 5 in., respectively, and the included angle between them is 45°. Determine the length of the third side.

SOLUTION

This problem is a check of Illustrative Problem 1.5. From the data given we have

$$
\begin{aligned}
a^2 &= b^2 + c^2 - 2bc\cos A \\
&= 6^2 + 5^2 - 2 \times 6 \times 5 \cos 45° \\
&= 36 + 25 - 60 \times 0.707 \\
&= 18.58 \\
a &= 4.31 \text{ in.}
\end{aligned}
$$

which checks Illustrative Problem 1.5.

ILLUSTRATIVE PROBLEM 1.7

The three sides of a triangle are 6, 11, and 16 in., respectively, as shown in Fig. 1.17. Determine the included angle between the 6 in. and 11 in. sides.

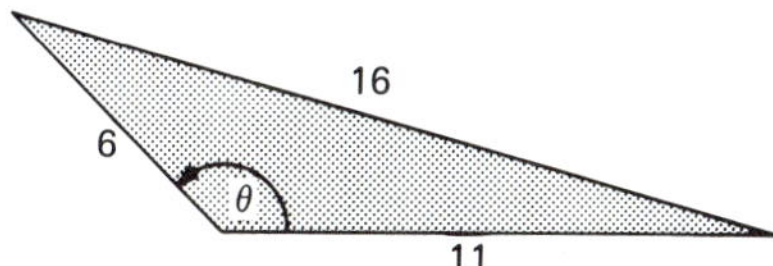

FIGURE 1.17   Illustrative Problem 1.7.

SOLUTION

Using the law of cosines, we have

$$16^2 = 11^2 + 6^2 - 2 \times 6 \times 11 \cos \theta$$

$$256 = 121 + 36 - 132 \cos \theta$$

$$132 \cos \theta = -256 + 121 + 36 = -99$$

$$\cos \theta = \frac{-99}{132} = -0.75$$

If the included angle is greater than 90°, the cosine is negative. Conversely, if the cosine is found to be negative, the included angle is greater than 90°.

Referring to Fig. 1.15(c), we have $\cos \theta = -\cos \phi$. Since $\theta + \phi = 180°$, $\cos \theta = -\cos (180 - \theta)$. The angle whose cosine is 0.75 is found from Table 1.1 by interpolation to be 41.4°. Therefore the angle $\theta = 180 - 41.4 = 138.6°$. An alternate way to obtain the angle if its cosine is negative is to look up the angle whose sign has the same positive value and add 90°. For this problem, the angle whose sine is 0.75 is 48.6°. The desired angle is 48.6 + 90 = 138.6°.

1.4 CLOSURE

The introductory concepts put forth in this chapter, as well as the brief review of some trigonometry, are intended to orient the student's thinking in a directed manner. The basic technique used in this chapter will be used throughout this book. We will first discuss and define our terms, then express these qualitative concepts in quantitative mathematical form, and finally apply these mathematical formulations to illustrative situations. Each of the three steps applied in order will help to develop each topic. The student will find that this procedure will lead to a good basic understanding of and an ability to apply the material of each section and chapter of this book.

REFERENCES

Bassin, M. G., S. M. Brodsky, and H. Wolkoff, STATICS AND STRENGTH OF MATERIALS, 3rd ed. New York: McGraw-Hill, 1979.

Beer, F. P., and E. R. Johnston, Jr., VECTOR MECHANICS FOR ENGINEERS—STATICS, 3rd ed. New York: McGraw-Hill, 1977.

Higdon, A., and W. B. Stiles, ENGINEERING MECHANICS, VOL. I—STATICS, 3rd ed. Englewood Cliffs, NJ: Prentice-Hall, 1968.

Levinson, I. J., INTRODUCTION TO MECHANICS, 2nd ed. Englewood Cliffs, NJ: Prentice-Hall, 1968.

Shames, I. H., ENGINEERING MECHANICS, VOL. I—STATICS, 2nd ed. Englewood Cliffs, NJ: Prentice-Hall, 1966.

Singer, F. L., ENGINEERING MECHANICS, 2nd ed. New York: Harper and Brothers, 1954.

White, H. E., MODERN COLLEGE PHYSICS, 5th ed. Princeton, NJ: D. Van Nostrand, 1966.

PROBLEMS

1.1 Find the sin, cos, and tan for:
   (a) 34°.
   (b) 89°.
   (c) 125°.

Basic Concepts

1.2   A right triangle has a hypotenuse of 13 in., and one side is 5 in. Determine the third
      side.

1.3   A right triangle has a hypotenuse of 21 in., and one angle is 20°. Determine the sides
      of the triangle.

1.4   A right triangle has sides of 3, 4, and 5 in. Determine the angles of the triangle.

1.5   The cos of an angle is -0.5. Determine the angle if it is in the second quadrant. What
      is the angle if it is in the third quadrant?

1.6   A ladder 35 ft long rests 18 ft high on a wall. How far from the wall is the base of the
      ladder (Fig. P 1.6)?

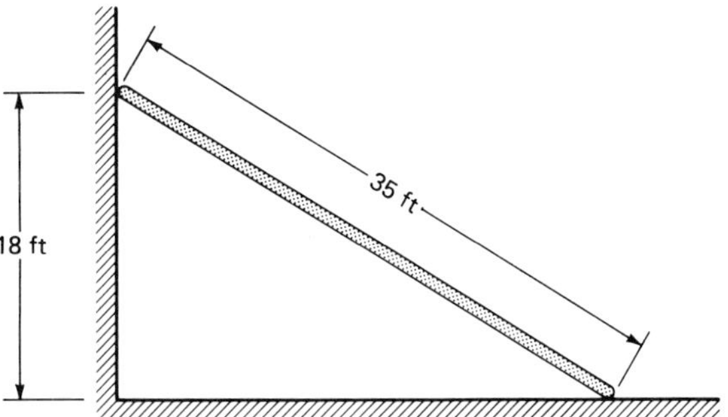

FIGURE P 1.6

1.7   A flag pole is 100 ft from an observer who sights the top of the flag pole to be at an
      angle of 50° from the horizontal. How high is the flagpole (Fig. P 1.7)?

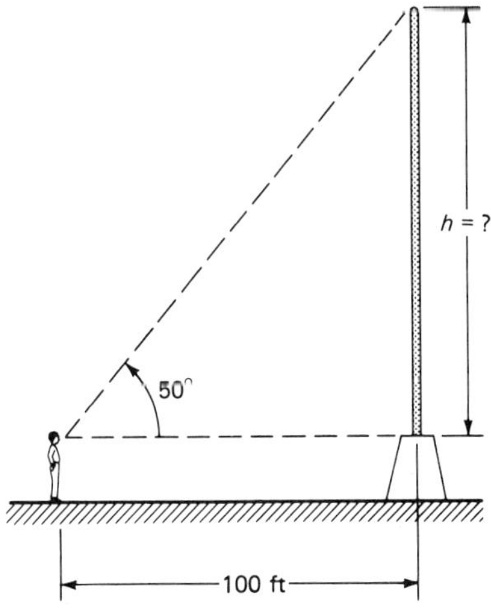

FIGURE P 1.7

1.8   A man walks 1 mi west, 2 mi north, 2 mi east, and 4 mi south. How far is he from his starting point (Fig. P 1.8)?

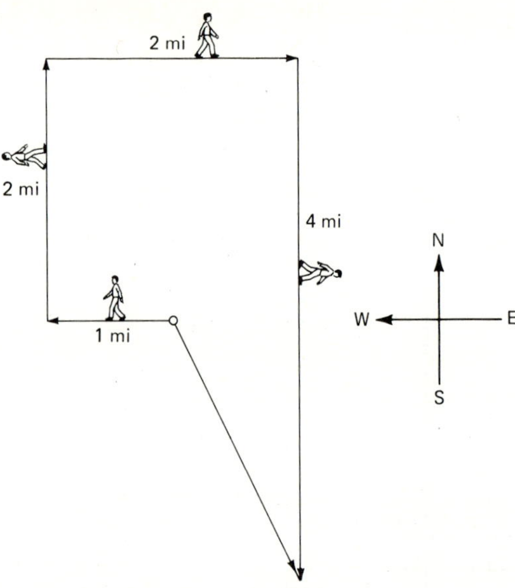

FIGURE P 1.8

1.9   In order to determine the width of a river, a base line of 100 ft is marked off on one side. On the other side of the river there is a large rock. When sighted from each end of the base line it is found that the lines of sight make angles of 60° and 75°, respectively, with the base line. Determine the width of the river (Fig. P 1.9).

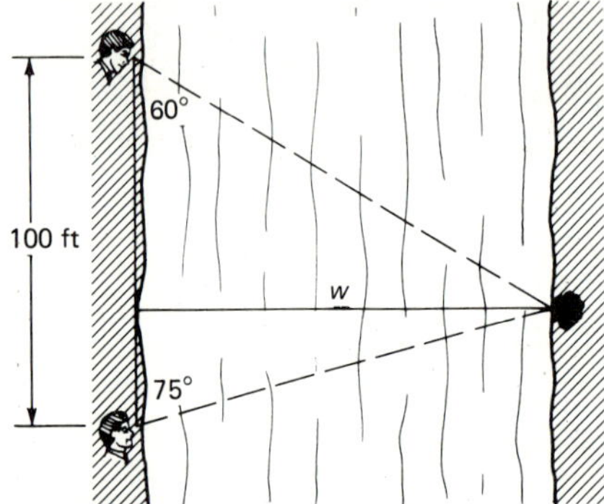

FIGURE P 1.9

1.10   A triangle has two angles of 30° and an included side of 15 in. Determine the sides of the triangle.

1.11   The shadow of a tall building is found to be 65 ft when the sun is 5° from directly overhead. Determine the height of the building (Fig. P 1.11).

1.12   Determine the sides of the triangle shown in Fig. P 1.12.

                                                                    Basic Concepts

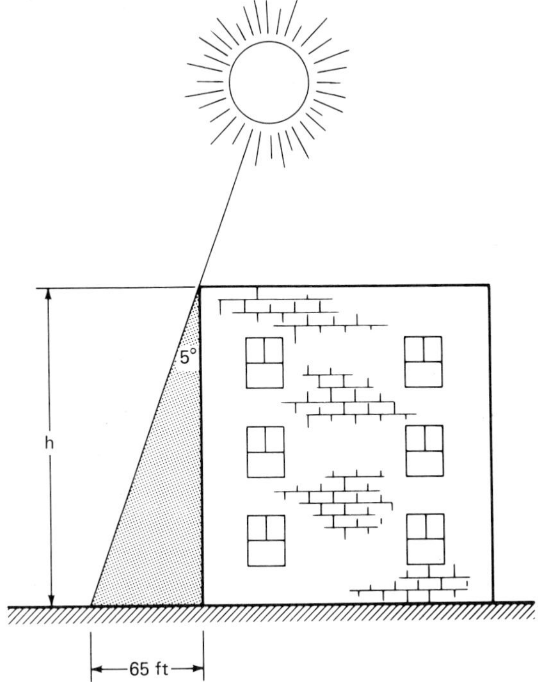

FIGURE P 1.11

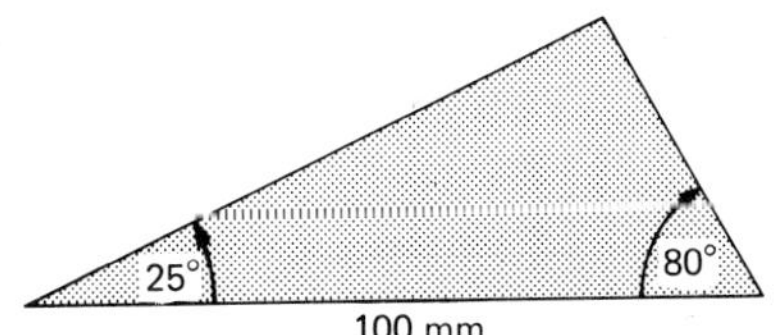

FIGURE P 1.12

1.13   Determine the angles of the triangle shown in Fig. P 1.13.

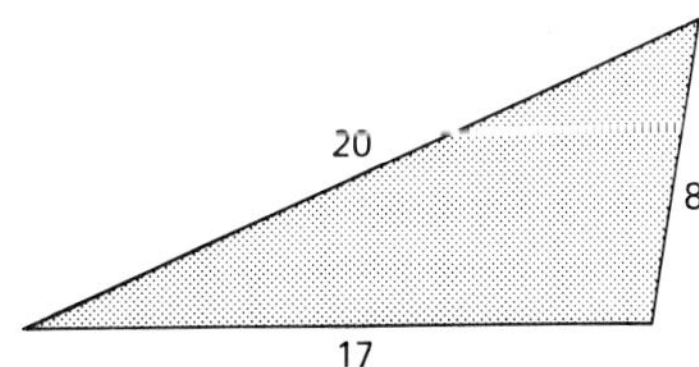

FIGURE P 1.13

1.14   Determine the unknown side of the triangle shown in Fig. P 1.14. Also determine the angles of the triangle shown.

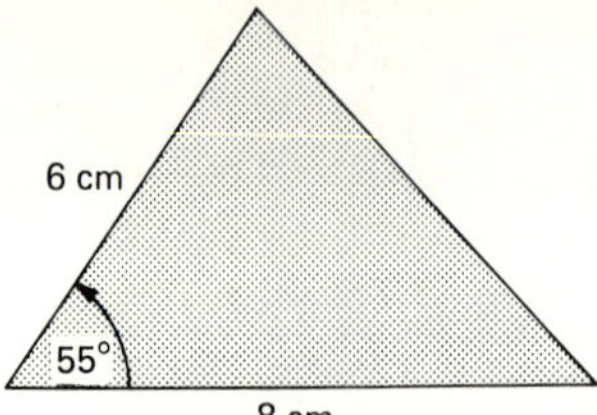

FIGURE P 1.14

1.15   The parallelogram will be used in the next chapter to obtain resultants of force systems and for vector addition. For the figure shown (Fig. P 1.15), determine R.

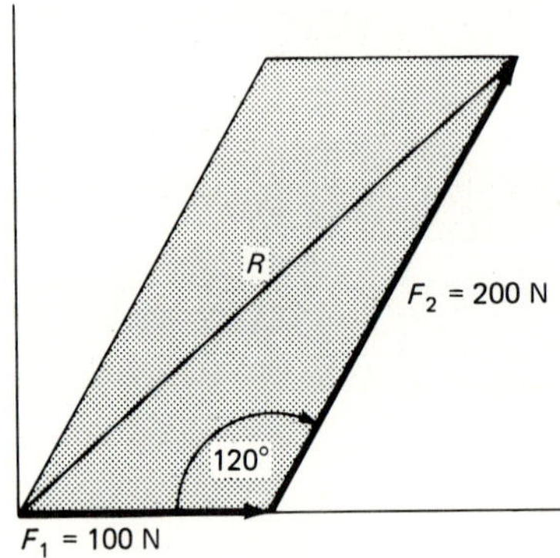

FIGURE P 1.15

1.16   In Fig. P 1.15 determine the angle that R makes with F_1.

1.17   In the parallelogram shown in Fig. P 1.17, $F_1 = 200$ N, $F_2 = 150$ N, and the angle (30°) is as shown. Determine R.

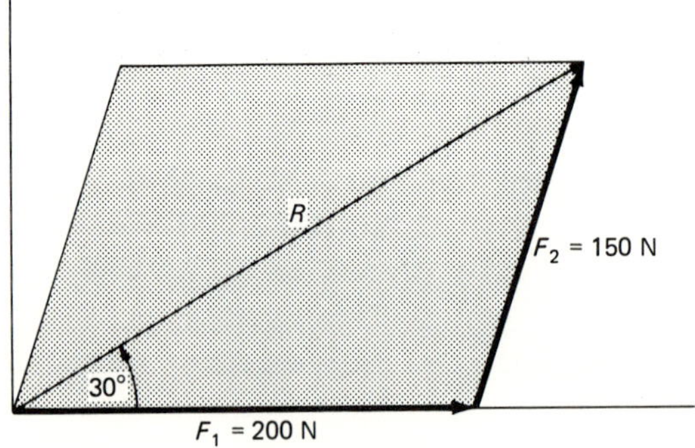

FIGURE P 1.17

1.18   Determine the angles of the triangle shown in Fig. P 1.18.

                                                                    Basic Concepts

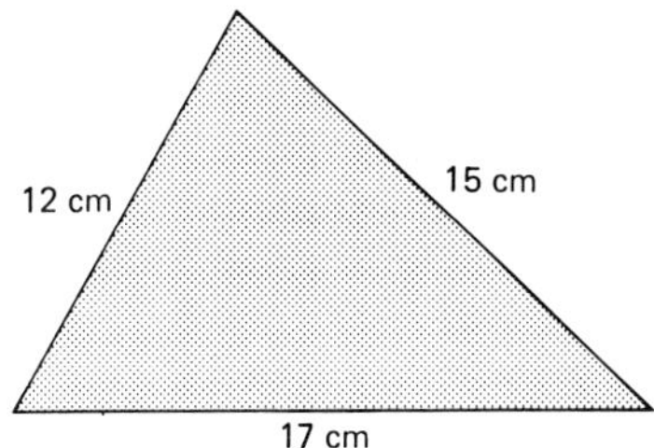

FIGURE P 1.18

1.19 A ship radios a distress signal to two shore stations which are located 150 mi apart. The first station finds that the ship lies 30° north of east, while the second station finds the ship to be located at 45° north of west. How far is the ship from each station (Fig. P 1.19)?

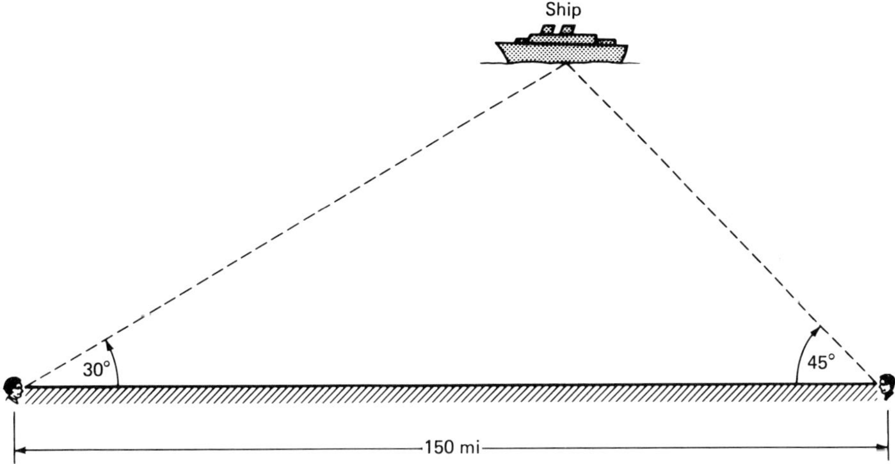

FIGURE P 1.19

Problems

2

Vectors and Force Systems

2.1 INTRODUCTION

In Chapter 1 we discussed some basic concepts of vectors and scalars, and a brief review of pertinent mathematics was undertaken. In this chapter we will use the concepts and tools developed in Chapter 1 to perform the operations of vector addition and subtraction and to determine the resultant of systems of vectors (forces). The resultant of a force system acting on a body is the single force that will produce the same effect on the body as the original force system. As we shall see later on in our study, it is often more convenient to consider the single resultant rather than each of the forces acting on a body. Also, as we shall find in Chapter 3, the equations of equilibrium for a force system acting on a body require the resultant of the force system to be zero.

2.2 VECTOR ADDITION OF COLLINEAR VECTORS

We have already distinguished between a scalar and a vector by noting that a scalar has magnitude only, whereas a vector is specified by both its magnitude and its direction. We can add scalars by the usual arithmetic processes with which we are all familiar. However, we cannot add scalars and vectors (apples and pears), and the usual arithmetic process of addition is not, in general, applicable to vectors. As a matter of fact, none of the usual arithmetic processes apply to similar vector operations such as addition, subtraction, and multiplication.

Let us now consider the case of a man running on a train in the same direction as the train is traveling. If the train is traveling at 40 mi/h relative to the ground, and the man travels at 10 mi/h relative to the train in the same direction as the train, we would readily say that the man is traveling at 50 mi/h relative to the ground. Similarly, if the man were running in a direction opposite to the direction of the train, we would then say that the man is going at 30 mi/h relative to the ground. While the foregoing appears to be straightforward, it is really necessary to look more closely at these two situations in order to establish certain ideas for our later studies. Since velocity is a vector, we can represent the velocity of the train by a vector whose length is proportional to the 40 mi/h speed and whose direction is that of the train. The arrowhead on the vector indicates the direction of motion of the train, as shown in Fig. 2.1(a). Similarly, the vector indicating the man's velocity relative to the train is also shown by a length proportional to the 10-mi/h speed and the direction of the train's motion. The resultant velocity of 50 mi/h is obtained by ''adding'' collinearly the 10-mi/h vector to the 40-mi/h vector, as shown in Fig. 2.1(b). We have added these two vectors by placing the two vectors in a line from tail to head of the train vector and adding the man's vector from the head of the train vector to the head of the man's vector, that is, tail to head plus tail to head.

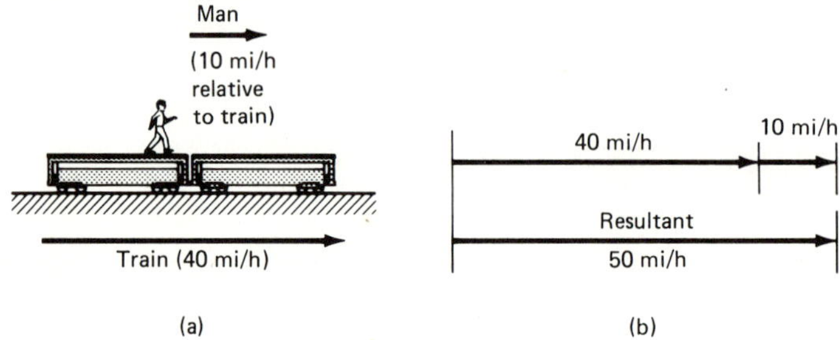

FIGURE 2.1   Vector addition of collinear vectors.

When the man's velocity is opposite to the train's velocity, we have the situation shown in Fig. 2.2. We again perform the addition process of tail to head plus tail to head, and in this case we find the results shown in Fig. 2.2(b). In this case we have performed the process of vector subtraction since the two vectors are in opposite directions.

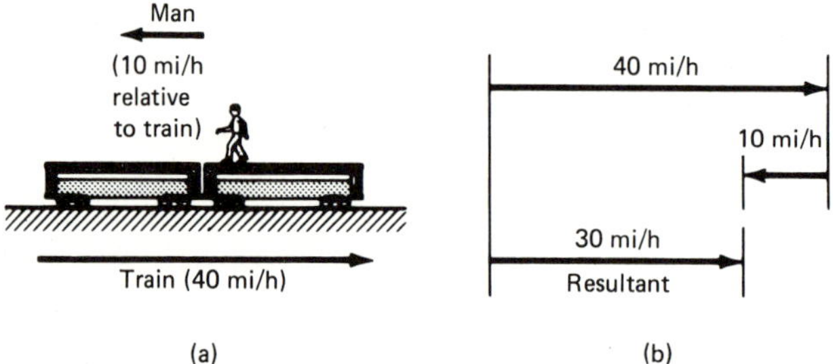

FIGURE 2.2   Vector subtraction of collinear vectors.

ILLUSTRATIVE PROBLEM 2.1

Two men start out from a given point, one travels due east, the other due west. If both go at speeds of 10 metres per hour (m/h), how far apart are they after 1 h?

                                                      Vectors and Force Systems

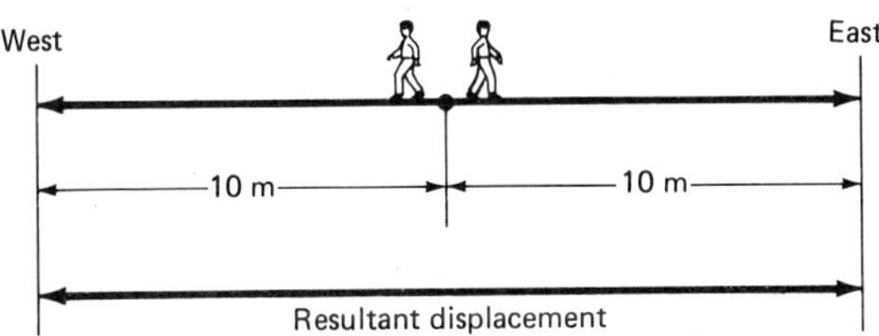

FIGURE 2.3   Illustrative Problem 2.1.

SOLUTION

From Fig. 2.3, where each directed displacement is plotted as a vector, we find the displacement to be 20 m. From the final position of the man who headed west, an observer would state that the other man is 20 m east of his position. From the final position of the man who headed east, the observer would find the other man to be 20 m west.

ILLUSTRATIVE PROBLEM 2.2

A body is acted upon by four forces, as shown in Fig. 2.4. Find the sum (resultant) of these forces.

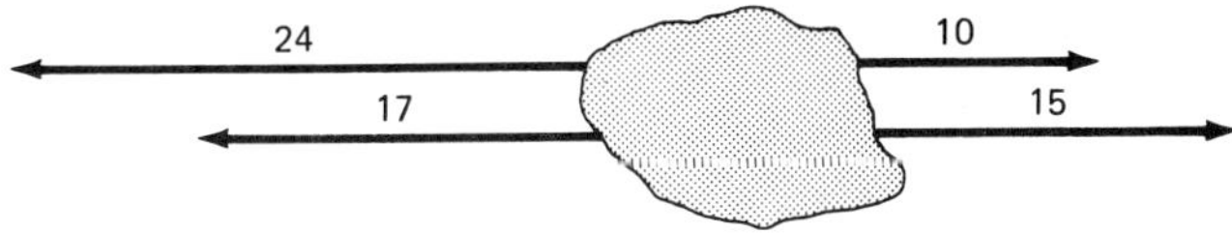

FIGURE 2.4   Illustrative Problem 2.2.

SOLUTION

If we arbitrarily assign as positive those vectors that point to the right and negative those that point to the left, we can algebraically add these collinear forces as

$$R = + 10 + 15 - 17 - 24 = -16 \text{ N}$$

Thus the resultant R is a vector 16 units in magnitude and pointed to the left.

2.3   THE PARALLELOGRAM LAW

For the general case of vector addition where the vectors are not collinear, there are several methods that can be used which are equivalent. All of these methods are based upon the *parallelogram law*. We can state the parallelogram law for the addition of two vectors as follows: *If from a common point two vectors are drawn and if a parallelogram is drawn having these vectors as sides, the diagonal of the parallelogram drawn from the common point to the opposite vertex is the vector that represents, in direction and magnitude, the sum (resultant) of the forces.*

The Parallelogram Law

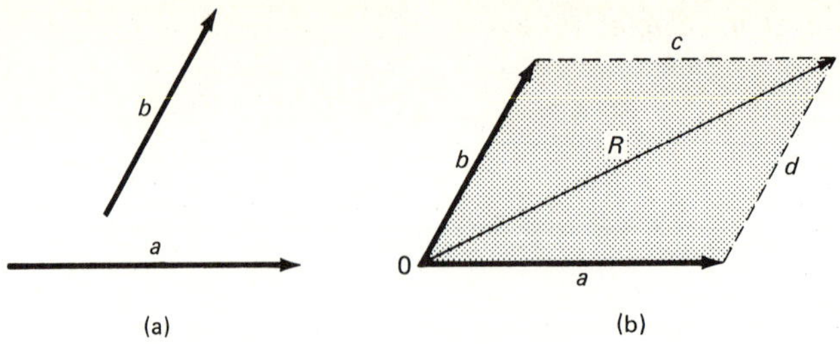

(a)   (b)

FIGURE 2.5   The parallelogram law.

In order to illustrate the use of the parallelogram law, let us refer to Fig. 2.5. From a common point 0 we lay off to scale in the proper directions the two vectors *a* and *b*. The parallelogram is then completed by drawing sides *c* and *d* parallel to *a* and *b*, respectively. The diagonal drawn from the starting point to the opposite vertex is the resultant *R* of the original two forces *a* and *b* and yields both the magnitude and direction of the resultant vector. It should be noted that as we have drawn the figure, we have not necessarily related the vertex 0 to the lines of action of each of the forces. Therefore the resultant that we have determined is a free vector parallel to, but not necessarily located on the line of action of the actual resultant of the two forces being added.

A force (vector) of 10 units acts horizontally to the right on a body. A second force of 5 units acts to the right and up from the horizontal at an angle of 45°. Determine the resultant of these two forces.

SOLUTION

We first lay off each vector from a common point to scale, as shown in Fig. 2.6 and complete the parallelogram. If care is taken, we can measure the resultant R to scale, and we can also measure the angle θ with a protractor in order to specify the direction of the resultant. If we use a mathematical solution, we can proceed in two ways. The first is to apply the law of cosines to the triangle consisting of the resultant R and the two sides 5 and 10. In doing this, note that the angle of the triangle that is of interest is not the angle 45° but rather its supplement 180° − 45°, as shown in Fig. 2.6. Therefore

$$R^2 = 5^2 + 10^2 - 2 \times 5 \times 10 \cos(180° - 45°)$$

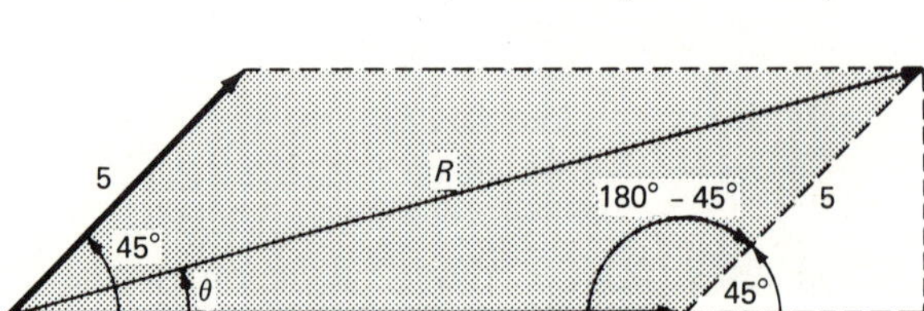

FIGURE 2.6   Illustrative Problem 2.3.

   Vectors and Force Systems

But we showed in Chapter 1 that cos $(180° - \theta) = -\cos \theta$. Thus

$$R^2 = 5^2 + 10^2 - 2 \times 5 \times 10 \, (-\cos 45°)$$

$$= 25 + 100 + 70.7 = 195.7$$

and

$$R = 13.99$$

We can find the angle θ by applying the law of sines to the triangle, which gives us

$$\frac{R}{\sin (180° - 45°)} = \frac{5}{\sin \theta}$$

and

$$\sin \theta = \frac{5}{R} \sin (180° - 45°) = \frac{5}{13.99} (\sin 135°) = 0.253$$

Thus

$$\theta = 14.6°$$

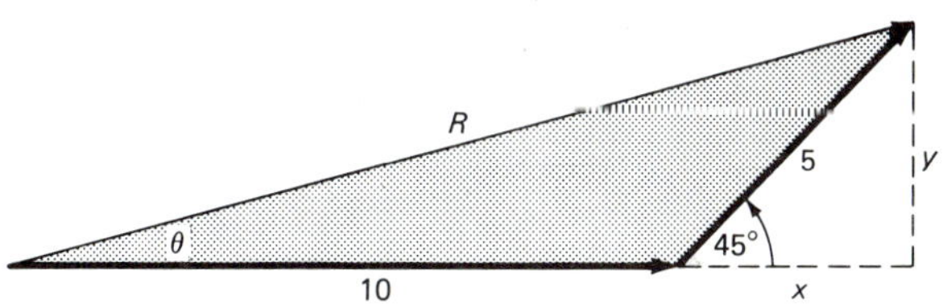

FIGURE 2.7   Illustrative Problem 2.3, alternate solution.

These same results can also be achieved by considering right triangles only. Figure 2.7 shows the construction simplified. We first determine the distances x and y using the sine and cosine of the exterior angle, which is 45° for this problem. Therefore

$$x = 5 \cos 45° = 3.535$$

$$y = 5 \cos 45° = 3.535$$

We now have to solve the right triangle whose horizontal dimension is 10 + 3.535 and whose vertical dimension is 3.535. Therefore

$$R^2 = 13.535^2 + 3.535^2 = 183.2 + 12.5 = 195.7$$

and

$$R = 13.99$$

The Parallelogram Law

Since

$$\sin \theta = \frac{3.535}{13.99} = 0.253$$

hence

$$\theta = 14.6°$$

It is obvious that either mathematical approach yields identical results. The student should use both for practice and to obtain facility in their usage.

Determine the resultant of the two forces shown in Fig. 2.8, using the parallelogram law.

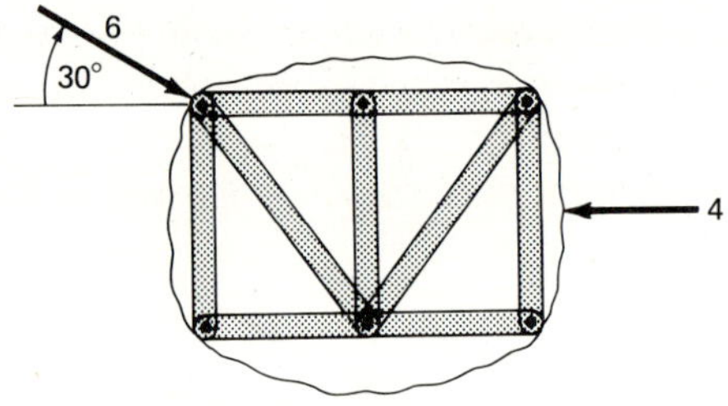

FIGURE 2.8   Illustrative Problem 2.4.

SOLUTION

We first complete the parallelogram as shown in Fig. 2.9. The resultant is obtained from the law of cosines as

$$R^2 = 4^2 + 6^2 - 2 \times 4 \times 6 \cos 30°$$

$$= 16 + 36 - 48(0.866) = 10.43$$

and

$$R = 3.23$$

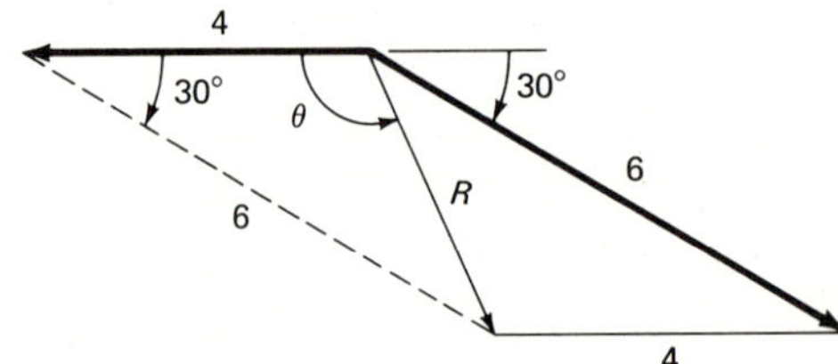

FIGURE 2.9   Illustrative Problem 2.4.

   Vectors and Force Systems

The angle θ is found from the law of sines,

$$\frac{3.23}{\sin 30°} = \frac{6}{\sin \theta}$$

Thus

$$\sin \theta = \frac{6 \sin 30°}{3.23} = 0.929$$

From Table 1.1, θ = 68.3°. However, since the angle is greater than 90°, θ = 180° − 68.3° = 111.7°. It is left as an exercise to verify the solution using the alternate right triangle method.

Since the parallelogram law can be used to find the resultant of any two vectors, it can be used to find the resultant of any number of vectors by applying it successively to pairs of vectors until the desired result is found. It should be noted that the parallelogram law as we have used it applies to the addition of two vectors. Where we desire to subtract one vector from another we can still use the parallelogram law, but we must modify our procedure. If it is desired to subtract one vector from another, *reverse its direction*, and then proceed as we did for addition. Therefore if we desired to subtract vector *b* from vector *a* in Fig. 2.5, we would reverse *b* and then proceed as shown in Fig. 2.10.

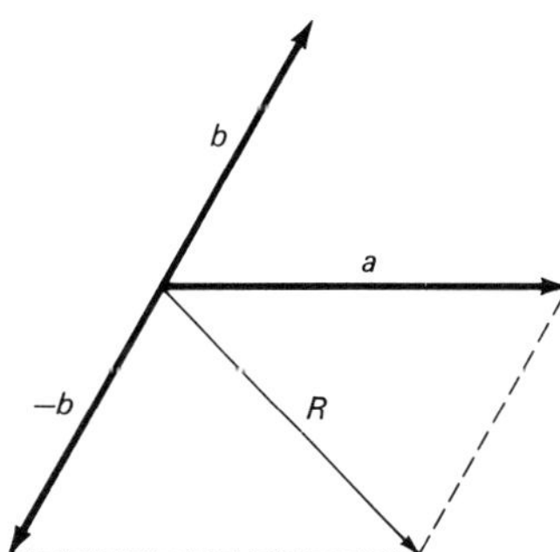

FIGURE 2.10   Vector subtraction.

ILLUSTRATIVE PROBLEM 2.5

Solve Illustrative Problem 2.3 if the 5-unit vector is to be subtracted from the 10-unit vector.

SOLUTION

As in Illustrative Problem 2.3, we first lay off the two vectors from the common point 0, as shown in Fig. 2.11. We now reverse the 5-unit vector and complete the parallelogram formed by the −5 and 10 vectors. Solving for R,

$$R^2 = 5^2 + 10^2 − 2 \times 5 \times 10 \cos 45°$$

$$= 125 − 70.7 = 54.3$$

The Parallelogram Law

Thus

$$R = 7.37$$

The angle θ is found from

$$\frac{R}{\sin 45°} = \frac{5}{\sin \theta}$$

$$\sin \theta = \frac{5 \sin 45°}{R} = \frac{5 \times 0.707}{7.37} = 0.480$$

and

$$\theta = 28.70°$$

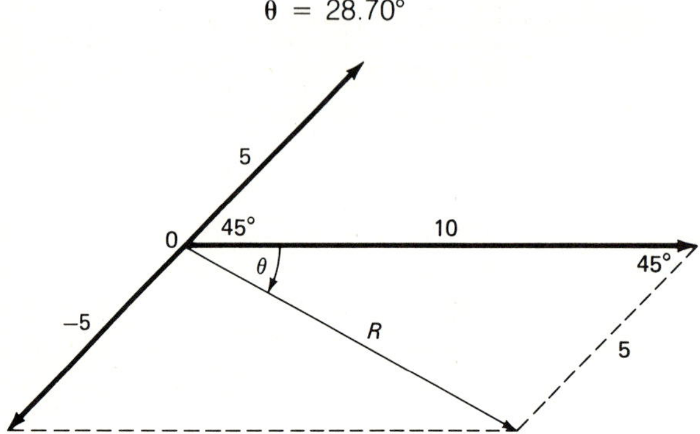

FIGURE 2.11   Illustrative Problem 2.5.

2.4   THE TRIANGLE LAW AND THE FORCE POLYGON

The Triangle Law

The *triangle law* of vector addition is a corollary of the material already presented in Sections 2.2 and 2.3. In Section 2.2 we concluded that by placing two collinear vectors tail to tip plus tail to tip yielded the resultant of this type of force system. This is a special case of the triangle law which can be stated as follows: *If two vectors are placed tail to tip, their resultant vector is the third side of the triangle, the direction of the resultant being from the tail of the first vector (origin) to the tip of the last vector.* We anticipated this law by the alternate solution to Illustrative Problem 2.3. Figure 2.12 shows two free vectors which are shown added by the triangle law. The mathematical solution of the addition of two vectors using the triangle law is identical to the procedures used to solve problems utilizing the parallelogram law. Also, the subtraction of vectors is handled exactly as before, that is, the sign of the vector being subtracted is reversed, and the procedure is as noted above.

                                            Vectors and Force Systems

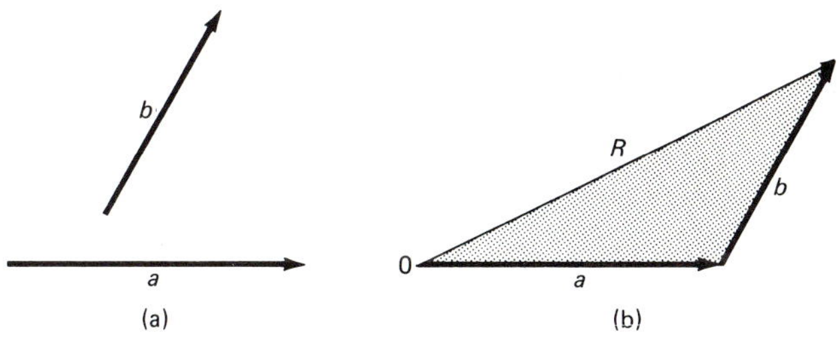

FIGURE 2.12   The triangle law.

ILLUSTRATIVE PROBLEM 2.6

Solve Illustrative Problem 2.4 using the triangle law.

SOLUTION

Referring to Fig. 2.8, we can construct the required triangle by adding tail to tip the 4-unit and 6-unit forces, as shown in Fig. 2.13.

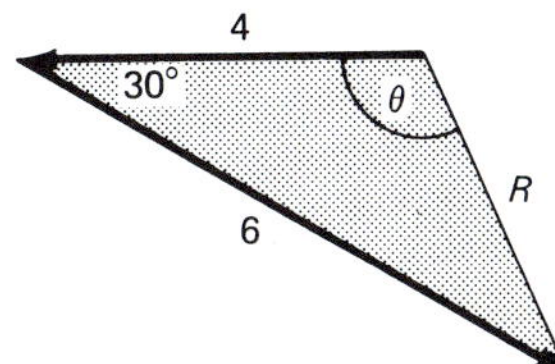

FIGURE 2.13   Illustrative Problem 2.6.

Applying the law of cosines, the solution proceeds exactly as for Illustrative Problem 2.4,

$$R^2 = 4^2 + 6^2 - 2 \times 4 \times 6 \cos 30°$$

$$= 16 + 36 - 48(0.866) = 10.43$$

and

$$R = 3.23$$

The angle θ is found from

$$\frac{4}{\sin \theta} = \frac{R}{\sin 30°}$$

Thus

$$\sin \theta = \frac{4 \sin 30°}{R} = \frac{4 \times 0.5}{3.23} = 0.619$$

and

$$\theta = 38.3°$$

or

$$\theta = 180° - 30° - 38.3° = 111.7°$$

The Triangle Law and the Force Polygon

SOLUTION

In this problem we are subtracting one vector from another. Our first step is to reverse the vector in question and then to apply the triangle law, as shown in Fig. 2.14. Once the triangle is completed, we once again apply the law of cosines to obtain the resultant and the law of sines to obtain the angular relation between the resultant and the forces. Proceeding,

$$R^2 = 5^2 + 10^2 - 2 \times 5 \times 10 \cos 45°$$

$$= 125 - 70.7 = 54.3$$

and

$$R = 7.37$$

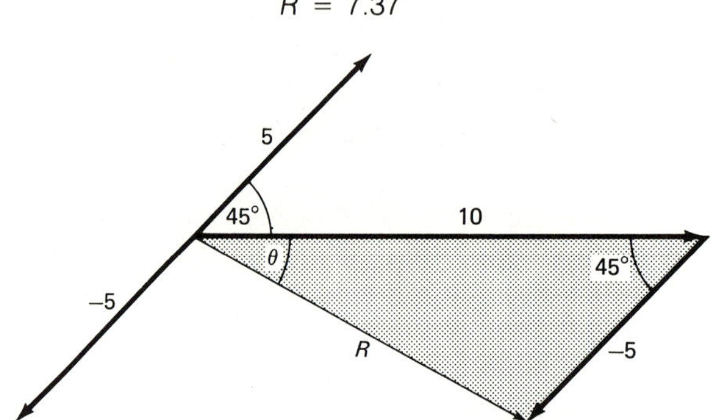

FIGURE 2.14   Illustrative Problem 2.7.

The angle θ is found from

$$\frac{R}{\sin 45°} = \frac{5}{\sin \theta}$$

and

$$\sin \theta = \frac{5 \sin 45°}{R} = \frac{5 \times 0.707}{7.37} = 0.480$$

Thus

$$\theta = 28.7°$$

The Polygon Method

The addition of three or more vectors using the parallelogram law is cumbersome in that successive pairs of vectors must be added until the final resultant is obtained. In principle, the method that is found to be more convenient, known as the *polygon method,* is simply an extension of the triangle law to more than two vectors. Consider the three free vectors shown in Fig. 2.15(a). We first add vectors *a* and *b* using the

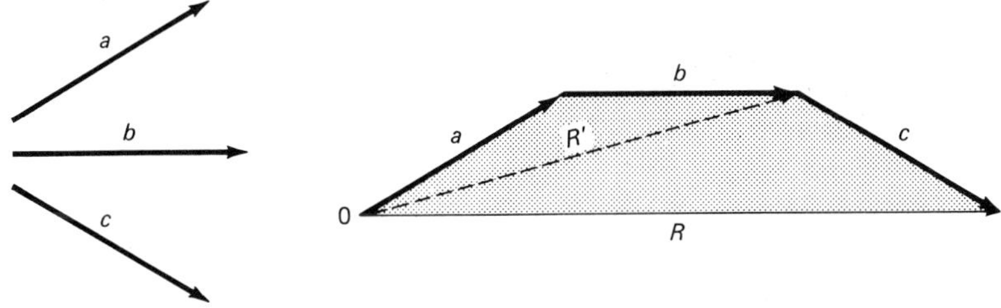

FIGURE 2.15   The polygon method.

triangle law to obtain the resultant R'. To R' we then add c to obtain the desired final resultant R. It will be noted that the intermediate step of obtaining R' could have been eliminated without changing the final result, and that the addition of the three vectors could have been obtained by joining tail to tip each vector successively to the other vectors, with the resultant being the vector that starts from the origin 0 and closes the figure. This procedure applies to any number of vectors, and the resultant R can be determined by dividing the polygon into any number of convenient triangles and calculating the sides and angles in turn. It is important to note that any convenient order of addition of the vectors using the tail-to-tip method can be used, and the resultant will always be the same. This is a general conclusion, which states that vectors follow the commutative rule of scalars, which is a basic rule of algebra, namely, that the order of addition of vectors does not affect the final result. Using capital letters to denote a vector, we can write this fact as

$$F_1 + F_2 = F_2 + F_1 \tag{2.1}$$

Figure 2.16 shows three vectors added by the polygon method to illustrate that the final resultant is independent of the order in which the vectors are added. The vectors a, b, and c are first added, so that $R = a + b + c$. The dashed lines show $R = a + c + b$, with R the same in both cases.

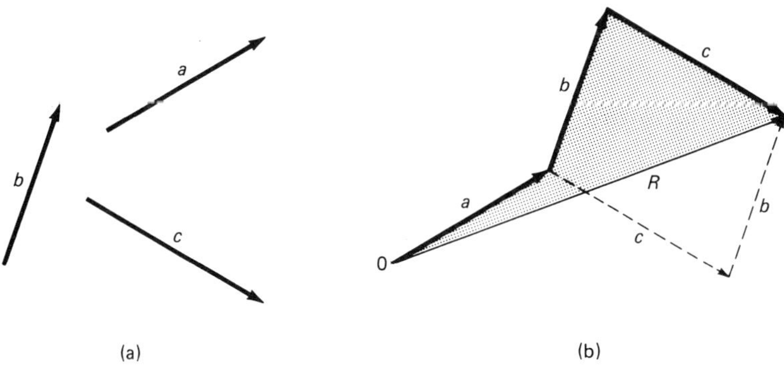

(a)

(b)

FIGURE 2.16   The polygon method further illustrated.

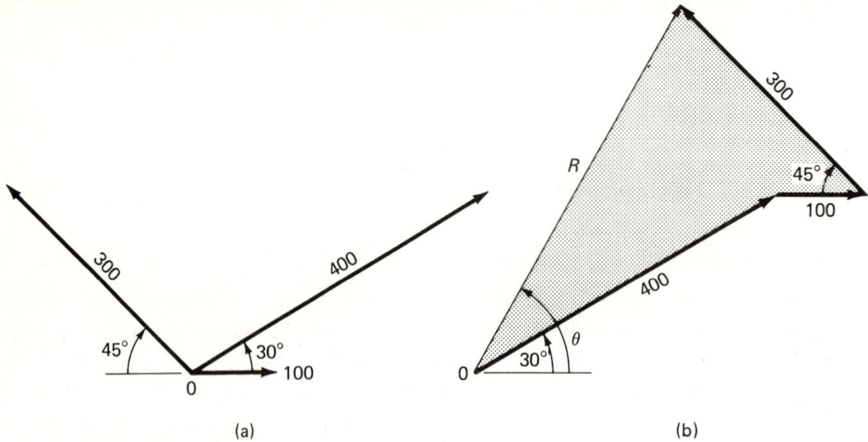

FIGURE 2.17   Illustrative Problem 2.8.

SOLUTION

Let us first draw the polygon to an appropriate scale, as indicated in Fig. 2.17(b). For this problem we have added the 100-unit vector to the 400-unit vector, and then we added the 300-unit vector to obtain R. We can measure R and the angle θ to complete the problem. We can also solve this problem by successively applying the law of cosines and the law of sines. As we shall see in the next section, the resultant can also be found in a systematic and somewhat easier manner. It is left as an exercise for the student to show that changing the order of addition does not change the resultant.

2.5   THE METHOD OF COMPONENTS

The most common and useful method of solving problems in mechanics is known as the *method of components*. For the present we will only be concerned with applying this method to obtaining the resultant of vector systems. Later on we shall apply it to force systems that are in equilibrium. This method considers that the resultant of a force system may be thought of as being the end result of the addition of two vectors or *components*. A resultant may be resolved into any pair of components, but for engineering purposes it is most convenient to use rectangular components, that is, components that are at right angles to each other, known as rectangular or orthogonal components. The process of breaking up a vector into components is known as *resolution*.

Let us first apply this method to the resolution of a single vector into two components. For our purpose we will use the conventional x–y axes (Cartesian coordinates), and x-directed quantities to the right and y-directed quantities vertically up will be taken arbitrarily as positive, while x and y quantities oppositely directed will be taken

                                                                 Vectors and Force Systems

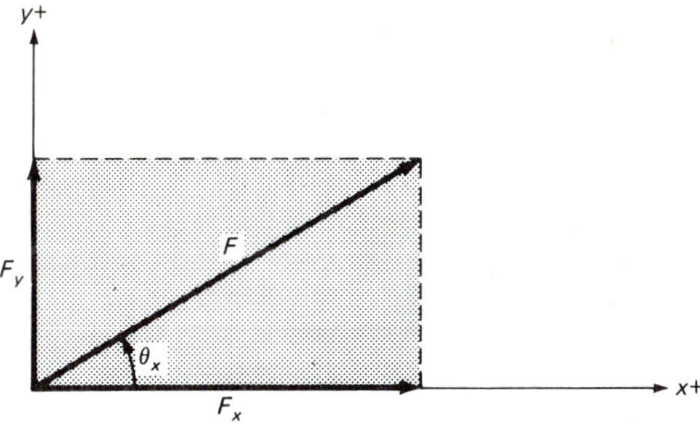

FIGURE 2.18   Resolution of a single vector into rectangular components.

as negative for consistent notation. Figure 2.18 shows an arbitrary vector F making an angle of θ_x with the x axis. We project F onto the x and y axes as shown to obtain the vectors F_x and F_y. Note that the vector addition of F_x and F_y will yield the vector F. From basic trigonometric relations we find

$$F_x = F \cos \theta_x$$

$$F_y = F \sin \theta_x$$

(2.2)

Note that our choice of x and y axes was arbitrary, and any choice of orthogonal axes can be chosen as x and y.

The importance of this method lies in the evident fact that the resultant can be obtained when the components are known. Thus

$$F = \sqrt{F_x^2 + F_y^2}$$

and

(2.3)

$$\tan \theta_x = \frac{F_x}{F_y}$$

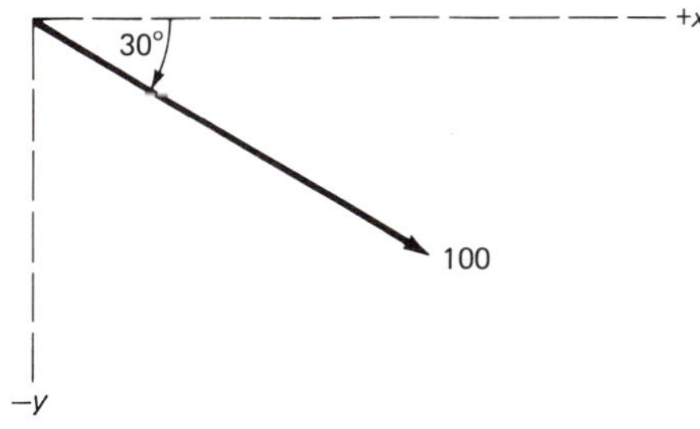

FIGURE 2.19   Illustrative Problem 2.9.

The Method of Components

SOLUTION

The solution to this type of problem requires us to note carefully the directions of the components in order to properly obtain their signs. For the vector shown in Fig. 2.19,

$$F_x = 200 \cos (30°) = 100(0.866) = 86.6$$

$$F_y = 200 \sin (-30°) = 100(-0.5) = -50$$

Notice that F_y is directed down in the negative y direction. This is taken care of by the sign of the trigonometric function of the angle (in this case $-30°$). It is also a good procedure to look at the vector and to note the direction of the component as a check on the mathematics.

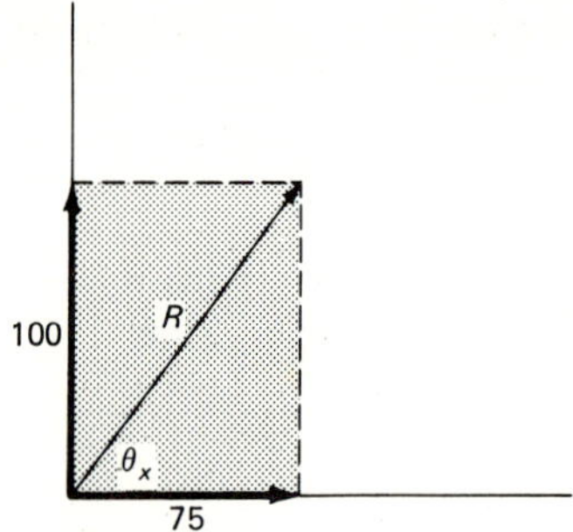

FIGURE 2.20   Illustrative Problem 2.10.

SOLUTION

We can apply Eqs. (2.3) directly. Performing the required algebra yields

$$R = \sqrt{75^2 + 100^2} = \sqrt{5625 + 10\,000}$$
$$= \sqrt{15\,625} = 125$$

and

$$\tan \theta_x = \frac{100}{75} = 1.333$$

Thus

$$\theta_x = 53.1°$$

                                                                 Vectors and Force Systems

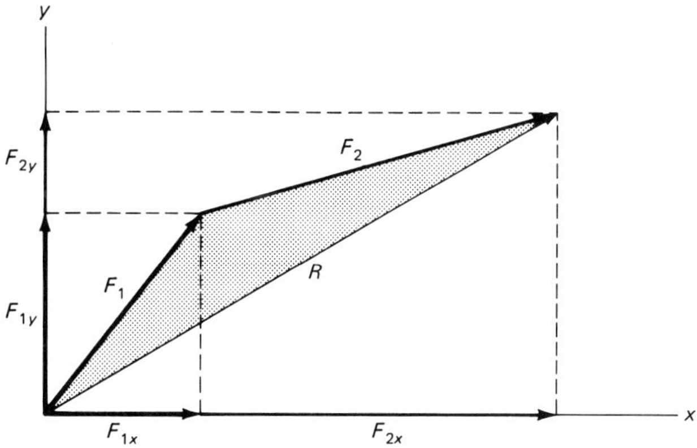

FIGURE 2.21   Addition of two vectors by the method of components.

Let us now apply the method of components to add the two vectors F_1 and F_2. Figure 2.21 shows these vectors added by the polygon method, tail to tip. On this diagram the x and y components of each of the vectors is shown as F_{1x}, F_{2x}, F_{1y}, and F_{2y}, respectively. Referring to Fig. 2.21 we see that the resultant R can be considered to be composed of two components consisting of $(F_{1x} + F_{2x})$ and $(F_{1y} + F_{2y})$ in the x and y directions. The finding of the resultant of these components is then given by Eqs. (2.4) as

$$R = \sqrt{(F_{1x} + F_{2x})^2 + (F_{1y} + F_{2y})^2}$$

$$\tan \theta_x = \frac{F_{1y} + F_{2y}}{F_{1x} + F_{2x}}$$

(2.4)

ILLUSTRATIVE PROBLEM 2.11

Add the two vectors shown in Fig. 2.22 using the method of components and find their resultant.

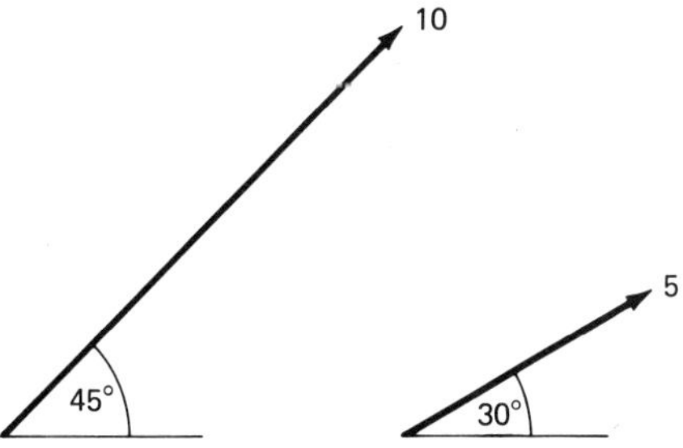

FIGURE 2.22   Illustrative Problem 2.11.

The Method of Components

SOLUTION

We first determine the x and y components of each vector. Thus for the 10-unit vector,

$$F_x = 10 \cos 45° = 10 \times 0.707 = 7.07$$

$$F_y = 10 \sin 45° = 10 \times 0.707 = 7.07$$

For the 5-unit vector,

$$F_x = 5 \cos 30° = 5 \times 0.866 = 4.33$$

$$F_y = 5 \sin 30° = 5 \times 0.50 = 2.50$$

We now algebraically add the x components to each other (since they are collinear) and the y components to each other to obtain the components of the resultant, R_x and R_y,

$$R_x = 7.07 + 4.33 = 11.40$$

$$R_y = 7.07 + 2.50 = 9.57$$

The resultant R is given by

$$R = \sqrt{11.40^2 + 9.57^2} = \sqrt{129.96 + 91.58}$$

$$= \sqrt{221.54} = 14.88$$

and

$$\tan \theta_x = \frac{9.57}{11.40} = 0.839$$

Thus

$$\theta_x = 40.0°$$

We can generalize the foregoing to the addition of any number of vectors by stating that the sum of the rectangular components of the vectors yields the rectangular components of the resultant of these vectors. Using the mathematical notation Σ to denote the sum of a quantity,

$$\Sigma F_x = F_{1x} + F_{2x} + F_{3x} + \ldots$$

$$\Sigma F_y = F_{1y} + F_{2y} + F_{3y} + \ldots$$

(2.5)

we have, for the resultant,

$$R_x = \Sigma F_x$$

$$R_y = \Sigma F_y$$

$$\tan \theta_x = \frac{\Sigma F_y}{\Sigma F_x}$$

(2.6)

Vectors and Force Systems

and

$$R = \sqrt{(\Sigma F_x)^2 + (\Sigma F_y)^2}$$

It is important to keep track of the algebraic signs of the vector components in order to obtain the correct solution to a problem. As is shown in the following illustrative problem, a tabular solution is sometimes a very effective method of keeping track of these quantities, especially when several vectors are involved.

ILLUSTRATIVE PROBLEM 2.12

Determine the resultant of the vectors shown in Fig. 2.23.

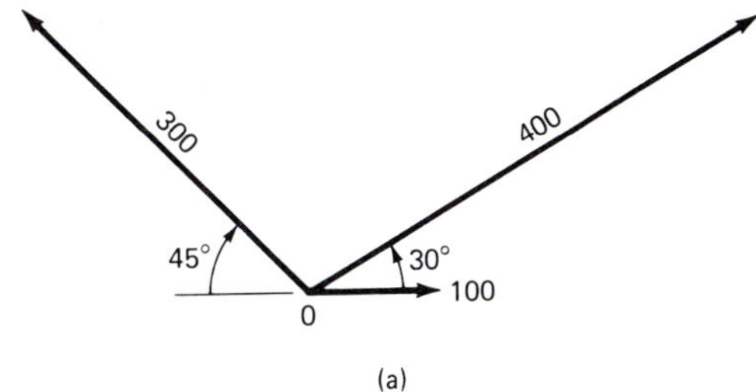

FIGURE 2.23   Illustrative Problem 2.12 (Repeated from Fig. 2.17(a).)

SOLUTION

In a formal manner let us first apply Eqs. (2.5) and (2.6) to the solution of this problem.

1. The 100 vector,

$$F_x = 100 \qquad F_y = 0$$

2. The 400 vector,

$$F_x = 400 \cos 30° = 400 \times 0.866 = 346.4$$

$$F_y = 400 \sin 30° = 400 \times 0.50 = 200.0$$

3. The 300 vector,

$$F_x = -300 \cos 45° = -300 \times 0.707 = -212.1$$

$$F_y = 300 \cos 45° = 300 \times 0.707 = 212.1$$

We now find the sum of the x and y components,

$$\Sigma F_x = 100 + 346.4 - 212.1 = 234.3$$

$$\Sigma F_y = 0 + 200 + 212.1 = 412.1$$

The Method of Components

Thus

$$R = \sqrt{(\Sigma F_x)^2 + (\Sigma F_y)^2} = \sqrt{234.3^2 + 412.1^2} = 474.0$$

$$\tan \theta_x = \frac{\Sigma F_y}{\Sigma F_x} = \frac{412.1}{234.3} = 1.759$$

and

$$\theta_x = 60.4$$

This solution could have been obtained using a tabular form as shown in Table 2.1.

TABLE 2.1   Illustrative Problem 2.12

Vector	Reference Angle θ_x	$F_x = F \cos \theta_x$	$F_y = F \sin \theta_x$
100	0°	100	0
400	30°	$400 \cos 30° = 400 \times 0.866 = 346.4$	$400 \sin 30° = 400 \times 0.50 = 200$
300	135°	$-300 \cos 45° = -300 \times 0.707 = -212.1$	$300 \sin 45° = 300 \times 0.707 = 212.1$
		$\Sigma F_x = 234.3$	$\Sigma F_y = 412.1$

These results are the same as obtained previously, and the rest of the problem proceeds in the same manner as before. The compact notation of the table as well as the systematic approach thus afforded tend to minimize mathematical errors.

ILLUSTRATIVE PROBLEM 2.13

Determine the resultant of the vectors shown in Fig. 2.24.

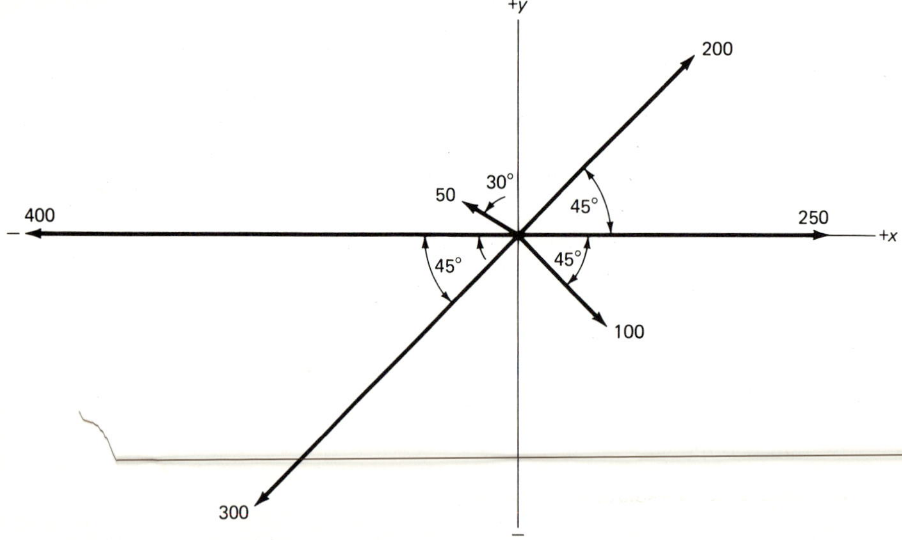

FIGURE 2.24   Illustrative Problem 2.13.

                    Vectors and Force Systems

To further illustrate the tabular method of solving a multivector problem, we first set up a table of the pertinent quantities, as shown in Table 2.2.

TABLE 2.2   Illustrative Problem 2.13

Vector	Reference Angle θ_x	$F_x = F\cos\theta_x$	$F_y = F\sin\theta_x$
250	0	$+250$	0
200	45°	$200\cos 45° = 200 \times 0.707 = 141.4$	$200\sin 45° = 200 \times 0.707 = 141.4$
50	150°	$-50\cos 30° = -50 \times 0.866 = -43.3$	$50\sin 30° = 50 \times 0.5 = 25.0$
400	180°	-400	0
300	225°	$-300\cos 45° = -300 \times 0.707 = -212.1$	$-300\sin 45° = -300 \times 0.707 = -212.1$
100	315°	$+100\cos 45° = 100 \times 0.707 = 70.7$	$-100\sin 45° = -100 \times 0.707 = -70.7$
		$\Sigma F_x = -193.3$	$\Sigma F_y = -116.4$

Thus

$$R = \sqrt{(\Sigma F_x)^2 + (\Sigma F_y)^2} = \sqrt{-193.3^2 + -116.4^2} = 225.6$$

$$\tan\theta_x = \frac{\Sigma F_y}{\Sigma F_x} = \frac{-116.4}{-193.3} = 0.602$$

and

$$\theta_x = 31.1°$$

With both the x and y components of the resultant being negative and the tangent being positive we have the resultant in the third quadrant. A quick reference as to the location of the resultant with respect to each quadrant can be obtained from Fig. 2.25.

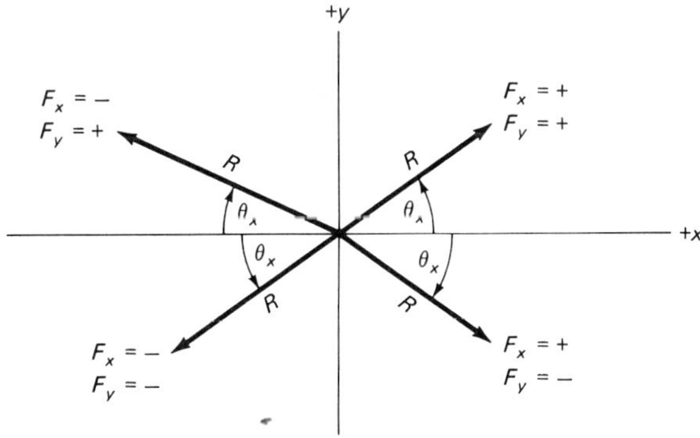

FIGURE 2.25   Location of resultant.

The Method of Components

2.6   THE MOMENT OF A VECTOR

The *moment of a vector* (for a force this is known as the *torque*) is qualitatively expressed as the tendency of the vector to cause rotation about some axis. Quantitatively we can define the moment of a vector to be the product of the vector and the perpendicular distance from the line of action of the vector to the axis in question. This concept is illustrated in Fig. 2.26, where the vector F is indicated to lie in a horizontal plane and the axis Y–Y is perpendicular to the plane. The moment of the vector F is therefore $F \times d$, where d is the perpendicular distance from the axis Y–Y to the force.

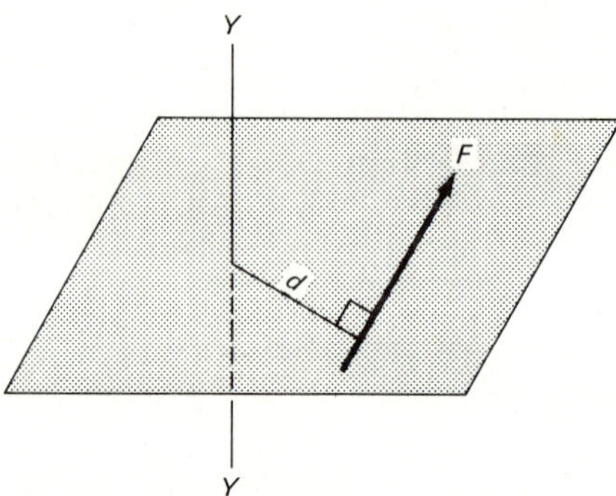

FIGURE 2.26   The moment of a vector.

From the foregoing definition, it is important to note two properties of a moment. The first concerns its physical units. If the vector is a force expressed in pounds, then the moment ($F \times d$) is expressed in units of foot pounds or inch pounds. The second property is that the sign of the moment about a given point must be stated. In other words, does the moment tend to produce clockwise or counterclockwise motion? In order to make this differentiation, it is conventional (and also quite arbitrary) to call a counterclockwise moment as a positive moment and a clockwise moment a negative moment. We will generally adhere to this conventional notation, even though the opposite convention is used by some engineers. The point to stress is that *in a given problem,* whichever convention is selected, *it must not be changed* and must be used consistently throughout, otherwise errors will occur in the solution of the problem.

ILLUSTRATIVE PROBLEM 2.14

Find the moment of the force shown in Fig. 2.27 about the axis O, which is perpendicular to the page.

SOLUTION

From O, a perpendicular is constructed to the line of action of the force (extended). The distance from O to the line of action of the force is $10 \times 0.707 = 7.07$ in. The moment of the force is therefore $7.07 \times 100 = 707$ in.·lb. Since this moment is clockwise, we indicate it to be -707 in.·lb.

                                          Vectors and Force Systems

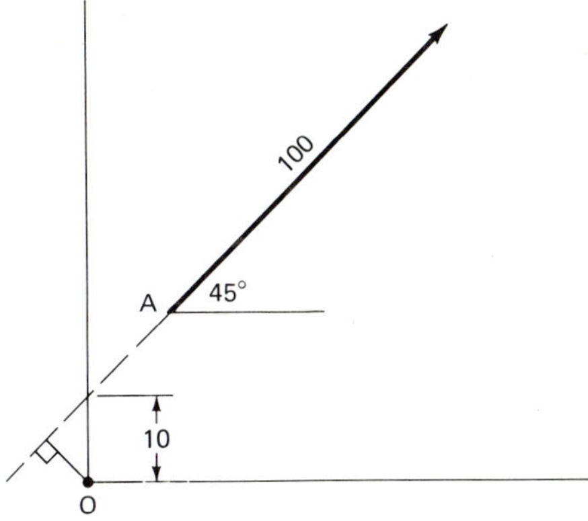

FIGURE 2.27   Illustrative Problem 2.14.

One of the most powerful principles in mechanics is known as the *principle of moments* or *Varignon's theorem*. As we shall use it, it can be stated that *the moment of a vector (force) about an axis is equal to the algebraic sum of the moments of its components about the same axis.** The importance of this theorem lies in the fact that it is often easier to determine the moment of a force about an axis by determining the sum of the moments of its components about the same axis.

ILLUSTRATIVE PROBLEM 2.15

Solve Illustrative Problem 2.14 using Varignon's theorem.

SOLUTION

It will be noted in Fig. 2.27 that point A of the 100-lb vector is not located in space. In order to illustrate certain important conclusions let us first assume that point A is located on the y axis, as shown in Fig. 2.28(a).

Resolving the 100-lb force into F_x and F_y gives us $F_x = 70.7$ and $F_y = 70.7$. The moment of F_y about O is zero since this vector intersects the axis through O and therefore has a zero moment arm. The moment of F_x about O is $70.7 \times 10 = 707$ in.·lb clockwise, or -707 in.·lb, which is the total moment and agrees with the conclusion reached in Illustrative Problem 2.14.

Let us now consider that point A is located as shown in Figure 2.28(b), where we have moved the force along its line of action. F_x and F_y remain each 70.7 lb. The moment of F_y about O is $70.7 \times 1 = 70.7$ in.·lb counterclockwise, or $+70.7$ in.·lb. The moment of F_x about O is $70.7 \times 11 = 777.7$ in.·lb clockwise, or -777.7 in.·lb. The net moment is the algebraic sum of these components or $-777.7 + 70.7 = -707$ in.·lb, which agrees with our earlier results.

*See Shames, I. H., *Engineering Mechanics*, vol. I—*Statics*, 2nd ed. Englewood Cliffs, N.J., Prentice-Hall, 1966, pp. 40, 41, for a proof of Varignon's theorem.

The Moment of a Vector

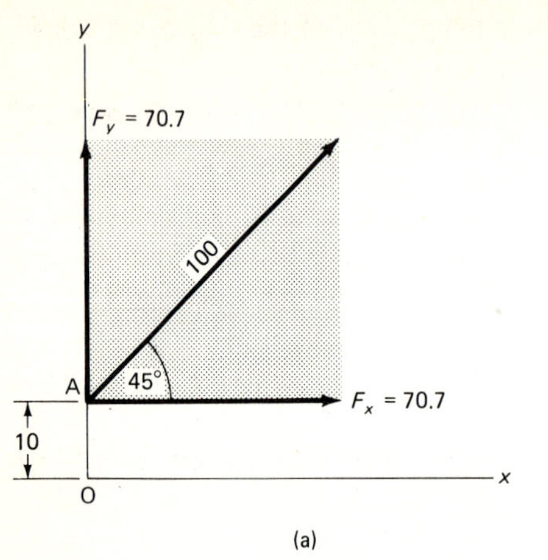

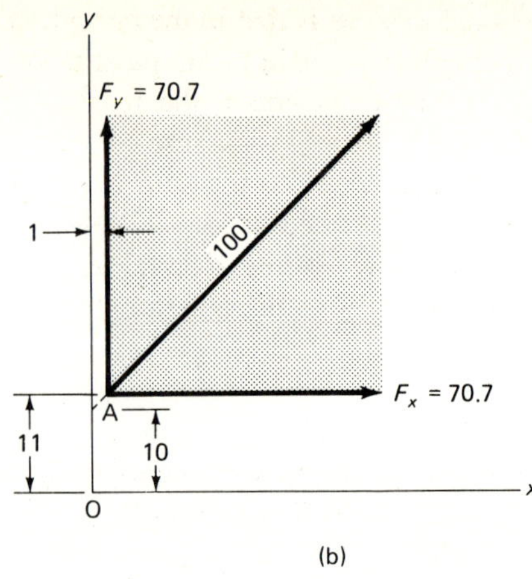

FIGURE 2.28   Illustrative Problem 2.15.

From Illustrative Problem 2.15 we can conclude that the moment of a force is zero if the line of action of the force (extended if necessary) intersects the axis about which moments are taken. We can also further conclude that the moment of a force about a given axis does not depend upon where along its line of action the force is located. The moment only depends upon the magnitude of the force and the geometric relation between the line of action of the force and the reference axis.

2.7   PARALLEL FORCE SYSTEMS

A system of coplanar parallel forces is one in which the lines of action of all of the coplanar forces are parallel. The resultant of this type of system is a vector R whose line of action is parallel to the lines of action of the forces in the system. The magnitude and sign of the resultant are simply found by adding each of the forces algebraically, as was done for the case of collinear vectors earlier in this chapter. The difference between this type of vector system and the concurrent systems which we have con-

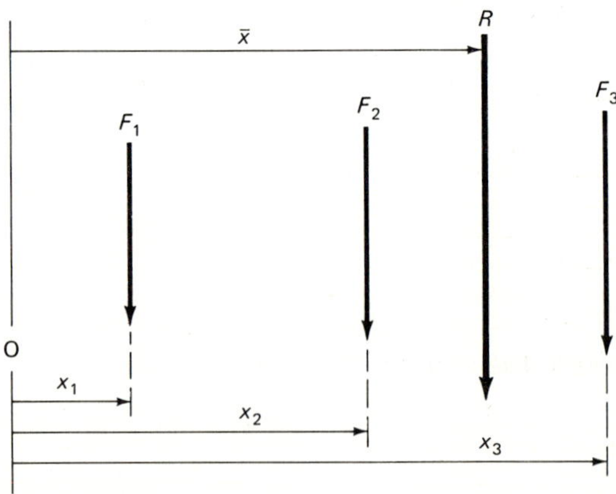

FIGURE 2.29   A parallel force system.

Vectors and Force Systems

sidered thus far is that in the concurrent system the position of the resultant is known by inspection, while in the parallel system it is not known.

In order to obtain the resultant of a parallel force system such as shown in Fig. 2.29, we will make use of Varignon's theorem, namely, that the moment of the resultant is equal to the sum of the moments of its components. As applied to the situation illustrated in Fig. 2.29, we first find the magnitude of the resultant by adding the individual vectors algebraically since the addition of parallel vectors is the same mathematically as the addition of collinear vectors. We therefore can write

$$R = F_1 + F_2 + F_3 \tag{2.7}$$

Applying Varignon's theorem yields the following relation for moments taken about the axis O (which is arbitrary):

$$R\bar{x} = F_1 x_1 + F_2 x_2 + F_3 x_3 \tag{2.8}$$

Equation (2.8) expresses the fact that the resultant has the same moment as the force system that it replaces. We shall use these conclusions later on when studying centroids and centers of gravity.

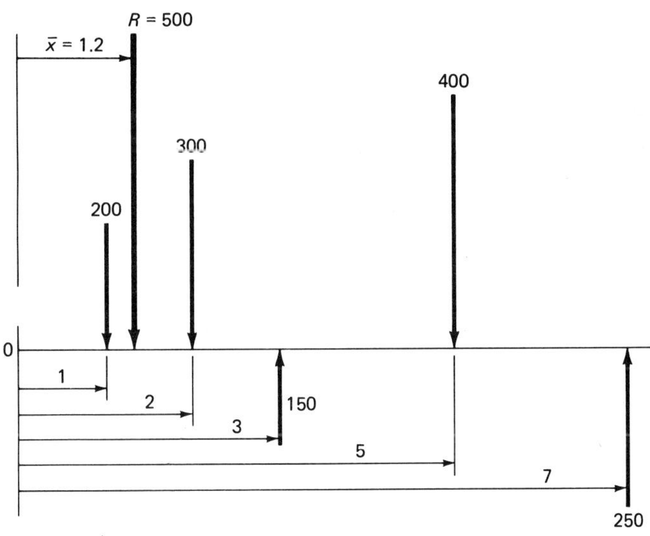

FIGURE 2.30   Illustrative Problem 2.16.

SOLUTION

Our first task is to assign an algebraic convention to the system. We shall use the conventional designation that up is positive, down is negative, and counterclockwise moments are positive. Applying Eq. (2.7) to this situation gives

$$R = -200 - 300 + 150 - 400 + 250 = -500$$

The resultant is a 500-lb force acting down. To find the line of action we take moments about the left side of the system and apply Eq. (2.8),

$$(-500)\,(\bar{x}) = -200\,(1) - 300\,(2) + 150\,(3) - 400\,(5) + 250\,(7)$$

$$\bar{x} = \frac{-200 - 600 + 450 - 2000 + 1750}{-500}$$

and

$$\bar{x} = 1.2$$

Notice that in the solution a downward acting force to the right of the axis O yields a clockwise $(-)$ moment, which is consistent with our convention.

A special case of the parallel force system occurs when the system is composed of two vectors, equal in magnitude but directed in opposite directions. For this force system the resultant is zero in magnitude, but there is a resultant moment.

This force system is known as a *couple*, and it is illustrated in Fig. 2.31. To determine the net moment produced by this couple let us evaluate its moment about the axis O. Taking moments of each force about O yields

$$M = F(d + x) - F(x) = Fd \qquad (2.9)$$

From this illustration we can now define the *moment of a couple* about any point or axis as the product of the magnitude of either force and the perpendicular distance between the lines of action of the forces. In keeping with our earlier convention, we will consider the moment of a couple to be positive if the sense of the rotation caused by the couple is counterclockwise. Notice that the moment of the couple Fd is inde-

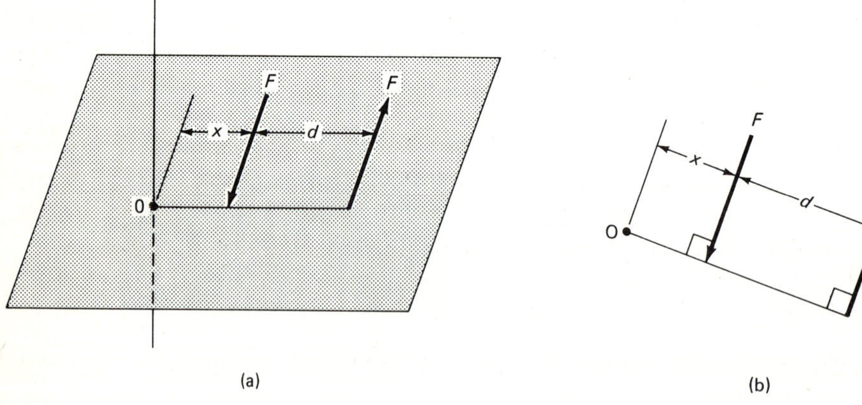

FIGURE 2.31   A couple.

                                                          Vectors and Force Systems

pendent of the location of the axis. Notice too that the resultant is a pure moment and can be replaced by any pair of oppositely directed forces that will yield the same moment. A couple is fully specified once the magnitude of its moment is given and the sense of rotation is stated. For the general case of forces in space it is also necessary to specify the orientation of the plane containing the couple.

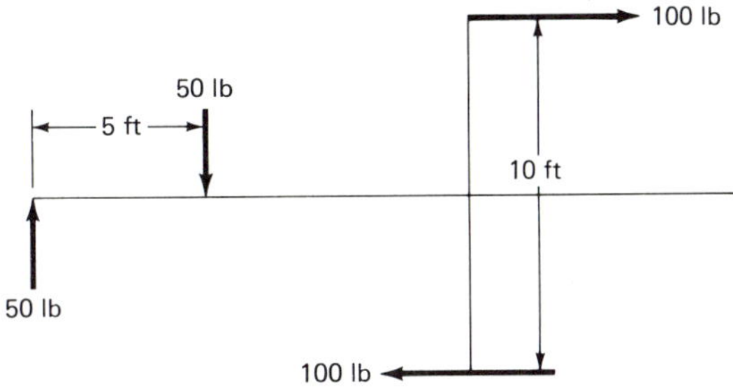

FIGURE 2.32   Illustrative Problem 2.17.

SOLUTION

Each set of forces forms a couple. Thus the 50-lb force system is a couple whose moment is $50 \times 5 = 250$ ft·lb clockwise (-250 ft·lb). The 100-lb system is also a couple whose sense is clockwise ($-$) and whose magnitude is $100 \times 10 = 1000$ ft·lb. The net moment is therefore $-250 + (-1000) = -1250$ ft·lb (clockwise). The net force is zero.

2.8   FORCES IN SPACE

All of the concepts that we have studied thus far have been applied to coplanar force systems. These same concepts are applicable to force systems that are not coplanar, that is, force systems in space. For the present we will restrict our attention to forces in space that are concurrent, that is, a force system in which the lines of action of the forces intersect in a common point. Let us first consider a single force in space and determine its x, y, and z components. In order to do this we first set up our coordinate system starting at the selected reference point O. The angles that the vector F makes with the x, y, and z axes are labeled θ_x, θ_y, and θ_z, respectively. Our vector F lies in the plane OACB (shaded) and from the geometry of Fig. 2.33 we can write

$$(OB)^2 = F_x^2 + F_z^2 \tag{2.10}$$

Since the triangle OBC is a right triangle, with OC (F) the hypotenuse, we also have

$$F^2 = F_y^2 + (OB)^2 \tag{2.11}$$

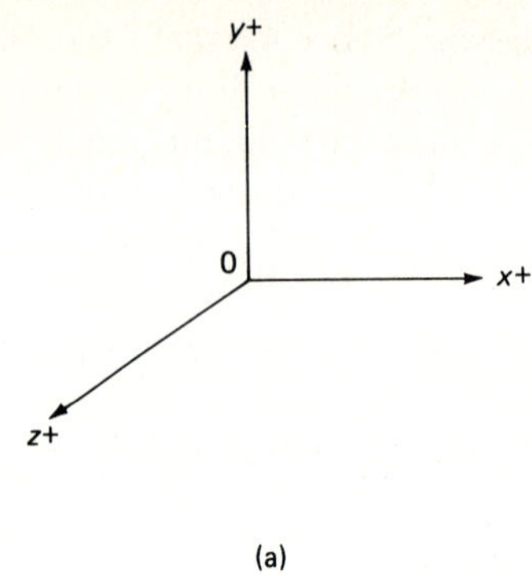

(a)

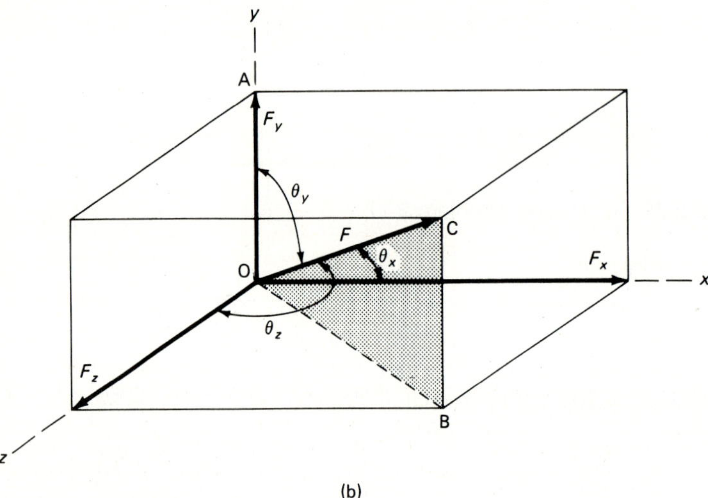

(b)

FIGURE 2.33 Components of a force in space.

Combining Eqs. (2.10) and (2.11),

$$F^2 = F_x^2 + F_y^2 + F_z^2 \qquad (2.12)$$

If we again refer to Fig. 2.33 we see that θ_x lies in the plane determined by F and the x axis. Similarly θ_y lies in the plane determined by F and the y axis, and θ_z lies in the plane determined by F and the z axis. Although these are shown in the figure, it is important to be aware that there is a distortion of the angles due to the perspective of the figure. We now consider triangle OAC, which is a right triangle. F is the hypotenuse, and the angle OAC is a right angle. F_y can be written for this triangle from basic trigonometric considerations as

$$F_y = F \cos \theta_y \qquad (2.13a)$$

   Vectors and Force Systems

We obtain similar expressions for F_x and F_z as

$$F_x = F \cos \theta_x \tag{2.13b}$$

$$F_z = F \cos \theta_z \tag{2.13c}$$

If we now combine Eqs. (2.12) and (2.13), we obtain

$$F^2 = F_x^2 + F_y^2 + F_z^2 = F^2 \cos^2 \theta_x + F^2 \cos^2 \theta_y + F^2 \cos^2 \theta_z \tag{2.14}$$

Based upon Eq. (2.14) we can conclude that

$$\cos^2 \theta_x + \cos^2 \theta_y + \cos^2 \theta_z = 1 \tag{2.15}$$

The cosine of the angles θ_x, θ_y, and θ_z are also called the *direction cosines*. From Eq. (2.13)

$$\cos \theta_x = \frac{F_x}{F}$$

$$\cos \theta_y = \frac{F_y}{F} \tag{2.16}$$

$$\cos \theta_z = \frac{F_z}{F}$$

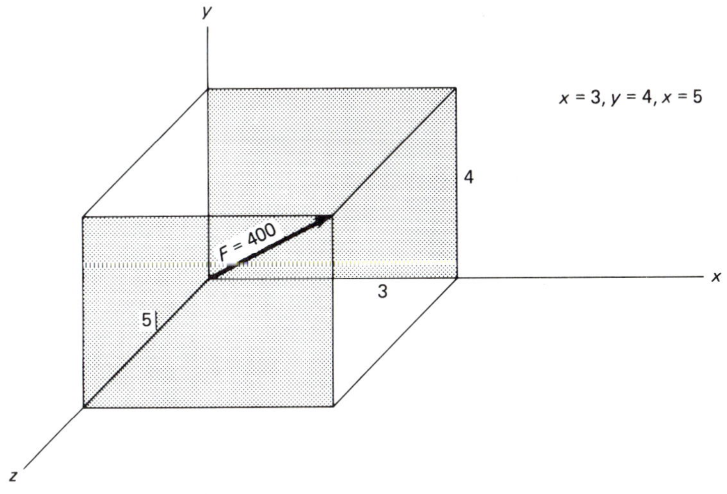

FIGURE 2.34   Illustrative Problem 2.18.

SOLUTION

We first determine the length of the line from the origin to the point $x = 3$, $y = 4$, $z = 5$ (tail to tip). This distance d is given by Eq. (2.12) as

$$d^2 = x^2 + y^2 + z^2 = 3^2 + 4^2 + 5^2$$

Hence

$$d = \sqrt{50} = 7.07$$

The direction cosines of the angles are

$$\cos \theta_x = \frac{3}{7.07} = 0.424 \qquad \theta_x = 64.9°$$

$$\cos \theta_y = \frac{4}{7.07} = 0.566 \qquad \theta_y = 55.5°$$

$$\cos \theta_z = \frac{5}{7.07} = 0.707 \qquad \theta_z = 45.0°$$

As a check,

$$\cos^2 \theta_x + \cos^2 \theta_y + \cos^2 \theta_z = 1$$

$$0.424^2 + 0.566^2 + 0.707^2 = 1$$

$$0.18 + 0.32 + 0.50 = 1.0 \text{ (checks)}$$

Continuing, we obtain the components F_x, F_y, and F_z as

$$F_x = F \cos \theta_x = 400 \times 0.424 = 169.6 \text{ N}$$

$$F_y = F \cos \theta_y = 400 \times 0.566 = 226.4 \text{ N}$$

$$F_z = F \cos \theta_z = 400 \times 0.707 = 282.8 \text{ N}$$

2.9 CLOSURE

The material of this chapter incorporates the material of Chapter 1 with the concepts of vector addition and the resultant of force systems. It is a prerequisite for the material that will be studied throughout our study of statics and strength of materials. While a large number of illustrative problems appear in the text material, the student is cautioned that a mere understanding of the text material is only part of the learning process—the solution of problems represents the other part. Each problem assigned should be carefully and neatly solved. Most students will find that a neatly drawn figure will help in visualizing and solving problems. Keeping the arithmetic processes neatly displayed also helps to avoid arithmetic errors. These brief cautions will be found to be equally valid for most technical courses.

REFERENCES

Bassin, M. G., S. M. Brodsky, and H. Wolkoff, STATICS AND STRENGTH OF MATERIALS, 3rd ed. New York: McGraw-Hill, 1979.

Vectors and Force Systems

Greene, E. G., Principles of Physics. Englewood Cliffs, NJ: Prentice-Hall, 1962.

Levinson, I. J., Introduction to Mechanics, 2nd ed. Englewood Cliffs, NJ: Prentice-Hall, 1968.

Sears, F. W., M. W. Zemansky, and H. D. Young, College Physics, 4th ed. Reading, MA: Addison-Wesley, 1977.

Shames, I. H., Engineering Mechanics, vol. I—Statics, 2nd ed. Englewood Cliffs, NJ: Prentice-Hall, 1966.

Singer, F. L., Engineering Mechanics, 3rd ed. New York: Harper and Brothers, 1975.

White, H. E., Modern College Physics, 5th ed. Princeton, NJ: D. Van Nostrand, 1966.

PROBLEMS

2.1   Two men start out from a given point, one heads north at 10 mi/h the other heads south at 25 mi/h. How far apart are they after 2 h?

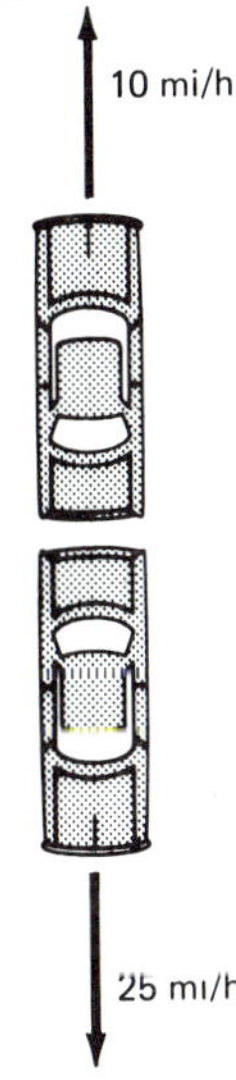

FIGURE P 2.1

2.2   A plane heads due west at an airspeed (relative to the air) of 450 km/h. The wind blows due east at 100 km/h. What is the plane's velocity relative to the earth?

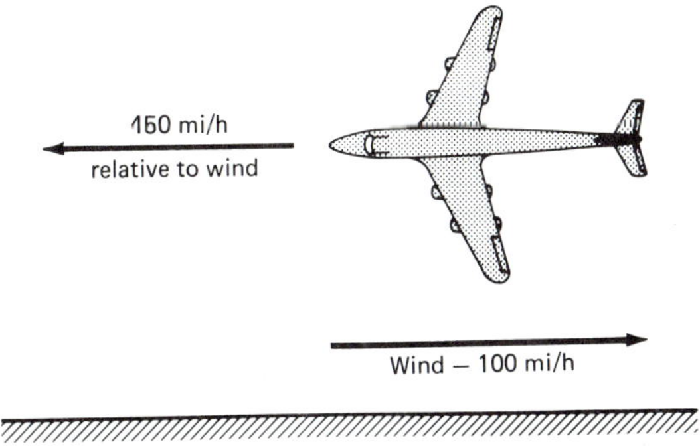

FIGURE P 2.2

Problems

2.3 A boat is traveling at 20 mi/h. A man runs at 8 km/h from the stern of the boat toward its bow. What is the man's velocity relative to the earth? If he runs from the bow to the stern at 8 km/h, what is his velocity relative to the earth?

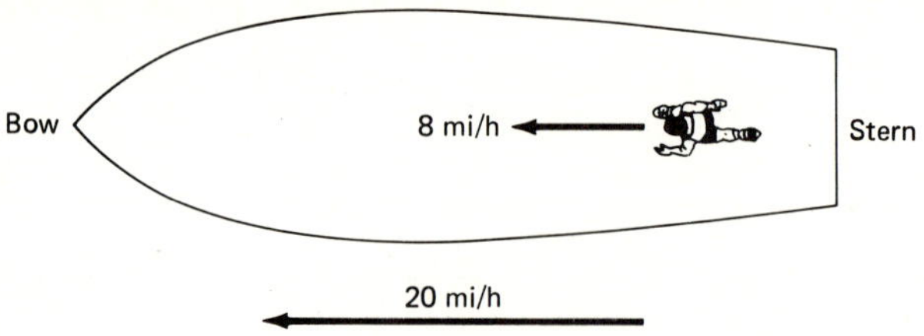

FIGURE P 2.3

2.4 A body is acted upon by the forces shown. Determine the resultant force acting on the body.

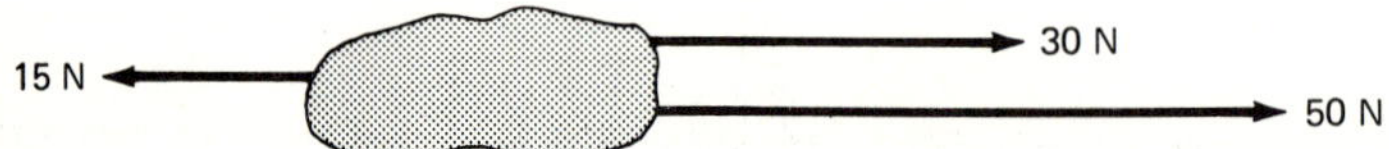

FIGURE P 2.4

2.5 Using the parallelogram law, determine the resultant of the forces shown.

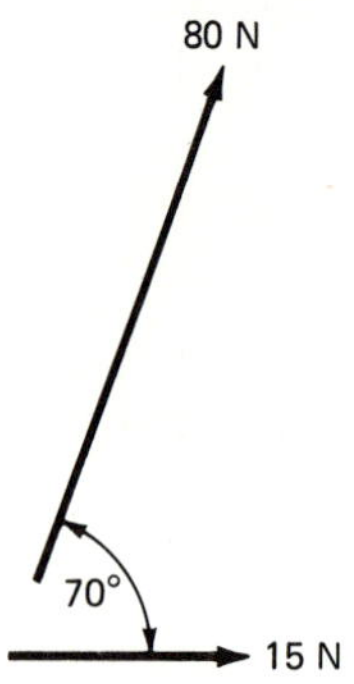

FIGURE P 2.5

2.6 Using the parallelogram law, determine the resultant of the forces shown.

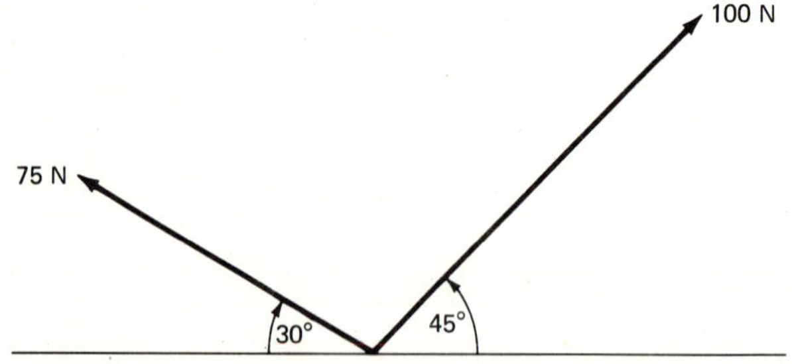

FIGURE P 2.6

 Vectors and Force Systems

2.7   Using the parallelogram law, determine the resultant of the forces shown.

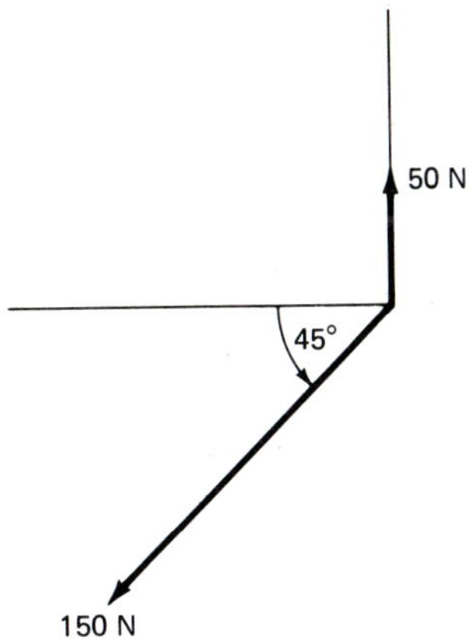

FIGURE P 2.7

2.8   A man can row across a still stream east to west at the rate of 5 mi/h. If the velocity of the stream is 3 mi/h from north to south, determine his velocity relative to the shore using the parallelogram law. Specify magnitude and direction.

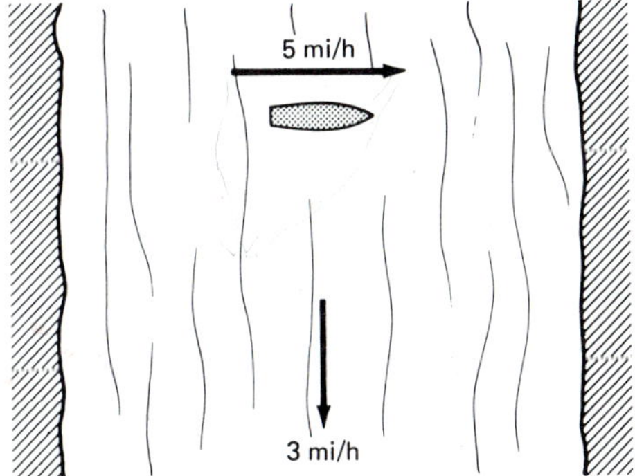

FIGURE P 2.8

2.9   Subtract a displacement of 5 km due south from a displacement of 6 km due east using the parallelogram law.

2.10   Subtract vector A from vector B using the parallelogram law.

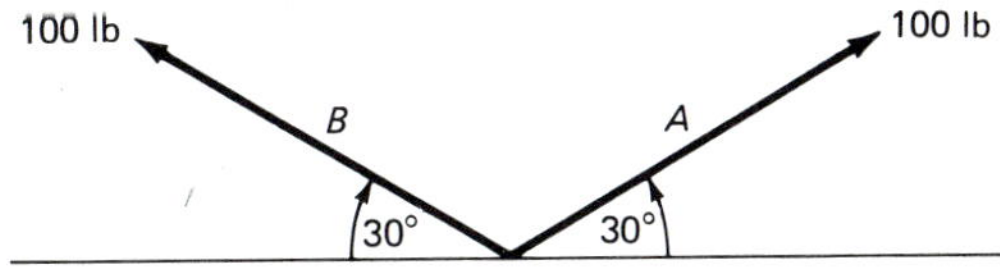

FIGURE P 2.10

2.11   Solve Problem 2.5 using the triangle law.

2.12   Solve Problem 2.6 using the triangle law.

Problems   61

2.13   Solve Problem 2.7 using the triangle law.

2.14   Solve Problem 2.8 using the triangle law.

2.15   Solve Problem 2.9 using the triangle law.

2.16   Solve Problem 2.10 using the triangle law.

2.17   A man walks 5 mi east and 3 mi north. How far is he from his starting point? At what angle to the easterly direction is he located?

2.18   Add the three vectors shown using the polygon method.

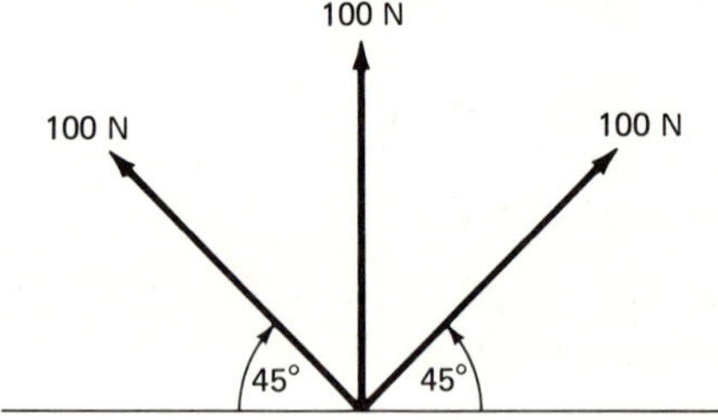

FIGURE P 2.18

2.19   If a woman walks 10 mi north, then 5 mi east, and finally 2 mi south, determine how far she is from her starting point.

2.20   Determine the resultant of the vectors shown using the polygon method.

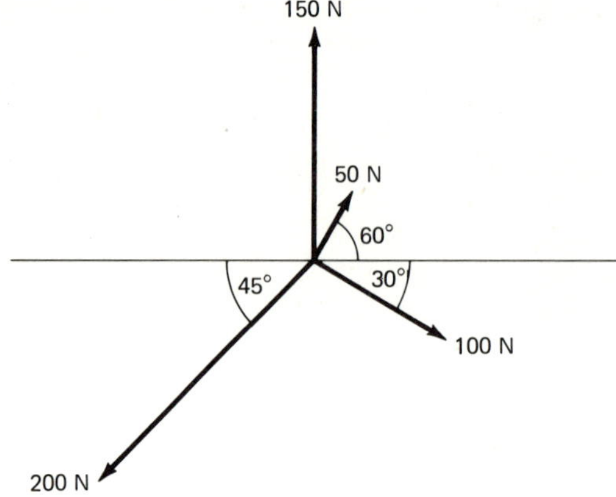

FIGURE P 2.20

2.21   Resolve the vectors shown into horizontal and vertical components.

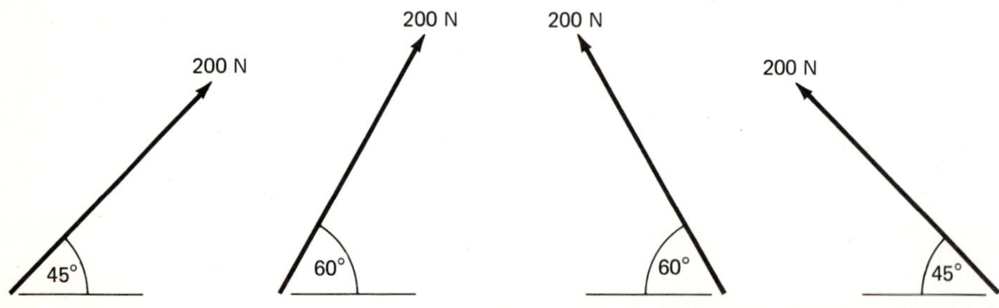

FIGURE P 2.21

2.22 Solve Problem 2.5 using the components method.

2.23 Solve Problem 2.6 using the components method.

2.24 Solve Problem 2.7 using the components method.

2.25 Solve Problem 2.8 using the components method.

2.26 Solve Problem 2.9 using the components method.

2.27 Solve Problem 2.10 using the components method.

2.28 Resolve a 100-lb force acting vertically downward into two components, one of which is directed 45° down from the horizontal toward the right and the other component is directed at right angle to the first component.

2.29 A force of 200 N acting vertically up is to be added to another force acting at an angle of 45° down from the horizontal. If their resultant is horizontal, determine the magnitude of the unknown force and also of the resultant.

2.30 Determine the resultant of the forces shown.

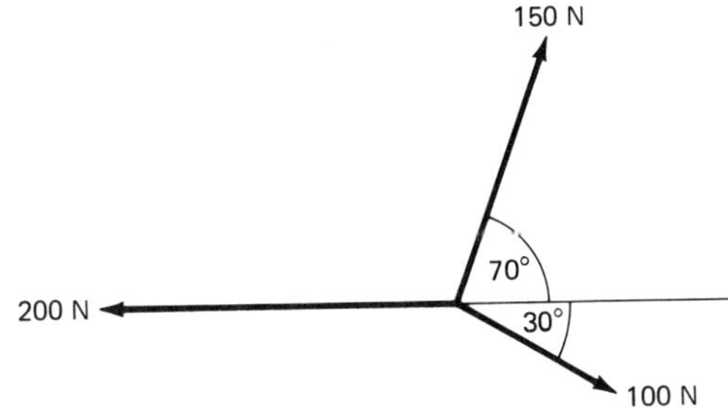

FIGURE P 2.30

2.31 Determine the resultant of the forces shown.

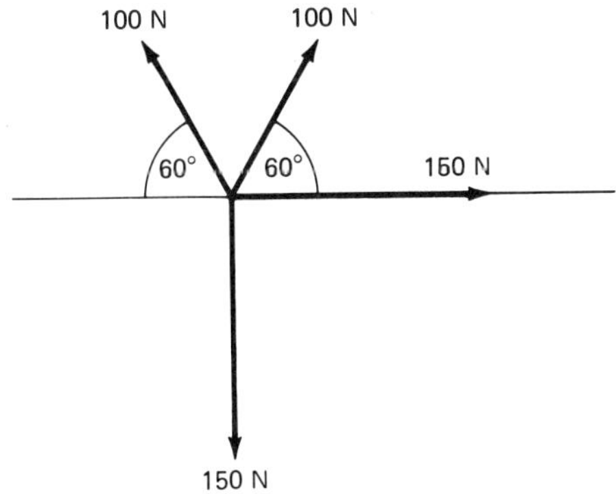

FIGURE P 2.31

Problems

63

2.32   Determine the moment with respect to O of the forces shown.

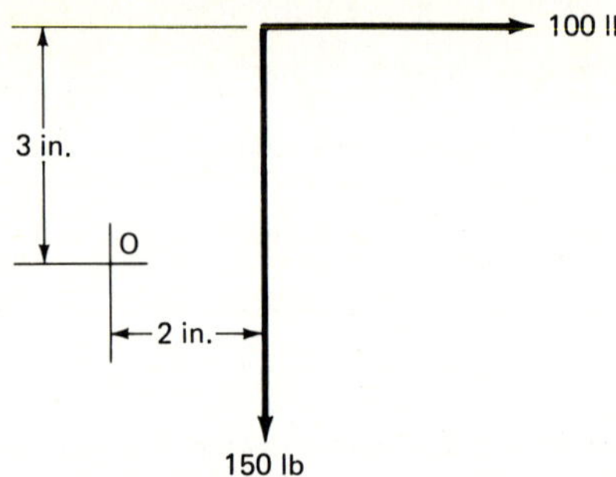

FIGURE P 2.32

2.33   Determine the moment of the force shown about the axis O.

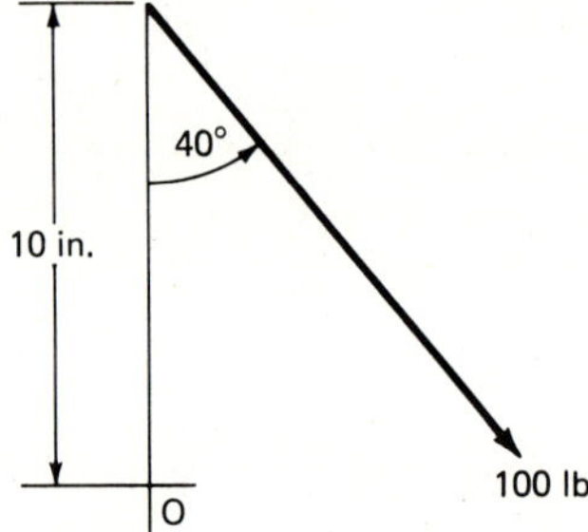

FIGURE P 2.33

2.34   Determine the location of the resultant of the force system shown.

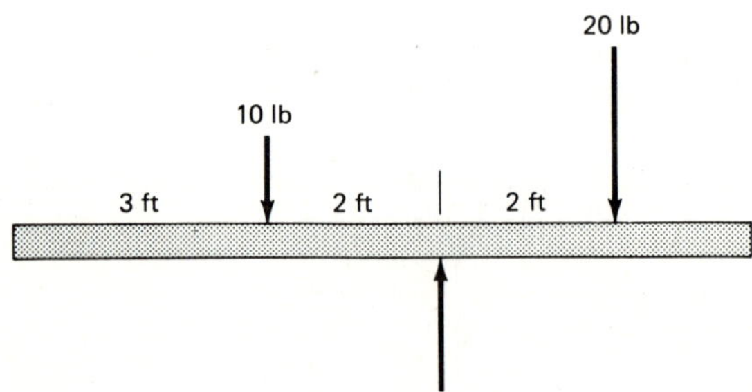

FIGURE P 2.34

2.35   Determine the resultant of the force system shown.

                                          Vectors and Force Systems

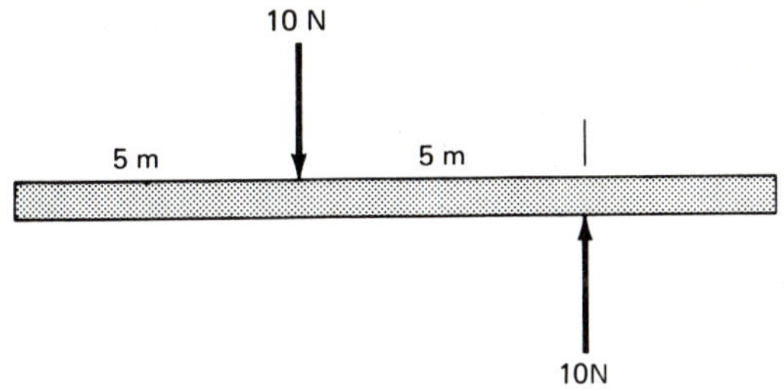

2.36 Determine the resultant of the force system shown.

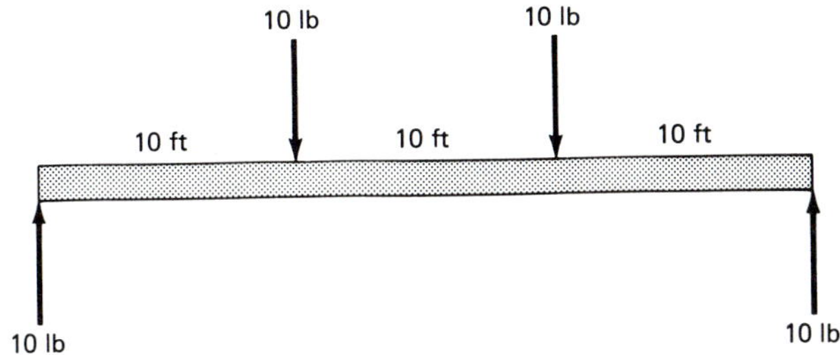

2.37 Replace the force shown by a force through point P and a moment, such that the resultant of the system remains the same.

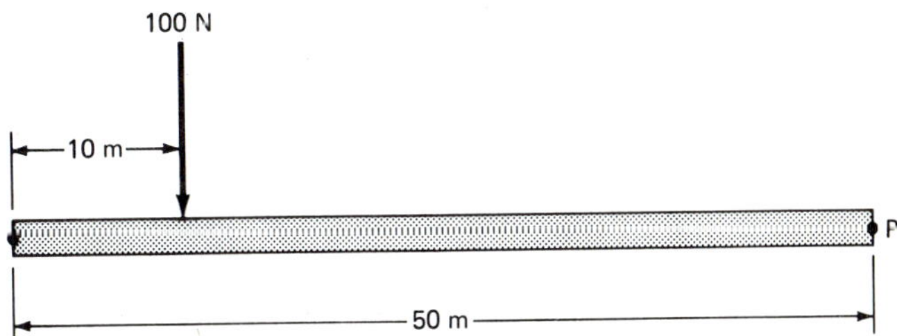

2.38 Determine the resultant of the force system shown.

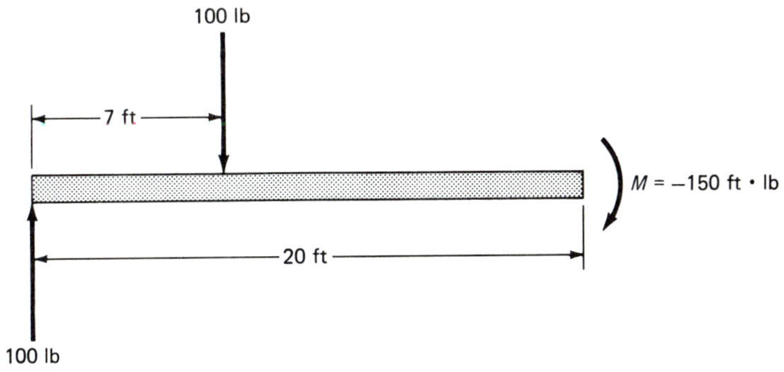

2.39 If the end moment shown in Figure P 2.38 is located at 10 ft along the beam, determine
the resultant of the force system shown.

2.40 Determine the resultant of the force system shown.

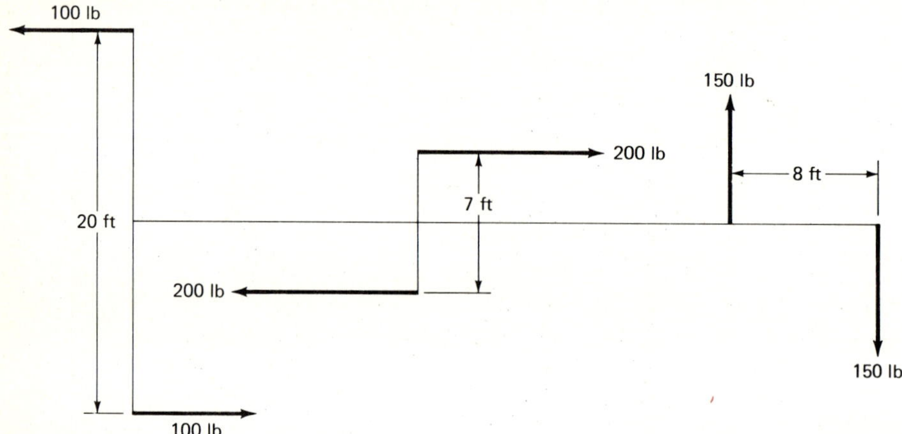

FIGURE P 2.40

2.41 A vector in space starts at the origin, and its tip is located at $x = 3$, $y = 5$, and $z = 7$. If the vector is 100 N, determine its x, y, and z components.

2.42 If the x, y, and z components of a vector are 100 N, 100 N, and 200 N, respectively, determine the magnitude and direction cosines of the resultant vector.

2.43 Resolve a 200-lb vector in space that starts at the origin and whose tip goes through the coordinates $x = -4$, $y = 5$, $z = -5$ into its x, y, and z components.

2.44 If the x, y, and z components of a vector are -400, -200, and $+100$ N, respectively, determine its magnitude and direction cosines.

Vectors and Force Systems

$$$$

3 Equilibrium

3.1 INTRODUCTION

In this chapter our study will concern itself with bodies that are in equilibrium. Our concept of equilibrium derives from Newton's first law of motion, which requires a body to be at rest or moving with a uniform velocity unless it is compelled to change its motion by the action of external forces acting on it. We therefore define equilibrium to be that state in which a body is at rest or moving with a uniform velocity. The converse of Newton's first law provides the necessary conditions for equilibrium to exist, namely, there must be no net external force or moment acting on the body. Therefore equilibrium requires the resultant of all external forces and all external moments acting on a body to be zero. We can therefore write

$$\Sigma F = 0 \tag{3.1}$$

$$\Sigma M = 0 \tag{3.2}$$

as the necessary conditions for equilibrium. It should be noted that Eqs. (3.1) and (3.2) are independent conditions. For our present study we will concern ourselves with bodies that are at rest and are said to be in static equilibrium.

3.2 THE FREE-BODY DIAGRAM

As noted in Section 3.1, a body is in equilibrium when the resultant of all of the external forces and moments acting on it are zero. This requires us to be able to accurately describe all of the forces acting on the body in order to obtain a correct solution to any problem. In order to do this, a sketch of the body is made that is free of all other bodies and on which all forces exerted by other bodies *on* it are shown. In effect the body (or portions of the body) in question is completely isolated from all contacting or attached bodies, and their effects are replaced by vectors representing the forces they exert on the body in question. The diagram is known as a *free-body diagram*. The free-body diagram is the most useful single tool in mechanics, and it must be thoroughly understood by the student. Its applications extend to the subjects of dynamics, strength of materials, fluid mechanics, thermodynamics, aerodynamics, and so on, and its importance as a conceptual tool cannot be overemphasized.

In applications the free-body diagram is constructed by making a sketch of the body (or portions of a body) in question. This sketch must show the body completely cut from all other bodies. Then the action on the body in question of each body removed by the cutting process is shown as a force. It is important to include all forces acting on the body, but by the same token the inclusion of nonexistent forces can lead to incorrect results.

In order to draw a free-body diagram it is necessary to correctly identify the action of the cut body or the body in question. The most common situations that occur will now be briefly considered so that when they occur, we may readily identify them in a given problem.

1. *The rope or cable* (Fig. 3.1). This member is considered to be perfectly flexible and only able to sustain tension along its axis.

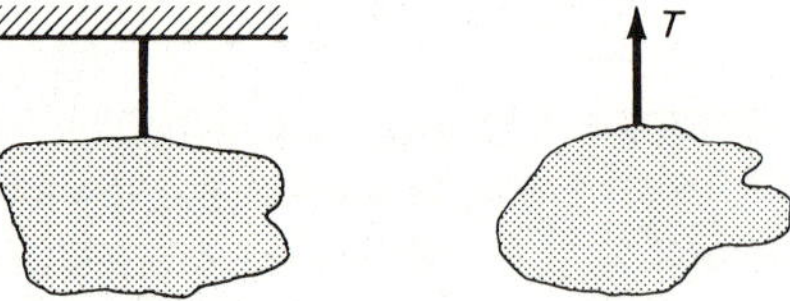

2. *Gravity* (Fig. 3.2). The action of the earth on a body is a force directed vertically downward through the center of gravity of the body and equal to its weight.

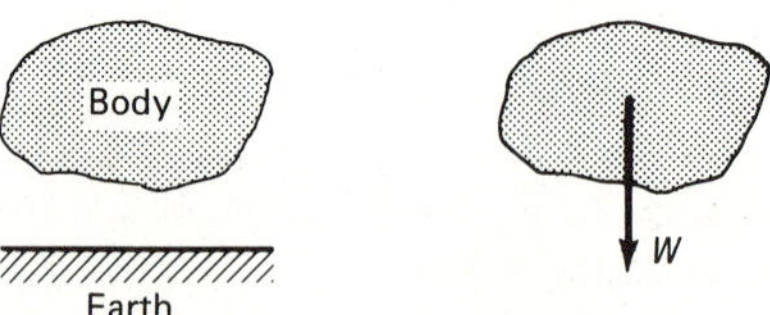

FIGURE 3.2 Gravity.

3. *Smooth surface* (Fig. 3.3). The action of a smooth surface is indicated by a single force perpendicular to the smooth surface since there is no frictional resistance to sliding along the surface.

                                                    Equilibrium

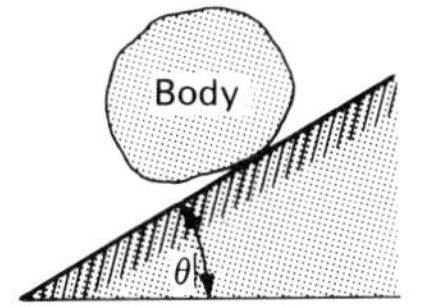
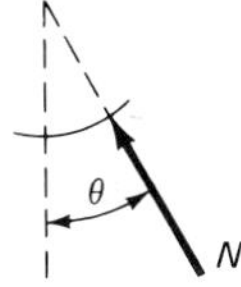

FIGURE 3.3   Smooth surface.

4. *Roller* (Fig. 3.4). The action of a roller is similar to the action of a smooth surface and is always represented as a force perpendicular to the surface on which the roller rests.

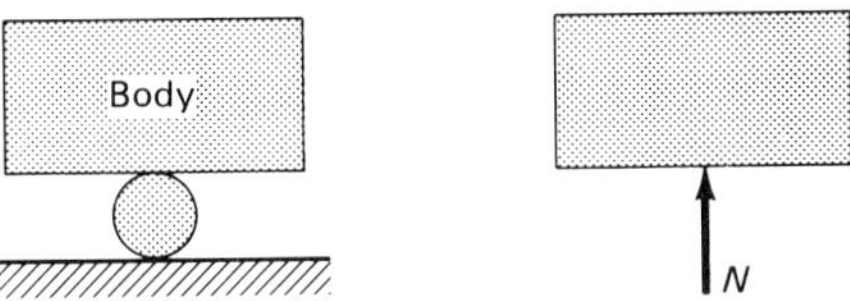

FIGURE 3.4   Roller.

5. *Smooth pin* (Fig. 3.5). The force exerted by a pin on a body is through the pin at some unknown angle. We shall usually show this force by denoting its two components.

FIGURE 3.5   Smooth pin.

6. *Rough surface* (Fig. 3.6). The same considerations hold as for the smooth pin.

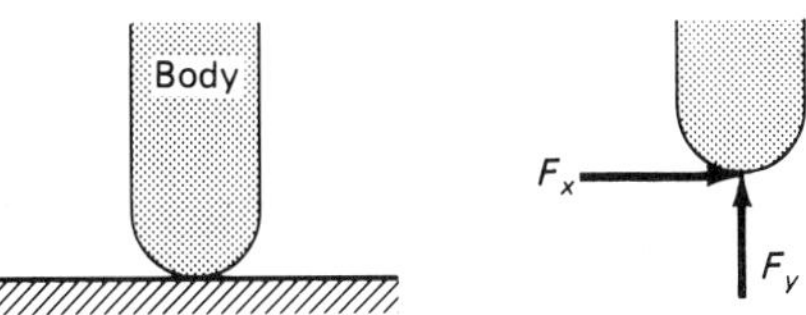

FIGURE 3.6   Rough surface.

7. *Fixed support* (Fig. 3.7). An unknown force and a couple usually shown as two components and a moment.

FIGURE 3.7   Fixed support.

The Free-Body Diagram

It is noted that one of Newton's laws states that for every action there is an equal but opposite reaction. Thus for each force on a body there is an equal, collinear, oppositely directed force acting on the body in contact with it. However, we once again emphasize that a free-body diagram only shows the forces acting on the body and not the forces it exerts on other bodies.

A block rests on a rough, inclined plane as shown in Fig. 3.8. Draw a free-body diagram of the block. The force F is parallel to the plane. The block tends to move up the plane.

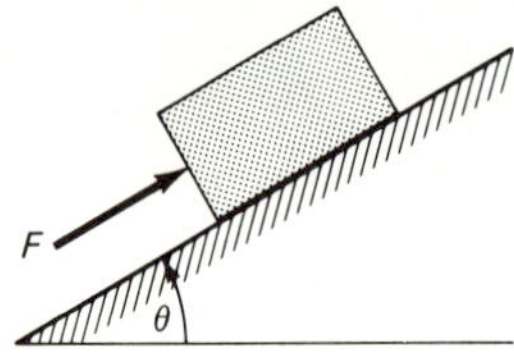

FIGURE 3.8   Illustrative Problem 3.1.

SOLUTION

We first draw the block and note that its weight acts vertically down through its center of gravity. Since the plane is rough, the reaction of the plane on the body will not be perpendicular to the plane (see also Chapter 5). Instead of drawing the reactive force of the plane on the block, we will show its components parallel to and perpendicular to the plane, denoting these as F_F and N. Figure 3.9 shows the complete free-body diagram. The choice of axes perpendicular to and parallel to the plane was arbitrary but, as will be seen later, quite convenient.

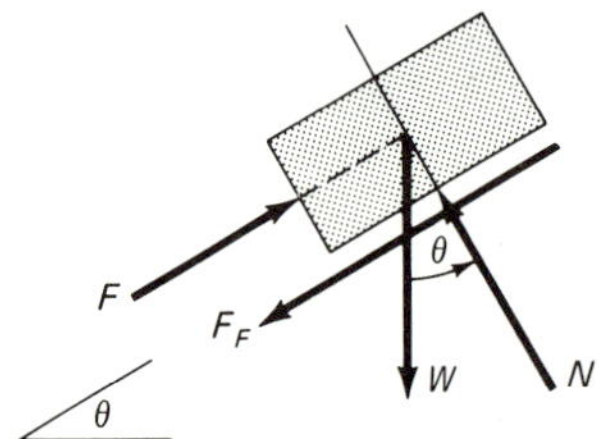

FIGURE 3.9   Illustrative Problem 3.1.

Draw a free-body diagram of the cylinder A and the rod BC. The cylinder weighs W N. Neglect the weight of the rod.

SOLUTION

Let us first consider the cylinder. It contacts the rod at D and the wall at E. At each of these points there is a force perpendicular to the surface (see item 4, roller). In addition, there is its weight W acting through its center vertically

                                                                 Equilibrium

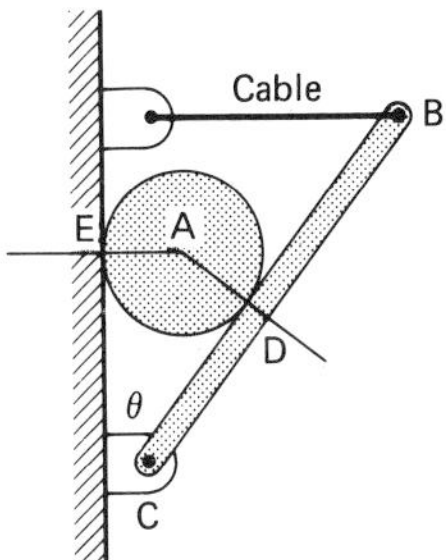

FIGURE 3.10 Illustrative Problem 3.2.

down. The free-body diagram of the cylinder is shown in Fig. 3.11(a). For the rod we have a cable attached at B, which exerts a horizontal tension T on the rod at B. At D there is the force due to the cylinder, which must be equal and opposite to the force D, which we showed to act on the cylinder. At the pinned end C we have the two components of the force (see item 5, smooth pin). The complete free-body diagram of the rod is shown in Fig. 3.11(b).

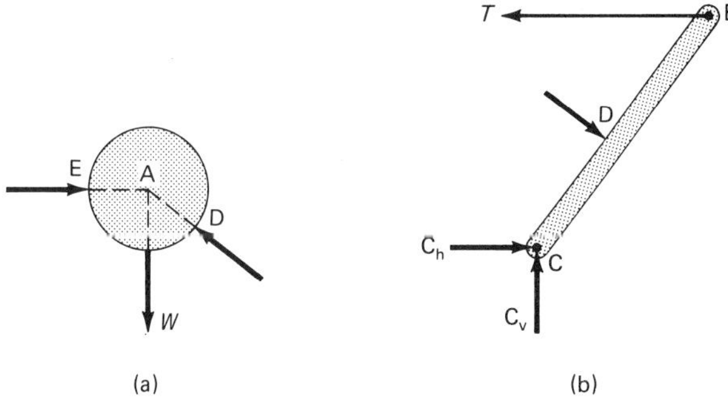

FIGURE 3.11 Illustrative Problem 3.2.

ILLUSTRATIVE PROBLEM 3.3

Draw a free-body diagram of each of the spheres shown in Fig. 3.12 if all of the contact surfaces can be assumed to be smooth.

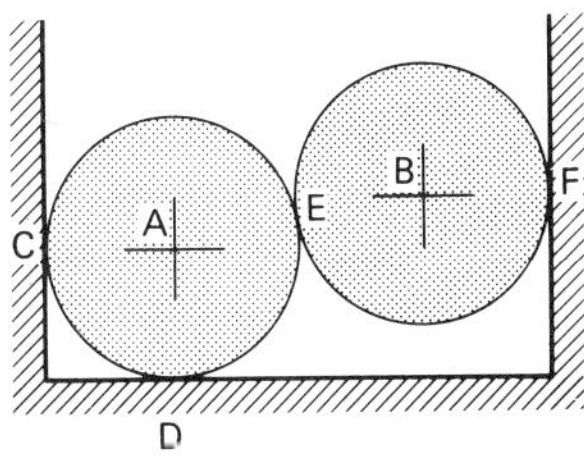

FIGURE 3.12 Illustrative Problem 3.3.

The Free-Body Diagram

SOLUTION

Since we are dealing with smooth surfaces, the contact force at each point of contact for each sphere must be normal to the surface of contact. Thus the force exerted by B on A (and A on B) must lie along the line of centers between the spheres. Figure 3.13 shows the complete free-body diagram for each sphere. Note that the force at E on body A is equal to but oppositely directed to the force at E on body B.

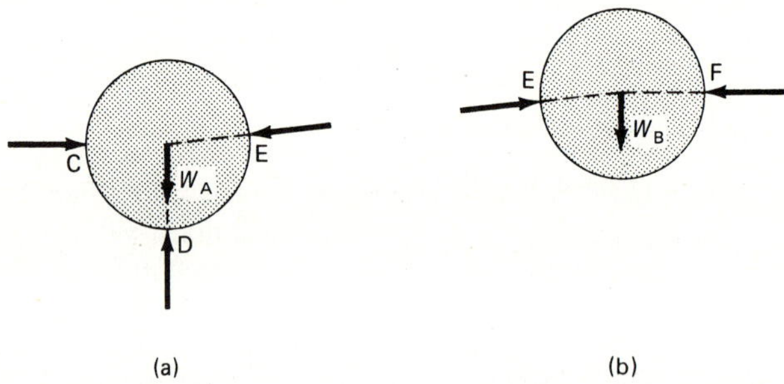

FIGURE 3.13   Illustrative Problem 3.3.

3.3   THE EQUATIONS OF EQUILIBRIUM—TWO-DIMENSIONAL FORCE SYSTEMS

We have already stated in Section 3.1 that equilibrium requires that the resultant of all external forces and all external moments acting on a body be zero, which can be written as

$$\Sigma F = 0 \tag{3.1}$$

$$\Sigma M = 0 \tag{3.2}$$

An alternate way of writing these equations is

$$R = \sqrt{(\Sigma F_x)^2 + (\Sigma F_y)^2} = 0 \tag{3.3}$$

$$\Sigma F_x = 0 \tag{3.4}$$

$$\Sigma F_y = 0 \tag{3.5}$$

as well as

$$\Sigma M_O = 0 \tag{3.6}$$

Equations (3.1) through (3.6) have been written for any coplanar force system, sometimes denoted as a two-dimensional force system. The term ΣM_O in Eq. (3.6) represents the summation of the moments about an axis perpendicular to the x–y plane.

There are four two-dimensional force systems that we will consider:

1.   The collinear force system

Equilibrium

2. The parallel force system
3. The concurrent force system
4. The nonconcurrent force system

The condition for equilibrium for two collinear forces can be written directly from a consideration of the conditions shown in Fig. 3.14. If the forces are collinear, they must be parallel, oppositely directed, and equal in magnitude.

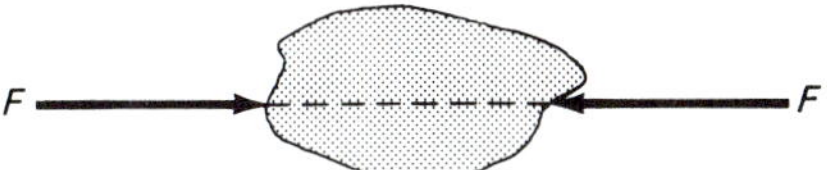

FIGURE 3.14   Collinear forces.

Conversely, if only two forces act on a body, they must be collinear. Also $\Sigma M_O = 0$ is obviously always satisfied.

ILLUSTRATIVE PROBLEM 3.4

A small weightless ball is held in place by two springs, as shown in Fig. 3.15(a). If one spring exerts a 100-N tensile force on the body, what force is exerted by the other spring?

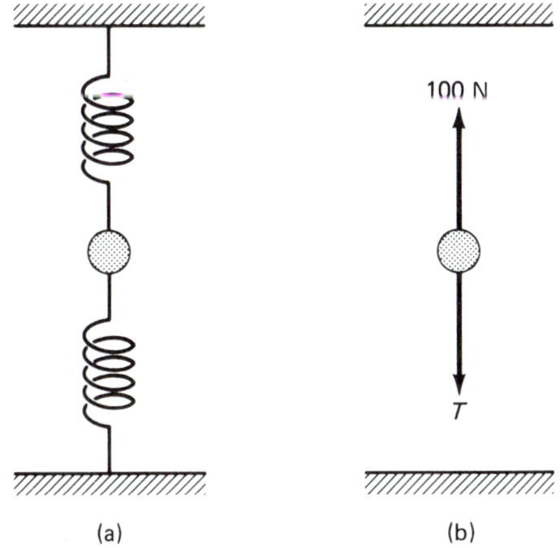

FIGURE 3.15   Illustrative Problem 3.4.

SOLUTION

Referring to Fig. 3.15(b), the free-body diagram, we write $\Sigma F_y = 0$ and obtain immediately $T = 100$ N. Since the forces are collinear, $\Sigma M_O = 0$ is automatically satisfied.

The parallel force system will be in equilibrium when its resultant is zero. By selecting the orthogonal axes so that the parallel forces are all pointed parallel to the direction of the axis, we can eliminate one of the directions, which reduces the equations

The Equations of Equilibrium

for equilibrium for this case to $\Sigma F_y = 0$, $\Sigma M_O = 0$. It is sometimes more convenient to replace the force summation by $\Sigma M_A = 0$, where O and A do not lie on a line parallel to the force system.

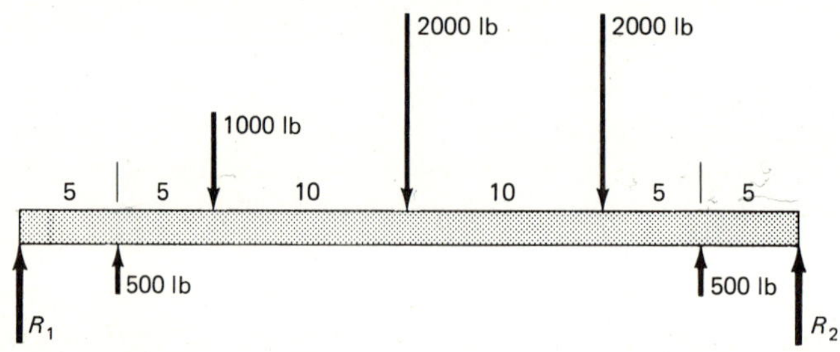

FIGURE 3.16   Illustrative Problem 3.5.

SOLUTION

For this case Fig. 3.16 can also be used as the free-body diagram of the beam. Using the condition that $\Sigma M_{R_1} = 0$ and $\Sigma M_{R_2} = 0$ will give us the unknown reactions as follows.

For $\Sigma M_{R_1} = 0$,

$$500 \times 5 - 1000 \times 10 - 2000 \times 20 - 2000 \times 30 + 500 \times 35 + 40R_2 = 0$$

$$2500 - 10\,000 - 40\,000 - 60\,000 + 17\,500 + 40R_2 = 0$$

$$40R_2 - 90\,000 = 0$$

$$R_2 = 2250 \text{ lb}$$

For $\Sigma M_{R_2} = 0$,

$$500 \times 5 - 2000 \times 10 - 2000 \times 20 - 1000 \times 30 + 500 \times 35 + 40R_1 = 0$$

$$2500 - 20\,000 - 40\,000 - 30\,000 + 17\,500 + 40R_1 = 0$$

$$40R_1 - 70\,000 = 0$$

$$R_1 = 1750 \text{ lb}$$

The requirement that $\Sigma F_y = 0$ can be used as a check on the calculation. Notice that $\Sigma F_y = 0$ is not an independent condition once the two moment equations are used. For $\Sigma F_y = 0$,

$$1750 + 500 - 100 - 200 - 200 + 500 + 2250 = 0 \text{ (checks)}$$

Equilibrium

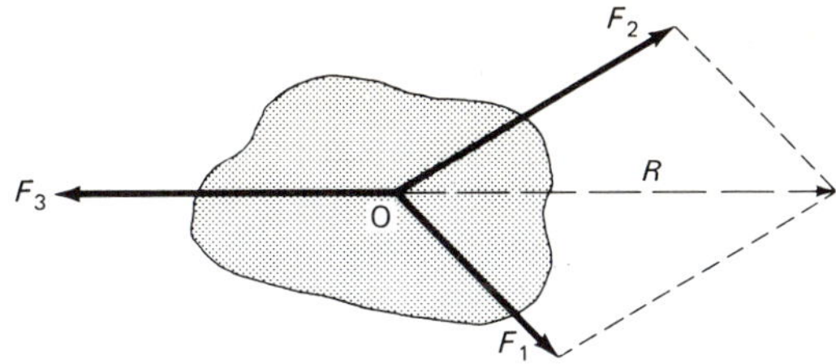

FIGURE 3.17 A concurrent, nonparallel, coplanar force system.

When a body in equilibrium is acted upon by three coplanar nonparallel forces, these forces must have lines of action that must intersect (extended, if necessary) at a single point; in other words, they must be concurrent. We can reach this conclusion by considering the three coplanar nonparallel forces shown in Fig. 3.17 which are in equilibrium. Using the parallelogram law, we can combine any two of these forces to obtain a single resultant R as shown. The system of R and F_3 now becomes a collinear force system in which we have already seen that equilibrium requires the two forces to be collinear, equal, and oppositely directed. Therefore R and F_3 must have common lines of action passing through the common point O. An alternate viewpoint is that if the forces did not have lines of action that extend through the common point O, then $\Sigma M_O = 0$ could not be met, and this condition for equilibrium would be violated.

ILLUSTRATIVE PROBLEM 3.6

A crane is constructed, as shown schematically in Fig. 3.18, with the cable C used to support the free end of the beam. Determine the tension in the cable and the reaction at pin A. Neglect the weight of the beam.

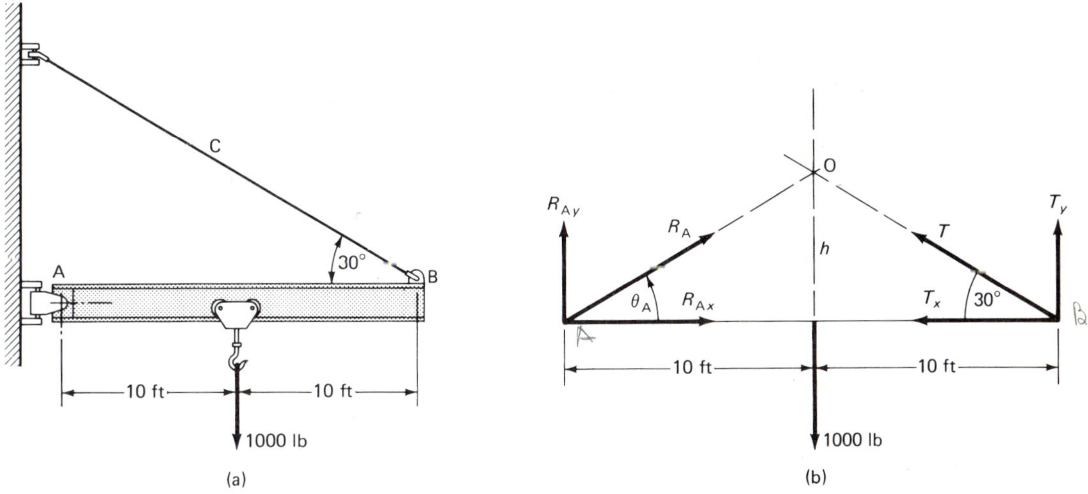

FIGURE 3.18 Illustrative Problem 3.6.

The Equations of Equilibrium

SOLUTION

We can solve this problem by several methods. If we consider the free-body diagram of Fig. 3.18(b), we notice that the system consists of three coplanar, nonparallel forces. Their lines of action must therefore be concurrent, and for the condition of equilibrium to exist, the resultant of the three forces must be zero. If we use the polygon law, this means that the three force vectors must give us a closed figure in order for the resultant to be zero. The 1000-lb force is given, and the direction of the cable tension is known. Since the forces are concurrent, the reaction R_A must pass through the intersection of the lines of action of the 1000-lb force and the cable tension at point O. This establishes the slope of the reaction at R_A. If this slope is drawn from the tail of the 1000-lb force, as shown in Fig. 3.19, the polygon is completed and T and R_A can be calculated or measured. We can carry our this calculation as follows.

From Fig. 3.18(b), tan θ_A = $h/10$; but $h/10$ = tan 30°. Therefore θ_A must also be 30°. This makes each of the angles in Fig. 3.19 equal to 60°, that is, the triangle is an equilateral triangle, and therefore all sides are equal to 1000. Thus R_A = 1000 lb and T = 1000 lb.

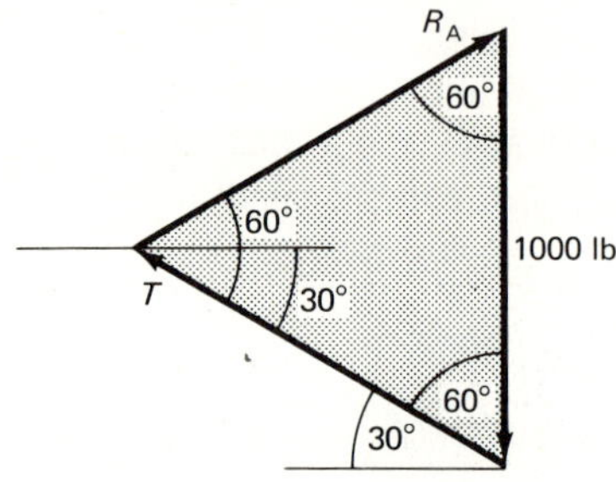

FIGURE 3.19 Force polygon for Illustrative Problem 3.6.

An alternate method is to use the equations of equilibrium and to use the components R_{Ax} and R_{Ay} of R_A, and T_x and T_y of T. We first take moments about A. For $\Sigma M_A = 0$,

$$-1000 \times 10 + 20T_y = 0$$

$$T_y = 500 \text{ lb}$$

But $T_y = T \sin 30° = 0.5T$. Thus $T = 500/0.5 = 1000$ lb.

From $\Sigma F_y = 0$,

$$R_{Ay} + T_y - 1000 = 0$$

$$R_{Ay} = 500 \text{ lb}$$

and from $\Sigma F_x = 0$,

$$R_{Ax} - T_x = 0 \qquad R_{Ax} = T_x$$

Equilibrium

But $T_x = T \cos 30° = 1000 \times 0.866 = 866$ lb. Therefore $R_{Ax} = 866$ lb and

$$R = \sqrt{R_{Ax}^2 + R_{Ay}^2} = \sqrt{500^2 + 866^2} = 1000 \text{ lb}$$

Both solutions check.

ILLUSTRATIVE PROBLEM 3.7

A 100-lb body is kept at rest on a smooth inclined plane by a force F, as shown in Fig. 3.20. Determine the magnitude of the force F and the normal force exerted by the plane on the body.

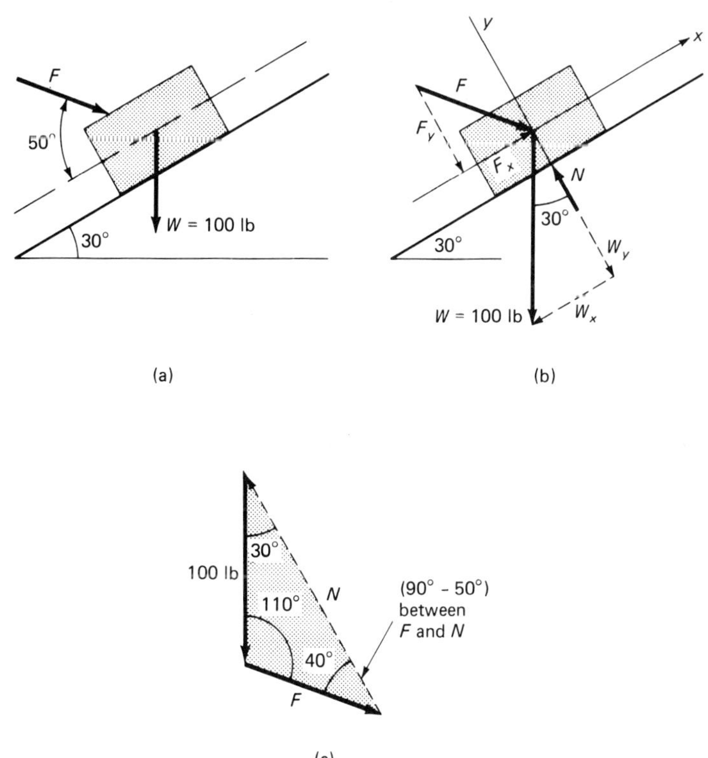

FIGURE 3.20   Illustrative Problem 3.7.

The Equations of Equilibrium

SOLUTION

Since the plane is specified to be smooth, the reaction of the plane on the body must be perpendicular to the plane, that is, the normal force exerted by the plane on the body is the *only* force the plane exerts on the body. We can set up the force polygon [(Fig. 3.20(c)] from the free-body diagram of Fig. 3.20(b) by noting that we are given the directions of the lines of action of the three forces and the magnitude of one of the forces. Therefore by applying the law of sines twice, we obtain

$$\frac{F}{\sin 30} = \frac{100}{\sin 40} \qquad F = \frac{100 \sin 30}{\sin 40} = \frac{100\,(0.5)}{0.643} = 77.8 \text{ lb}$$

and

$$\frac{N}{\sin 110} = \frac{100}{\sin 40} \qquad N = \frac{100 \sin 110}{\sin 40} = \frac{100\,(0.940)}{0.643} = 146.2 \text{ lb}$$

An alternate solution to Illustrative Problem 3.7 can be obtained by using $\Sigma F_x = 0$, $\Sigma F_y = 0$, where we will now select as the axes, axes that are parallel to and perpendicular to the plane, as shown in Fig. 3.20(b). Referring to Fig. 3.20(b), we show the components of each force along these axes and apply the equations of equilibrium. For $\Sigma F_x = 0$,

$$F \cos 50 - W \sin 30 = 0$$

$$0.643F = 100 \times 0.5$$

$$F = 77.8$$

For $\Sigma F_y = 0$,

$$N - F \sin 50 - 100 \cos 30 = 0$$

$$N - 77.8 \times 0.766 - 100 \times 0.866 = 0$$

$$N = 146.2$$

These values check the previous solution. Use of the force polygon reduces some of the algebra and provides a visual relation between the forces, which is sometimes very helpful.

The final coplanar force system that we will consider is the coplanar, noncurrent, nonparallel force system. This is the most general two-dimensional force system, and in principle the others can be obtained from it. However, having treated the other two-dimensional force systems, we can now readily treat this one. The requirements for equilibrium for this type of system is that Eqs. (3.4), (3.5), and (3.6) (or their equivalents) be satisfied. Regardless of the set of equations we may use, it should be noted that there are only at most three independent equations from statics to satisfy the requirement that a nonconcurrent coplanar force system is in equilibrium.

Equilibrium

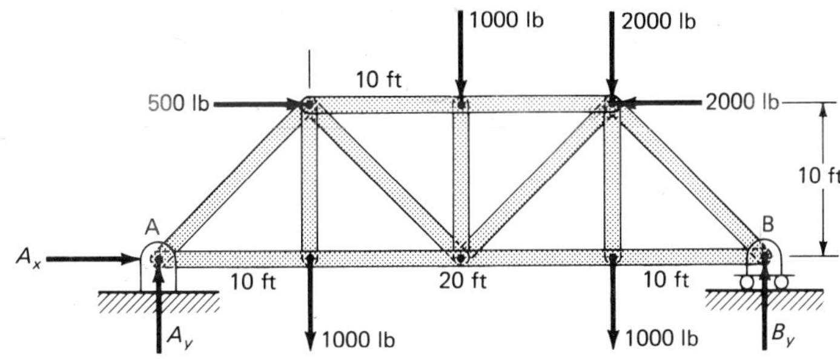

FIGURE 3.21   Illustrative Problem 3.8.

SOLUTION

At B the rollers give us a vertical force only, while at A the force has both horizontal and vertical components. By using A as the axis for moments, we can eliminate two variables and simplify the problem. Then for $\Sigma M_A = 0$,

$$-500 \times 10 - 1000 \times 10 - 20 \times 1000 - 30$$

$$\times 2000 - 1000 \times 30 + 2000 \times 10 + 40B_y = 0$$

$$B_y = 2625$$

Using $\Sigma F_y = 0$,

$$A_y - 1000 - 1000 - 2000 - 1000 + 3125 = 0$$

$$A_y = 2375$$

and $\Sigma F_x = 0$,

$$A_x + 500 - 2000 = 0$$

$$A_x = 1500$$

As a check, taking moments about B, for $\Sigma M_B = 0$,

$$-2000 \times 10 - 1000 \times 10 - 2000 \times 10$$

$$- 1000 \times 20 - 1000 \times 30 + 500 \times 10 + 40A_y = 0$$

$$A_y = 2375 \text{ (checks)}$$

Note that $\Sigma M_B = 0$ does not give us an independent condition—it does permit us to check one of the other equations.

The Equations of Equilibrium

A beam is supported by a pin at one end and rests on a smooth surface at the other end. Neglect the weight of the beam and determine the reaction at both ends of the beam.

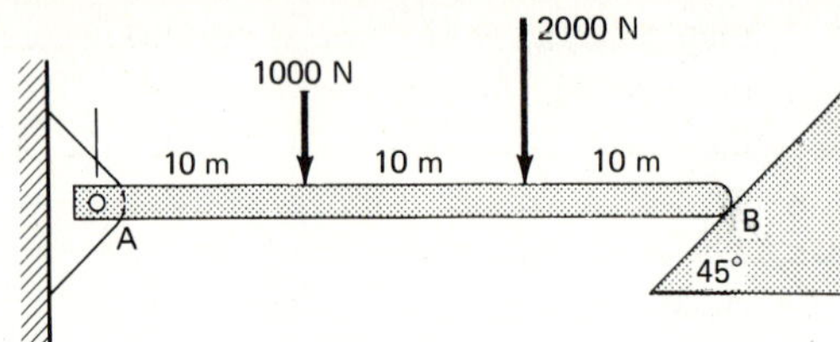

FIGURE 3.22   Illustrative Problem 3.9.

SOLUTION

Our first task is to draw the free-body diagram of the beam. This problem lends itself to the use of two moment equations, one at A and one at B. Therefore for $\Sigma M_B = 0$

$$-10 \times 2000 - 20 \times 1000 + 30 A_V = 0$$

$$A_V = 1333$$

for $\Sigma M_A = 0$

$$-10 \times 1000 - 20 \times 2000 + N \times 30 \cos 45° = 0$$

$$-10\,000 - 40\,000 - 21.21N = 0$$

$$N = 2357$$

and for $\Sigma F_x = 0$

$$A_H - 2357 \times 0.707 = 0$$

$$A_H = 1667 \text{ N}$$

As a check, for $\Sigma F_y = 0$

$$1333 - 1000 - 2000 + 2357 \times 0.707 = 0 \text{ (checks)}$$

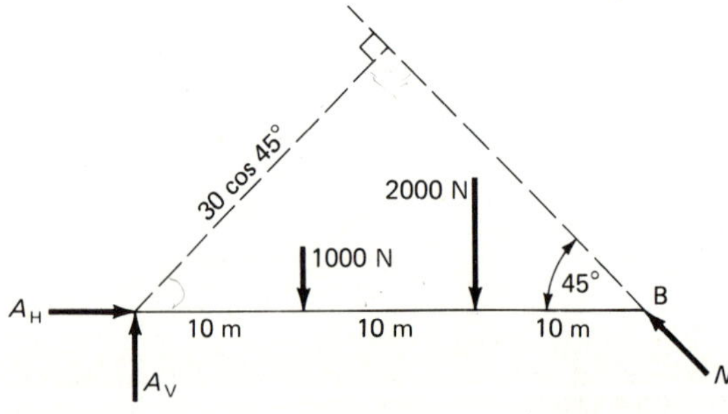

FIGURE 3.23   Free-body diagram for Illustrative Problem 3.9.

                                                                 Equilibrium

From the previous work in this section and the illustrative problems we can set up a general procedure to be used as a guide to solving problems in mechanics.

1. Read and understand the problem. Too frequently students do not fully understand the stated problem and, in essence, solve one that they make up.
2. Draw a free-body diagram (or diagrams) as needed for solving the problem.
3. The number of free-body diagrams required in step 2 is obtained from the number of independent equations expressing the equilibrium condition that can be written.
4. Solve the equations of equilibrium for the unknown required.
5. Wherever possible, check the results from another viewpoint, that is, a force or moment equation.

This procedure will be found to be most helpful, and with time and practice will become one of the professional tools that the student will acquire.

3.4   NONCOPLANAR FORCE SYSTEMS

Equilibrium requires the resultant forces and moments on a body to be zero. Noncoplanar force systems add another dimension to the two-dimensional system that we studied in Section 3.3, and it is therefore necessary to include this additional dimension in our equations of equilibrium. The complete requirements for the equilibrium of noncoplanar force systems are

$$\Sigma F_x = 0$$

$$\Sigma F_y = 0$$

$$\Sigma F_z = 0$$

$$\Sigma M_x = 0 \tag{3.7}$$

$$\Sigma M_y = 0$$

$$\Sigma M_z = 0$$

Equations (3.7) give us six independent conditions that can be used to determine at most six unknowns in a given problem. Usually it is not necessary to use all six equations.

ILLUSTRATIVE PROBLEM 3.10

Given the parallel force system shown in Fig. 3.24, determine the tensions in the rods supporting the plate. The weight of the plate, 1000 lb, is assumed to act at its center.

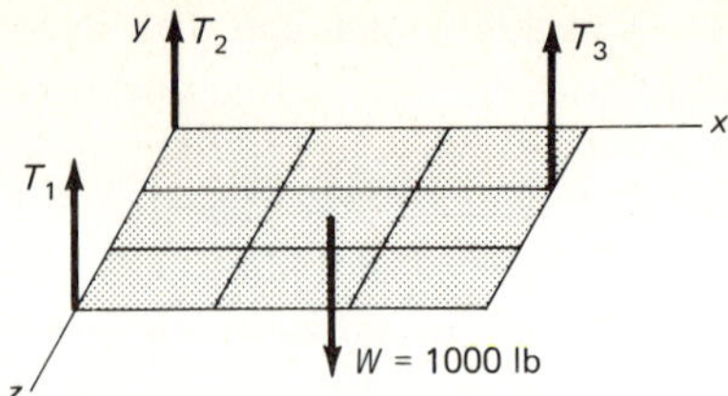

FIGURE 3.24   Illustrative Problem 3.10.

SOLUTION

The forces shown all act parallel to the y axis, and we accordingly do not need to perform a force summation in the x and z directions. We will take moments about the z and x axes as well as a force summation in the y direction. Proceeding, for $\Sigma M_z = 0$,

$$-1000 \times 1.5 + T_3 \times 3 = 0$$

$$T_3 = 500 \text{ lb}$$

For $\Sigma M_x = 0$,

$$3T_1 + T_3 - 1.5 \times 1000 = 0$$

but $T_3 = 500$, so

$$T_1 = 333 \text{ lb}$$

For $\Sigma F_y = 0$,

$$T_1 + T_2 + T_3 = 1000$$

Since $T_3 = 500$ and $T_1 = 333$,

$$333 + T_2 + 500 - 1000 = 0$$

$$T_2 = 167 \text{ lb}$$

ILLUSTRATIVE PROBLEM 3.11

A frame is to support the load shown in Fig. 3.25. Assuming the 500-lb load to be parallel to the z axis and the 1000-lb load to be vertical, determine the force in each member.

SOLUTION

Since members BC and BD lie in the x–z plane, they have no y components. Also member AB lies in the x–y plane and has no z component. The free-body diagram of the structure is shown in Fig. 3.25(b). Let us first use a force summation in the y direction. For $\Sigma F_y = 0$,

$$-1000 + A_y = 0$$

$$A_y = 1000 \text{ lb}$$

Equilibrium

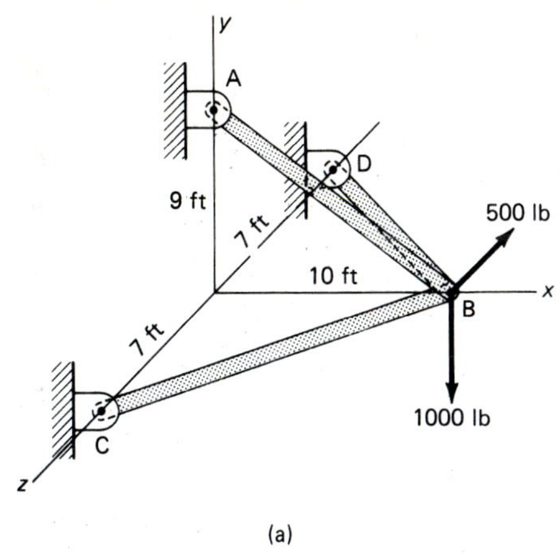

(a)

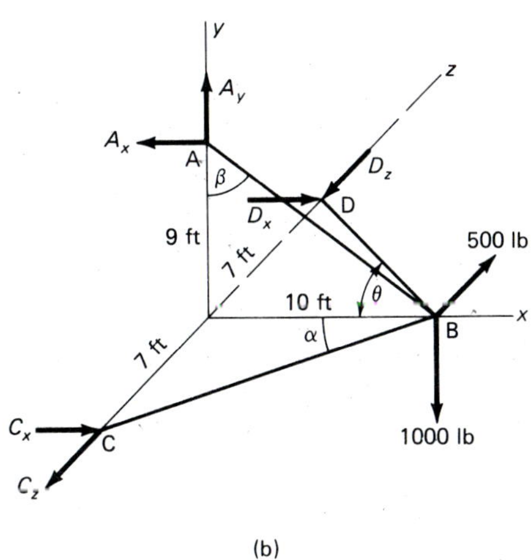

(b)

FIGURE 3.25   Illustrative Problem 3.11.

Next let us take moments about the axis CD (z axis). Notice that forces at C and D have zero moments. Therefore, for $\Sigma M_z = 0$,

$$9A_x - 1000 \times 10 = 0$$

$$A_x = 1111 \text{ lb}$$

If we now select an axis through point C parallel to the y axis and take moments, for $\Sigma M_y = 0$ at C,

$$-7A_x - 500 \times 10 + 14D_x = 0$$

$$-7777 - 5000 + 14D_x = 0$$

$$D_x = 912.6 \text{ lb}$$

Noncoplanar Force Systems

Now for $\Sigma F_x = 0$,

$$C_x + D_x - A_x = 0$$

$$C_x + 912.6 - 1111 = 0$$

$$C_x = 198.4 \text{ lb}$$

Therefore the resultant force at R_A is

$$R_A = \sqrt{A_x^2 + A_y^2} = \sqrt{1111^2 + 1000^2} = 1495 \text{ lb}$$

Since $D_x = 912.6$ lb, $R_D = D_x/\cos\theta$, and

$$\cos\theta = \frac{10}{\sqrt{7^2 + 10^2}} = \frac{10}{12.2} = 0.82$$

we find

$$R_D = \frac{912.6}{0.82} = 1113 \text{ lb}$$

Also since $C_x = 1984$, $R_C = C_x/\cos\alpha$, and

$$\cos\alpha = \frac{10}{\sqrt{7^2 + 10^2}} = 0.82$$

we find

$$R_C = \frac{198.4}{0.82} = 242 \text{ lb}$$

As a check,

$$R_A = \frac{A_x}{\sin\beta} = \frac{1111}{10/\sqrt{10^2 + 9^2}} = 1495 \text{ (checks)}$$

3.5   CLOSURE

The free-body diagram provides us with the tool necessary to solve problems in mechanics. It must be thoroughly understood before undertaking more advanced studies. Also, when applying the conditions for equilibrium, a great deal of the work can be simplified by the judicious choice of axes and summations. Often a moment summation will replace a force summation to yield a simpler condition. This type of reasoning may eliminate the need to solve simultaneous algebraic equations, which can be time-consuming and can lead to arithmetic errors.

The free-body concept, introduced in 1856, still remains the single most useful and necessary concept for problem solving in mechanics. The only way to really learn and understand the free-body concept is to do as many practice problems as possible and to make an overt effort in each problem to carefully draw the applicable free-body diagram or diagrams required.

                                                      Equilibrium

REFERENCES

Bassin, M. G., S. M. Brodsky, and H. Wolkoff, STATICS AND STRENGTH OF MATERIALS, 3rd ed. New York: McGraw-Hill, 1979.

Greene, E. G., PRINCIPLES OF PHYSICS. Englewood Cliffs, NJ: Prentice-Hall, 1962.

Levinson, I. J., INTRODUCTION TO MECHANICS, 2nd ed. Englewood Cliffs, NJ: Prentice-Hall, 1968.

Sears, F. W., M. W. Zemansky, and H. D. Young, COLLEGE PHYSICS, 4th ed. Reading, MA: Addison-Wesley, 1977.

Shames, I. H., ENGINEERING MECHANICS, VOL. I—STATICS, 2nd ed. Englewood Cliffs, NJ: Prentice-Hall, 1966.

Singer, F. L., ENGINEERING MECHANICS, 3rd ed. New York: Harper and Brothers, 1975.

White, H. E., MODERN COLLEGE PHYSICS, 5th ed. Princeton, NJ: D. Van Nostrand, 1966.

PROBLEMS

3.1 Draw the free-body diagram of the bar shown. Neglect its weight.

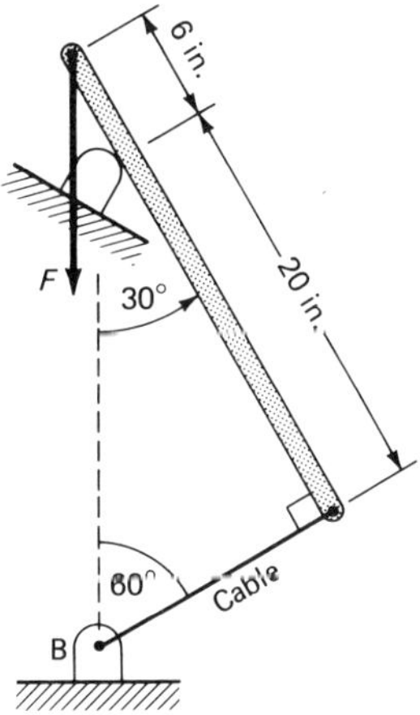

FIGURE P 3.1

3.2 A ladder rests against a wall as shown. If the vertical wall only is smooth, draw the free-body diagram for the ladder. Assume that its weight W acts at its center.

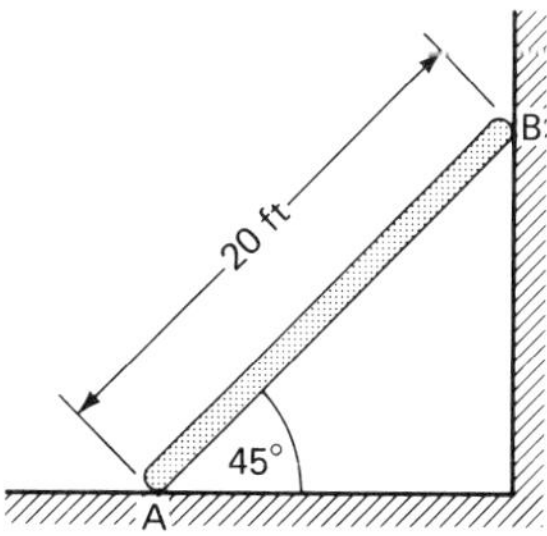

FIGURE P 3.2

Problems

3.3   Neglecting friction, draw the free-body diagrams for the figure shown. The weight of
the cylinder can be taken to act at its center. Neglect the weight of the bar.

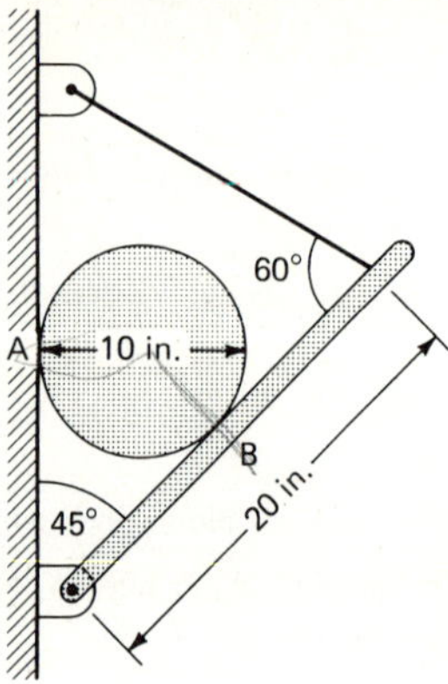

FIGURE P 3.3

3.4   Draw a free-body diagram for the cantilever beam shown.

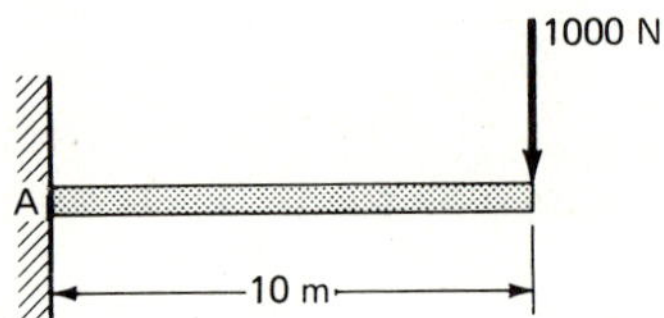

FIGURE P 3.4

3.5   Draw a free-body diagram for the cantilever beam of Problem 3.4 if a 1000-lb force
makes an angle of 30° with the vertical.

3.6   Assuming the bar to be weightless, draw the required free-body diagram for the situation
shown.

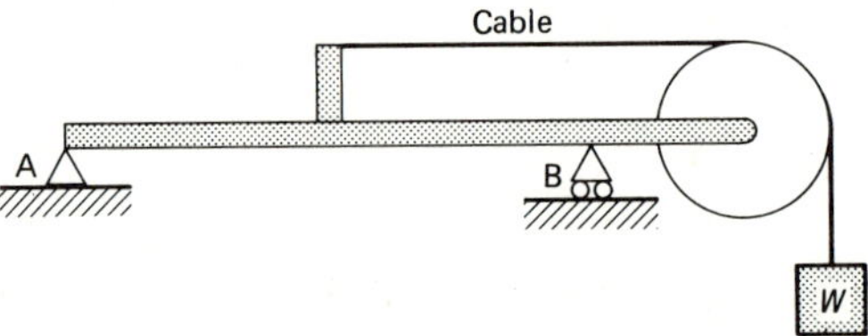

FIGURE P 3.6

3.7   Determine the reactions at the supports for the beam shown. Neglect its weight.

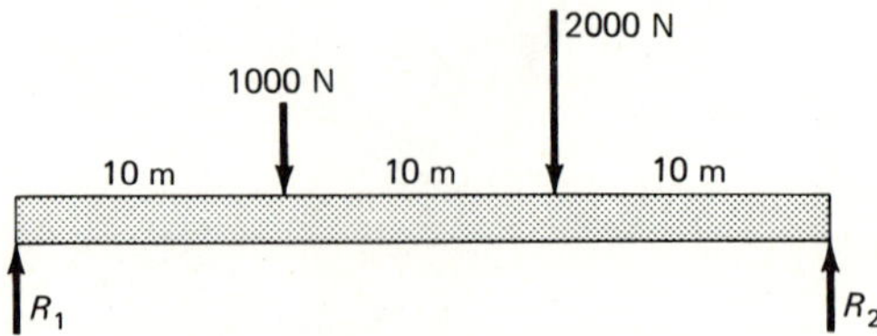

FIGURE P 3.7

                                                                    Equilibrium

3.8   If the weight of the beam in Problem 3.7 is 250 N, determine R_1 and R_2. Assume that the weight acts at the center of the beam.

3.9   If a counterclockwise moment of 20 000 N·m is acting on the beam in Problem 3.7, determine R_1 and R_2.

3.10   Neglecting friction and the weight of the pulleys, determine the tension in the cable shown.

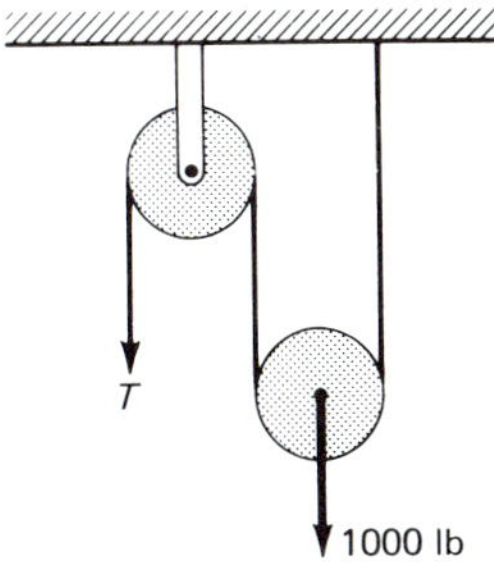

FIGURE P 3.10

3.11   Determine the reactions R_1, R_2, and R_3. Neglect beam weights.

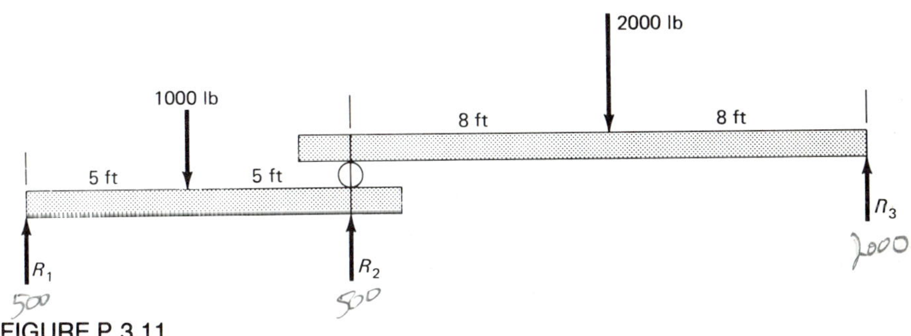

FIGURE P 3.11

3.12   Determine the tension in each cable for the figure shown.

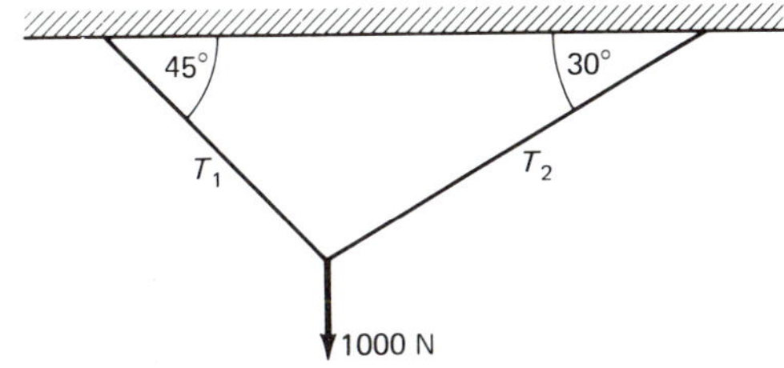

FIGURE P 3.12

Problems   **87**

3.13   What force is required to start the cylinder to roll over the block as shown? The cylinder
weighs 500 lb.

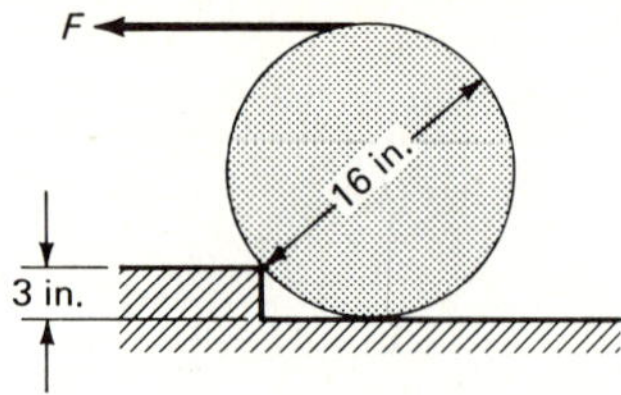

FIGURE P 3.13

3.14   Determine the reactions at A, B, C, and D if the wall surfaces are smooth.

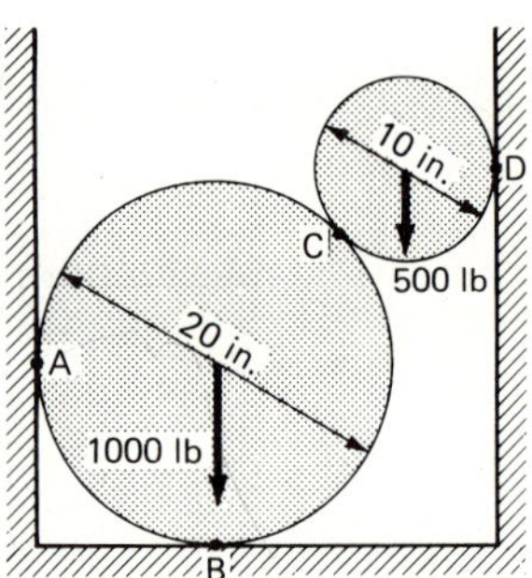

FIGURE P 3.14

3.15   Determine the forces at A and B. Neglect the weight of the bars.

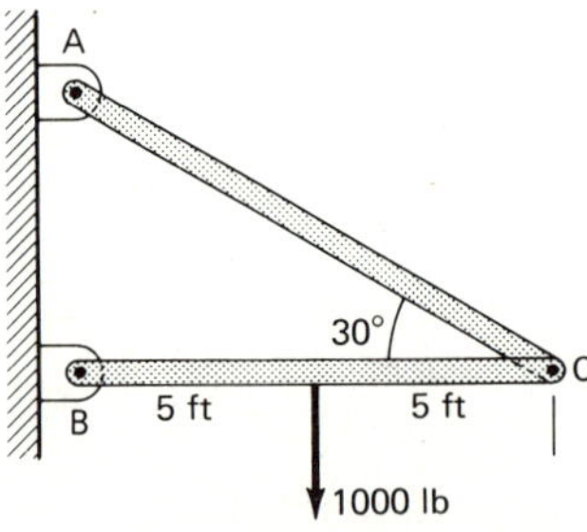

FIGURE P 3.15

3.16   Determine the reactions at A and B for the ladder of Problem 3.2 if the ladder weighs
100 lb.

                                                            Equilibrium

3.17   What forces act at A and B?

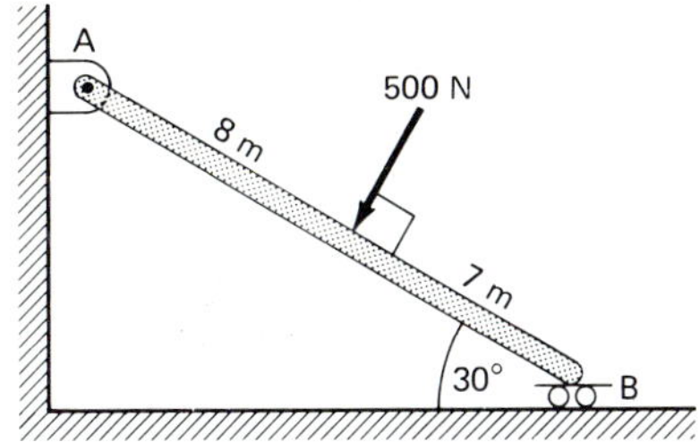

FIGURE   P 3.17

3.18   If the 500-N force in Problem 3.17 is vertical, determine the reactions at A and B.

3.19   If F in Problem 3.1 is 1000 lb, determine the tension in the cable.

3.20   Determine the tension in the cable of Problem 3.3. The cylinder weighs 600 lb. Neglect the weight of the bar.

3.21   A driver can pull with 50 lb on the parking brake handle of a car. What tension can be exerted on the cable and what is the reaction at the pivot A?

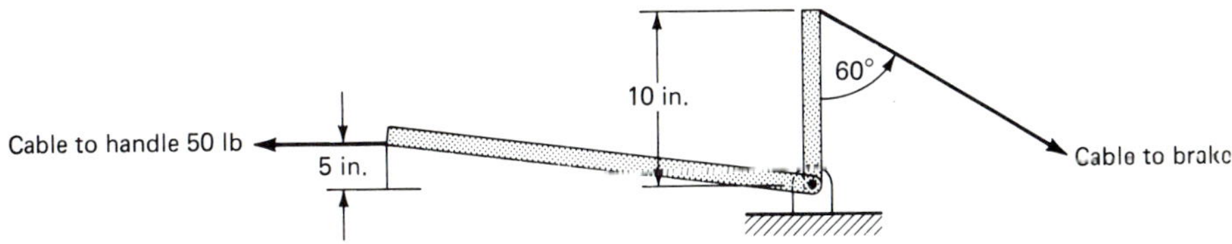

FIGURE P 3.21

3.22   Determine the reactions at A and B for the structure shown. Neglect the weight of each member.

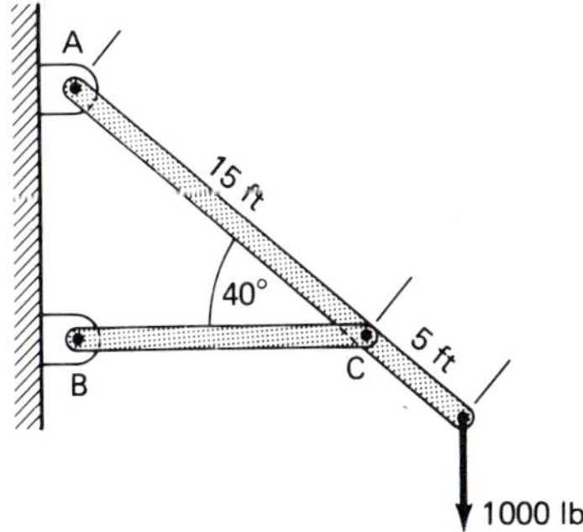

FIGURE P 3.22

Problems   89

3.23 Determine the support reactions for the truss shown.

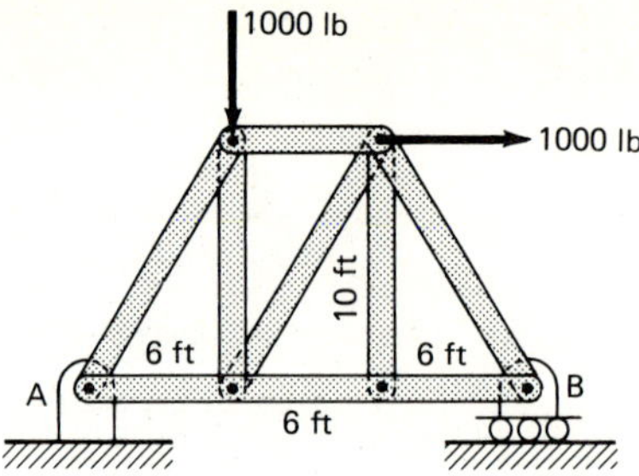

FIGURE P 3.23

3.24 A cable is used to support a hinged beam in the horizontal position. If the weight of the beam is negligible, determine the tension in the cable and the support reaction.

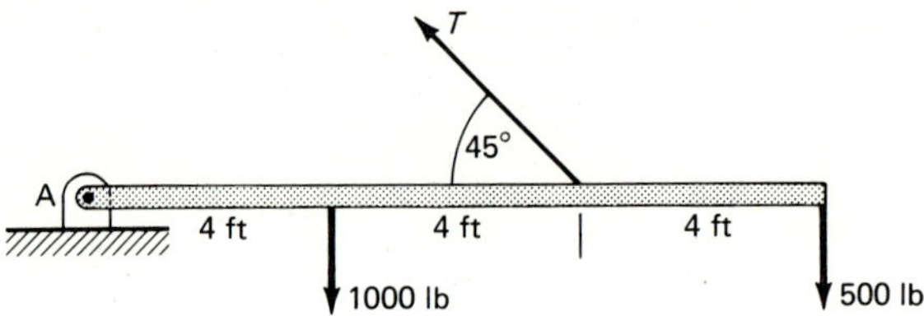

FIGURE P 3.24

3.25 Solve Illustrative Problem 3.10 if in addition to the weight there is a 2000-lb force acting downward at coordinates $x = 3$, $z = 3$.

3.26 Solve Illustrative Problem 3.11 if the 500-lb force is deleted.

3.27 Determine the reaction at A and B for the shaft shown. Neglect pulley and shaft weights. Both pulleys have the same diameter, and the bearings at A and B are smooth.

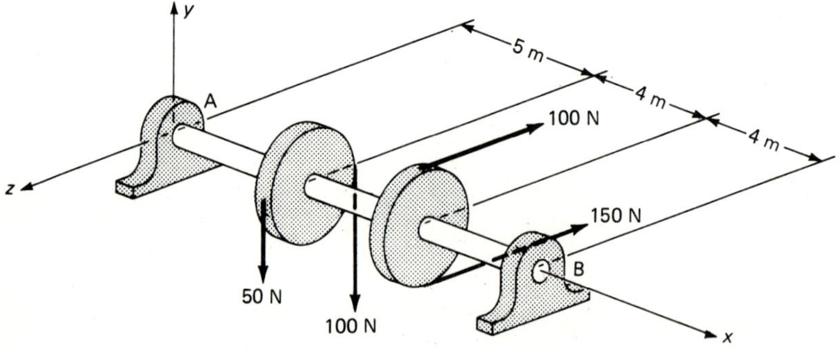

FIGURE P 3.27

Equilibrium

3.28 A flag pole is braced as shown. If there is a ball joint at the base (three force components) and the weight is neglected, determine the cable tensions and the force at the ball joint due to the 1000-lb pull.

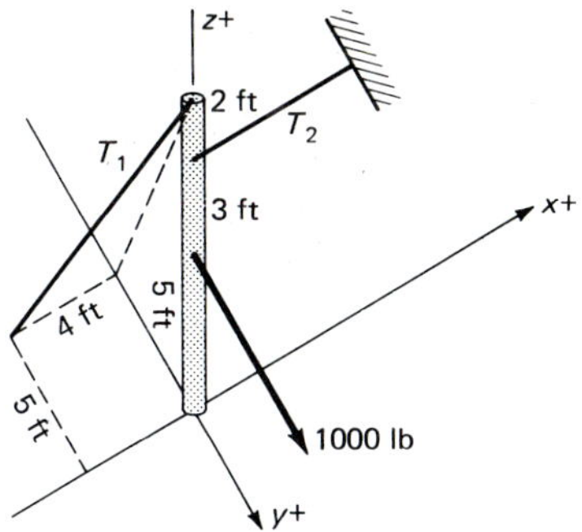

FIGURE P 3.28

4 Structures

4.1 INTRODUCTION

In Chapter 3 we were concerned about the conditions required to keep a system in equilibrium. By taking the free body to be essentially a single rigid body we were able to ascertain the external requirements on the body in order for it to be in equilibrium. A structure can be considered to consist of a number of uniform members fastened together in order to safely support the loads imposed upon it. Our present concern will be to determine the force in each member of the structure. A truss is a structural assemblage of members joined at their ends to form a structure capable of supporting larger loads or spanning larger distances than can be accomplished by any single member of the truss. The ends of the members of a truss are considered to be joined by pins, and all loads are considered to act at the ends of the members. Due to this action the members of a truss are considered to be subjected to forces directed axially along the member. In many cases such as roof trusses and bridge trusses the centerlines of all members and of all loads and reactions lie essentially in one plane, and these trusses are known as plane trusses.

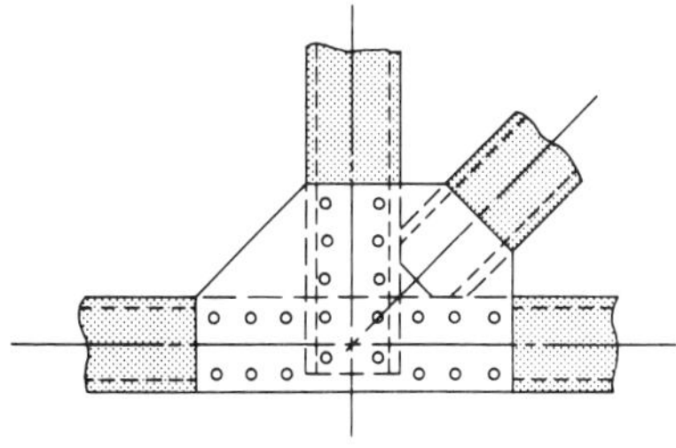

FIGURE 4.1   A riveted gusseted structural joint.

FIGURE 4.2   Some typical structural details.

When riveted or welded connections are used to join the members of a truss, it is often assumed that the connection is a pin joint. This assumption is usually satisfactory if due care is taken when the structure is designed for the centerlines of the members to intersect at a common point at each joint. This type of construction is shown in Fig. 4.1, where the members are joined by riveting using a gusset plate. Figure 4.2 shows some typical structural connections.

4.2   THE METHOD OF JOINTS

In determining the forces acting on the members of a truss, certain simplifying assumptions will be made. These assumptions will be used regardless of the method of calculation used to analyze the structure. We have already briefly mentioned some of these assumptions and we will now formally note the following.

1. The members of a truss are connected by pins at their ends. Each member is assumed to be rigid, and the pins are assumed to fit perfectly.
2. The external loads and reactions on the members of the truss act only at the pins, which are at the ends of the members.
3. The members of the truss and the external loads are coplanar. The force system on each truss member is a coplanar system. Therefore the forces at

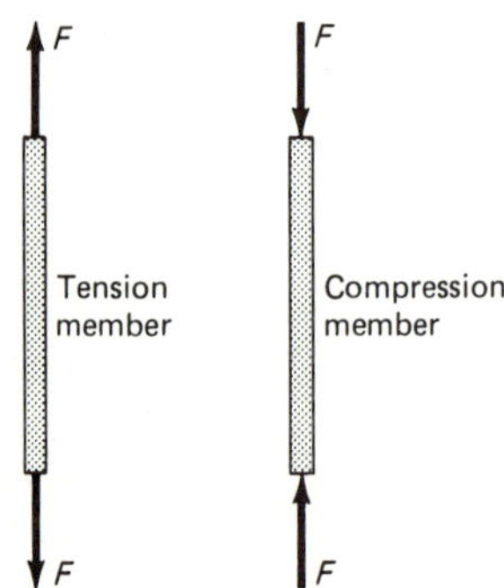

FIGURE 4.3   Tension and compression illustrated.

                                                                                                    Structures

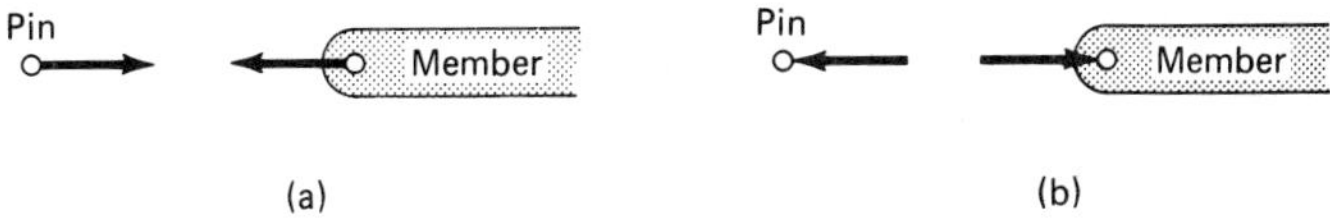

FIGURE 4.4   Tension and compression convention on pin joints.

each end of the member must be collinear, equal in magnitude, and opposite in direction, since the member is in equilibrium. These members are called two-force members since they are in equilibrium under the action of two forces.

4. The weight of the members of a truss can be neglected since the forces, due to the weights, are invariably much smaller than the external loads on the system.

Since the above assumptions require a truss member to be loaded axially, it is necessary to adopt a convention regarding the direction of these axial forces. Where the member is subjected to an axial load that causes the member to stretch or elongate, we will state that the member is in tension, while axial loads that cause the member to contract will be denoted as compressive loads. Since every action causes an equal and opposite reaction, the effect on the joints will be opposite to that shown in Fig. 4.3. Thus a member in tension causes a pull on the joints, while a member in compression causes a push on the end joints. This convention is illustrated in Fig. 4.4.

The method of joints utilizes the fact that each member of the truss is subjected to a coplanar concurrent force system that must be in equilibrium. For such a system the equations of equilibrium reduce to two independent equations. This method basically consists of considering a pin connection and cutting all of the members attached to the pin to obtain a free-body diagram of the pin. Thus it is important (necessary) to start at a joint having only two members in which the forces are unknown. This joint is then completely determined, using the equations of equilibrium. The remaining joints are determined successively until the forces in all members of the truss are completely determined. The sequence of selection of the joints must be such that at the joint selected there will be no more than two unknown forces. It should also be noted from Fig. 4.4 that we have selected the convention that tension is indicated by an arrow *away* from a pin and compression is indicated by an arrow *toward* the pin. During the calculation it is often necessary to assign a given direction to a force acting on a member. In the event that a negative value is obtained from the calculation, the assumed direction is incorrect and must be reversed. The following illustrative problem will serve to illustrate the method of joints, and afterward we shall summarize the procedure used for convenience.

ILLUSTRATIVE PROBLEM 4.1

Determine the force in each member of the truss shown in Fig. 4.5.

SOLUTION

The first step in the solution is to obtain the reactions at joints A and C. To accomplish this, a free-body diagram of the entire truss is shown in Fig. 4.6. Since

The Method of Joints   95

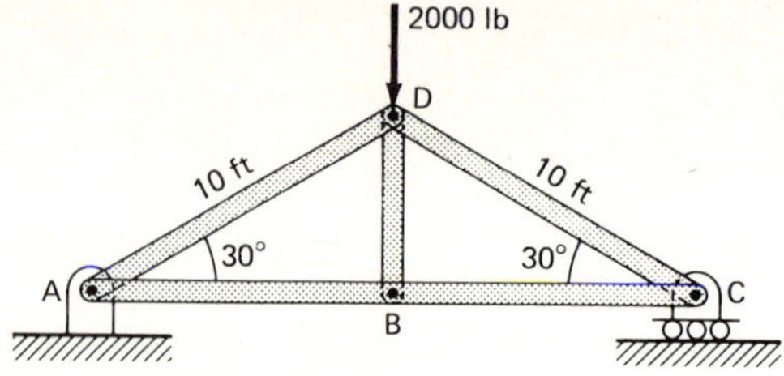

FIGURE 4.5   Illustrative Problem 4.1.

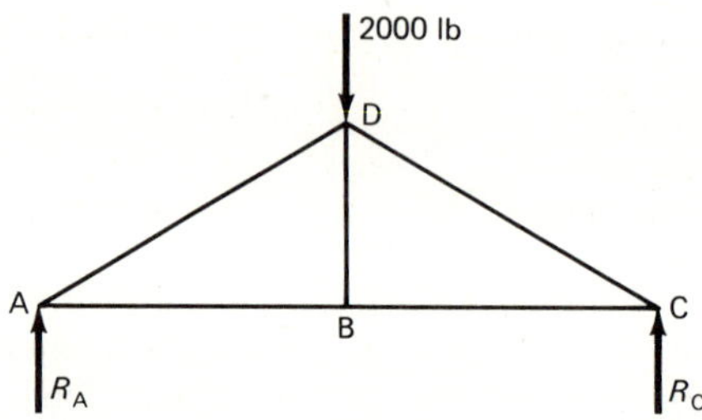

FIGURE 4.6   Illustrative Problem 4.1, entire truss.

the reaction at C is vertical (roller joint), the reaction at A is also vertical. Either by symmetry or by taking moments about A, we readily find $R_A = R_C = 1000$ lb. We can now consider the joint at A, since there will be only two unknown forces at this joint. We first draw a free-body diagram of joint A, as shown in Fig. 4.7. The direction of the forces AB and AD are indicated by the arrows at the joint. These could be arbitrarily assumed, but at this joint there is enough information to be able to correctly assign the directions of the forces. The vertical components of AD must be opposite to R_A, requiring the arrow on AD to be toward the joint. We now find that AD will have a horizontal component to the left, requiring AB to be a force on the joint acting to the right, as shown in Fig. 4.7. Setting up the equations of equilibrium, for $\Sigma F_y = 0$,

$$1000 - AD \sin 30° = 0$$

$$1000 - AD \times 0.5 = 0$$

and

$$AD = 2000 \text{ lb}$$

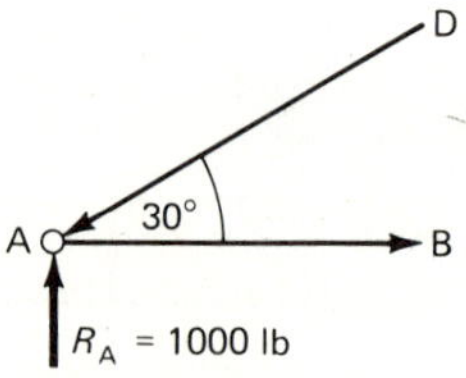

FIGURE 4.7   Illustrative Problem 4.1, joint A.

Structures

As shown, AD is in compression. For $\Sigma F_x = 0$,

$$AB - AD \cos 30° = 0$$

$$AB = AD \cos 30° = 2000 \times 0.866 = 1732 \text{ lb in tension}$$

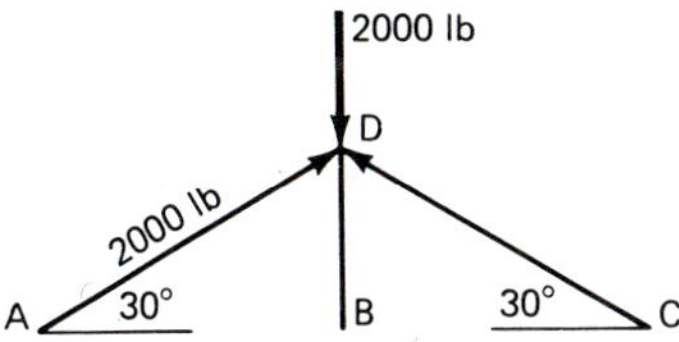

FIGURE 4.8   Illustrative Problem 4.1, joint D.

We now proceed to joint D since there will be two unknown forces at this joint. A free-body diagram of the joint is shown in Fig. 4.8 where it will be noted that we must correctly show that AD is in compression by an arrow *toward* joint D. Setting up the equations of equilibrium, for $\Sigma F_x = 0$,

$$AD \cos 30° - CD \cos 30° = 0$$

$$AD = CD = 2000 \text{ lb in compression}$$

We could anticipate this result by noting the symmetry of the truss.
For $\Sigma F_y = 0$,

$$AD \sin 30° + CD \sin 30° - 2000 + DB = 0$$

$$2000 \times 0.5 + 2000 \times 0.5 \quad 2000 + DB - 0$$

$$DB = 0$$

This result DB = 0 is interesting, since it would indicate that member DB is not needed (redundant). However, such members do serve to stiffen the structure. We could have considered joint B rather than joint D to obtain this same result, as shown in Fig. 4.9. Notice that BA and BC are collinear and have no vertical components. Therefore DB must be zero to meet $\Sigma F_y = 0$. Also, since BA and BC are collinear, and as the only forces acting on B in the horizontal direction, BA = BC and must be directed in the opposite sense to satisfy $\Sigma F_x = 0$.
The problem is now completely solved with the following results:

$$AD = CD = 2000 \text{ lb compression } (C)$$

$$AB = BC = 1732 \text{ lb tension } (T)$$

and

$$DB = 0$$

$$R_A = R_C = 1000 \text{ lb}$$

FIGURE 4.9   Illustrative Problem 4.1, joint B.

As a further illustration of the method of joints, consider the following problem.

ILLUSTRATIVE PROBLEM 4.2

Determine the forces in each member of the truss shown in Fig. 4.10(a).

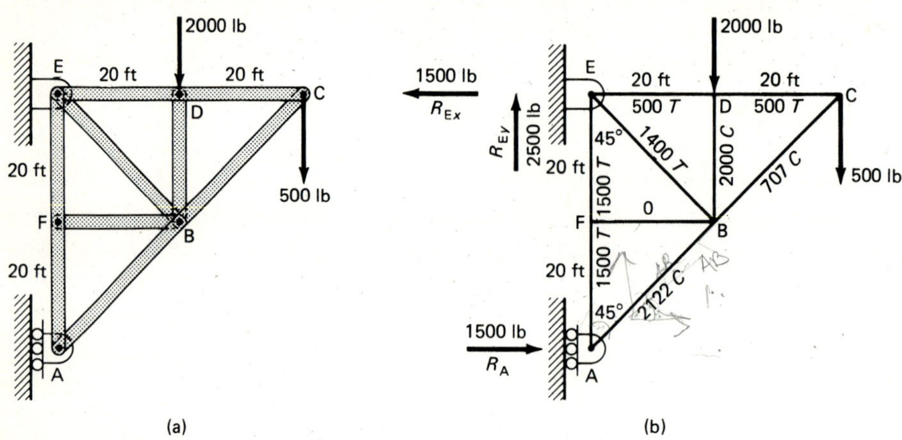

FIGURE 4.10   Illustrative Problem 4.2.

SOLUTION

As in Illustrative Problem 4.1, we first determine the support reactions. Taking moments about E,

$$R_A \times 40 - 2000 \times 20 - 500 \times 40 = 0$$

$$R_A = 1500 \text{ lb}$$

The reaction at E will have two components R_{Ex} and R_{Ey}. From $\Sigma F_x = 0$, $R_{Ex} = -R_A = -1500$ lb and $R_{Ey} = 2000 + 500 = 2500$ lb up, as shown in Fig. 4.10(b). The first joint to consider is joint A. From the equilibrium conditions, for $\Sigma F_x = 0$,

$$1500 - AB \cos 45° = 0$$

$$AB = \frac{1500}{\cos 45°} = 2122 \text{ lb compression}$$

FIGURE 4.11   Illustrative Problem 4.2, joint A.

Structures

For $\Sigma F_y = 0$,

$$AF - AB \sin 45° = 0$$

$$AF = AB \times 0.707 = 2122 \times 0.707 = 1500 \text{ lb tension}$$

Consideration of joint F, shown in Fig. 4.12, leads to the immediate conclusion that FB = 0, and FE = FA = 1500 lb tension.

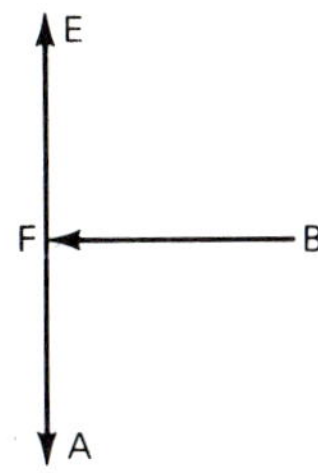

FIGURE 4.12   Illustrative Problem 4.2, joint F.

The next joint with two unknown forces is joint E as shown in Fig. 4.13. Using the conditions of equilibrium once again, for $\Sigma F_x = 0$,

$$ED + EB \cos 45° - 1500 = 0$$

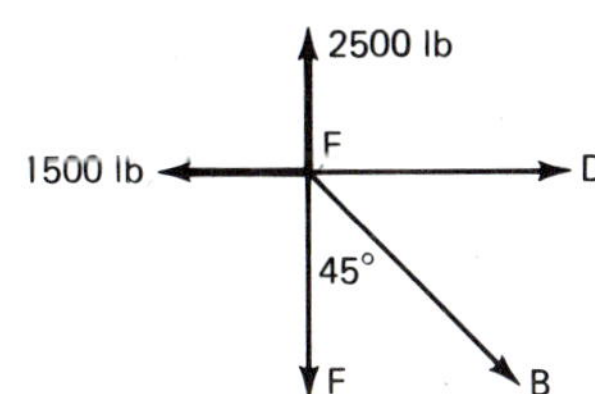

FIGURE 4.13   Illustrative Problem 4.2, joint E.

For $\Sigma F_y = 0$,

$$2500 - EF - EB \sin 45° = 0$$

Putting in known values,

$$ED + 0.707EB - 1500 = 0$$

$$2500 - 1500 - 0.707EB = 0$$

$$0.707EB = 1000$$

$$EB = 1414 \text{ lb tension}$$

$$ED + 1000 - 1500 = 0$$

$$ED = 500 \text{ lb tension}$$

We now consider joint D, as shown in Fig. 4.14, which has two unknown forces.

The Method of Joints

We obtain for $\Sigma F_y = 0$,

$$DB - 2000 = 0$$

$$DB = 2000 \text{ lb compression}$$

$$DC - DE = 0$$

$$DC = DE = 500 \text{ lb tension}$$

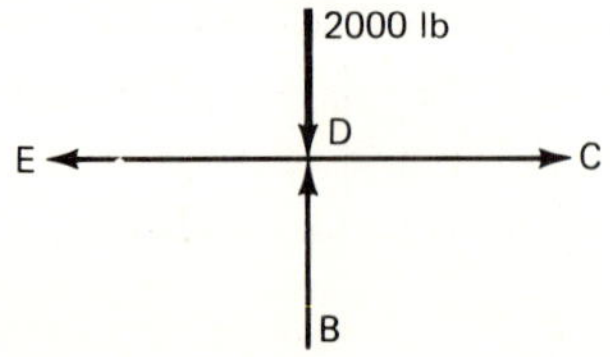

FIGURE 4.14   Illustrative Problem 4.2, joint D.

The last joint that has to be considered is joint C, as shown in Fig. 4.15. Once again the conditions of equilibrium give us $\Sigma F_x = 0$,

$$BC \sin 45° - DC = 0$$

$$BC = \frac{DC}{\sin 45°} = \frac{500}{0.707} = 707 \text{ lb compression}$$

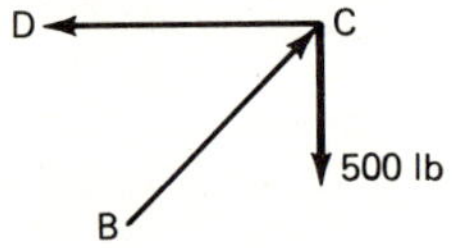

FIGURE 4.15   Illustrative Problem 4.2, joint C.

Figure 4.10(b) shows all of the forces obtained from the complete solution of the truss.

We can summarize the procedure used in the two illustrative problems considered thus far as follows.

1. Determine the reactions acting on the truss.
2. Select a joint at which there are only two unknown forces.
3. Draw a free-body diagram of the joint selected in step 2. Note that tension is shown by an arrow away from a joint; compression is denoted by an arrow toward a joint.
4. Apply the conditions of equilibrium to the joint, that is, $\Sigma F_x = 0$ and $\Sigma F_y = 0$, and solve for the unknown forces. A negative value of a force indicates a direction opposite to the assumed direction.
5. Select the next joint with two unknown forces and repeat the foregoing procedure.
6. Continue until forces in all members are obtained.

                                                                    Structures

    # THE METHOD OF SECTIONS

The method of joints will always yield a solution to the problem of obtaining the forces in the members of a truss. However, it is obvious that the procedure is long and tedious. One of the reasons for this is that only two of the three conditions of equilibrium are used. In order to take advantage of the third condition, namely, $\Sigma M = 0$, we consider the truss to consist of sections, each of which is in equilibrium and acted upon by a system of nonconcurrent coplanar forces. As we shall see, this method permits us to obtain the force in any member without going from joint to joint until the desired member is reached. The only qualification to the method is that the free-body diagram of the section cannot have more than three unknown forces, since the equations of equilibrium permit us to write only three simultaneous independent equations.

ILLUSTRATIVE PROBLEM 4.3

Determine the forces in members DC and BC of the truss discussed in Illustrative Problem 4.2 and shown in Fig. 4.16.

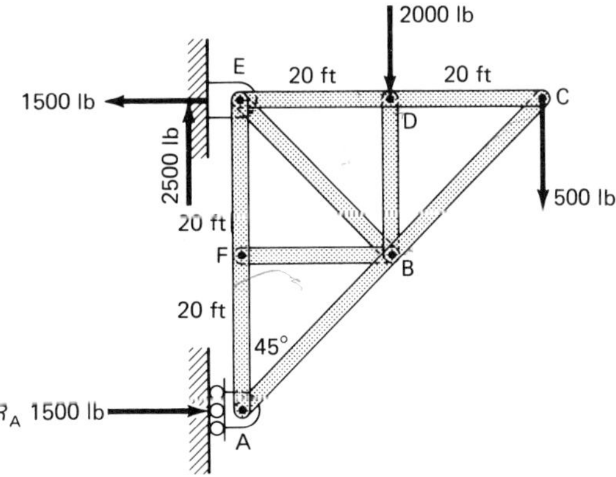

FIGURE 4.16   Illustrative Problem 4.3.

SOLUTION

We first obtain the reactions at A and E, as was done in Illustrative Problem 4.2. We now cut the truss at section X–X and isolate the left side as illustrated in Fig. 4.17. By selecting the axis for moments conveniently, we can apply $\Sigma M = 0$ to obtain one of the unknown forces. For this case let us take an axis through D for the moments. The distance from D to the line of action of BC is $20 \cos 45° = 14.14$ ft. Therefore, for $\Sigma M_D = 0$,

$$1500 \times 40 - 2500 \times 20 - BC \times 14.14 = 0$$

$$BC = 707 \text{ lb compression}$$

This result checks with that of Illustrative Problem 4.2. Continuing, let us now take

The Method of Sections

moments about B. For $\Sigma M_B = 0$

$$1500 \times 20 + 1500 \times 20 - 2500 \times 20 - DC \times 20 = 0$$

$$DC = 500 \text{ lb tension}$$

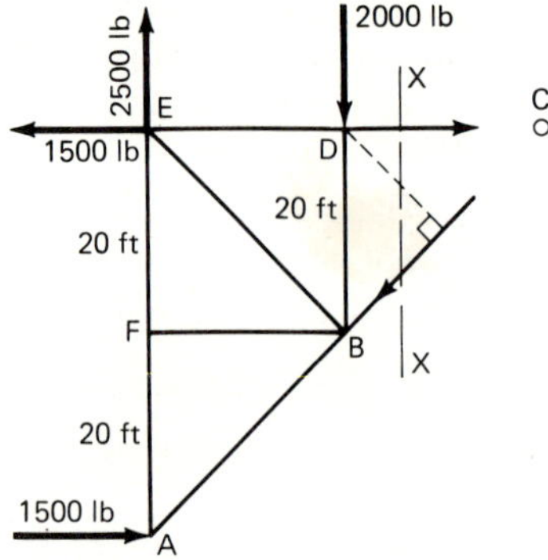

FIGURE 4.17   Illustrative Problem 4.3.

It is obvious that we obtained these results with much less effort than by use of the method of joints.

ILLUSTRATIVE PROBLEM 4.4

Determine the forces in members BC, BE, and AE in the truss shown in Fig. 4.18.

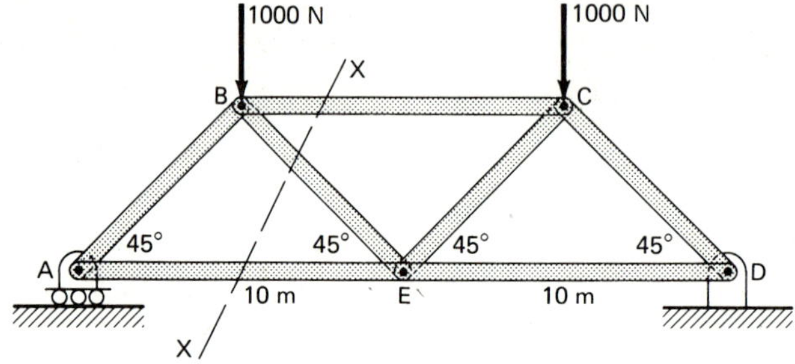

FIGURE 4.18   Illustrative Problem 4.4.

SOLUTION

From symmetry we have the reactions at A and D = 1000 N. Using plane X–X as the cutting plane gives us Fig. 4.19, which shows the free-body diagram of the left sections of the truss. Taking moments about B, for $\Sigma M_B = 0$,

$$-1000 \times 5 + AE \times 5 = 0$$

$$AE = 1000 \text{ N tension}$$

                                    Structures

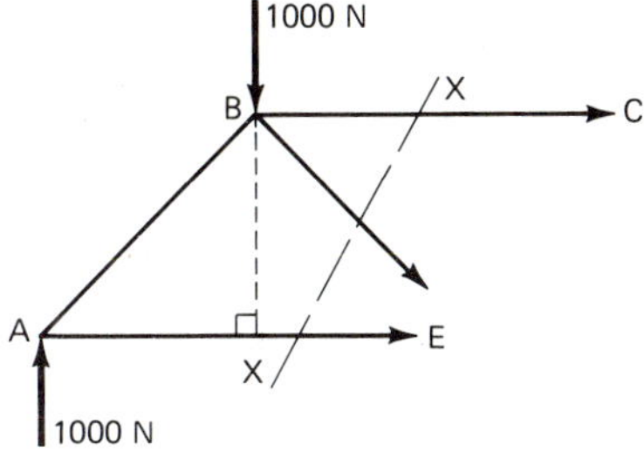

FIGURE 4.19 Illustrative Problem 4.4.

Using $\Sigma F_y = 0$,

$$1000 - 1000 - BE \sin 45° = 0$$

$$BE = 0$$

Notice again that by symmetry BE (and CE) are redundant force members, which carry no load but are necessary for the stability of the structure.

Taking moments about A, for $\Sigma M_A = 0$,

$$-5 \times 1000 - BC \times 5 = 0$$

$$BC = -1000 \text{ N}$$

The minus sign of BC indicates that we assumed its direction opposite in sense to the correct direction. Therefore BC is a 1000 N compressive force.

4.4 FRAMES

The truss was considered to be a structure composed of two force members pinned at each end. When one or more members of a structure are acted upon by three or more forces, we call this type of structure a *frame*. In a frame the forces will not generally be directed along the axis of the members and we must now consider the forces as either having unknown directions or represent them by two unknown orthogonal components. While none of the principles used to solve frame problems are new or different, certain considerations will be emphasized, principally the principle of action and reaction.

Consider a pin-connected joint used to connect two members, as shown in Fig. 4.20. Each member exerts forces on the pin, but the pin is in equilibrium.

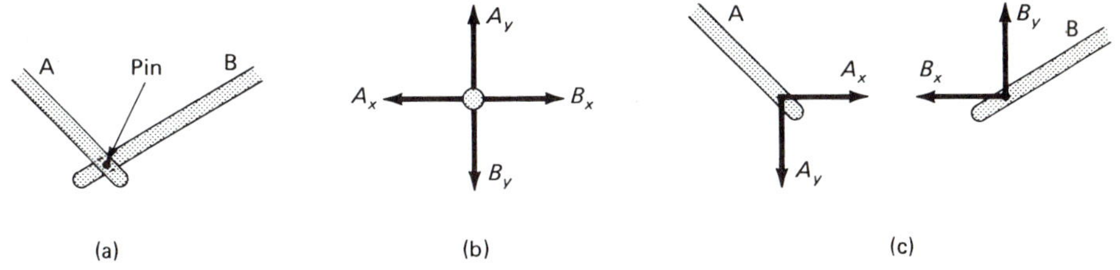

FIGURE 4.20 Forces on a pin.

Frames

Therefore we note that the components A_x and A_y of A at the pin are equal and opposite to those of B at the pin. Therefore when members are separated (for purposes of analysis), the forces (or their components) exerted by the pin on *each* member must be represented to act in opposite directions in each member, as shown in Fig. 4.20(c). In order to obtain the correct solution to the problem of a frame it is necessary for the forces to be represented correctly on interconnecting bodies. In the event that a force or its component is found to be negative, the magnitude of the force will be correct, but its sense will be opposite to that originally assumed. However, the sign must be changed on each member at an interconnecting pin in order to be correct. Changing the direction on one member only will lead to an incorrect result.

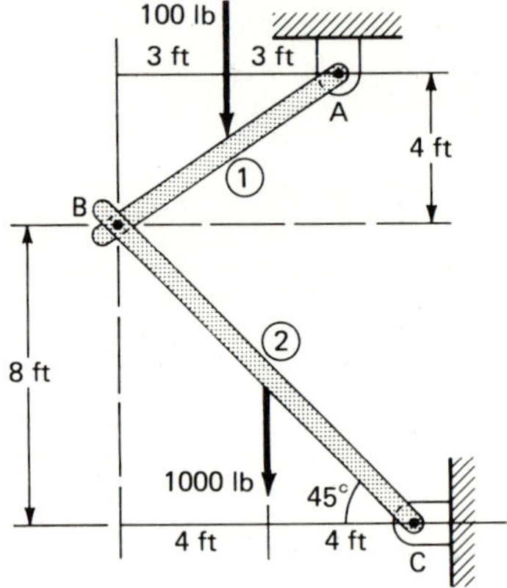

FIGURE 4.21   Illustrative Problem 4.5.

SOLUTION

Let us draw a free-body diagram of each member. Notice that at each joint there is an equal but opposite force on the two connecting members. The components of each force are shown.

Using Fig. 4.22(a) and taking moments about B gives, for $\Sigma M_B = 0$,

$$-1000 \times 4 - C_x \times 8 - C_y \times 8 = 0$$

$$C_x + C_y = -500 \text{ lb}$$

From Fig. 4.22(c) we take moments about A, for $\Sigma M_A = 0$,

$$100 \times 3 + 1000 \times 2 - C_x \times 12 - C_y \times 2 = 0$$

$$12C_x + 2 C_y = +2300 \text{ lb}$$

                                                                                    Structures

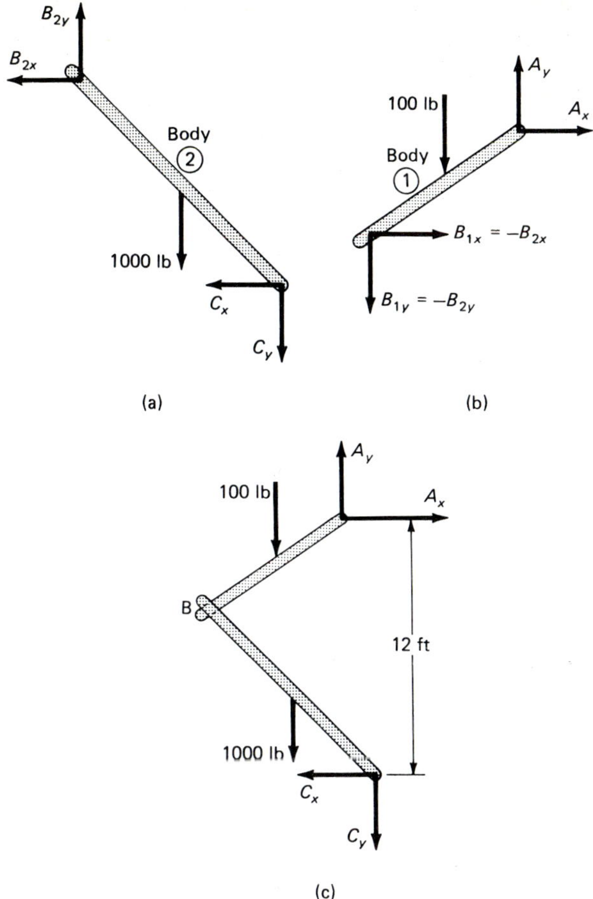

FIGURE 4.22   Free-body diagrams for Illustrative Problem 4.5.

Solving simultaneously,

$$C_x = 330 \text{ lb} \qquad C_y = -830 \text{ lb}$$

Notice that C_y is negative, which indicates that it is opposite in sense to the direction assumed.

We can now consider the free-body diagram of Fig. 4.22(a), which has two unknowns, B_{2x} and B_{2y}. Therefore, for $\Sigma F_y = 0$,

$$-1000 + 830 + B_{2y} = 0$$

$$B_{2y} = 170 \text{ lb (up, as shown)}$$

For $\Sigma F_x = 0$,

$$B_{2x} + C_x = 0$$

$$B_{2x} = -C_x = -330 \text{ lb (to the right, opposite to direction shown)}$$

The last member to be considered is shown by the free-body diagram of Fig. 4.22(b), and there are now two unknowns, A_y and A_x. For $\Sigma F_y = 0$,

$$A_y - 100 - 170 = 0$$

where $B_{1y} = -B_{2y}$, and

$$A_y = 270 \text{ lb}$$

For $\Sigma F_x = 0$,

$$A_x + B_{1x} = 0$$

but $B_{1x} = -B_{2x} = -330$, and

$$A_x = 330 \text{ lb}$$

As a check on the solution let us take moments about B in Fig. 4.22(b). For $\Sigma M_B = 0$,

$$-100 \times 3 - 4 \times 330 + 270 \times 6 = 0 \text{ (checks)}$$

As another check, for $\Sigma M_C = 0$ in Fig. 4.22(a),

$$1000 \times 4 - B_{2y} \times 8 + B_{2x} \times 8 = 0$$

$$1000 \times 4 - 170 \times 8 - 330 \times 8 = 0 \text{ (checks)}$$

4.5   CLOSURE

The material in this chapter is really an extension of the material of Chapter 3 on equilibrium. The areas of departure are in the assumptions made in order to deal with truss problems and the action and reaction that occurs in members of frames. While these items are somewhat novel, we find that the solution to structural problems depends for the most part on the construction of the correct free-body diagram(s) for the given conditions. As we found in Chapter 3, the time and care taken to think out and construct the free-body diagram repays itself in setting up a problem correctly. While the algebra of problems of structures is somewhat long and tedious, it will be found that with experience some of the work can be simplified. Symmetry, the choice of axes, the moment center used, and so on, all greatly lessen the work and decrease the chances for error. It is therefore strongly suggested for these problems (as well as for all problems in engineering) that a little forethought and understanding brought to bear *before* ''doing'' a problem will lessen the work and contribute to satisfaction in one's ability to solve problems correctly.

REFERENCES

Bassin, M. G., S. M. Brodsky, and H. Wolkoff, STATICS AND STRENGTH OF MATERIALS. 3rd ed. New York: McGraw-Hill, 1979.

Beer, F. P., and E. R. Johnston, Jr., VECTOR MECHANICS FOR ENGINEERS—STATICS, 3rd ed. New York: McGraw-Hill, 1977.

Higdon, A., and W. B. Stiles, ENGINEERING MECHANICS, VOL. I—STATICS, 3rd ed. Englewood Cliffs, NJ: Prentice-Hall, 1968.

Levinson, I. J., INTRODUCTION TO MECHANICS, 2nd ed. Englewood Cliffs, NJ: Prentice-Hall, 1968.

Meriam, J. L., MECHANICS, PART I—STATICS, 2nd ed. (SI version) New York: John Wiley, 1975.

Seely, F. B., and N. E. Ensign, ANALYTICAL MECHANICS FOR ENGINEERS, 3rd ed. New York: John Wiley, 1941.

Shames, I. H, ENGINEERING MECHANICS, VOL. I—STATICS, 2nd ed. Englewood Cliffs, NJ: Prentice-Hall, 1966.

Singer, F. L., ENGINEERING MECHANICS, 3rd ed. New York: Harper and Brothers, 1975.

PROBLEMS

4.1   Determine the force in each member of the truss shown using the method of joints.

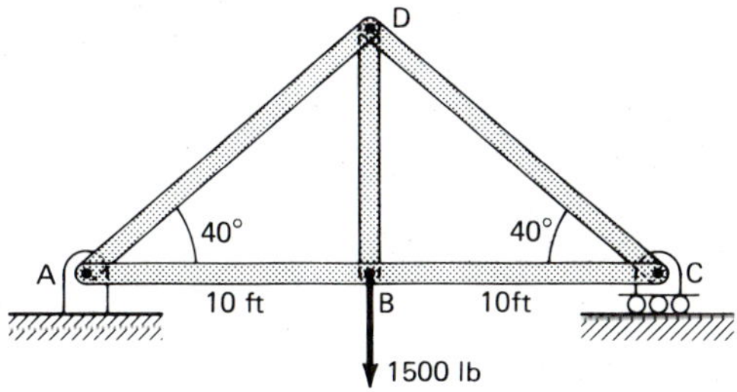

FIGURE P 4.1

4.2   If both ends of the truss of Problem 4.1 are pinned, determine the reactions of A and C.

4.3   Determine the force in each member of the truss shown. Use the method of joints.

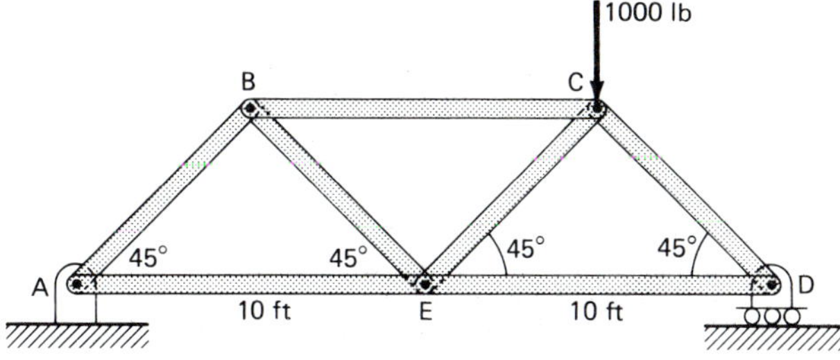

FIGURE P 4.3

4.4   Solve Problem 4.3 using the method of sections.

Problems   107

4.5 Determine the forces in members GC, BC, and BG in the truss shown. Use the method of joints.

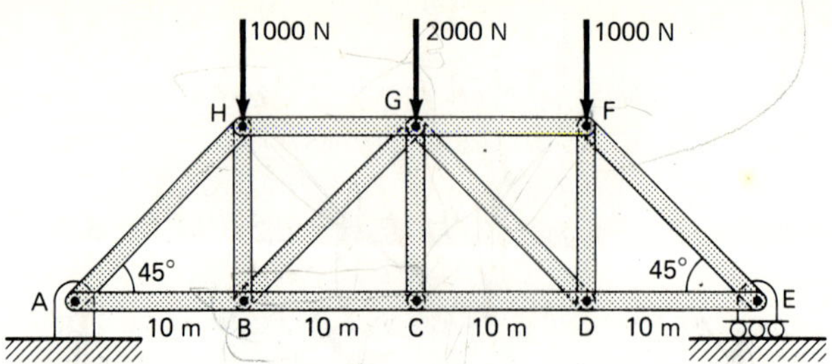

FIGURE P 4.5

4.6 Solve Problem 4.5 using the method of sections.

4.7 Determine the force in each member of the truss shown, using the method of joints.

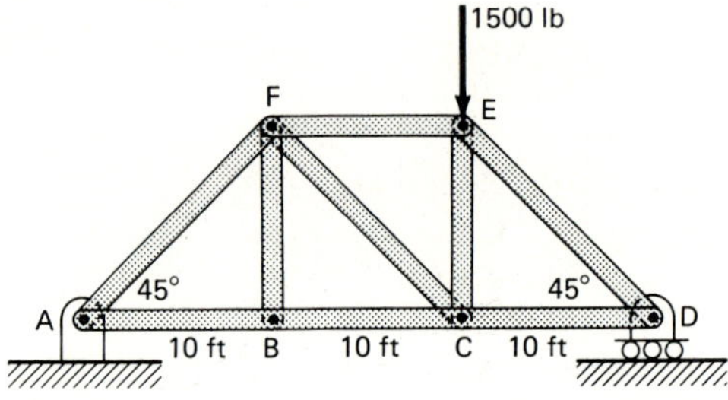

FIGURE P 4.7

4.8 Using the method of sections, determine the force in members FE, FC, and BC of Fig. P 4.7.

4.9 Determine the forces in the members of the truss shown.

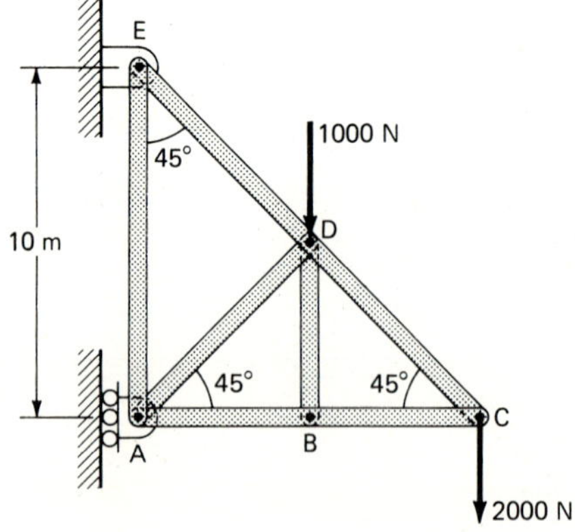

FIGURE P 4.9

Structures

4.10   Determine the reactions at A for the structure shown.

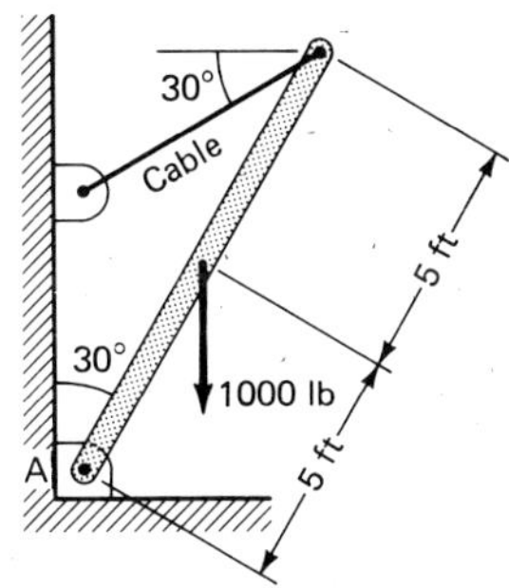

FIGURE P 4.10

4.11   Show that all of the internal members of the truss shown carry zero load.

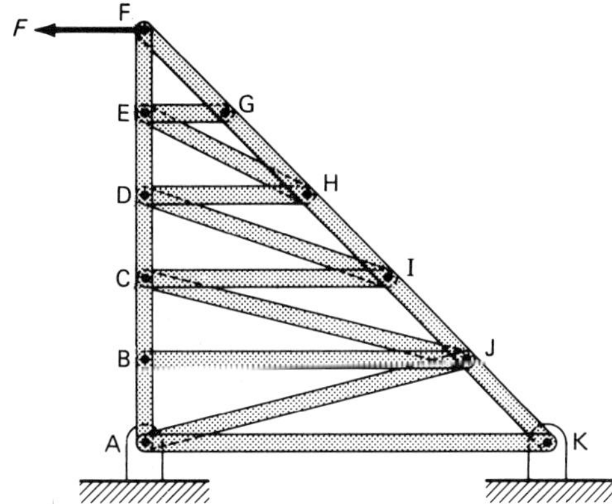

FIGURE P 4.11

4.12   Determine the forces in members EF, FC, and BC for the truss shown, using the method of sections.

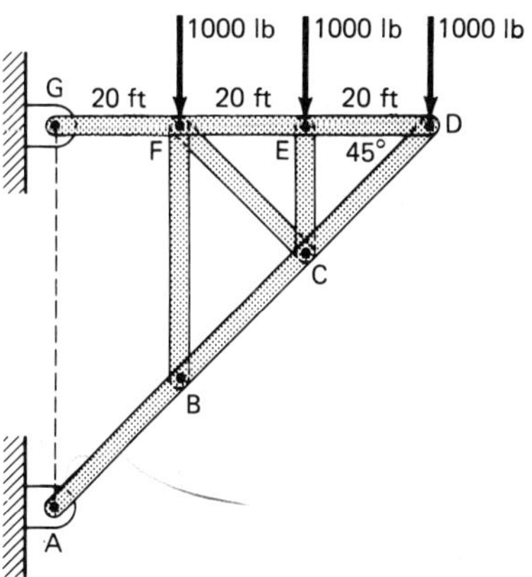

FIGURE P 4.12

Problems   109

4.13 Using the method of sections, determine the forces in members JI, JC, and BC.

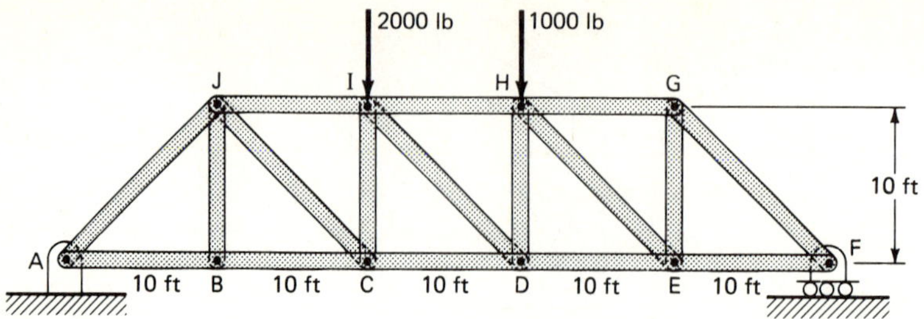

FIGURE P 4.13

4.14 Using the method of sections determine the forces in members IH, ID, and CD in Fig. P 4.13.

4.15 Using the method of sections determine the forces in members HG, HE, and DE in Fig. P 4.13.

4.16 Determine the forces at pin B for the frame shown.

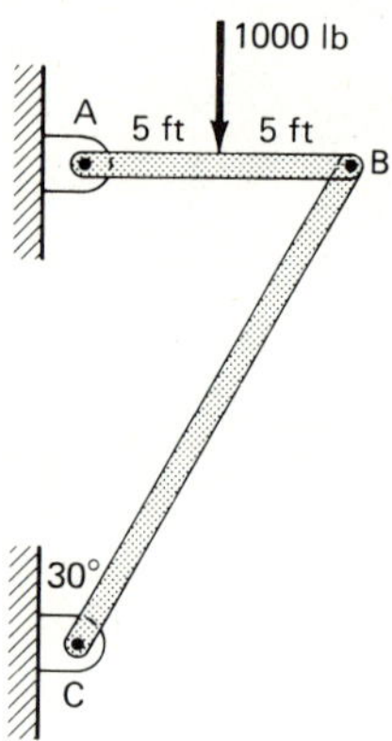

FIGURE P 4.16

4.17 Determine the forces at pins B, C, and D for the frame shown.

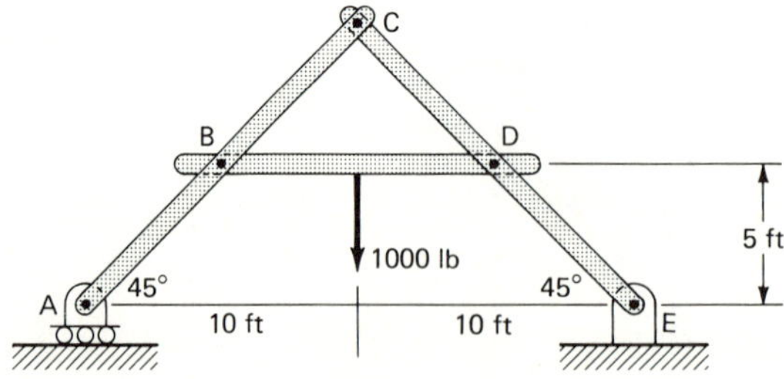

FIGURE P 4.17

110   Problems

4.18   Determine the forces at pin B which act on bar AC.

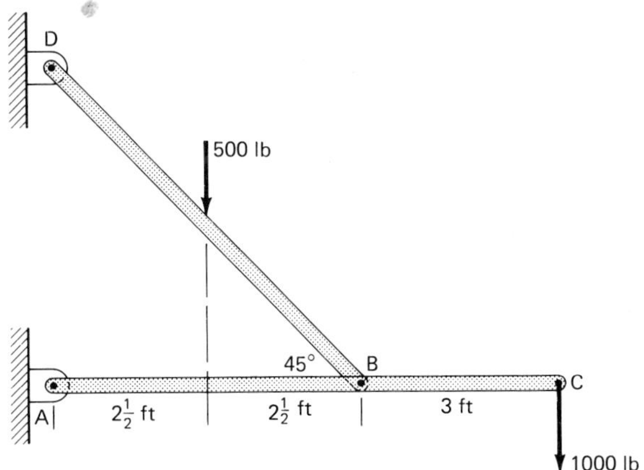

FIGURE P 4.18

4.19   Determine the forces at pins A and D acting on bars AC and DB, respectively, of Fig. P 4.18.

4.20   If a horizontal force of 500 lb acting to the left on Fig. P 4.18 replaces the vertical 500-lb force, determine the forces at pin B which act on bar AC.

5
Friction

5.1 INTRODUCTION

Most people do not consider that the absence of friction would make our daily lives impossible. We could not walk; vehicles could not start and stop; nails, screws, and bolts would not hold; nothing would stay on any surface that was not perfectly horizontal; motion, once started, would go on endlessly, and so on. We are more aware though that in many instances the role of friction is to create wear and heat, and in this sense it is destructive. To minimize these frictional effects, lubrication is most frequently used in the form of grease, oil, and so on. The cost of the effects of friction in machines is enormous, and a continual effort must be made to minimize these effects. The total elimination of friction is not possible, and this fact is embodied in the physics concept that perpetual motion is impossible.

The concepts to be studied in this chapter are basic and somewhat simplified. This is unfortunate, but it does express the fact that the present knowledge of friction is far from satisfactory, and that much more work must be done before our quantitative ability to predict frictional effects reaches a fully satisfactory status.

5.2 FRICTION THEORY

Whenever one body *tends* to slide over another or actually slides, a force exists which tends to oppose the impending or actual motion. The origin of these forces is largely due to the irregularities of the surfaces in contact with one another. Figure 5.1 depicts these irregularities greatly magnified. Where the surfaces are in contact, forces of adhesion exist, and we may consider these contacts to be points of microscopic welds. Also, in order for the surfaces to start to move relative to each other, it is necessary

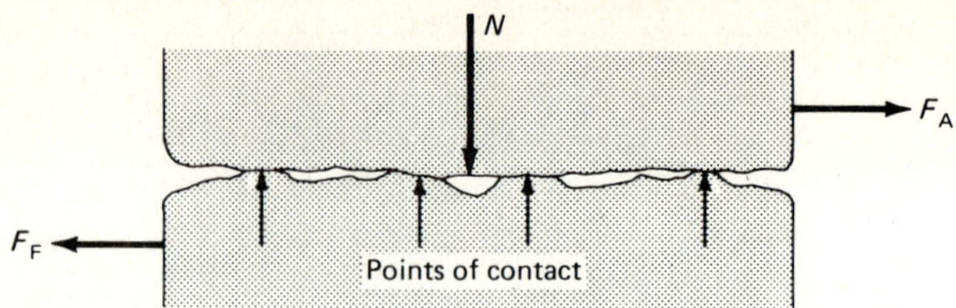

FIGURE 5.1   Friction.

for the meshing irregularities to lift or shear. Thus it is expected that it would require a greater force to start one body moving relative to another than to keep it moving once movement occurs. Also, the force to start motion first is a function of the character of the surfaces in contact, that is, the degree of roughness of the two surfaces.

If we now plot the applied force F_A shown in Fig. 5.1 against the frictional force F_F, we find that up to that point of impending motion the conditions of equilibrium require the frictional force F_F to equal the applied force. This is shown in Fig. 5.2, where applied force is plotted against frictional force. Since both are equal, this portion of the graph is a straight line at 45° to the horizontal, as shown. For a given pair of surfaces of given materials and roughness, it is found that the maximum value that the static frictional force attains is proportional to the *normal* (perpendicular) force between the surfaces. We can write proportionality as

$$F_S = \mu_S N \qquad (5.1)$$

where

F_S = maximum static frictional force developed

N = normal force between surfaces

μ_S = proportionality constant known as the coefficient of static friction

Some typical values of the coefficient of static friction are tabulated in Table 5.1. It should be noted that these are typical values and can only be used as a guide since the character of the surfaces in contact causes the coefficient of static friction to vary greatly.

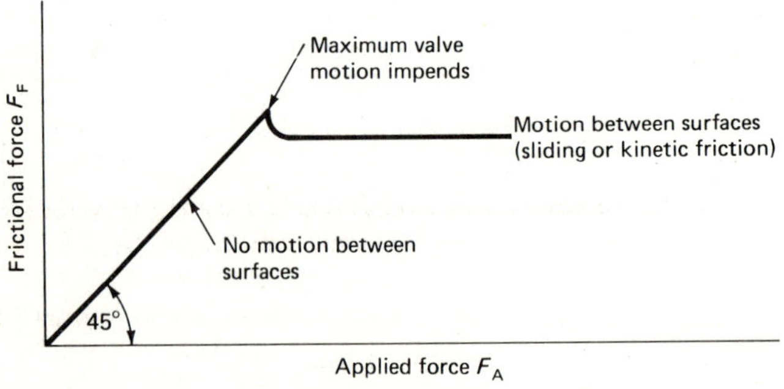

FIGURE 5.2   Static and kinetic friction.

Friction

TABLE 5.1 Coefficients of Friction—Typical Values

Materials	Coefficient of Static Friction	Coefficient of Kinetic Friction
Steel on steel	0.76	0.50
Steel on aluminum	0.61	0.47
Steel on copper	0.53	0.36
Steel on brass	0.51	0.44
Zinc on cast iron	0.85	0.21
Copper on cast iron	1.05	0.29
Cast iron on cast iron	1.10	0.15
Glass on glass	0.94	0.40
Copper on glass	0.68	0.53
Teflon on Teflon	0.04	0.04
Teflon on steel	0.04	0.04
Wood on wood	0.25–0.50	0.48
Metal on wood	0.20–0.60	

Once motion starts (breakaway), the resistance to motion decreases to a value less than the maximum value F_S. Two physical events occur at the breakaway point. The first is the breaking of the microwelds at the points of contact, and the second is that there is less meshing of the surface irregularities. Both effects cause the maximum static frictional force developed at the point of impending motion to decrease until the kinetic friction value is reached. The value of the kinetic friction resisting force F_K is *always* the maximum value that can be developed between the surfaces in contact. Although at low speeds and at very high speeds there is a change in the kinetic friction, it is usual to assume that the kinetic friction force F_K is constant in value, depending only on the normal force between the surfaces in contact and the materials of these surfaces. Equation (5.1) can be applied to kinetic friction by altering the subscripts to read

$$F_K = \mu_K N \tag{5.2}$$

where

F_K = kinetic friction force

N = normal force between surfaces in contact

μ_K = coefficient of kinetic friction

Very often the subscripts S and K are deleted, and Eqs. (5.1) and (5.2) are written simply as

$$F = \mu N \tag{5.3}$$

Table 5.1 also gives typical values for the coefficient of kinetic friction.

We can summarize some of the foregoing into the so-called laws of friction, which are essentially empirical in nature and are not fundamental (as are Newton's laws of motion) as follows.

1. Friction always tends to oppose the impending or actual motion of bodies relative to each other.
2. Up to its maximum value, static frictional forces develop values equal to the force acting to cause relative motion of the two surfaces.
3. The value of the static friction force developed is a function of the materials, the character (roughness) of the surfaces, and the normal force between the surfaces.
4. The maximum value of the static friction force developed between two surfaces is directly proportional to the normal force between the surfaces, and the proportionality constant is known as the coefficient of static friction.
5. Once motion starts, the resisting frictional force is assumed to reach its maximum value, which is somewhat less than the maximum value of the static frictional force and is assumed to be constant and independent of the relative velocities between the surfaces. The value of the kinetic frictional force is proportional to the normal force between the surfaces, and the proportionality constant is known as the coefficient of kinetic friction.
6. The general relation between frictional force and normal force can be written as $F = \mu N$.

ILLUSTRATIVE PROBLEM 5.1

A body weighing 100 N rests on a horizontal surface. If the coefficient of static friction is 0.3, determine the force required to cause motion to impend.

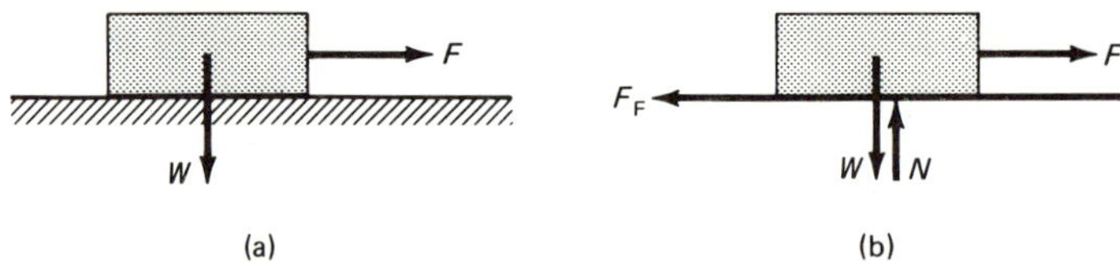

FIGURE 5.3   Illustrative Problem 5.1.

SOLUTION

Under the action of the forces shown in Fig. 5.3(a), we note that the system is in equilibrium. A free-body diagram of the body is shown in Fig. 5.3(b), where N is the normal force between body and surface, and F_F is the frictional force acting in a direction to oppose impending motion. Since the body is in equilibrium, a force summation yields

$$W - N = 0$$
$$F_F - F = 0$$

Friction

We also have the fact that $F_F = \mu_S N$. Solving these three equations gives us $W = N$ and $F_F = F$. Therefore

$$F = \mu_S W = 0.3 \times 100 = 30 \text{ N}$$

Note that friction *always opposes* impending or actual motion.

Illustrative Problem 5.1 also leads us to a consideration of the forces exerted by the horizontal surface on the body. Redrawing Fig. 5.3(b), we have the free-body diagram of the block shown in Fig. 5.4, but now we show the resultant of the friction force and the normal force to be R. The angle that the resultant R makes with the normal force is designated as ϕ. The magnitude of this angle ϕ varies from zero to a maximum value when motion impends. This angle ϕ is known as the angle of friction, and it is the angle of static friction when motion impends. Referring to Fig. 5.4 we have tan $\phi = F_F/N$. However, we have defined $F_F = \mu_S N$. Therefore

$$\mu_S = \tan \phi \tag{5.4}$$

which, expressed in words, states that the coefficient of static friction is equal to the tangent of the angle of static friction.

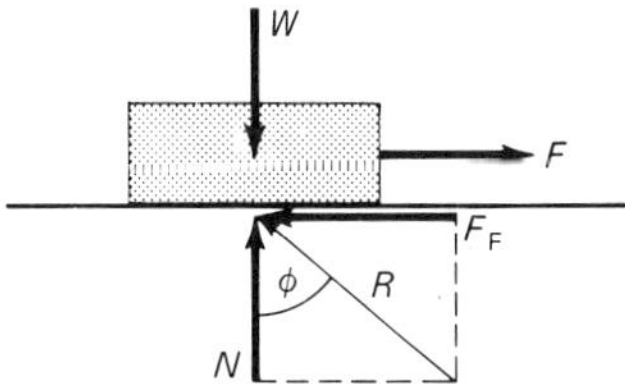

FIGURE 5.4   The angle of friction.

ILLUSTRATIVE PROBLEM 5.2

Determine the force F required to cause motion to impend in the situation shown in Fig. 5.5.

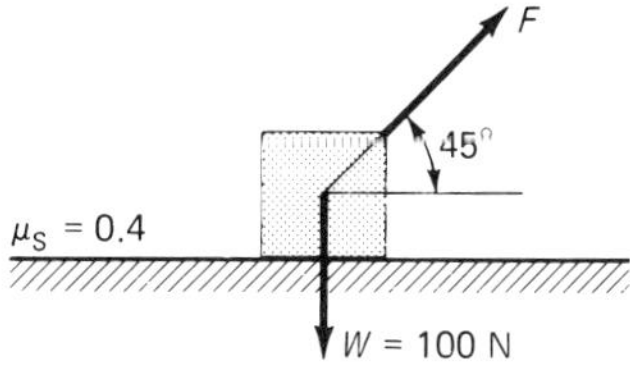

FIGURE 5.5   Illustrative Problem 5.2.

SOLUTION

Drawing a free-body diagram with the components of F shown, yields

Fig. 5.6. A summation of the forces in the horizontal direction and the condition of equilibrium gives

$$F_F - F_x = 0$$

and the summation of the forces in the vertical direction must be zero. Therefore

$$W - F_y - N = 0$$

and finally,

$$F_F = \mu_S N$$

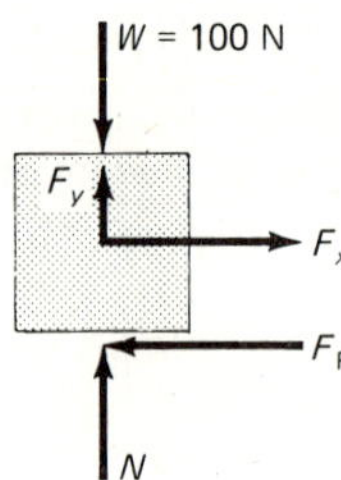

FIGURE 5.6   Illustrative Problem 5.2.

From the summation of vertical forces we have

$$100 - F \sin 45° - N = 0$$

From the horizontal force summation we have

$$F_F - F \cos 45° = 0$$

but

$$F_F = 0.4N$$

Therefore

$$0.4N - F \cos 45° = 0$$

or

$$N = \frac{F \cos 45°}{0.4}$$

Substitution into the vertical force summation equation,

$$100 - F \sin 45° - \frac{F \cos 45°}{0.4} = 0$$

Simplifying,

$$F \left(\sin 45° + \frac{\cos 45°}{0.4} \right) = 100$$

$$F(0.707 + 1.768) = 100$$

$$F = 40.4 \text{ N}$$

It should be noted that the angle that the force F makes with the horizontal gives us a vertical force component, which decreases the value of the normal force N from its maximum value of the weight, which we found for the case of a horizontal force. In this problem $N = 100 - F \sin 45° = 71.4$ N, not the 100-N weight.

5.3   THE INCLINED PLANE

Let us start this portion of our discussion by considering that a body is placed upon a horizontal plane which is hinged at its forward edge so that it can be pivoted up, as shown in Fig. 5.7. As the angle that the plane makes with the horizontal is increased, there is no motion of the body on the plane, until at some angle ϕ we find the block to be on the verge of sliding down the plane. The angle of the plane at which motion of the block impends is known as the *angle of repose*. Since the body is in equilibrium under the action of the forces acting on it, we can draw a free-body diagram of the body and apply the conditions of equilibrium to it. Figure 5.8 shows a free-body diagram of the body on the plane at the position of impending motion.

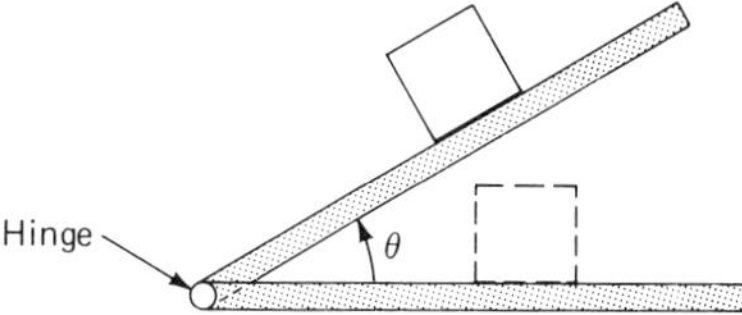

FIGURE 5.7   A body on an inclined plane.

Since the resultant of F_F and N must equal W, $R = W$, and the angle between the normal N and the resultant R is ϕ. Thus when motion impends, the angle of inclination of the plane is equal to the angle ϕ, which we have called the angle of repose for this case or the angle of friction for the case of a horizontal plane surface. This gives us a relatively simple method of determining the coefficient of static friction, since $\mu_S = \tan \phi$.

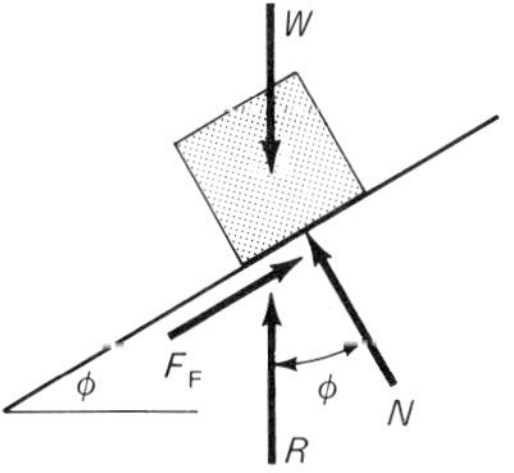

FIGURE 5.8   Free-body diagram of a body on an inclined plane.

ILLUSTRATIVE PROBLEM 5.3

A body weighing 100 lb rests on an inclined plane that makes an angle of 30°

with the horizontal. Will a force of 40 lb acting parallel to and up the plane cause motion to occur? If so, will the body move up or down the plane?

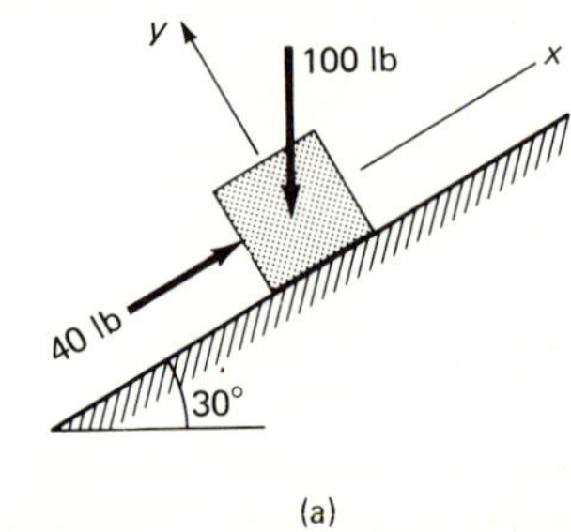

(a)

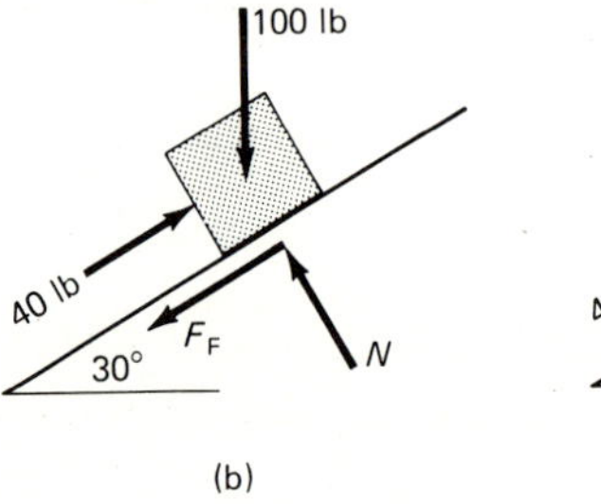

(b)

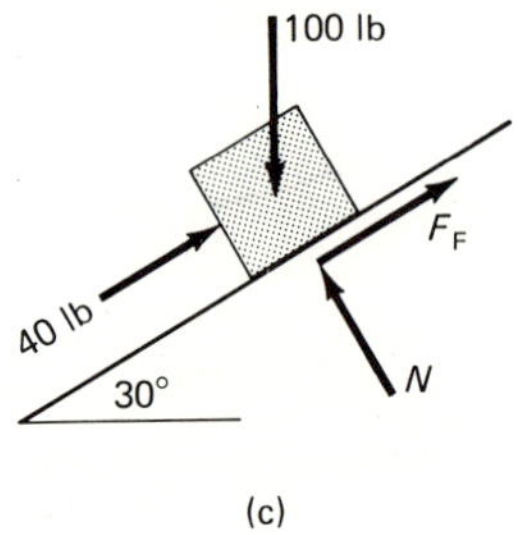

(c)

FIGURE 5.9   Illustrative Problem 5.3.

SOLUTION

Let us first assume that the body is in equilibrium and that the force of friction tends to oppose motion up the plane. The free-body diagram for this case is shown in Fig. 5.9(b). From conditions of equilibrium, using axes parallel to and perpendicular to the plane, we have

$$40 - F_F - 100 \sin 30° = 0$$

$$N - 100 \cos 30° = 0$$

Solving these equations,

$$N = 100 \cos 30° = 100 \times 0.866 = 86.6 \text{ lb}$$

and

$$F_F = -100 \sin 30° + 40$$

$$F_F = -100 \times 0.5 + 40 = -50 + 40 = -10 \text{ lb}$$

The negative value of F_F indicates that the body will not move up the plane since the plane cannot generate a frictional force in the direction of motion. Let us therefore investigate whether the body can move down the plane by considering the force system shown in Fig. 5.9(c). Writing the equations of equilibrium

Friction

for the forces parallel to and perpendicular to the plane gives us

$$40 + F_F - 100 \sin 30° = 0$$

$$N - 100 \cos 30° = 0$$

Therefore,

$$F_F = 100 \sin 30° - 40 = 100 \times 0.5 - 40 = 50 - 40 = 10 \text{ lb}$$

and

$$N = 100 \cos 30° = 100 \times 0.866 = 86.6 \text{ lb}$$

The plane can generate a maximum frictional force of $\mu_s N = 0.4 \times 86.6 = 34.6$ lb, which is in excess of the value of 10 lb if the body were just on the verge of starting to slide down the plane. We therefore conclude that the body will not slide up or down the plane—it will remain stationary under the action of the forces shown in Fig. 5.9.

SOLUTION

For this problem, we refer to Fig. 5.9(b) with a force F replacing the 40-lb force. Writing the equations of equilibrium in an x, y coordinate system parallel to and perpendicular to the plane gives us

$$F - F_F - 100 \sin 30° = 0$$

$$N - 100 \cos 30° = 0$$

At this point we also invoke

$$F_F = 0.4 N$$

Solving

$$N = 100 \cos 30° = 100 \times 0.866 = 86.6$$

$$F_F = 0.4 \times 86.6 = 34.6$$

$$F - 34.6 - 100 \sin 30° = 0$$

$$F = 34.6 + 100 \sin 30° = 34.6 + 100 \times 0.5 = 84.6$$

Therefore the least force required to cause motion of the body up the plane is 84.6 lb.

Another form of inclined plane is the wedge. The wedge is basically an inclined plane that is used to produce small motions in a body or as a means of applying large forces to another body. The basic action of the wedge depends upon friction, and the principles studied in the earlier paragraphs can be applied directly to this machine. The following illustrative problem indicates the method of solution of this type of problem.

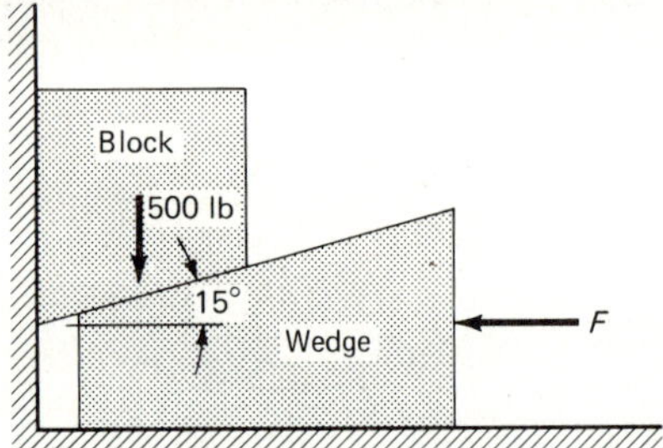

FIGURE 5.10   Illustrative Problem 5.5.

SOLUTION

Our first step is to draw the free-body diagram of each block. On each free body we show the components of each force in the horizontal and vertical directions. It is important to note that the reaction on the block by the wedge gives rise to forces that are equal and opposite on block and wedge. Also, the normal and frictional components of the forces on the inclined surface are shown for convenience in Fig. 5.11.

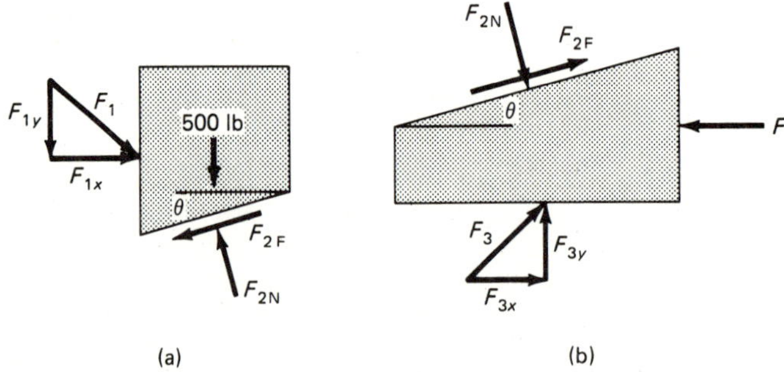

FIGURE 5.11   Illustrative Problem 5.5.

Applying the conditions for equilibrium to the block, for $\Sigma F_y = 0$,

$$-F_{1y} + F_{2N} \cos \theta - F_{2F} \sin \theta - 500 = 0 \tag{A}$$

For $\Sigma F_x = 0$,

$$F_{1x} - F_{2F} \cos \theta - F_{2N} \sin \theta = 0 \tag{B}$$

also

$$F_{1y} = \mu F_{1x} \tag{C}$$

$$F_{2F} = \mu F_{2N} \tag{D}$$

Using Eqs. (C) and (D) in Eqs. (A) and (B) gives Eqs. (A') and (B'), where Eq. (B) has been multiplied by μ:

$$-\mu F_{1x} + \frac{F_{2F}}{\mu}(\cos \theta) - F_{2F} \sin \theta - 500 = 0 \tag{A'}$$

$$\mu F_{1x} - \mu F_{2F} \cos \theta - F_{2F} \sin \theta = 0 \tag{B'}$$

Adding Eqs. (A') and (B'),

$$F_{2F}\left(\frac{\cos \theta}{\mu} - \sin \theta - \mu \cos \theta - \sin \theta\right) - 500 = 0$$

Setting $\theta = 15°$ and $\mu = 0.2$, gives

$$F_{2F} = 121.4 \text{ lb}$$

Therefore

$$F_{2N} = \frac{F_{2F}}{\mu} = \frac{121.4}{0.2} = 607.0 \text{ lb}$$

We can now consider the free-body diagram of the wedge and proceed as before. For $\Sigma F_y = 0$,

$$-F + F_{2F} \cos \theta + F_{2N} \sin \theta + F_{3x} = 0 \tag{E}$$

For $\Sigma F_x = 0$

$$-F_{2N} \cos \theta + F_{2F} \sin \theta + F_{3y} = 0 \tag{F}$$

also

$$F_{2F} = \mu F_{2N} \tag{G}$$

From above,

$$F_{2F} = 121.4 \text{ lb}$$

$$F_{2N} = 607.0 \text{ lb}$$

and since

$$F_{3x} = \mu F_{3y}$$

From Eq. (F),

$$F_{3y} = F_{2N} \cos \theta - F_{2F} \sin \theta \tag{H}$$

The Inclined Plane **123**

Substituting and solving,

$$-F + F_{2F} \cos \theta + \mu F_{2N} \cos \theta + F_{2N} \sin \theta - \mu F_{2F} \sin \theta = 0$$

Inserting the constants of the problem,

$$-F + 121.4 \times 0.966 + 0.2 \times 607 \times 0.966$$
$$+ 607 \times 0.259 - 0.2 \times 121.4 \times 0.259 = 0$$

and

$$F = 385.5 \text{ lb}$$

Using the method of components to solve this problem is a tiring exercise in algebra and can easily lead to arithmetic errors. An alternate solution for this problem can be obtained by using the polygon method, which is much simpler. We first redraw the free-body diagram of Fig. 5.11 in a more convenient form, as shown in Fig. 5.12. Notice that the forces F_1, F_2, and F_3 are shown inclined from their normals by an amount equal to the friction angle ϕ, where $\phi = \arctan \mu$. Drawing the polygon of forces for each body gives us Fig. 5.13.

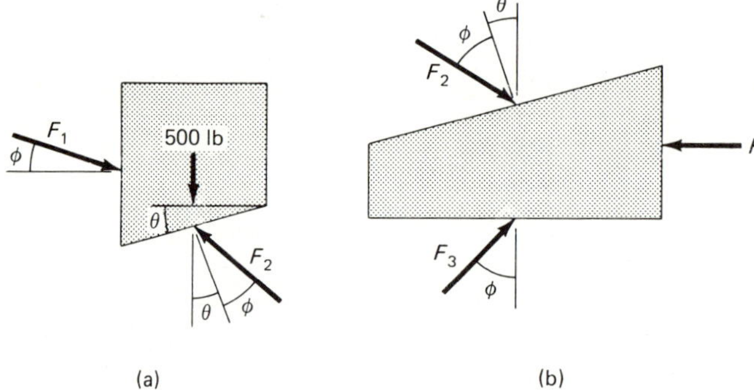

FIGURE 5.12   Alternate solution of Illustrative Problem 5.5.

F_2 is obtained from Fig. 5.13(a), using the law of sines,

$$\frac{500}{\sin (90° - 2\phi - \theta)} = \frac{F_2}{\sin (90° + \phi)}$$
$$F_2 = \frac{500 \sin (90° + \phi)}{\sin (90° - 2\phi - \theta)}$$

Since $\mu = 0.2$, $\tan \phi = 0.2$, $\phi = 11.3°$, and θ is given as 15°,

$$F_2 = \frac{500 \sin (90 + 11.3)}{\sin [90 - 2(11.3) - 15]} = \frac{500 (\sin 101.3)}{\sin 52.4} = \frac{500 \sin 78.7}{\sin 52.4}$$
$$F_2 = 500 \times \frac{0.981}{0.792} = 619.3 \text{ lb}$$

Note that our earlier solution gives

$$F_2 = \sqrt{121.4^2 + 607^2} = 619.0$$

Friction

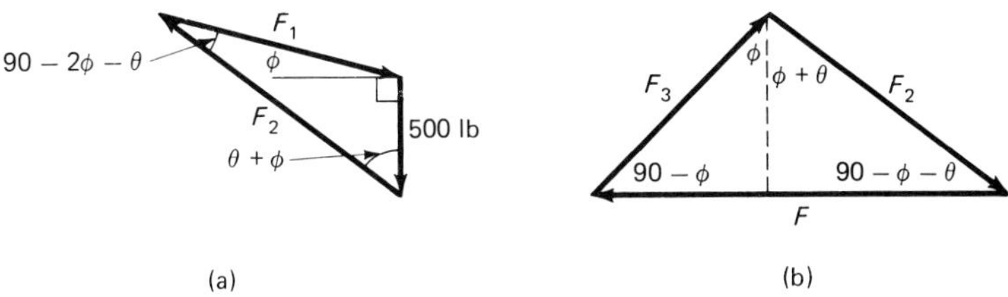

FIGURE 5.13   Force polygon for Illustrative Problem 5.5.

We now solve the triangle of Fig. 5.13(b), using the law of sines,

$$\frac{F}{\sin(\phi + \phi + \theta)} = \frac{F_2}{\sin(90° - \phi)}$$

$$F = \frac{619.3 \sin(11.3 + 11.3 + 15)}{\sin(90 - 11.3)}$$

$$= \frac{619.3 \sin 37.6}{\sin 78.7} = \frac{619.3 \times 0.610}{0.981} = 385.0 \text{ lb}$$

The effort required to solve this type of problem is obviously considerably less when using the alternate method, which yields the same results as using the direct algebraic approach.

5.4   ROLLING FRICTION

When a wheel or roller rolls along any surface, the surface will yield and be deformed under and adjacent to the wheel. This action causes a resistance to the motion of the wheel, and this resistance is known as rolling friction. If the wheel did not cause yielding of the surface, a situation would occur as shown in Fig. 5.14. In this case the weight of the wheel passes through the contact point, and the resisting force F also passes through the same point, since the sum of the moments about the contact point must be zero. We must therefore conclude that a wheel rolling on a horizontal, unyielding surface would encounter no resistance (since $F = 0$) and once set in motion would roll endlessly, giving us a perpetual motion machine.

For the case of a wheel on a yielding surface we have the situation shown with exaggerated friction in Fig. 5.15. When the wheel is pulled forward by the force F, it must climb out of the groove which was created by the yielding of the surface. In

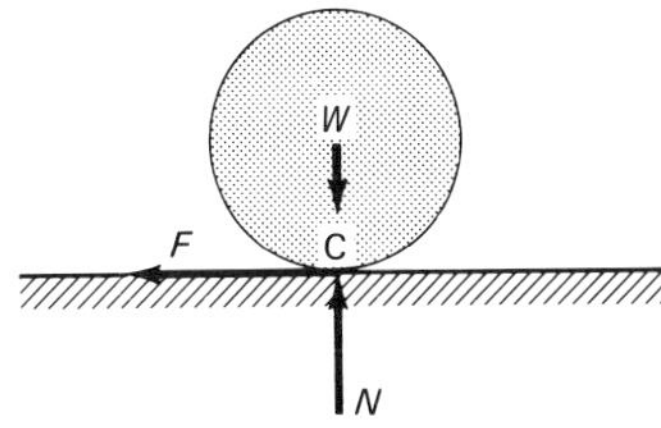

FIGURE 5.14   Wheel on an unyielding horizontal surface.

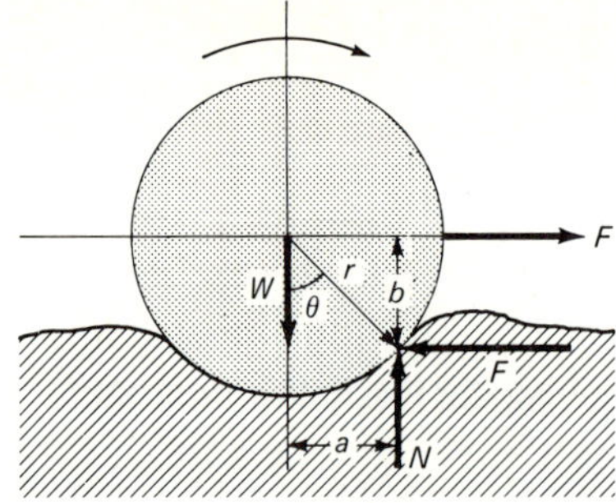

FIGURE 5.15   Wheel on yielding horizontal surface.

order to do this, the applied force F must create a moment about the contact point equal to the moment of the weight of the roller about this point. Taking moments about the point of contact, we have

$$Fb = Wa \qquad (5.5)$$

or

$$F = \frac{Wa}{b} \qquad (5.5a)$$

However, the depth of indentation is small, and for practical purposes we can taken $b \cong r$. Therefore

$$F = \frac{Wa}{r} \qquad (5.6)$$

Equation (5.6) is the defining equation for rolling resistance, and the distance a is known as the *coefficient of rolling resistance*. Since the dimension of a is usually expressed in inches or millimetres, it differs from our earlier concept of a coefficient of friction that is dimensionless, that is, F/N. Unfortunately there is a lack of infor-

TABLE 5.2   Coefficients of Rolling Resistance

Material	a (in.)	a (mm)
Annealed steel on annealed steel	0.02	0.51
Hardened steel on hardened steel	0.0030–0.0040	0.08–0.10
Steel on wood	0.06–0.10	1.52–2.54
Steel on soft ground	3.0–5.0	76.2–127.0
Tires on good road	0.02–0.022	0.51–0.56
Tires on mud road	0.04–0.06	1.02–1.52
Ball bearings	0.0008–0.0012	0.02–0.03
Cylindrical roller bearings	0.002–0.004	0.051–0.102
Steel wheels on steel rails	0.006	0.152

Friction

mation regarding the laws governing rolling resistance, but two assumptions are generally made which are open to question. The first is that the distance a is a constant for a given wheel radius and material. We do know that the distance a is a function of the elastic and plastic properties of the materials in contact, the roughness of these surfaces, and the speed of travel of the wheel. The second assumption usually made is that for a given material the distance a is constant for all sizes of rollers. From these assumptions it is apparent that results obtained from Eq. (5.6) using tabulated values for the coefficient of rolling resistance cannot be expected to yield completely satisfactory results. Table 5.2 gives some values for the coefficient of rolling resistance which should be considered to be approximate. Variations from these values can exceed 100% of the tabulated data shown.

SOLUTION

From Table 5.2 we can take a to be 0.006 in. Therefore for the locomotive F will be

$$F = \frac{197\,660 \times 0.006}{79/2} = 30.0 \text{ lb}$$

For sliding, $F = \mu N$. Therefore

$$F_{sliding} = 0.3 \times 197\,660 = 59\,300 \text{ lb}$$

It should be noted that the magnitude of the frictional force required to overcome sliding friction is almost 2000 times greater than that required to overcome rolling friction.

5.5 BELT FRICTION

When a belt or band is wrapped around a pulley or drum, the friction developed between belt and pulley can be used either to transmit power or, as in the case of brakes, to retard motion. In order to analyze the forces acting in this situation it is necessary to consider the case of impending slippage of a belt on a pulley, and to apply the equations of equilibrium. Before proceeding it should be noted that if the pulley were perfectly smooth, the tension on the belt would be the same on each side of the pulley, since no friction would be developed between belt and pulley. Therefore when friction exists, we can anticipate that the tensions on both sides of the pulley differ.

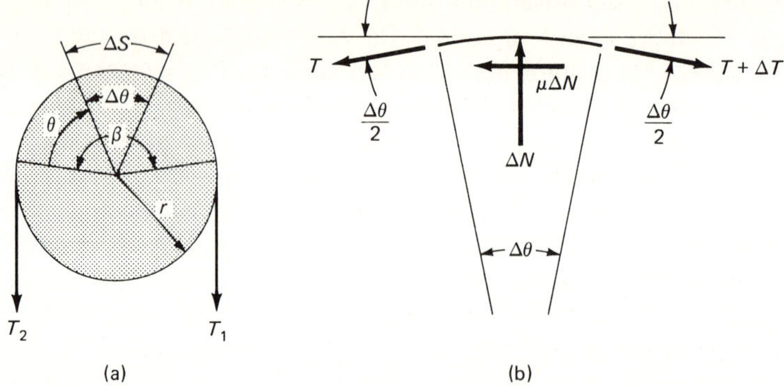

FIGURE 5.16 Belt friction.

Figure 5.16 shows a belt wrapped around a pulley with the angle of wrap (in radians) conventionally denoted by β. We shall assume the pulley to be stationary and the belt to be on the verge of slipping. With $T_1 > T_2$ the belt will cause impending motion of the pulley to be in the clockwise direction. In order to establish the relationship relating the tensions, angle of wrap, and coefficient of friction, it is necessary to consider the free-body diagram of a small segment of the belt, which we will designate as ΔS, that is, a small element of belt length wrapped around the pulley for an angle $\Delta\theta$. The normal force acting on this small belt segment is designated as ΔN, and the frictional force acting to oppose motion is denoted by $\mu\Delta N$.

The procedure at this time is to set up axes in the radial and transverse directions and then to apply the conditions of equilibrium to the components of the forces acting on the system. From these equations a differential equation is obtained, which can be solved by methods beyond the scope of our study.* The final result obtained for the case where centrifugal effects on the belt are neglected is

$$\frac{T_1}{T_2} = e^{\mu\beta} \quad \text{or} \quad \ln\frac{T_1}{T_2} = \mu\beta \quad \text{or} \quad \ln T_1 - \ln T_2 = \mu\beta \qquad (5.7)$$

where

e = base of natural logarithms, = 2.718

μ = coefficient of friction

β = angle of wrap, in radians (1 rad $\cong$ 57.3°)

ln = natural logarithm

Since friction always tends to oppose impending motion, the larger force must basically

*See Shames, I. H., ENGINEERING MECHANICS, VOL. I—STATICS, 2nd ed. Englewood Cliffs, NJ: Prentice-Hall, 1966.

Friction

equal the smaller force plus the summation of the frictional forces. Thus in Eq. (5.7) the sense of the larger force is determined by the conditions of the problem.[†]

In order to determine the torque (moment) that a belt can transmit we consider Fig. 5.16(a) and take moments about the center of the pulley. Therefore

$$\text{torque} = T_1 r - T_2 r = (T_1 - T_2)r \qquad (5.7a)$$

SOLUTION

From Eq. (5.7a) we have

$$\text{torque} = 1000 = (T_1 - T_2)8 \text{ in.·lb}$$

Therefore

$$T_1 - T_2 = \frac{1000}{8} = 125 \text{ lb}$$

From Eq. (5.7),

$$\frac{T_1}{T_2} = e^{\mu\beta} = e^{0.3\pi} = e^{0.942} = 2.57$$

and

$$T_1 = 2.57 T_2 \text{ lb}$$

Substituting this into the previous relation yields

$$2.57 T_2 - T_2 = 125$$

$$1.57 T_2 = 125$$

$$T_2 = 79.6 \text{ lb}$$

$$T_1 = 2.57 \times 79.6 = 204.6 \text{ lb}$$

[†]For a more general discussion of belt problems see Granet, I., "Graphs Help Solve Belt Problems." *Design News*, August 7, 1972.

Belt Friction

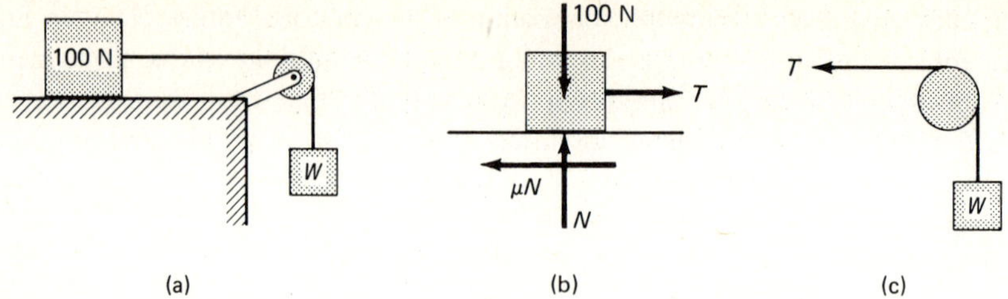

FIGURE 5.17   Illustrative Problem 5.8.

SOLUTION

We first consider the free-body diagram of the block shown in Fig. 5.17(b). For equilibrium,

$$T = \mu N = 0.3 \times 100 = 30 \text{ N}$$

The angle of wrap of the cable is 90° ($\pi/2$ rad). Therefore

$$\frac{W}{T} = e^{\mu\beta} = e^{0.3\,(\pi/2)} = e^{0.471} = 1.602$$

Therefore

$$W = T \times 1.602 = 30 \times 1.602 = 48.1 \text{ N}$$

5.6   SQUARE-THREADED SCREWS

A screw is an inclined plane wrapped around a cylinder as shown in Fig. 5.18(a). It can also be considered to be a helix with a constant lead. While screws can and do have a multiplicity of cross-sectional shapes, we will be interested in the square thread. For a single-threaded screw we define the pitch of the screw to be p, the distance between similar points on adjacent threads. In one turn of a nut it will advance a distance along the screw equal to the pitch. Thus the inclined plane is defined by

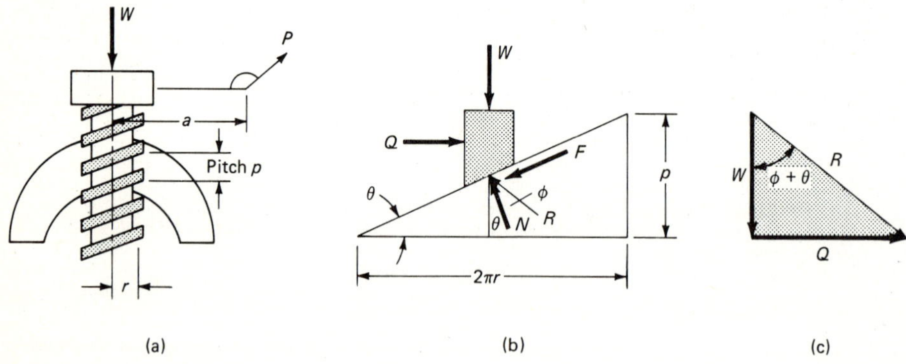

FIGURE 5.18   A square-threaded screw.

Friction

Fig. 5.18(b), where the base of the helix is $2\pi r$, and the rise is the pitch p, r being the mean radius of the screw thread. If we assume that the thread is used to lift the weight W, we obtain the free-body diagram shown in Fig. 5.18(b). The simplest way to determine the force Q required to make the weight W go up the inclined plane is to use the force triangle shown in Fig. 5.18(c). From this we obtain Q directly as

$$Q = W \tan (\phi + \theta) \qquad (5.8)$$

where $\tan \phi = \mu$, the coefficient of friction. If we now take moments about the central vertical axis of the screw, we obtain

$$Pa = Qr \qquad (5.9)$$

and

$$P = \frac{Wr}{a} \tan (\phi + \theta) \qquad (5.10)$$

If the body is being lowered, the force triangle will yield

$$Q = W \tan (\phi - \theta) \qquad (5.11)$$

and

$$P = \frac{Wr}{a} \tan (\phi - \theta) \qquad (5.12)$$

Let us now ask the question as to whether, once the load is raised, will releasing the applied moment cause the screw to unwind and therefore lower the load? As we have seen [Eqs. (5.10) and (5.11)], raising or lowering the load is equivalent to changing the friction force from one side of the normal to the plane to the other. The screw will be on the verge of self-unwinding if $\phi = \theta$ and will remain in place (self-locking) if $\phi > \theta$. If $\phi < \theta$, the screw will unwind itself, and a torque of $Wr \tan (\theta - \phi)$ must be applied to prevent unwinding.

ILLUSTRATIVE PROBLEM 5.9

A square-threaded jackscrew is used to raise a weight of 200 lb. The pitch of the screw os 0.25 in., and the mean radius of the screw is 1/2 in. If the coefficient of friction is 0.3, determine the torque necessary to raise the weight. Is the screw self-locking? What torque is required to lower tho scrcw?

SOLUTION

The tangent of the angle θ is the pitch p divided by $2\pi r$. Therefore

$$\tan \theta = \frac{0.25}{2\pi \times 0.5} = 0.0795$$

$$\theta = 4.55°$$

We can immediately answer the second question: The screw is self-locking since

$\mu > \tan \theta$. The torque to raise the weight is Qr. From Eq. (5.8), multiplied by r, we have

$$Qr = Wr \tan (\phi + \theta)$$

Since $\tan \phi = 0.3$ and $\phi = 16.7°$,

$$Qr = 200 \times 0.5 \tan (16.7 + 4.55) = 200 \times 0.5 \tan 21.25$$

$$= 200 \times 0.5 \times 0.389 = 38.9 \text{ in.·lb}$$

The torque to lower the screw is found from Eq. (5.11), multiplied by r, as

$$Qr = Wr \tan (\phi - \theta)$$

$$= 200 \times 0.5 \tan (16.7 - 4.55) = 200 \times 0.5 \tan 12.15$$

$$= 200 \times 0.5 \times 0.215 = 21.5 \text{ in.·lb}$$

5.7   CLOSURE

The number of situations where friction enters into a problem are infinite since all real physical events involve friction in some manner or form. In this chapter we have tried to illustrate the principles that are involved and how they are applied to several situations that are of interest from an applied viewpoint. Undoubtedly students can think of many situations from their own experience which could also be used to illustrate the application of the principles of friction to real situations. In most instances a brief study of the situation will serve to show how the principles studied in this chapter can be applied to those situations. The key to solving all problems involving friction is the free-body diagram plus the simple statement that friction always opposes impending or actual motion.

REFERENCES

Bassin, M. G., S. M. Brodsky, and H. Wolkoff, STATICS AND STRENGTH OF MATERIALS, 3rd ed. New York: McGraw-Hill, 1979.

Granet, I., "Graphs Help Solve Belt Problems." *Design News,* August 7, 1972.

Levinson, I. J., INTRODUCTION TO MECHANICS, 2nd ed. Englewood Cliffs, NJ: Prentice-Hall, 1968.

Meriam, J. L., MECHANICS, PART I—STATICS, 2nd ed. New York: John Wiley, (SI Version) 1975.

Sears, F. W., and M. W. Zemansky, COLLEGE PHYSICS, 4th ed. Reading, MA: Addison-Wesley, 1977.

Seely, F. B., and N. E. Ensign, ANALYTICAL MECHANICS FOR ENGINEERS, 3rd ed. New York: John Wiley, 1941.

Semat, H., FUNDAMENTALS OF PHYSICS, 4th ed. New York: Holt, Rinehart and Winston, 1966.

Shames, I. H., ENGINEERING MECHANICS, VOL. I—STATICS, 2nd ed. Englewood Cliffs, NJ: Prentice-Hall, 1966.

Friction

Singer, F. L., ENGINEERING MECHANICS, 2nd ed. New York: Harper and Brothers, 1975.

White, H. E., MODERN COLLEGE PHYSICS, 5th ed. Princeton, NJ: D. Van Nostrand, 1966.

PROBLEMS

5.1   A 50-lb block is placed on an inclined plane. If the coefficient of friction between the block and the plane is 0.25, determine the angle of the plane at the instant that the block starts to slide down the plane.

5.2   A 100-lb body is placed on a plane that causes it to be on the verge of slipping when the plane is tilted 25° with respect to the horizontal. Determine the coefficient of friction between body and plane.

5.3   A body weighing 300 lb is placed on an inclined plane which makes an angle of 20° with the horizontal. If a force of 200 lb acting parallel to and up the plane causes the body to be on the verge of starting up the plane, determine the coefficient of friction between body and plane.

5.4   A force of 25 lb parallel to the plane of Problem 5.3 and facing down the plane causes the body to be on the verge of starting down the plane. Determine the coefficient of friction between body and plane.

5.5   What force F will cause motion to impend in the figure shown?

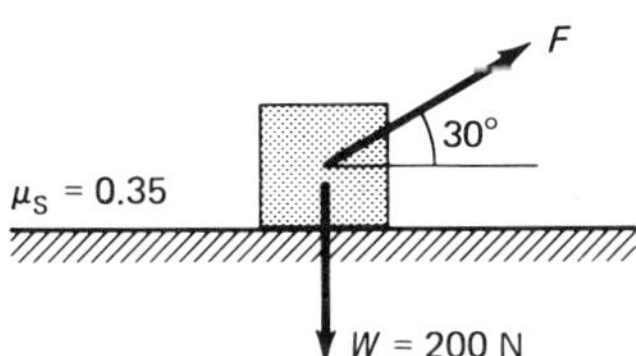

FIGURE P 5.5

5.6   Find the force necessary to cause the lower weight to move to the right if $\mu_S = 0.3$ for all surfaces.

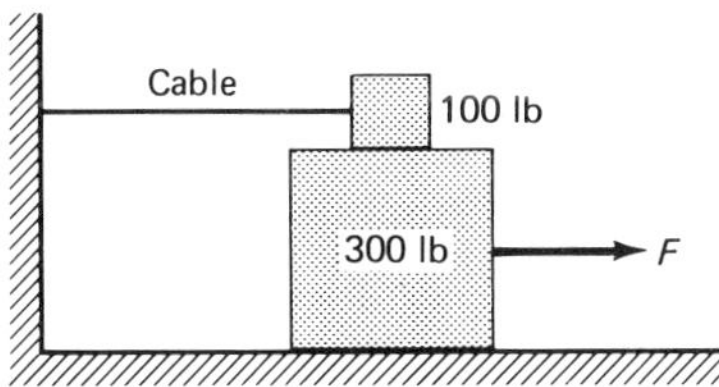

FIGURE P 5.6

5.7   If the cable in Fig. P 5.6 makes an angle of 15° with the horizontal, determine F.

5.8 What force F is required to cause the upper block to just start to move? Assume that $\mu_S = 0.3$ for all surfaces, except that the vertical wall may be considered to be perfectly smooth.

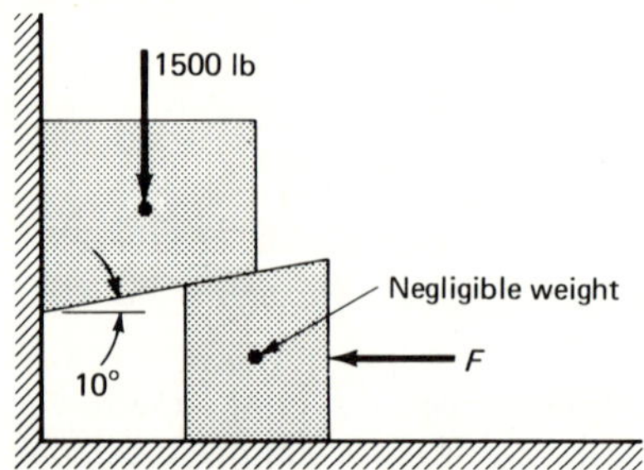

FIGURE P 5.8

5.9 If the vertical wall in Problem 5.8 has a coefficient of friction of 0.1, determine F.

5.10 Compare the horizontal force required to cause a 5000-lb block to start to slide as compared to the force required if it is placed on 1-ft diameter rollers. Assume annealed steel on steel for the materials.

5.11 What force is required to start the 100-lb cylinder over the step?

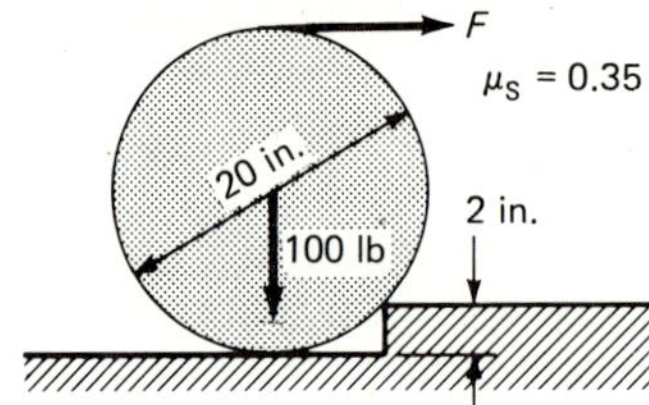

FIGURE P 5.11

5.12 A weight of 100 lb hangs over a pulley. What force is necessary to just keep the weight from moving? Assume that $\mu_S = 0.25$.

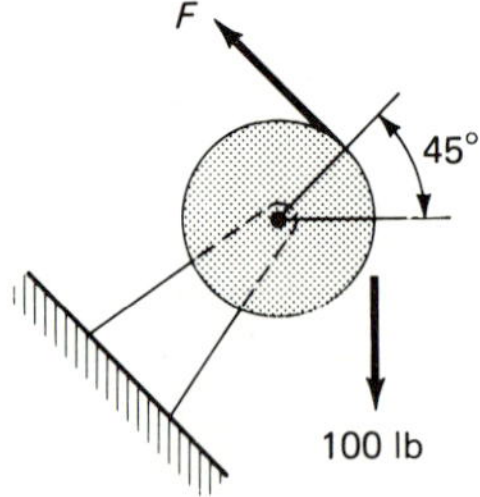

FIGURE P 5.12

5.13 If in Problem 5.12 the rope is wound for an additional complete turn so that $360° + 45°$ is the angle of contact, determine F.

                                                                 Friction

5.14   If a belt is wrapped around a shaft for half a turn (180°), determine the torque that it can exert on a 20-in. diameter shaft. Assume that the restraining cable can sustain a force of 500 lb and $\mu_S = 0.3$.

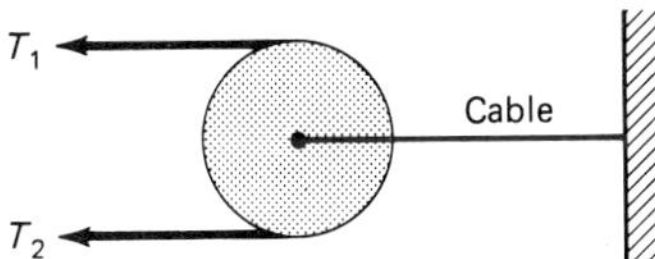

FIGURE P 5.14

5.15   A man wraps a rope around a pole for two complete turns. If he can pull on one end with a force of 40 lb, what load can he restrain on the other end? Use $\mu_S = 0.25$.

5.16   What weight will cause the 200-lb block to start to move? Use $\mu_S = 0.25$ for all surfaces.

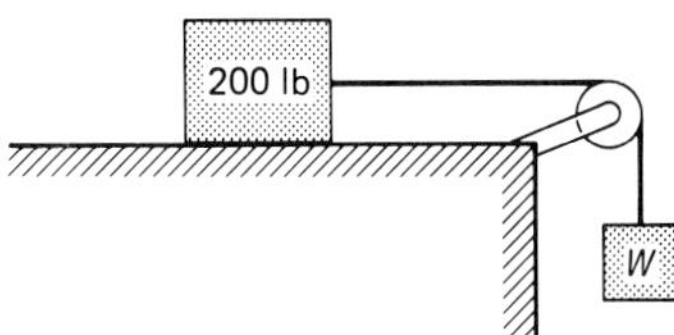

FIGURE P 5.16

5.17   A square-threaded screw has a mean radius of 0.75 in. If it is used to raise a weight of 500 lb, and the pitch is 3/8 in. per turn, determine the torque necessary to raise the weight. Use $\mu_S = 0.25$.

5.18   If the weight in Problem 5.17 is to be lowered, determine the torque required.

5.19   If the screw in Problem 5.17 is lubricated so that $\mu_S = 0.002$, determine whether it is self-locking.

5.20   If the screw in Problem 5.19 is not self-locking, determine the torque necessary to prevent the weight from going down of its own accord.

6

Concepts of Stress
and Strain

6.1   INTRODUCTION

The objectives of any course in *strength of materials* are to study how materials behave when subjected to differing loadings and restraints and to be able to predict this behavior in a given situation. The ultimate goal of this study is to relate the external forces on a body to the internal forces and deformations of that body. In most instances we will not be able to calculate the exact distribution of the internal resisting forces in the body, but by the use of the equations of equilibrium, we can usually obtain the resultant of these forces. A review of statics, especially that portion concerning the equilibrium of bodies subjected to various external forces, is strongly suggested at this time.

The purposes of this chapter are to define certain terms, such as stress and strain, to illustrate these definitions, and to apply the free-body concept and these definitions to some common simple situations.

6.2   STRUCTURAL LOADS

Since all structures, structural members, machine elements, and so on, can be subjected to various types of loading, such as static loads, dynamic loads, and repeated loads, all of which may act over wide temperature ranges, let us first discuss load categories. For our present purposes we will separate loadings into two broad classifications: *static loading* and *dynamic loading*.

The category of static loading can be further broken down into three subdivisions, namely, *continuous loading, gradually* or *slowly applied loading,* and *repeated gradually applied loading*. The term *continuous load* is used to characterize a load that remains on a member for a long time period. The weight of a structure or a tank subjected to internal pressure for a long period of time is a common example of this type of loading. A *gradually* or *slowly applied load* is one that slowly builds up to its maximum value and does not cause shock or vibration when it is applied. The last type of static load is the *repeated gradually applied load*. In this load classification are included those loads that are gradually applied but repeated a large number of times. The repeated gradually applied load is important because it can cause failure under a load that would be safe if it were applied once or only a few times. Failure of structures due to repeated loads usually occurs catastrophically.

When the application of a load is much more rapid than it is for those loads that are classified as static loads, appreciable shock or vibration can occur. This type of load is known as a *dynamic* or *impact load* and its application can cause internal forces in the structure to momentarily exceed those that occur when the same load is applied gradually. In addition, the deflection of the structure will also momentarily exceed the deflection caused by an equal gradually applied load. As an example, the gradual placing of a weight on the end of a cantilever beam will cause the beam to deflect and gradually come to maximum value. However, if a weight of the same magnitude is dropped on the end of the beam, the maximum deflection will be found to be several times greater than for the same gradually applied load, even if the height of drop is small.

In the following chapters of this book, we shall be concerned with static loads and, in particular, continuous or slowly applied loads. Structures or machine members that are subjected to either repeated or dynamic loads will not be part of our study. The reactions of structures to dynamic loads is a study area involving considerations beyond the scope of this text.

6.3 STRESS

The application of an external force to a solid causes internal resisting forces to exist within the body whose resultant will be equal in magnitude but opposite in direction to the applied force. A bar subjected to a longitudinal axial force that tends to elongate the bar is said to be *in tension,* while a bar subject to a longitudinal axial force that tends to decrease the axial dimension of the bar is said to be *in compression*. Figure 6.1 shows short bars in tension and compression. Due to the applied axial forces, internal resisting forces which are continuously distributed over the cross section of the bar are shown on the *free-body diagrams*. If the applied force passes through the *centroid* (center of gravity) of the cross section of the bar, the resisting forces will be uniformly distributed over the cross section of the bar. This rule is quite general and holds with only minor exceptions.

When comparing the resisting loads of geometrically similar members, we will find it more convenient to use as a basis of comparison the resistive load per unit area. For a uniform distribution of internal resistive forces, we simply express the resistive force per unit area in terms of the applied loading as

$$S = \frac{F}{A} \tag{6.1}$$

                                    Concepts of Stress and Strain

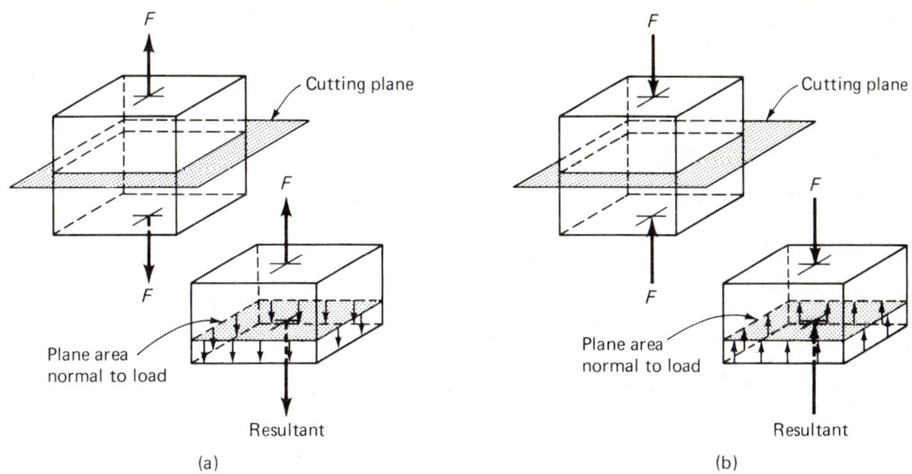

FIGURE 6.1 Tension and compression loads. (a) Tension. (b) Compression.

where S is the stress, A is the cross-sectional area of the bar, and F is the applied force. With the force expressed in pounds and the area expressed in square inches, the stress is expressed in pounds per square inch (psi).

In SI units, the force is expressed in newtons and the area is expressed in square metres, giving us the unit of stress as newtons per square metre, which is called the pascal (Pa).

For the loading conditions shown in Fig. 6.1, we note that the area is perpendicular (*normal*) to the direction of the applied force, and the stress defined by Eq. (6.1) is called a *normal stress*.

ILLUSTRATIVE PROBLEM 6.1

The body shown in Fig. 6.2 is subjected to an axial tensile load of 40 kN. Determine the stress at sections A–A, B–B, and C–C.

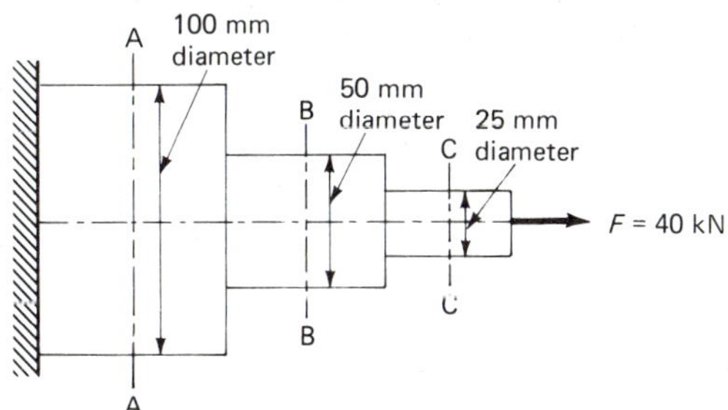

FIGURE 6.2 Illustrative Problem 6.1.

SOLUTION

Each of the sections must be capable of carrying the full load of 40 kN. The area at each of the sections is tabulated, and the stress at each section is F/A, or 40 kN/A, or 40 000/A Pa.

Section	Diameter (m)	Area (m²)	Stress $= \dfrac{40\,000}{\text{Area}}$
A–A	100×10^{-3}	7.85×10^{-3}	5.1 MPa
B–B	50×10^{-3}	1.96×10^{-3}	20.4 MPa
C–C	25×10^{-3}	4.91×10^{-4}	81.5 MPa

ILLUSTRATIVE PROBLEM 6.2

A uniform circular rod is hung vertically with one end free and the other end fixed. If the rod weighs w lb/ft, determine the stress at sections A–A, B–B, C–C, D–D, and E–E if the cross-sectional area of the rod is A in.²

SOLUTION

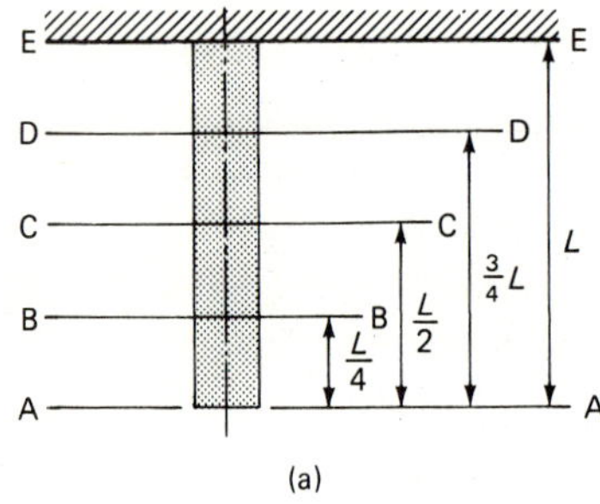
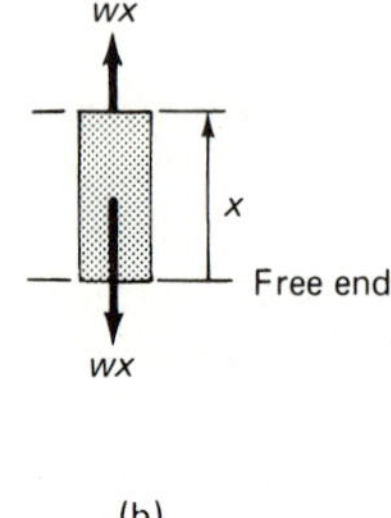

FIGURE 6.3   Illustrative Problem 6.2.

The stress in the bar arises from the weight of the portion of the bar hanging below the section in question, as shown in Fig. 6.3(b). Thus we can tabulate the loads at each section as follows to obtain the stress on the section. As will be seen, the stress varies from zero at the free end to a maximum value at the fixed end. This type of stress variation in a member occurs frequently.

Section	Weight Hanging on Section	Stress $= \dfrac{F}{A}$
A–A	0	0
B–B	$\dfrac{wL}{4}$	$\dfrac{wL}{4A}$
C–C	$\dfrac{wL}{2}$	$\dfrac{wL}{2A}$
D–D	$\dfrac{3wL}{4}$	$\dfrac{3wL}{4A}$
E–E	wL	$\dfrac{wL}{A}$

                                         Concepts of Stress and Strain

The tensile and compressive stresses discussed in the previous paragraphs arise from internal resistive forces within the body when the body is subjected to external forces. Another direct stress occurs when there is contact between two bodies. This external force is known as *bearing,* and the contact pressure between the two bodies is known as a *bearing stress.* Some examples of bearing stresses are shown in the situation illustrated in Fig. 6.4. Bearing occurs between the post and the plate, the plate and the footing, and the footing and the soil. The bearing stress in each case is the load *F* divided by the normal area carrying the load.

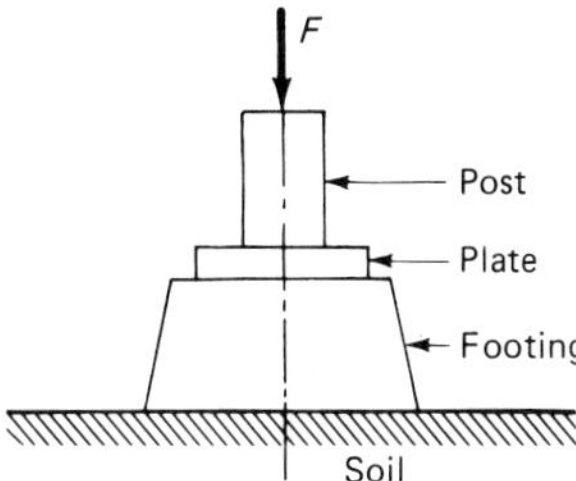

FIGURE 6.4 Bearing stress.

ILLUSTRATIVE PROBLEM 6.3

If the post in Fig. 6.4 is a 150-mm × 150-mm member, the plate is 250 mm × 250 mm, and the base of the concrete footing is 750 mm × 750 mm, determine the bearing stresses when the structure must carry 450 kN.

SOLUTION

Bearing occurs at the three sections noted in the preceding paragraph. The bearing areas are noted, and the bearing stress is 450 kN/A.

Location	Dimensions (m)	Bearing Area (m^2)	Bearing Stress F/A
Post–plate	$150\,(10)^{-3} \times 150\,(10)^{-3}$	2.25×10^{-2}	20.0 MPa
Plate–footing	$250\,(10)^{-3} \times 250\,(19)^{-3}$	6.25×10^{-2}	7.2 MPa
Footing–soil	$750\,(10)^{-3} \times 750\,(10)^{-3}$	5.625×10^{-1}	800 kPa

Tensile and compression loads are resisted by areas that are perpendicular to the direction of the applied load. Shearing stresses occur when the force being resisted acts in the plane of the reacting area. Several representative situations that give rise to shear stresses are shown in Fig. 6.5. In each case, the shearing stress is the shear force divided by the area being sheared. Mathematically,

$$S_s = \frac{F}{A} \tag{6.2}$$

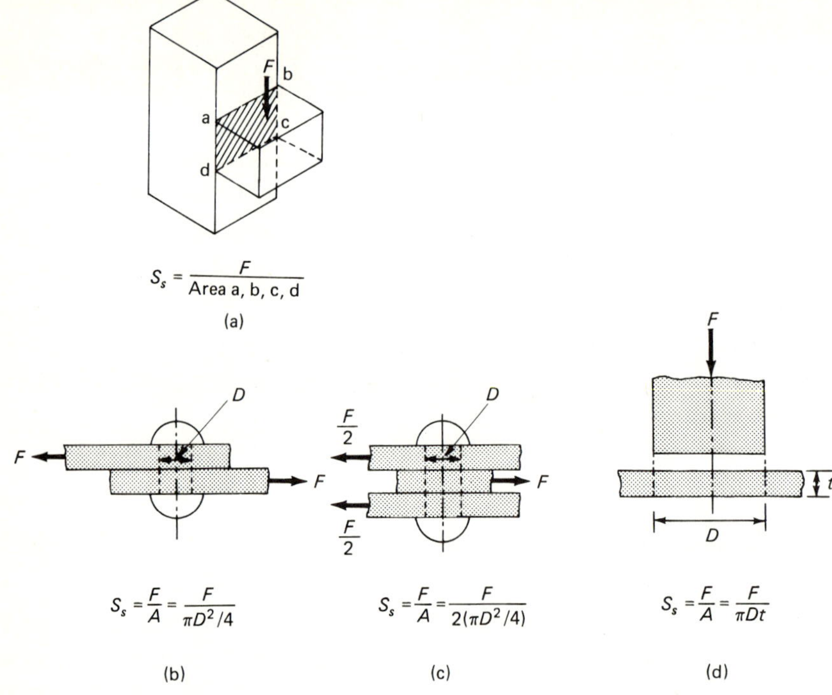

$$S_s = \frac{F}{\text{Area } a, b, c, d}$$

(a)

$$S_s = \frac{F}{A} = \frac{F}{\pi D^2/4}$$

(b)

$$S_s = \frac{F}{A} = \frac{F}{2(\pi D^2/4)}$$

(c)

$$S_s = \frac{F}{A} = \frac{F}{\pi D t}$$

(d)

FIGURE 6.5   Shear stresses.

Figure 6.5(a) and (b) shows the shearing force being resisted by a single shear area. The situation shown in Fig. 6.5(c) illustrates a case of double shear, that is, two areas resist the shear in the bolt being sheared. Figure 6.5(d) shows a punch shearing out a cylindrical disk from a plate, with the shear area being the outside surface of the disk.

ILLUSTRATIVE PROBLEM 6.4

The eyebar shown in Fig. 6.6 must carry a load of 45 kN. Calculate the shear stress in the bolt.

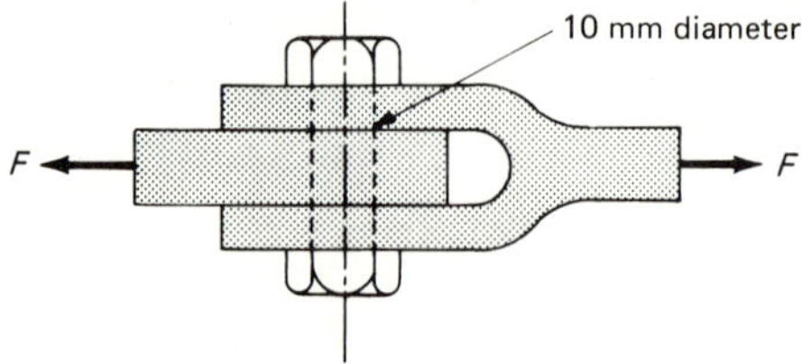

FIGURE 6.6   Illustrative Problem 6.4.

                                        Concepts of Stress and Strain

SOLUTION

The bolt is in double shear, and the shear area is $2(\pi D^2/4) = 2(\pi/4)\,(10 \times 10^{-3})^2 = 1.57 \times 10^{-4}\ \text{m}^2$. The shear stress is given by

$$S_s = \frac{F}{A} = \frac{45\ 000}{1.57 \times 10^{-4}} = 286.6\ \text{MPa}$$

ILLUSTRATIVE PROBLEM 6.5

A punch is used to punch a 1-in. diameter hole in a plate ¼ in. thick. If the ultimate shearing stress in the material is 50 000 psi, determine the necessary shearing force.

SOLUTION

Referring to Fig. 6.5(d), the shearing area is $\pi Dt = \pi(1)(¼) = 0.785\ \text{in.}^2$. Since $S_s = F/A$, $F = 0.785 \times 50\ 000 = 39\ 300$ psi.

6.4   STRAIN

The application of an external force gives rise not only to internal resistive forces in a body, but also to internal displacements in the body, which in the aggregate give rise to an elongation of the body if the external force is a tensile force. Similarly, a compressive force causes a shortening of the dimension of the body in the direction of the force. When discussing the elongation (or shortening) of a body, we will use several terms whose meaning must be clearly understood. The *deformation* will be designated by the symbol δ, and it will be taken to be the total change in a dimension due to an applied force. The *elongation* of the bar per unit of original length is designated by the symbol ϵ and is expressed as

$$\epsilon = \frac{\delta}{L_o} \tag{6.3}$$

where L_o is the original length. The elongation per unit length is called the *strain*.

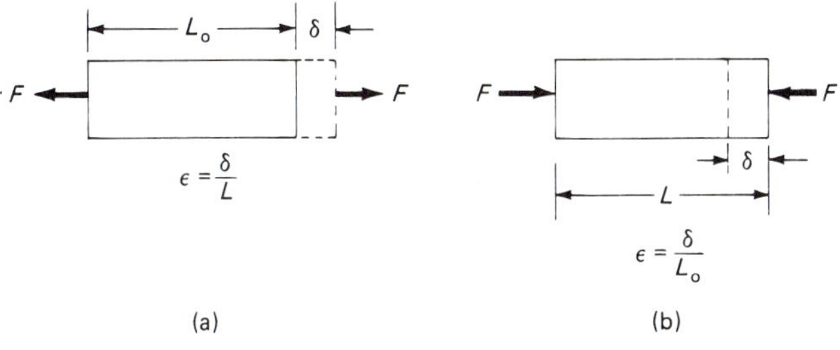

FIGURE 6.7   Strain. (a) Tension. (b) Compression.

Strain is a dimensionless quantity, usually expressed as inches per inch or metres per metre. These definitions are shown in Fig. 6.7 for tension and compression. It should be noted that the strain defined by Eq. (6.3) is an average value.

SOLUTION

By definition the strain is

$$\epsilon = \frac{\delta}{L_o} = \frac{0.0024}{2.0} = 0.0012 \text{ in./in.}$$

which is the same as metres per metre.

Shearing Strain

When a member is subjected to a shear stress, it undergoes a deformation that is somewhat different from that described for the cases of tension and compression. Rather than an elongation or shortening, shearing stress causes an angular deformation of the body. Thus a rectangular section becomes a parallelogram. The *shearing strain,* as shown in Fig. 6.8, can be defined as the total deformation divided by the height of the element L when the angle φ is small. Thus

$$\frac{\delta_s}{L} = \tan \varphi \cong \varphi \tag{6.4}$$

(Since the angle φ is small, $\tan \varphi \cong \varphi$, where φ is the angle expressed in radians.) Therefore

$$\varphi = \frac{\delta_s}{L} \tag{6.5}$$

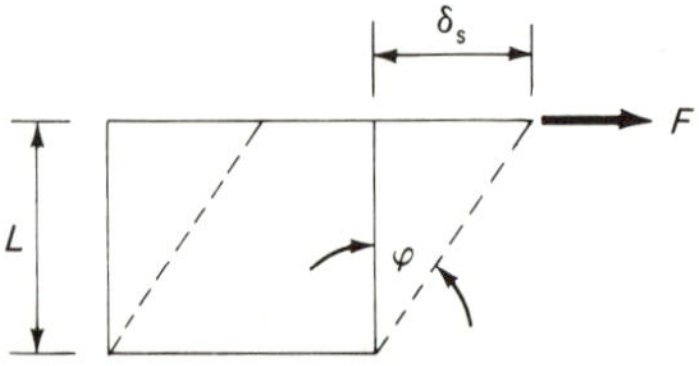

FIGURE 6.8   Shear strain.

                                            Concepts of Stress and Strain

SOLUTION

Using Eq. (6.5),

$$\varphi = \frac{\delta_s}{L} = \frac{0.005}{2.5} = 0.002 \text{ rad} \left(\frac{m}{m}\right)$$

6.5 HOOKE'S LAW

If a member is subjected to a specific stress, it will undergo a specific strain. If this strain vanishes, that is, the body returns to its original dimensions upon the removal of the stress, the action is said to be *elastic*. However, if upon the removal of the stress the body does not return to its original dimensions and there is a residual strain, the action is said to be *inelastic*. Almost 300 years ago Robert Hooke observed that stress is directly proportional to strain. Figure 6.9 shows a typical stress–strain curve for a material such as steel. As we can see from this curve, stress and strain are linearly proportional up to a point, known as the *proportional limit*. Past that point additional stress causes inelastic action to occur. All of the material in the subsequent chapters of this book will utilize the proportionality of stress and strain, which is known as *Hooke's law*. We can mathematically express the proportionality of stress and strain as

$$S = E\epsilon \tag{6.6}$$

where E is the constant of proportionality relating stress and strain. The proportionality constant E is sometimes known as *Young's modulus*, or more commonly the *modulus of elasticity*. Since stress is expressed in units of force per unit area and ϵ is nondimensional (inches per inch or metres per metre), E has the same units as S, namely,

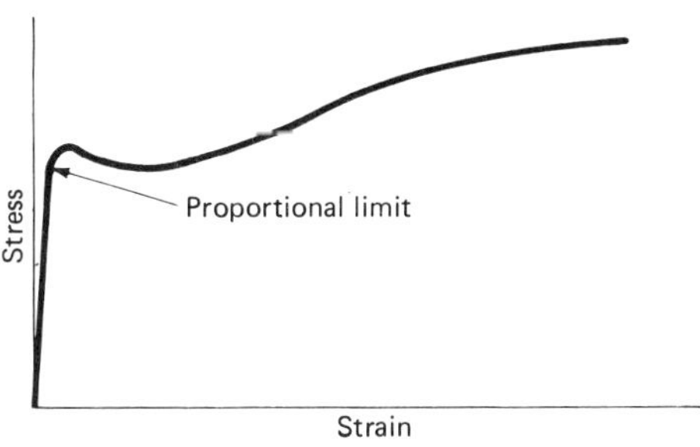

FIGURE 6.9 Typical stress–strain curve for steel.

TABLE 6.1   Modulus of Elasticity and Rigidity for Some Common Materials

Material	E(GPa)	E(psi)	G(GPa)	G(psi)
Carbon steel	210	30×10^6	85	12×10^6
Alloy steel	210	30×10^6	85	12×10^6
Cast iron (grey)	105	15×10^6	40	6×10^6
Aluminum	70	10×10^6	28	4×10^6
Brass	105	14×10^6	40	6×10^6
Copper	115	17×10^6	40	6×10^6

psi or Pa. It is sometimes more convenient to express Eq. (6.6) in terms of the total deformation δ, the length of the member L, the cross-sectional area A, and the applied load F. Noting that $S = F/A$ and $\epsilon = \delta/L$,

$$\frac{F}{A} = E\frac{\delta}{L} \tag{6.7}$$

or

$$\delta = \frac{FL}{EA} \tag{6.8}$$

A similar proportionality exists between shear stress and shear strain. It can be written as

$$S_s = G\varphi \tag{6.9}$$

where G is the shear modulus of elasticity (more commonly known as the *modulus of rigidity*). Table 6.1 gives some typical values of E and G for some common materials.

ILLUSTRATIVE PROBLEM 6.8

A steel rod is used to support a weight of 2 tons. If the allowable stress is 20 000 psi, determine the diameter of the rod and its elongation if it is 2 ft long. The modulus of elasticity is 30×10^6 psi for steel.

SOLUTION

The tensile stress is given by $S = F/A$. Therefore $A = F/S$. Using the numerical data of the problem in consistent units,

$$A = \frac{2000 \times 2}{20\ 000} = 0.2 \text{ in.}^2 \quad [1.29 \times 10^{-4} \text{ m}^2]$$

$$\frac{\pi D^2}{4} = 0.2$$

and

$$D = 0.505 \text{ in.} \quad [1.283 \times 10^{-2} \text{ m}]$$

                                    Concepts of Stress and Strain

The elongation may be found using either Eq. (6.6) or Eq. (6.8):

$$\epsilon = \frac{S}{E} = \frac{20\,000}{30 \times 10^6} = 0.67 \times 10^{-3} \text{ in./in.} \quad [\text{m/m}]$$

$$\delta = \epsilon L = 0.67 \times 10^{-3} \times 2 \times 12 = 16 \times 10^{-3}$$

$$= 0.016 \text{ in.} \quad [4.06 \times 10^{-4} \text{ m}]$$

Alternately,

$$\delta = \frac{FL}{EA} = \frac{2000 \times 2 \times 24}{30 \times 10^6 \times 0.2} = 0.016 \text{ in.} \quad [4.06 \times 10^{-4} \text{ m}]$$

6.6 POISSON'S RATIO

As was noted earlier, when a bar or member is subjected to a direct stress, that bar changes its axial dimension by an amount directly proportional to the magnitude of the applied stress—if the stress is tensile, then the bar elongates; if the stress is compressive, the bar contracts. In addition to this direct effect, it is found that the bar simultaneously undergoes a change in its lateral dimension. Thus for the bar shown in Fig. 6.10 the length will increase from L to $L + \epsilon L$, while the sides contract from their original dimensions of W and h. If we now denote the ratio of the lateral strain to the longitudinal strain as μ, we have

$$\mu = \frac{\epsilon_W}{\epsilon_y} \quad \text{and} \quad \mu = \frac{\epsilon_h}{\epsilon_y} \tag{6.10}$$

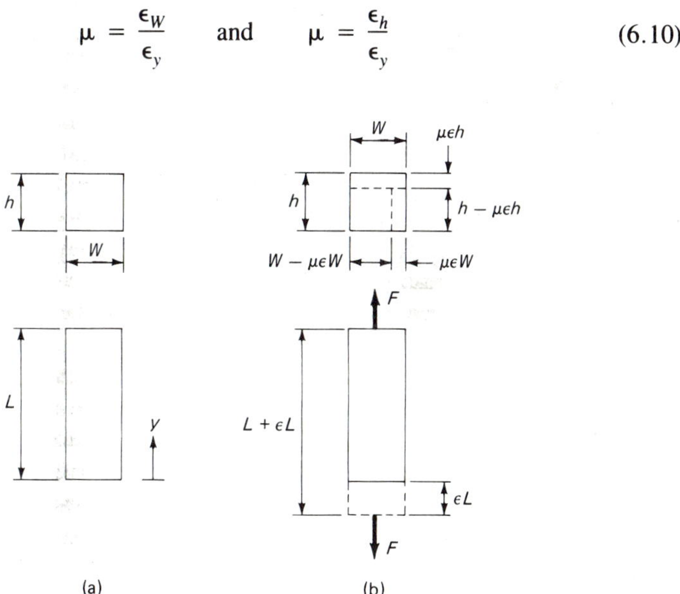

FIGURE 6.10   Lateral strain due to longitudinal stress. (a) Unstressed bar. (b) Stressed bar.

Material	μ
Aluminum	0.34
Aluminum alloys	0.32
Copper	0.35
Iron	0.28
Carbon steel	0.28
Magnesium	0.33
Titanium	0.34

This ratio is known as *Poisson's ratio*, and for steel it is usually taken to be approximately 0.3. Typical values of μ for some materials are given in Table 6.2.

It is important to note that if the longitudinal strain causes extension, the lateral strain will be compressive, that is, if the applied force causes the bar to elongate in the direction of the applied force, the bar will contract in the other two directions. Similarly, if a bar is compressed (shortened) in the longitudinal direction, it will undergo an increase in its lateral dimensions.

ILLUSTRATIVE PROBLEM 6.9

A steel bar 1 in. × 1 in. × 10 in. is subjected to an axial tensile stress of 20 000 psi (Fig. 6.11). Determine the axial elongation, the lateral contraction, the final volume, and the volumetric strain. Use $E = 30 \times 10^6$ psi and $\mu = 0.3$.

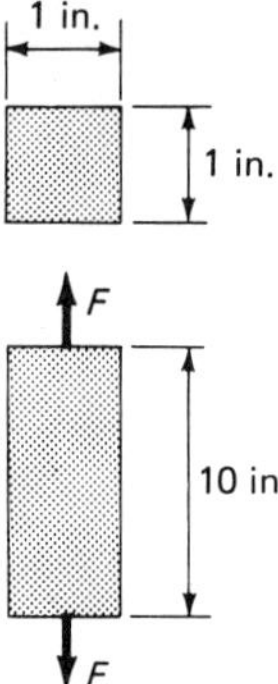

FIGURE 6.11   Illustrative Problem 6.9.

SOLUTION

The axial stress is 20 000 psi. Therefore the axial strain ϵ is $S/E = 20\ 000/(30 \times 10^6) = 0.000\ 667$ in./in. The lateral strain is $\mu\epsilon = 0.3(0.000\ 667) = 0.000\ 20$ in./in. The elongation of the 10-in. length is $10 \times 0.000\ 667 = 0.006\ 67$ in. The contraction of each side is $1 \times 0.000\ 20 = 0.000\ 20$ in. The final dimensions of the bar are 10.006 67 in., 0.9998 in., and 0.9998 in.

Remember that even though the strains appear as small numbers, they are sig-

                                    Concepts of Stress and Strain

nificant and are not to be ignored. Also, the foregoing discussion presumes that the material being deformed is not stressed beyond its proportional limit.

When a body is subjected to stresses in more than one direction, the strains may be found by considering each stress separately and determining the effect of each stress separately. The total effect is then the algebraic sum of the separate effects, with attention to the fact that, by convention, tension causes a positive deformation, and compression causes a negative deformation.

An important relation that connects the three constants E, G, and μ can be derived from theoretical considerations. Such a relation is important since the direct determination of μ is quite difficult due to the necessity of measuring small lateral contractions accurately. This connecting equation is

$$G = \frac{E}{2(1 + \mu)} \tag{6.11}$$

ILLUSTRATIVE PROBLEM 6.10

Using the values of E and G from Table 6.1, calculate μ for carbon steel.

SOLUTION

From Eq. (6.11), $\mu = (E/2G) - 1$. Using the values of $E = 30 \times 10^6$ psi and $G = 12 \times 10^6$ psi,

$$\mu = \frac{30 \times 10^6}{2 \times 12 \times 10^6} - 1$$

$$= 0.25$$

The value for carbon steel from Table 6.2 is 0.28.

6.7 STATICALLY INDETERMINATE STRESSES

Structures or members whose loadings and supporting conditions give rise to more conditions than can be solved using the three equations of statics for equilibrium, are classified as being *statically indeterminate*. In order to solve this type of situation, it is necessary to have additional information, such as the conditions at a support, to fully solve the problem. The method of approach to the solution of problems of this type is best illustrated by examining a specific problem.

ILLUSTRATIVE PROBLEM 6.11

A steel rod 1 in. in diameter is placed inside of a brass cylinder, as shown in Fig. 6.12. Both are initially 3 in. long. A 5000-lb weight is placed on the assembly. Determine the force, stress, and contraction of each member.

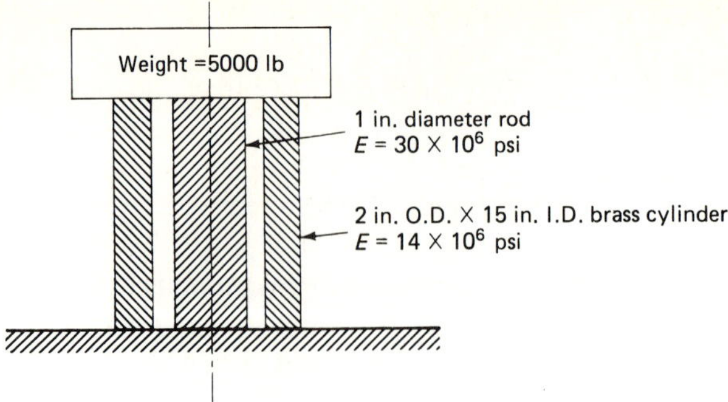

FIGURE 6.12   Illustrative Problem 6.11.

SOLUTION

Let us designate the force in the steel as F_S and the force in the brass as F_B. The necessary equation from statics for equilibrium is

$$F_S + F_B = 5000$$

The second condition of the problem is that the deflection of each member is the same. Therefore using δ_S as the contraction of the steel and δ_B as the contraction of the brass,

$$\delta_S = \delta_B$$

In terms of Hooke's law,

$$\frac{F_S L_o}{A_S E_S} = \frac{F_B L_o}{A_B E_B}$$

or

$$F_S = F_B \left(\frac{A_S}{A_B}\right)\left(\frac{E_S}{E_B}\right) = F_B \left[\frac{\pi(1)^2/4}{\pi(2^2 - 1.5^2)/4}\right]\left[\frac{30 \times 10^6}{14 \times 10^6}\right] = 1.22\, F_B$$

Using this information in the static equilibrium condition,

$$1.22 F_B + F_B = 5000$$

and

$$F_B = 2248 \text{ lb}$$

$$F_S = 2752 \text{ lb}$$

The corresponding stresses are $S_B = 2248/[(\pi/4)(2^2 - 1.5^2)] = 1636$ psi, and $S_S = 2752/[(\pi/4)(1)^2] = 3505$ psi. Since both members are originally and finally

Concepts of Stress and Strain

the same length, $\epsilon_S = \epsilon_B$. As a check,

$$\epsilon_S = \frac{S_S}{E_S} = \frac{3505}{30 \times 10^6} = 116.8 \times 10^{-6}$$

$$\epsilon_B = \frac{1636}{14 \times 10^6} = 116.8 \times 10^{-6}$$

which is satisfactory. The total deflection of each member $\delta = \epsilon L_o = 116.8 \times 10^{-6} \times 3 = 350 \times 10^{-6} = 0.000\,350$ in. (contraction).

6.8 TEMPERATURE STRESSES

If a bar is not restrained in any way, it will be found that an increase in temperature will cause an increase in its dimensions, and conversely, a decrease in temperature will cause a decrease in its dimensions. It is usual to describe the dimensional change due to temperature changes in terms of the change in a linear dimension. Thus the change in length of a bar ΔL is directly proportional to both the temperature change of the bar Δt and the original length of the bar L_o. The constant of proportionality is called the *linear coefficient of expansion* and is usually denoted by the symbol α. Referring to Fig. 6.13, we can write

$$\Delta L = \alpha L_o \, (\Delta t) \tag{6.12}$$

or

$$L_f = L_o + \alpha L_o \, (\Delta t) = L_o \, (1 + \alpha \Delta t) \tag{6.13}$$

The coefficient of linear expansion α is therefore the change in length per unit length for a one-degree change in temperature. Units for α are in./in.·°F, or m/m·°C, or mm/mm·°C. Some typical values of the linear coefficient of expansion for several materials are given in Table 6.3, and a more complete table will be found in the Appendix.

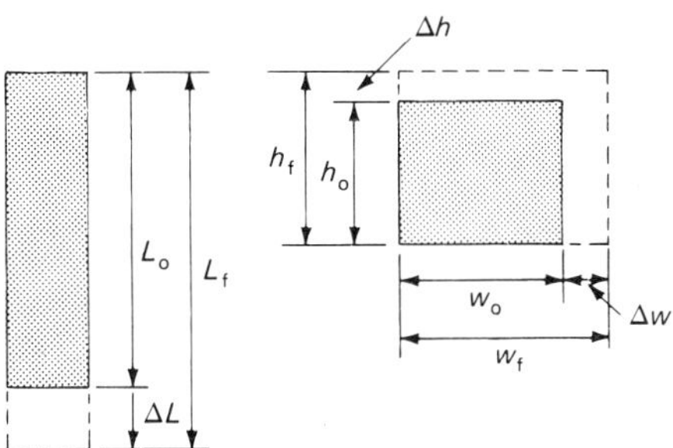

FIGURE 6.13 Thermal expansion of a bar.

Material	Coefficient α	
	(in./in.·°F)	(mm/mm·°C)
Aluminum	12.8×10^{-6}	23.0×10^{-6}
Brass	10.4×10^{-6}	18.7×10^{-6}
Copper	9.3×10^{-6}	16.7×10^{-6}
Iron	6.7×10^{-6}	12.1×10^{-6}
Steel (mild)	6.5×10^{-6}	11.7×10^{-6}
Steel (stainless)	9.9×10^{-6}	17.8×10^{-6}

If a bar is restrained and cannot move freely, stresses will be set up in the bar when it undergoes a change in temperature. Consider the bar shown in Fig. 6.14. If it were not restrained, it would expand a distance $\alpha L_0(\Delta t)$ due to a temperature increase of Δt. However, the rigid walls prevent this from occurring, and the final length of the bar must equal the initial length of the bar. We can consider the return of the bar from its expanded position to its initial position to have been accomplished by an axial compressive force on the bar, which caused a stress in the bar of magnitude S. From Hooke's law,

$$S = E\epsilon = E \frac{\text{total deflection}}{\text{initial length}} \tag{6.14}$$

and using $\alpha L_0(\Delta t)$ as the total deflection,

$$S = E \frac{\alpha L_0(\Delta t)}{L_0} = E\alpha(\Delta t) \tag{6.15}$$

From Eq. (6.15) we note that the stress in the bar, when the bar is restrained from freely expanding, is independent of the dimensions of the bar and is only a function of the temperature change, the modulus of elasticity, and the coefficient of linear expansion. Thus for a restrained, mild steel bar the stress for a temperature rise of 1 °C is $11.7 \times 10^{-6} \times 210 \times 10^9 = 2.457$ MPa; for a 100 °C rise it is 245.7 MPa. In conventional units, the corresponding stress for a temperature rise of 1 °F is $6.5 \times 10^{-6} \times 30 \times 10^6 = 195$ psi; for a 100 °F rise it is 19 500 psi.

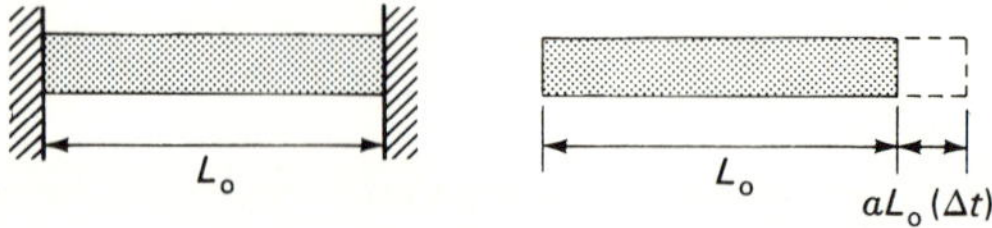

FIGURE 6.14   Restrained bar subject to a temperature increase.

Concepts of Stress and Strain

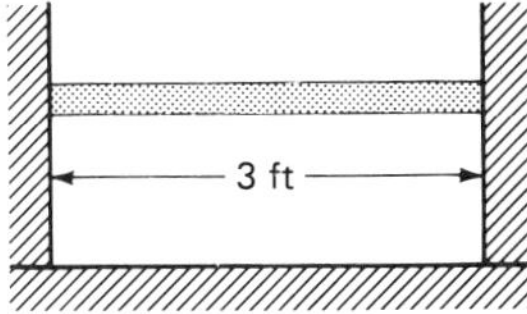

FIGURE 6.15   Illustrative Problem 6.12.

SOLUTION

The total elongation of the rod, if it is unrestrained, is

$$\Delta L = L_o\alpha(\Delta t) = 3 \times 12 \times 6.5 \times 10^{-6} \times 100 = 0.0234 \text{ in.}$$

Since the walls move 0.01 in., the rod must be compressed by a force causing a deflection of 0.0234 in. − 0.0100 in. = 0.0134 in. Therefore

$$S = E\epsilon = 30 \times 10^6 \times \frac{0.0134}{36} = 11\ 170 \text{ psi}$$

Thus far we have considered the case of temperature stresses that arise when a member is prevented from expanding freely due to imposed external conditions. When a structure is composed of more than one material, temperature stresses can occur due to the differences in the expansion between the materials. This condition is shown by the situation illustrated in Fig. 6.16. In this instance we have bars of different materials whose upper ends are free to move. However, the very large rigid member, whose weight is W, joining the upper end of the bars, can only move parallel to itself. Due to this action, one bar will elongate due to a temperature rise less than the amount of expansion that would occur if it were unrestrained, while the other end will elongate

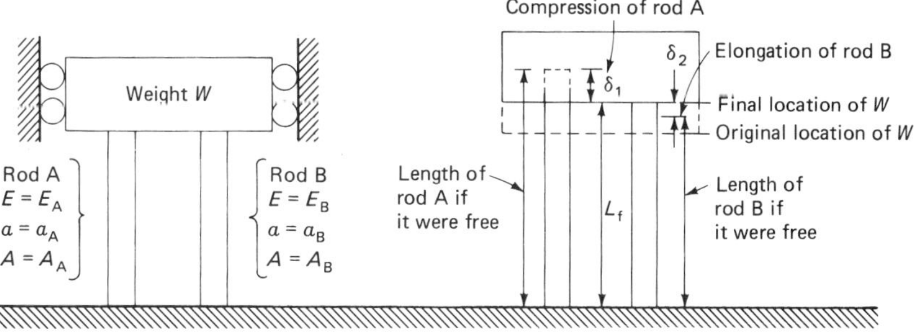

FIGURE 6.16   System with two different materials.

more than the amount it would expand for the same temperature rise if it too were unrestrained. The final lengths of both bars will be the same due to the constraint imposed by the upper rigid body. The elongation of rod A, if it is free, is $\alpha_A L_o(\Delta t)$ and that of rod B is $\alpha_B L_o(\Delta t)$. The difference in these lengths is made zero by compressing bar A an amount that we shall denote by δ_1 and elongating bar B by an amount δ_2. The force in rod A plus the force in rod B must support the weight W. Thus initially,

$$F_A + F_B = W \tag{6.16}$$

and finally

$$F'_A + F'_B = W \tag{6.17}$$

Also, the extension of rod A due to the temperature increase minus the decrease in its length due to the increased compressive load in rod A must equal the extension of rod B plus the increase in its length due to the decrease in the compressive load in rod B. Therefore,

$$\alpha_A L_o(\Delta t) - \frac{(F'_A - F_A)L_o}{E_A A_A} = \alpha_B L_o(\Delta t) - \frac{(F'_B - F_B)L_o}{E_B A_B} \tag{6.18}$$

Equations (6.17) and (6.18) yield the conditions necessary to solve the problem. Note that F'_B is greater than F_B, giving rise to a negative term on the right-hand side of Eq. (6.18).

ILLUSTRATIVE PROBLEM 6.13

If rod A is a 1-in. diameter stainless-steel rod and rod B is a 2-in. diameter mild steel rod, determine the stresses in each due to a 50 °F temperature rise if they support a weight W of 10 000 lb. Assume both rods to have been the same length before W was placed on them, and that W was constrained to stay horizontal. Also take $E = 30 \times 10^6$ for both rods. Use α from Table 6.3.

SOLUTION

When the weight is initially placed on the rods, both rods will deflect by an equal amount. Thus $\delta_1 = \delta_2$, or

$$\frac{F_A L_o}{E_A A_A} = \frac{F_B L_o}{E_B A_B}$$

Since E_A equals E_B,

$$F_A = \frac{A_A}{A_B} F_B = \left[\frac{\pi(1)^2/4}{\pi(2)^2/4} \right] F_B = \tfrac{1}{4} F_B$$

and both rods must support the total weight of 10 000 lb,

$$F_A + F_B = 10\,000$$

      Concepts of Stress and Strain

Therefore

$$\tfrac{1}{4}F_B + F_B = 10\ 000$$

Thus,

$$F_B = 8000\ \text{lb}$$

$$F_A = 2000\ \text{lb}$$

When the temperature increases by 50 °F, the expansion of rod A is equal to $\alpha_A L_o(\Delta t)$, or

$$\alpha_A L_o(\Delta t) = 9.9 \times 10^{-6} \times L_o \times 50 = 495 \times 10^{-6} L_o \ \text{in.}$$

The mild steel rod, rod B, expands

$$\alpha_B L_o(\Delta t) = 6.5 \times 10^{-6} \times L_o \times 50 = 325 \times 10^{-6} L_o \ \text{in.}$$

Denoting the new forces in the rods as F_A' and F_B',

$$F_A' + F_B' = 10\ 000 \qquad \text{or} \qquad F_A' = 10\ 000 - F_B'$$

Using the condition that the final lengths of both bars are equal [Eq. (6.18)],

$$\alpha_A L_o(\Delta t) - \frac{(F_A' - F_A)L_o}{F_A A_A} = \alpha_B L_o(\Delta t) - \frac{(F_B' - F_B)L_o}{E_B A_B}$$

Substituting numerical data and canceling out the common L_o term yields

$$495 \times 10^{-6} - \frac{(F_A' - 2000)}{E_A A_A} = 325 \times 10^{-6} + \frac{(8000 - F_B')}{E_B A_B}$$

Substituting for F_A' and letting $E = E_A = E_B$,

$$495 \times 10^{-6} - 325 \times 10^{-6} = \frac{8000 - F_B'}{E_A A_A} + \frac{8000 - F_B'}{E_B A_B}$$

$$= \frac{(8000 - F_B')}{E}\left[\frac{1}{A_A} + \frac{1}{A_B}\right]$$

$$= \frac{5}{30\pi}(8000 - F_B')$$

Simplifying,

$$F_B' = 8000 - 170 \times 10^{-6} \times 30 \times 10^6 \times \frac{\pi}{5} = 4796\ \text{lb}$$

and

$$F_A' = 5204\ \text{lb}$$

Temperature Stresses

Notice that the load in rod B has decreased since that rod has expanded less than rod A, while the load in rod A increased. The final stresses are F/A for each rod, which yield

$$S_A = \frac{F'_A}{A_A} = \frac{5204}{\pi(1)^2/4} = 6626 \text{ psi}$$

$$S_B = \frac{F'_B}{A_B} = \frac{4796}{\pi(2)^2/4} = 1527 \text{ psi}$$

6.9 STRESSES IN CYLINDERS AND SPHERES

When a fluid is contained, it exerts forces on the walls of the container that stress the material of the container. As an extension of the principles discussed in this chapter, let us consider the problem of evaluating the stresses in thin cylinders and spheres subjected to an internal pressure. To evaluate these stresses we shall assume that the ratio of the thickness to the diameter of the cylinder or sphere is small and that the stresses developed in these pressure vessels are uniform over the cross section of the vessel. By *thin* we mean that the ratio of thickness to diameter is 0.1 or less.

With these assumptions in mind, let us consider the cylinder shown in Fig. 6.17. Due to the internal pressure, there is a stress in the circumferential direction S_2 and a stress in the longitudinal direction S_1 [as shown in Fig. 6.17(a)]. If the pipe is cut perpendicular to the longitudinal axis (and far from the ends), the resultant free-body diagram will be the one shown in Fig. 6.17(b). The total resisting force in the cylinder will be the stress S_1 multiplied by the area over which the stress acts. Thus the resisting force is $S_1(2\pi Rt)$ since the material's area is $2\pi Rt$. The applied load is due to the pressure in the tube over the tube area. The applied load is therefore $p\pi R^2$. For equilibrium, these forces must be equal. Therefore

$$S_1 2\pi Rt = p\pi R^2$$

or

$$S_1 = \frac{pR}{2t} \tag{6.19}$$

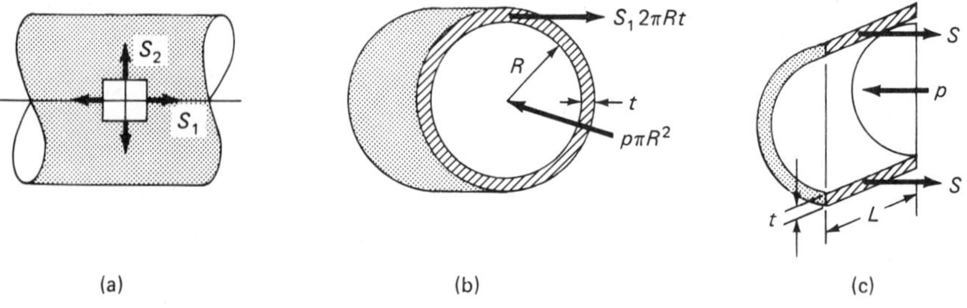

FIGURE 6.17   Thin cylinder under internal pressure.

Concepts of Stress and Strain

By considering a diametrical cut through the cylinder shown in Fig. 6.17(c), S_2 can be evaluated in an analogous manner. The area is $2tL$, where L is the pipe length. Acting on the pipe is the internal force component in the horizontal direction, which is $p2RL$, where $2RL$ is the projected area of the tube. Thus,

$$2tLS_2 = p2RL$$

and

$$S_2 = \frac{pR}{t} \tag{6.20}$$

From Eqs. (6.19) and (6.20) we can see that S_2 is twice S_1. The student should also note that these stresses were arrived at by assuming the stress distribution in the material of the cylinder to be uniform.

Using considerations beyond the scope of this text, we will find it possible to calculate the stresses in thick cylinders. Figure 6.18 shows the ratio of the stresses calculated for thick cylinders to the results obtained from Eq. (6.20) as a function of the outer radius R_o and the inner radius R_i. Our earlier definition of thin was a ratio of thickness to diameter as 0.1 or less. This corresponds to $t/R_i = 0.2$, which from Fig. 6.18 indicates a 10% error in calculating the stress using the thin-cylinder formula. As t/R_i increases, the error in using the thin-cylinder formula increases, as shown in Fig. 6.18.

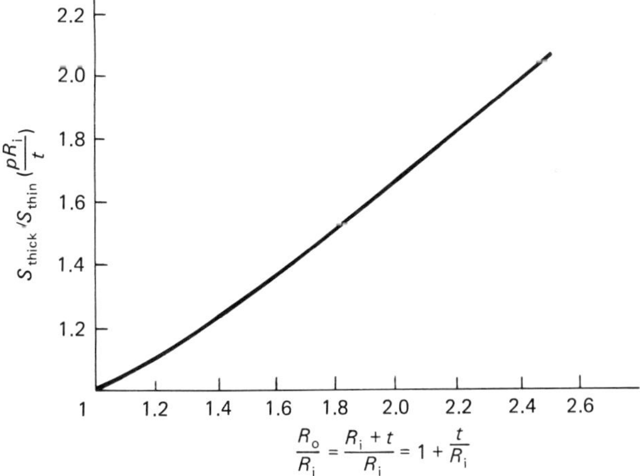

FIGURE 6.18　Ratio of stresses on a thick cylinder to stress calculated from thin-cylinder formula.

Let us now consider a thin sphere subjected to an internal pressure. In this sphere S_1 will be equal to S_2 by symmetry. As can be seen from Fig. 6.19, the stress S_2 acts over an area equal to $2\pi Rt$. The sum of the force components in the horizontal direction is once again just the product of the pressure multiplied by the projected area, that is, $p\pi R^2$. Therefore

$$S_2 = 2\pi Rt = p\pi R^2$$

Stresses in Cylinders and Spheres　　　　　　　　　　　　　　　157

or

$$S_2 = S_1 = \frac{pR}{2t} \tag{6.21}$$

Equation (6.21) yields the result that S_2 in a thin sphere is half of S_2 in a thin cylinder when both are subjected to the same internal pressure. The student is cautioned to consult applicable codes before attempting to design cylinders and spheres for either internal or external pressure.

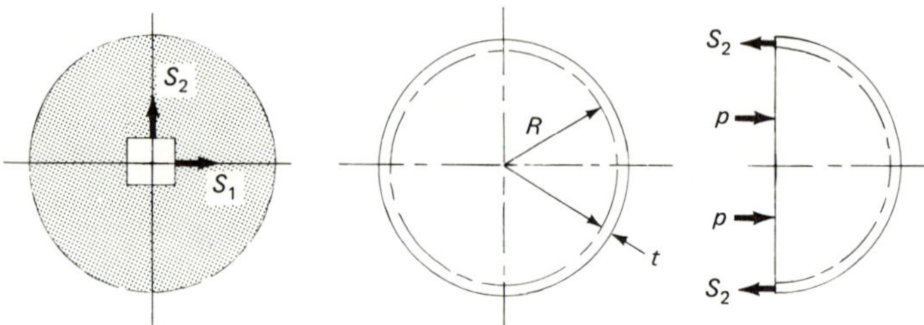

FIGURE 6.19   Thin sphere under internal pressure.

ILLUSTRATIVE PROBLEM 6.14

A 48-in. diameter steel pipe, 0.25 in. thick, carries water under an internal pressure of 100 psi. Compute the circumferential stress in the steel. If the pressure is raised to 250 psi and the allowable unit stress in the steel is 15 000 psi, what thickness of steel is required?

SOLUTION

The circumferential stress is S_2. Thus from Eq. (6.20),

$$S_2 = \frac{pR}{t} = \frac{100 \times 24}{0.25} = 9600 \text{ psi}$$

$$t = \frac{pR}{S_2} = \frac{250 \times 24}{15\ 000} = 0.4 \text{ in.}$$

6.10   CLOSURE

The concepts of stress and strain defined and illustrated in this chapter, Hooke's law, and the free-body diagram provide the foundations for the study of "strength of materials." It is therefore most important for the student to master this material fully before proceeding further into the subject. Similarly precautionary statements will be found at the closure of other chapters, but no other chapter will contain more essential material than this chapter. We may use the old saying that "a child must crawl before it walks and it must walk before it runs" to place this chapter in its proper perspective—having mastered it, the student is now prepared to walk.

                                    Concepts of Stress and Strain

One further suggestion is in order. When doing problems, draw a sketch and put all the given pertinent data on that sketch. Then draw the necessary free-body diagram (or diagrams). Apply the conditions of equilibrium to the free-body diagrams and use the known boundary conditions (such as equal deflection) and the stress–strain relations to solve the problem. Such an organized approach to problem solving in this course will prove to be most helpful in both understanding and solving problems.

REFERENCES

American Institute of Steel Construction, Manual of Steel Construction, 8th ed. 1980.

Bassin, M. E., S. M. Brodsky, and H. Wolkoff, Statics and Strength of Materials, 3rd ed. New York: McGraw-Hill, 1979.

Bruhn, E. F., Analysis and Design of Flight Vehicle Structures. Cincinnati: Tri-State Offset Co.

Case, J., and A. H. Chilver, Strength of Materials. London: Edward Arnold.

Granet, I., Fluid Mechanics for Engineering Technology. Englewood Cliffs, NJ: Prentice-Hall, 1981.

Jensen, A., and H. H. Chenoweth, Applied Strength of Materials, 3rd ed. New York: McGraw-Hill, 1971.

Levinson, I. J., Mechanics of Materials, 2nd ed. Englewood Cliffs, NJ: Prentice-Hall, 1970.

Panlilio, F., Elementary Theory of Structural Strength. New York: John Wiley, 1963.

Seely, F. B., and J. O. Smith, Advanced Mechanics of Materials, 2nd ed. New York: John Wiley, 1952.

Singer, F. L., Strength of Materials, 2nd ed. New York: Harper and Row, 1962.

Timoshenko, S., and D. H. Young, Elements of Strength of Materials, 5th ed. New York: D. Van Nostrand, 1968.

PROBLEMS

6.1   A steel bar having a 1-in. diameter is to support a load of 50 000 lb. Determine the stress in the bar.

6.2   A steel bar having a 25-mm diameter is to support a load of 220 kN. Determine the stress in the bar.

6.3   A short solid cylinder is to support a compressive load of 105 kN. If the allowable stress of the material is 35 MPa, determine the required diameter of the cylinder.

6.4   A ¾-in. diameter bolt is used to connect two plates. If the load F is 5000 lb, determine the shear stress in the bolt.

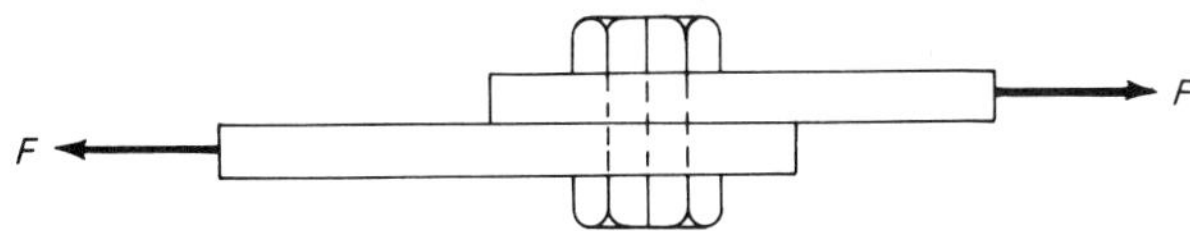

FIGURE P 6.4

6.5   Determine the average stress at sections A–A, B–B, and C–C.

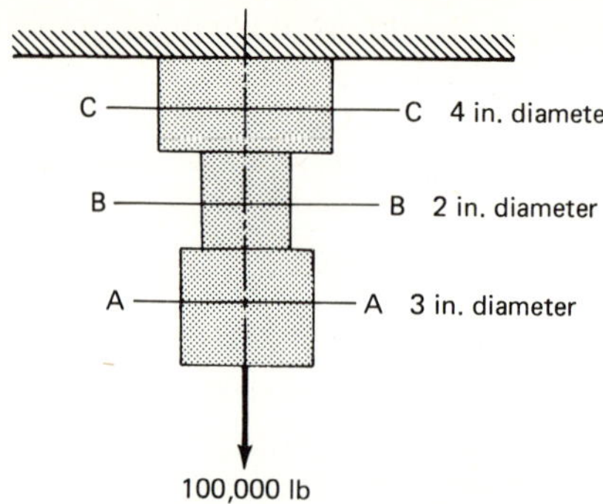

FIGURE P 6.5

6.6   A steel bar, 10 ft long, is to support a tensile force of 10 000 lb. Determine the stress in the bar at the built-in end of the bar—neglect the weight of the bar.

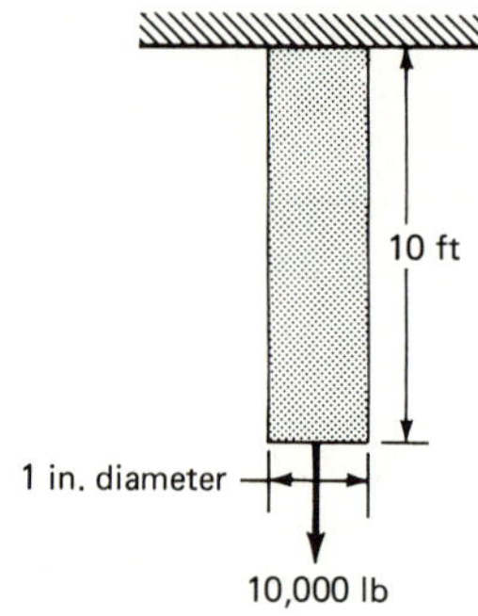

FIGURE P 6.6

6.7   Two bars, one aluminum and the other steel, are arranged as shown. Determine the total change in length when these bars are subjected to an axial force of 45 kN. E_{al} = 70 GPa and E_{steel} = 210 GPa.

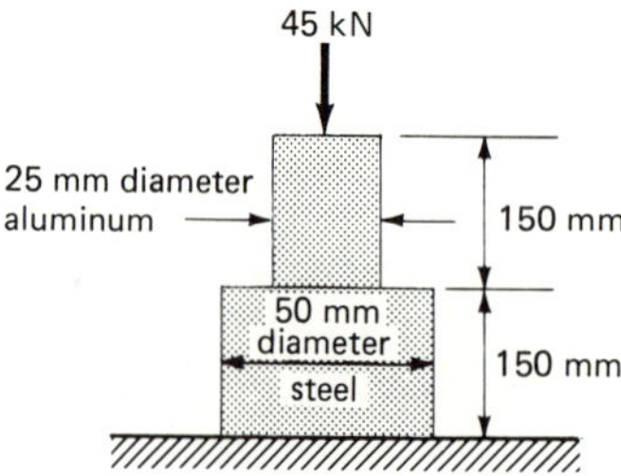

FIGURE P 6.7

6.8   If the allowable shear stress in steel is taken to be 20 000 psi, what load can the bolt carry in the configuration shown?

160   Concepts of Stress and Strain

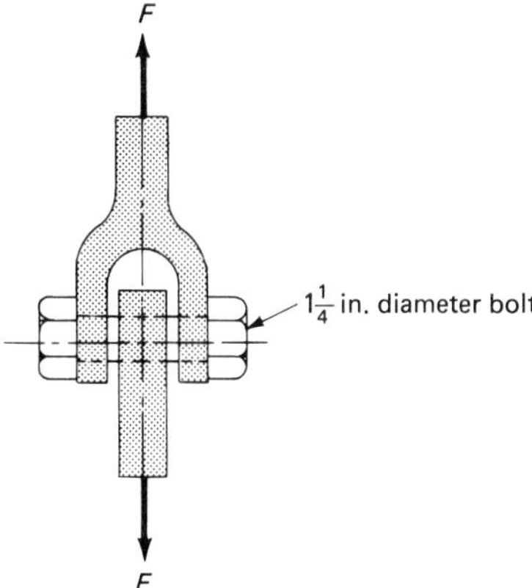

FIGURE P 6.8

6.9 Determine the maximum tensile stress in the configuration shown.

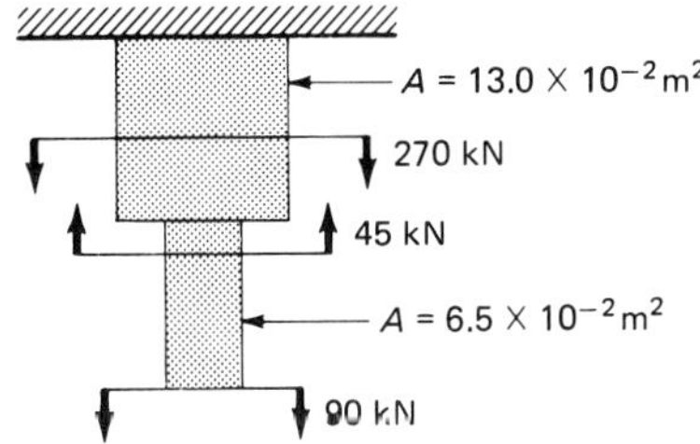

FIGURE P 6.9

6.10 A steel bolt carries a tensile load of 5000 lb. Determine the tensile stress in the unthreaded portion of the bolt and in the bottom of the threaded portion of the bolt, and also determine the shear stress at the root of the thread if the nut is ½ in. long.

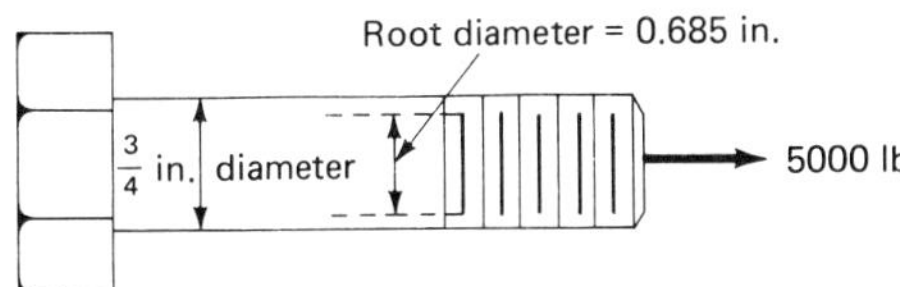

FIGURE P 6.10

6.11 A steel bolt carries a tensile load of 22 kN. Determine the tensile stress in the unthreaded portion of the bolt and in the bottom of the threaded portion of the bolt, and also determine the shear stress at the root of the thread if the nut is 12 mm long.

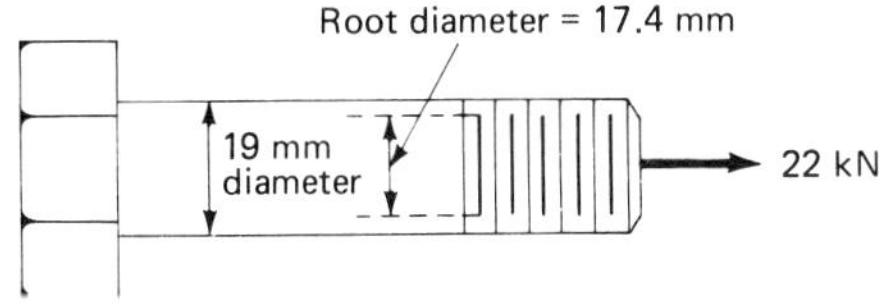

FIGURE P 6.11

Problems

6.12 It is desired to punch a 1-in. diameter hole in a steel plate ³⁄₁₆ in. thick. Determine the force on the punch if the shear stress is 52 000 psi.

6.13 A steel plate ⅝ in. thick and 12 in. wide is to be sheared into two 6-in. pieces. What shearing force is required if the shearing stress is 40 000 psi?

6.14 An S7 × 20 structural shape is to be used as a short column. If the column rests on a 12-in. × 12-in. plate, which in turn rests on a concrete footing whose bearing surface is 24 in. × 24 in., determine the stresses caused by an axial compressive load of 100 000 lb in each member.

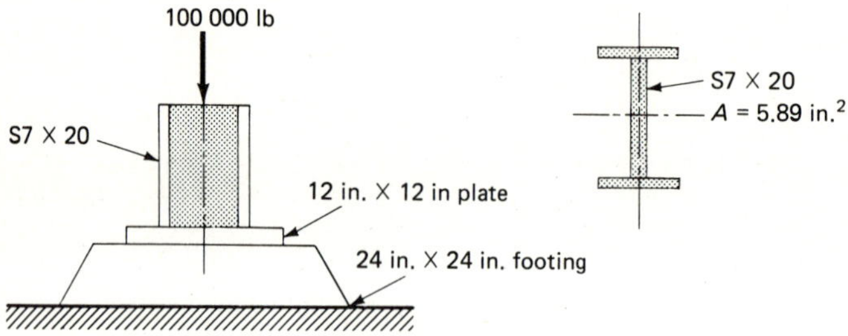

FIGURE P 6.14

6.15 A 4-in. Schedule 40 pipe (standard weight pipe) having an outer diameter of 4.500 in. and an inner diameter of 4.026 in. is subjected to an axial compressive load of 75 000 lb. Using $E = 30 \times 10^6$ psi, calculate the shortening of the pipe if it is 10 ft long.

6.16 Determine the change in length of the steel shaft shown due to an axial tension load of 100 000 lb. Use $E = 30 \times 10^6$ psi.

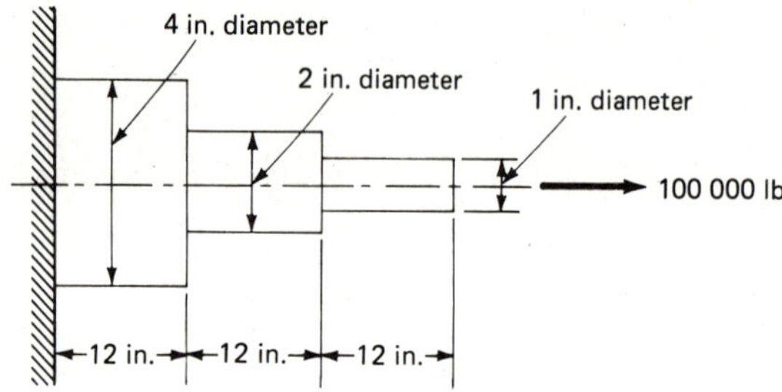

FIGURE P 6.16

6.17 A flange coupling is designed to support a twist (torque) of 7000 in·lb. If the mean bolt radius is 5 in. and there are four ½-in. diameter bolts, determine the shearing stress in the bolts.

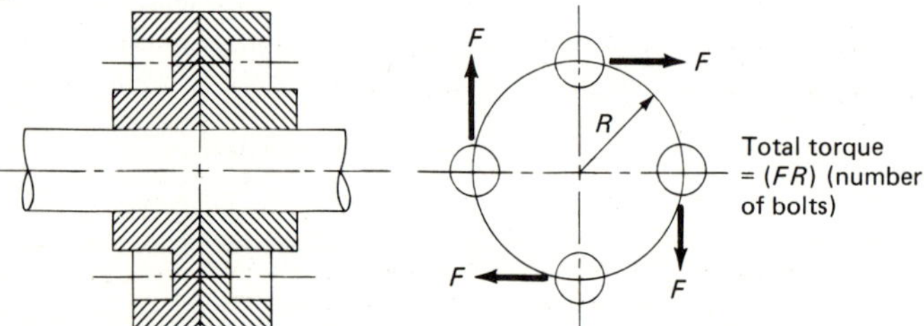

FIGURE P 6.17

                                    Concepts of Stress and Strain

6.18   A glued joint is made between a 100-mm × 300-mm board and a large table as shown. Determine the shear stress on the joint when it is subjected to the load shown.

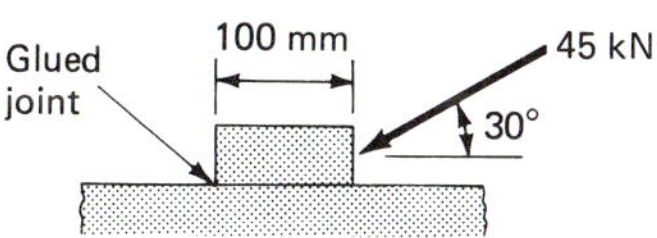

FIGURE P 6.18

6.19   A shaft is used to transmit a torque of 2100 N·m. A key 12 mm wide × 125 mm long is used to couple a wheel to the shaft. Determine the shear stress in the key.

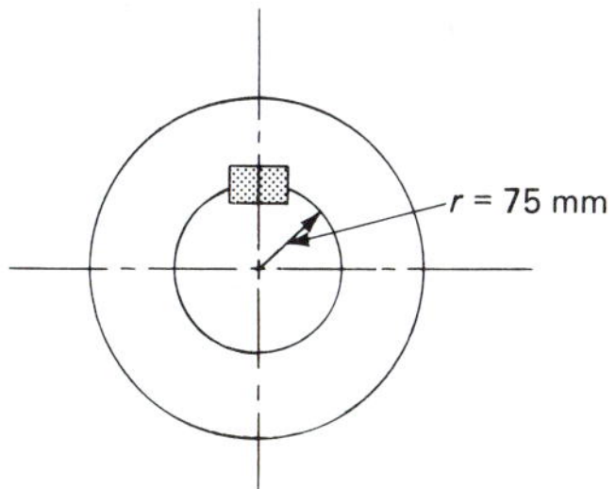

FIGURE P 6.19

6.20   A bar ½ in. in diameter and 10 in. long elongates 0.0085 in. when subjected to an axial load of 5000 lb. Compute the stress, strain, and E for the bar.

6.21   A 4-in. Schedule 40 pipe (standard weight pipe) having an outer diameter of 4.500 in. and an inner diameter of 4.026 in. is subjected to an axial compressive load of 300 kN. Using E = 210 GPa, calculate the shortening of the pipe if it is 3 m long.

6.22   A compression member consists of a short length of 4-in. Schedule 40 pipe having an outer diameter of 4.5 in. and an inner diameter of 4.026 in. If it is filled with concrete and supports an axial load of 45 kN, determine the load taken by the steel and concrete. Also determine the stress in each material. E_s = 210 GPa and E_c = 70 GPa.

6.23   What is the change of length of the composite bar shown due to the axially applied load? All bars have the same diameter, 25 mm.

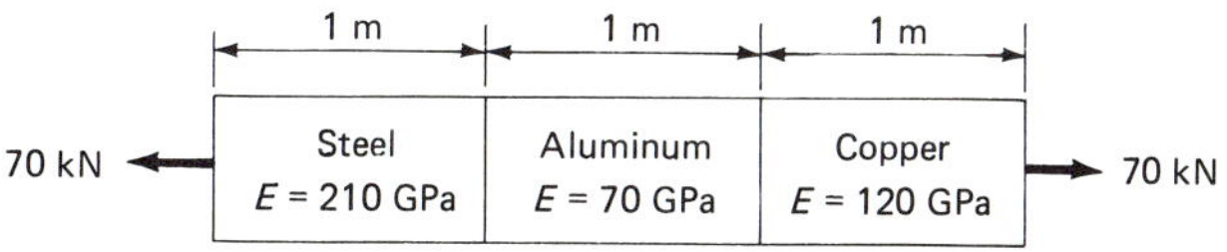

FIGURE P 6.23

Problems   163

6.24   Determine the change in length of the steel shaft shown due to an axial tension load of 450 kN. Use $E = 210 \times 10^9$ Pa.

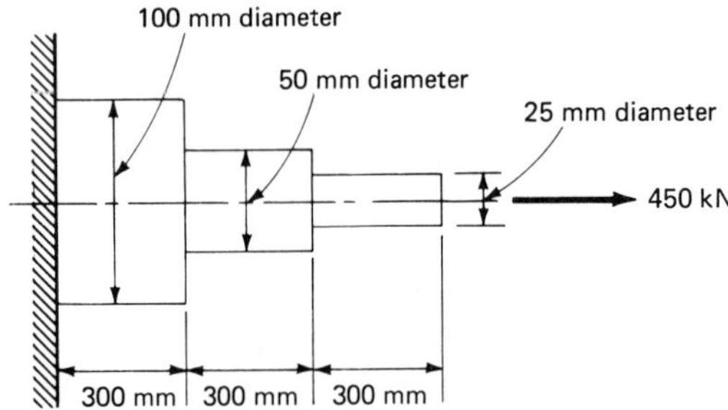

FIGURE P 6.24

6.25   Using the data in Table 6.1 calculate μ for aluminum, iron, and copper. Compare your results with the values given in Table 6.2.

6.26   A short steel bar is subjected to a transverse shear load of 5000 lb. If the block has a cross-sectional area of 2.0 in.^{2} and is 2 in. high, determine its total shear deformation and its shearing strain. Use $G = 12 \times 10^6$ psi.

6.27   A steel rod having a diameter of 2 in. is subjected to an axial tensile load of 100 000 lb. If $\mu = 0.28$ and $E = 30 \times 10^6$ psi, determine the change in diameter of the rod.

6.28   Determine the change in cross-sectional area for the steel rod in Problem 6.27.

6.29   If the total elongation of a bar is 0.5 mm when subjected to a load of 50 kN, determine the stress in the bar. $E = 210$ GPa and the bar had an initial length of 0.5 m.

6.30   The modulus of rigidity of a material is 80.0×10^9 Pa, and the modulus of elasticity is 202×10^9 Pa. Determine μ.

6.31   If the modulus of rigidity of a material is 11×10^6 psi and its modulus of elasticity is 30×10^6 psi, determine μ.

6.32   Using the data in Table 6.1 calculate μ for aluminum, iron, and copper. Compare your results with the values given in Table 6.2. (Use SI values.)

6.33   A short steel bar is subjected to a transverse shear load of 22 kN. If the block has a cross-sectional area of 0.0013 m^2 and is 25 mm high, determine its total shear deformation and its shearing strain. Use $G = 84 \times 10^9$ Pa.

6.34   A short bar is subjected to a transverse load of 30 kN. If the cross-sectional area of the block is 0.0018 m^2 and it is 40 mm high, determine its total shear deformation and its shearing strain if $G = 30$ GPa.

6.35   A steel rod, 1 in. in diameter, is placed between two rigid walls. If the rod is initially 2 ft long, determine its compressive stress after a temperatue increase of 80 °F. Use $E = 30 \times 10^6$ psi and $\alpha = 6.5 \times 10^{-6}$ in./in.

6.36. What total elongation will occur in a steel rail initially 25 ft long due to a 100 °F temperature change? Use $E = 30 \times 10^6$ psi and $\alpha = 6.5 \times 10^{-6}$ in./in.

6.37   If the axial stress in the rail described in Problem 6.36 is to be limited to 5000 psi, determine the required clearance between the rails.

6.38   Two rods, one of steel and the other of an unknown material, are initially at the same temperature. An increase of temperature of 75 °F causes the unknown material to expand

                                        Concepts of Stress and Strain

so that a line drawn across the tops of the rods makes an angle of 0.000 188 rad with the horizontal. If the rods were initially 12 in. long and were separated by 1 ft, determine α for the unknown material. α for steel can be taken as 6.5×10^{-6}.

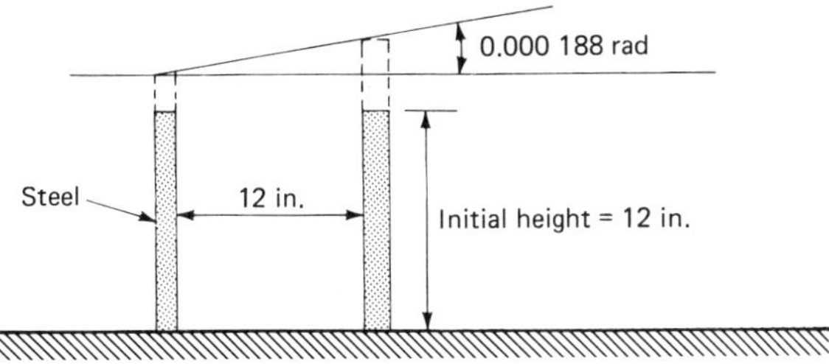

FIGURE P 6.38

6.39   A steel rod 25 mm in diameter is placed between two rigid walls. If the rod is initially 1 m long, determine its compressive stress after a temperature increase of 45 °C. Use $E = 210$ GPa and $\alpha = 11.7 \times 10^{-6}$ m/m·°C.

6.40   An aluminum rod 12.5 mm in diameter is placed between two rigid walls. If the rod is initially 500 mm long, determine its compressive stress after its temperature increases 20 °C. $E = 70$ GPa and $\alpha = 23.0 \times 10^{-6}$ m/m·°C.

6.41   A steel wire having a diameter of 2.5 mm is found to be 8 m long when subjected to a tensile force of 350 N at 20 °C. How long will it be at 38 °C when subjected to a force of 140 N in tension? Use $E = 210$ GPa and $\alpha = 12.6 \times 10^{-6}$ m/m·°C.

6.42   A 4-in. Schedule 40 pipe is used as a short compression member. At 20 °C it is 250 mm long. It is loaded by a compressive load of a magnitude to cause it to be 250 mm long at 38 °C. Determine the magnitude of the compressive load if $E = 210 \times 10^{9}$ Pa and $\alpha = 12.6 \times 10^{-6}$ m/m·°C.

6.43   A weight of 100 000 lb is to be carried by a steel rod in a copper cylinder. Determine the forces and stresses in each member. $E_{steel} = 30 \times 10^{6}$ psi and $E_{copper} = 10 \times 10^{6}$ psi.

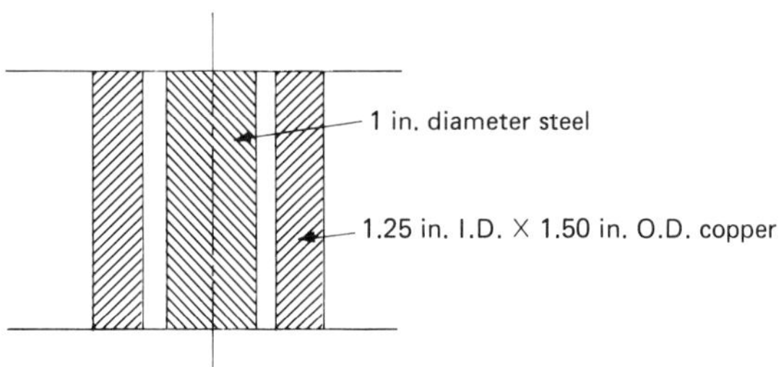

FIGURE P 6.43

6.44   A 1-in. diameter steel bolt has 10 threads per inch and a copper sleeve around it as shown. If the nut on the bolt is turned $\frac{1}{10}$ of a turn, determine the stress in each material. $E_{steel} = 30 \times 10^{6}$, $E_{copper} = 10 \times 10^{6}$.
(*Hint:* The final force in each member is the same. Also, the elongation of the bolt must

equal the shortening of the sleeve due to the unopposed turning of the nut minus the elongation of the bolt.)

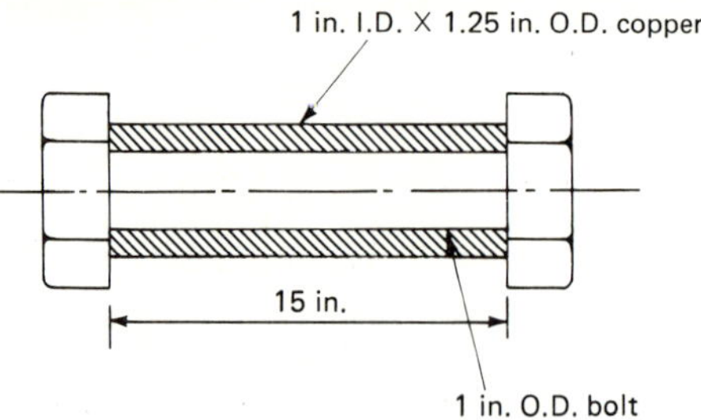

FIGURE P 6.44

6.45 Two rods are mounted between rigid walls as shown. Determine the stress in each bar due to a temperature drop of 85 °F. $\alpha_{steel} = 6.5 \times 10^{-6}$ in./in., $E_{steel} = 30 \times 10^6$ psi, $\alpha_{copper} = 9.3 \times 10^{-6}$ in./in., $E_{copper} = 16 \times 10^6$ psi.
(*Hint:* The force in each member is the same. Also the sum of the final lengths of the members must equal 12 in.)

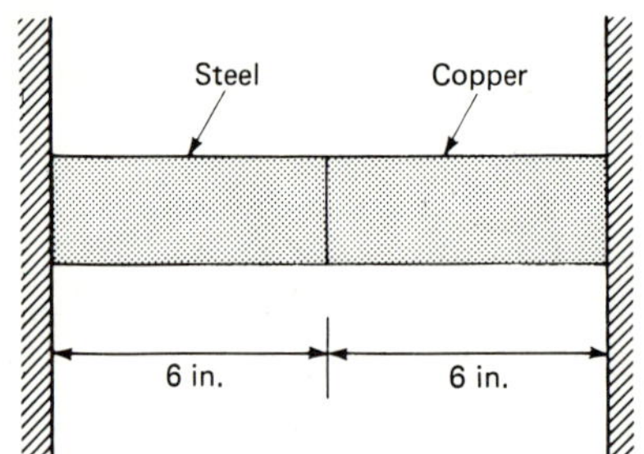

FIGURE P 6.45

6.46 A thin-walled cylinder is subjected to an internal pressure of 400 psi. If the internal diameter is 20 in. and the allowable working stress is 18 000 psi, calculate the required wall thickness.

6.47 To what internal pressure can a sphere be subjected if it is made of ¼-in. thick plate and has a diameter of 24 in.? Assume an allowable working stress of 16 000 psi.

6.48 A spherical vessel 6 m in diameter is to contain a fluid at 700 kPa. If the allowable stress in tension is 70 MPa, determine the required thickness of the sphere.

6.49 A spherical shell having an inside diameter of 6 m is made of 9 mm plate. Find the tensile stress in the shell if it contains a fluid at 525 kPa.

6.50 A cylindrical tank, 18 in. in diameter, has a flat cover plate held on by eight ½-in. diameter bolts. If the allowable stress is 15 000 psi in the bolts, to what pressure can the plate be subjected? Assume that bolt tension governs.

6.51 Using the results of Problem 6.50, determine the thickness of the tank if the allowable stress in the tank is the same as that of the bolts.

                                        Concepts of Stress and Strain

6.52 A cylindrical tank is made with a 20-in. inner diameter and a 25-in. outer diameter. Calculate the stress in this vessel using the mean diameter and a pressure of 1000 psi. Compare your results with the value obtained from Fig. 6.18.

7

Mechanical Properties of Materials

7.1 INTRODUCTION

The process of selecting a material for a given application is complex, and therefore it is quite difficult to simply write down a prescription for selection that would be generally applicable. The purpose of this chapter is to discuss briefly material properties that are generally needed, common test methods used to obtain data, and the results of such tests for certain selected materials. The study of the internal structure of a particular material, the effects of alloying, heat treating, or cold working, as well as other methods of tailoring the properties of materials, are most properly the concern and province of a course in physical metallurgy. However, the first consideration in selecting a material is usually that the strength of the material is adequate for the specified situation. Since more than one material can usually be found that would be suitable for a given application, it becomes the designer's function to select the most suitable material, using considerations of cost, availability, ease of handling, life, corrosion resistance, machinability, and so on.

The mechanical properties of materials and tests which will be discussed briefly in this chapter represent only a portion of the information that a designer needs to know. However, they are important in that they usually represent the starting point in the selection process.

7.2 THE TENSILE TEST

The *tensile test* is the most common test applied to materials. It is also the most important of the mechanical tests used to obtain data on the properties of materials. The test is usually performed by slowly and steadily applying a tensile load to a test

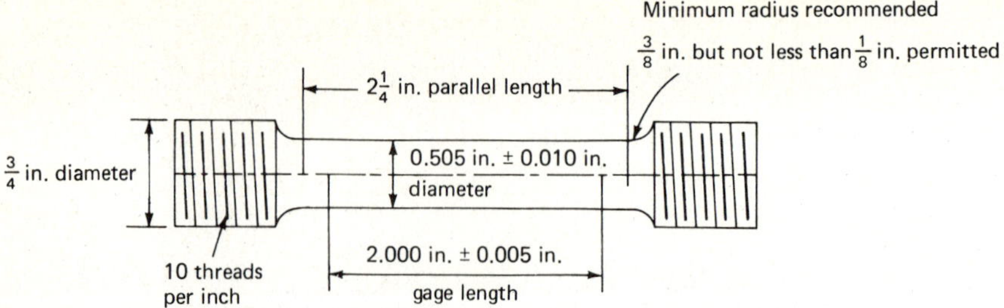

FIGURE 7.1 Standard tensile test specimens.

specimen at room temperature. The test is usually performed on a standard test specimen in order to standardize the results. One common type of test specimen is shown in Fig. 7.1. Other types of test specimens are specified by the American Society for Testing and Materials (ASTM) for machines having different types of grips, for wire, for sheet metal, and so on. The ASTM standards should be consulted for more detailed information on standardized test specimens. The test specimen must be accurately machined so that it is symmetrical, and extreme care must be exercised to mount it in the test machine so that only axial loads, as differentiated from bending loads, will be imposed on the specimen. For the test specimen shown in Fig. 7.1, it will be noted that two light prick point marks are shown located 2.000 ± 0.005 in. apart. This distance is known as the *gage length*. After the sample has been loaded in tension and is broken, the two halves of the test specimen are held firmly together, and the distance between the marks is again measured. The *percent elongation* is defined as the change in length divided by the original length × 100. Thus,

$$\text{percent elongation} = \left(\frac{L_f - L_o}{L_o}\right) \times 100 \tag{7.1}$$

where L_f is the final gage length and L_o is the initial gage length. Percent elongation is also used as a measure of the *ductility* of a material, that is, the ability of a material to be drawn into a wire or tube or to be forged. In referring to ductility in terms of percent elongation, we must also necessarily state the gage length since percent elongation varies with gage length. This is due to the fact that a large part of the total strain occurs in the necked down portion of the gage length just before fracture (see Fig. 7.9) and within 1 in. of the fracture. For example, a specimen of material having a gage length of 6.000 in. might have a final length of 7.8 in., which would give a percent elongation of (7.8 − 6)/6 × 100 = 30%. A 2.000-in. gage length of the same material might have as a final length 2.8 in., or a percent elongation of (2.8 − 2)/2 × 100 = 40%.

Figure 7.2 shows a modern universal testing machine with an automatic stress–strain recorder. The testing machine is a device in which the specimen can be accurately loaded at rates that are standardized. One standard requirement is that the speed of the testing machine cross-head should not exceed $\frac{1}{16}$ in. per inch of gage length per minute up to the yield point of the material, and it should not exceed $\frac{1}{2}$ in. per inch of gage length per minute from the yield point to the rupturing point of the material.

                    Mechanical Properties of Materials

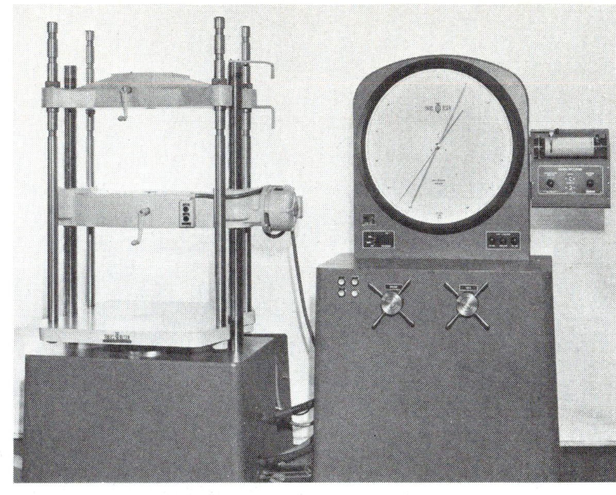
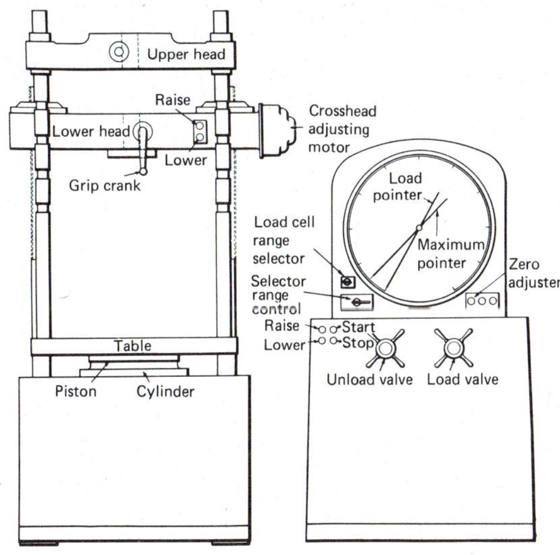

FIGURE 7.2   Universal testing machine. (Courtesy of Tinius Olsen Testing Machine Company.)

The reliability of data derived from any test depends upon the accuracy of the test instrumentation, the recorder, and the test machine. The measurement of the strain of the test specimen is made with an instrument known as an *extensometer*. Figure 7.3 shows six types of extensometers, all of which can be used to obtain the specimen strain. The extensometer is clamped to the test specimen and, as the specimen stretches or compresses, the core of a differential transformer moves proportionally to this displacement to produce a displacement voltage and a direction signal. This signal is subsequently amplified and is used to activate a servo motor, which rotates the recorder drum (2) and a cam (3), as shown in Fig. 7.4. As the cam turns, it moves the cam follower and the attached cores of the three differential transformers (one for each magnification range). This core motion produces an opposite and equal signal. The signal for the selected magnification range travels to the servo amplifier. Thus the pen

The Tensile Test

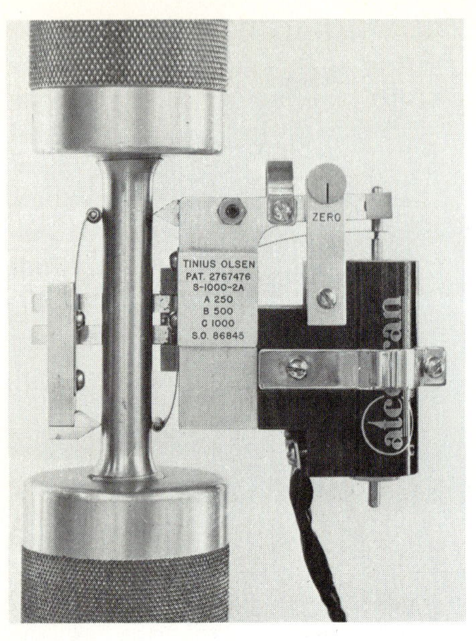

(a)

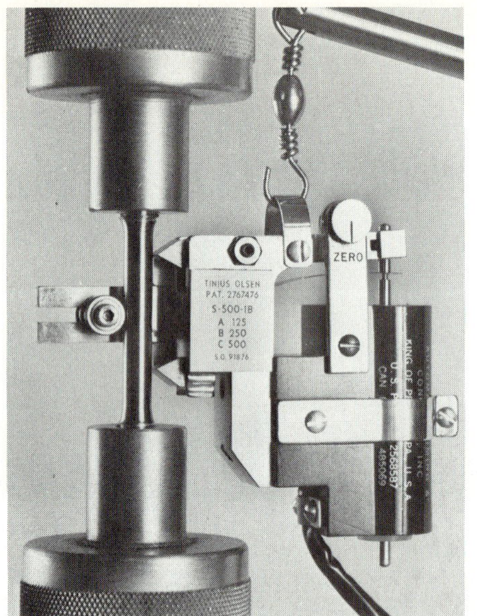

(b)

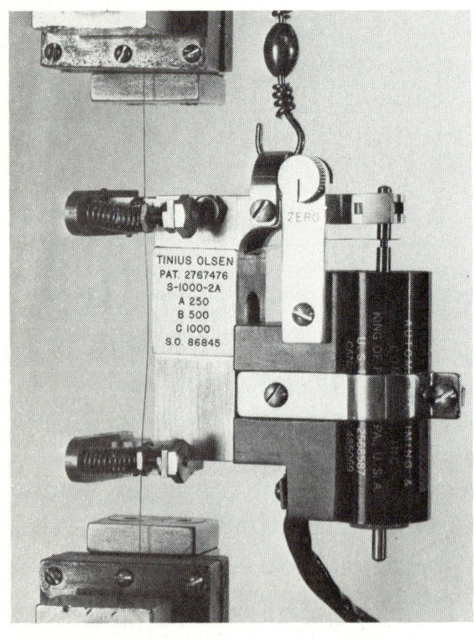

(c)

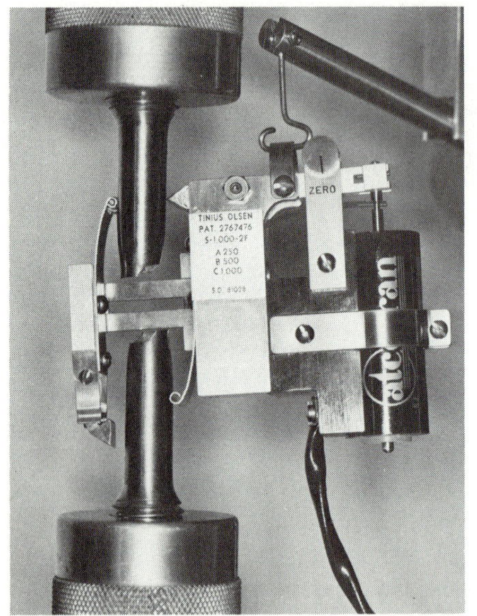

(d)

Mechanical Properties of Materials

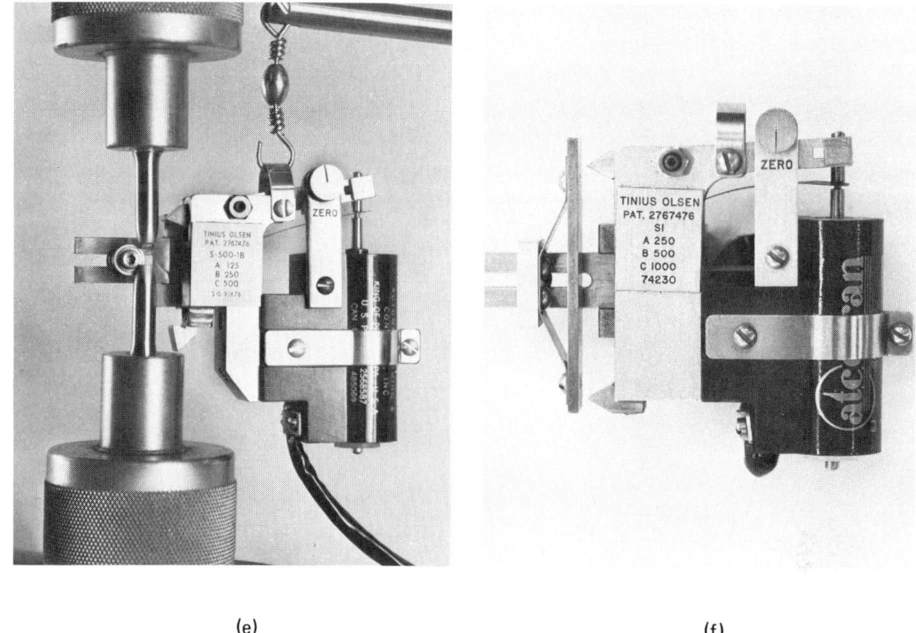

(e)   (f)

FIGURE 7.3 Extensometers. (a) Averaging type—Knife edges on opposite sides of specimen provide a measure of the average amount of strain between gage points. (b) Nonaveraging type—Provides measurement of strain between knife edges on same side of specimen. (c) Film clamps (nonaveraging only)—Spring-activated film clamps permit strain measurements of thin materials. (d) Breakaway extensometer, averaging type—Lower knife edge pivots free of specimen when elongation exceeds measuring range or specimen breaks. (e) Breakaway extensometer, nonaveraging type—Same as (d) but measures elongation between knife edges on the same side of specimen. (f) Sheet metal attachment (nonaveraging only)—Used for thin flat specimens. (All photos courtesy of Tinius Olsen Testing Machine Company.)

line drawn by the rotation of the recorder drum indicates the exact specimen strain at any instant. The signal restores the null balance when the opposing signals are of equal intensity.

While the recorder drum is being rotated in direct proportion to specimen strain, the load-activated pen mechanism moves the pen point across the drum a distance that is in direct proportion to the increasing or decreasing stress (load) on the specimen. For all practical purposes, both of these indicating processes are instantaneous. As a result of the simultaneous rotation of the recorder drum and the movement of the recorder pen, a highly magnified and extremely accurate stress–strain diagram, which can be studied in detail without further transposition, is produced.

The Tensile Test

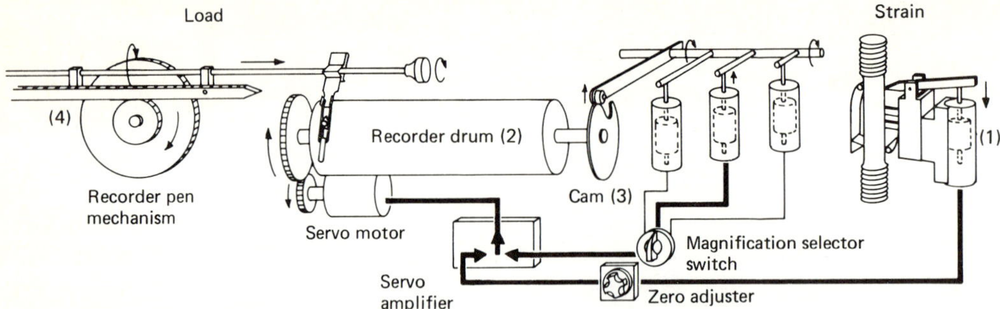

FIGURE 7.4 Automatic stress–strain recording machine. (Courtesy of Tinius Olsen Testing Machine Company.)

7.3   THE STRESS–STRAIN CURVE IN TENSION

Having tested a specimen, we find it necessary to reduce the observed data to meaningful engineering results. The most useful engineering presentation of the data obtained is the *engineering stress–strain diagram*. This diagram is a plot of strain as the abscissa, and stress as the ordinate. The *stress* is defined as the load in pounds divided by the cross-sectional area of the specimen at the start of the test. As the test proceeds, the actual cross-sectional area decreases, and at high stresses this reduction in area become appreciable. It should be noted that the stress based upon the initial area is not the true stress, but it is generally used and the calculated stress in load-carrying members is almost universally based on this original area. The strain used is the elongation of a unit length of the test specimen taken over the gage length. Figure 7.5 shows a typical engineering stress–strain diagram and the diagram obtained for the same material when based upon the actual area of the specimen. That the actual stress curve is one that continually rises indicates an ever-increasing applied stress up to the point of rupture, while the curve based upon the original area of the specimen shows a maximum stress and then a *decrease* in stress up to the rupture point. We shall examine all of these occurrences in some detail later on in this section.

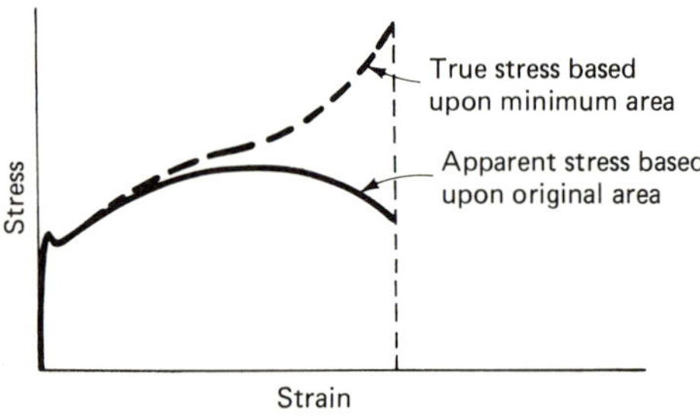

FIGURE 7.5 Apparent and true stresses.

                                    Mechanical Properties of Materials

Total Applied Load F (lb)	Stress (F/A_o) (psi)	Elongation in Gage Length (in.)	Strain ($\Delta L/L$) (in./in.)	Least Diameter (in.)	Area A (in.^{2})	True Stress (psi)
1 100	5 500	0.001	0.000 125	0.5020	0.1979	5 550
1 950	9 850	0.002	0.000 250	0.5015	0.1975	9 870
2 570	12 480	0.003	0.000 375	0.5012	0.1973	13 020
3 230	16 310	0.004	0.000 500	0.5012	0.1973	16 370
3 800	19 200	0.005	0.000 625	0.5012	0.1973	19 240
4 600	23 200	0.006	0.000 750	0.5012	0.1973	23 300
5 300	26 800	0.007	0.000 875	0.5012	0.1973	26 820
6 150	31 050	0.008	0.001 000	0.5012	0.1973	31 170
6 600	33 300	0.009	0.001 125	0.5012	0.1973	33 420
7 300	36 850	0.010	0.001 250	0.5012	0.1973	37 000
7 600	38 400	0.011	0.001 375	0.5012	0.1973	38 500
7 500	37 850	0.012	0.001 500	0.5012	0.1973	38 000
7 450	37 600	0.0135	0.001 688	0.5012	0.1973	37 700
7 560	38 200	0.014	0.001 750	0.5012	0.1973	38 300
7 600	38 400	0.015	0.001 875	0.5011	0.1972	38 500
7 780	39 300	0.016	0.002 000	0.5010	0.1971	39 420
7 700	38 900	0.0505	0.006 310	0.4990	0.1956	39 350
8 100	40 900	0.100	0.023 500	0.4965	0.1936	41 800
8 000	40 400	0.150	0.018 750	0.4960	0.1932	41 400
8 700	43 900	0.200	0.025 000	0.4960	0.1932	45 000
9 300	46 900	0.250	0.031 250	0.4940	0.1917	48 500
9 860	49 800	0.300	0.037 500	0.4925	0.1909	51 700
10 330	52 200	0.360	0.045 000	0.4900	0.1886	54 800
10 980	55 400	0.460	0.057 500	0.4870	0.1863	58 800
11 850	59 800	0.600	0.082 500	0.4810	0.1817	65 200
12 340	62 300	0.860	0.107 500	0.4740	0.1765	69 800
12 450	62 900	1.060	0.132 500	0.4660	0.1706	73 000
12 620	63 700	1.260	0.157 500	0.4570	0.1640	76 900
12 760	63 900†	1.460	0.182 500	0.4390	0.1514	83 600
9 980	50 400	1.660	0.207 500			
9 840	49 700‡	1.670	0.208 750	0.3350	0.0881	111 500

*(Initial diameter 0.502 in.; initial area of section, A_o 0.198 in.^{2}; initial gage length 8 in.) †Specimen begins to neck, Ultimate tensile strength. ‡Failure.

Table 7.1* gives the data recorded during a tension test conducted on a mild steel specimen. The specimen had an initial diameter of 0.502 in., an initial area A_o of 0.198 in.^{2}, and a gage length of 8 in.

The total applied load and the elongation of the specimen in the gage length are tabulated in Table 7.1. From the test data and the known size of the test specimen the

*Table 7.1 and Figs. 7.6 and 7.7 are reproduced with permission from D. S. Clark, *Engineering Materials and Processes.* International Textbook Co.

The Stress–Strain Curve in Tension

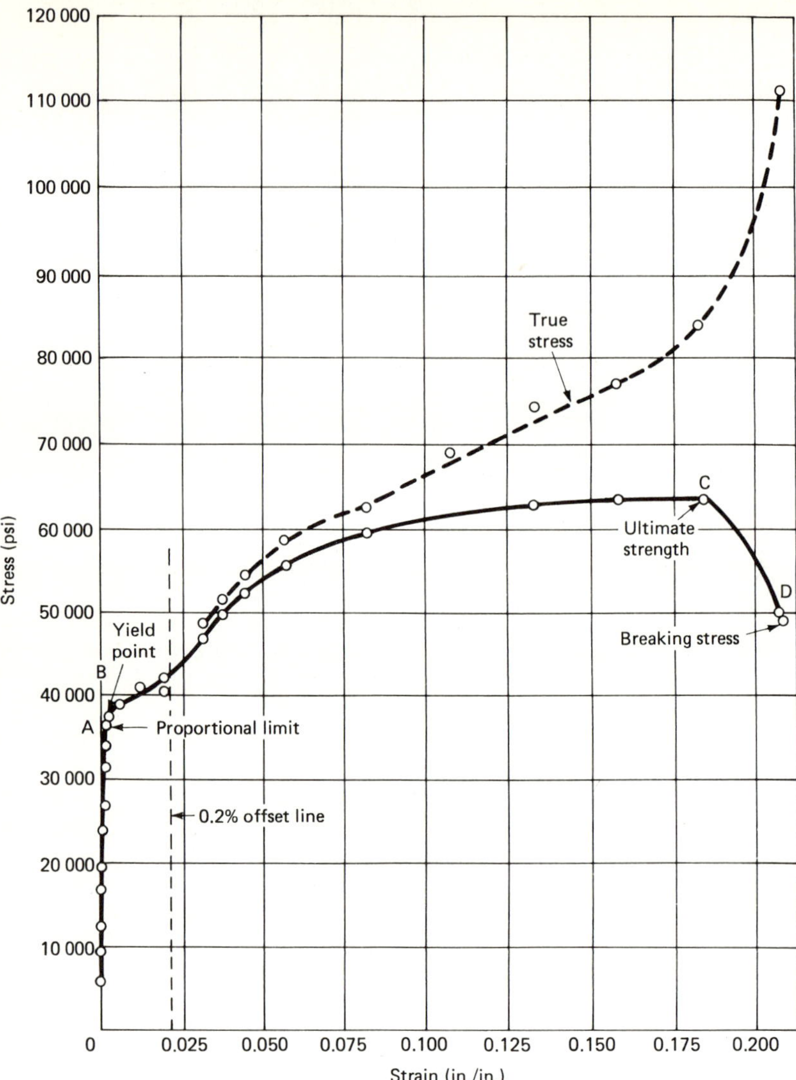

FIGURE 7.6   Stress–strain diagram for mild steel.

rest of the table is completed. The stress–strain diagram has been plotted in Fig. 7.6 with both the engineering stress and true stress curves shown. Note that the stress–strain curve begins with an almost vertical straight line when plotted on a scale to show the strain up to the breaking point. In order to show this initial section in detail, Fig. 7.7 has been plotted. This curve shows that the extensometer had a zero correction of 0.001 in./in., that is, all strains should be corrected by adding 0.001 in./in. to the tabulated values. Basing our calculations upon Table 7.1 and Figs. 7.6 and 7.7, we can obtain much significant information. These are defined and illustrated for the data of Table 7.1.

                                                     Mechanical Properties of Materials

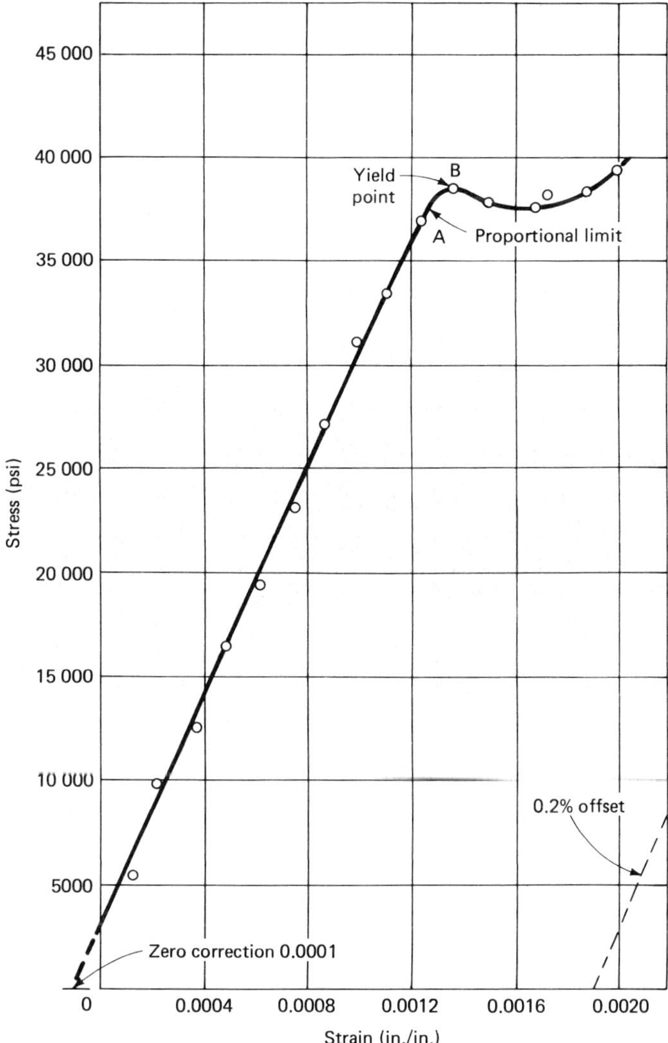

FIGURE 7.7   Portion of stress–strain diagram for mild steel.

Proportional Limit

Hooke's law is a statement of the proportionality of stress and strain. The *proportional limit* is defined as the greatest stress that a material is capable of developing without deviation from Hooke's law of stress–strain proportionality. In order to determine the proportional limit, it is necessary to use very sensitive extensometers to detect the slightest deviation from a straight line in the tensile test diagram.

Other methods of defining the proportional limit have been proposed. These usually define the proportional limit as that stress at which a certain defined permanent set takes place. Thus a permanent strain of 0.001% and a permanent strain of 0.01%

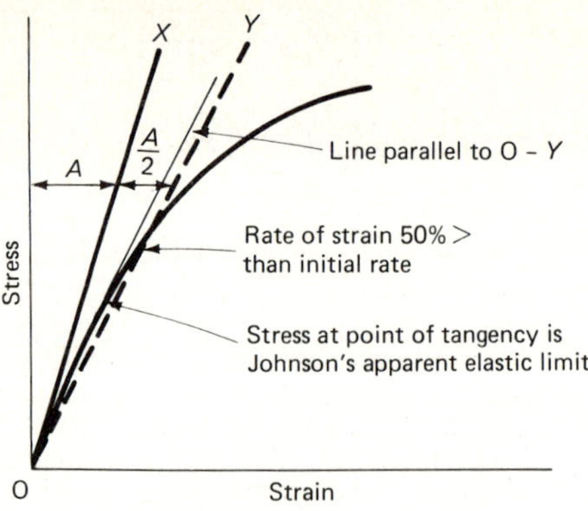

FIGURE 7.8 Construction for Johnson's apparent elastic limit.

have been used to obtain the proportional limit. Certain materials, such as cast iron and copper, do not have the straight line portion of the stress–strain curve as does steel. For these materials a value known as *Johnson's apparent elastic limit* is sometimes used instead of the proportional limit. Johnson's apparent elastic limit is defined as that stress at which the rate of deformation is 50% greater than the initial rate of deformation. Figure 7.8 shows the construction for Johnson's apparent elastic limit. The proportional limit for the data shown in Fig. 7.7 occurs at point A, and its value is 37 500 psi.

Modulus of Elasticity

The mechanical property that defines the resistance of a material in the elastic range is called *stiffness*; and for ductile materials it is measured by the value called the *modulus of elasticity* (or Young's modulus), which is designated by the capital letter E. Referring to Fig. 7.7, we note that the first part of the diagram is a straight line, which indicates a constant ratio between stress and strain over this range. The numerical value of this ratio is referred to as the modulus of elasticity E. E is therefore the slope of the initial straight portion of the stress–strain diagram, and its numerical value is obtained by dividing the stress in pounds per square inch by the strain, which in nondimensional. Thus E has the same units as stress, namely, pounds per square inch.

When the straight-line portion of the curve passes through the origin, E can be determined simply by dividing any stress value along the straight line by the corresponding strain. As we have already noted, the data shown in Fig. 7.7 show that a zero correction is needed due to the offset of the intercept. For cases such as this, E can be found by taking two values of stress and two values of strain far enough from the origin to eliminate any errors. Thus

$$E = \frac{S_1 - S_2}{\epsilon_1 - \epsilon_2} \tag{7.2}$$

                                          Mechanical Properties of Materials

For Fig. 7.7 we can obtain E from two points C and D as

$$E = \frac{30\ 000 - 13\ 750}{0.0010 - 0.0004} = 27.1 \times 10^6 \text{ psi}$$

We can check this by using the data at a strain (apparent) of 0.0010. The stress corresponds to 30 000 psi, and the true strain is $0.0010 + 0.0001$. Thus E is

$$E = \frac{30\ 000}{0.0010 + 0.0001} = 27.3 \times 10^6 \text{ psi}$$

It should be noted that the value of 27.3×10^6 psi for E is from a single test of one specific steel specimen.

Yield Strength

When the load on the test specimen is increased beyond the proportional limit, a stress level is reached where the material continues to elongate without an increase of load. The *yield point* is defined as the stress at which a marked increase in strain occurs without a concurrent increase in applied stress. Point B on Fig. 7.7 is the yield point. More correctly, point B is known as the *upper yield point*, and after it is reached, the force resisting deformation decreases due to the yielding of the material. After the initial yielding, the stress reaches a relatively constant lower value, while the deformation process continues. This latter value of stress is known as the *lower yield point*. The lower yield point is usually taken to be the true material characteristic to be used as the basis for the determination of working stresses.

Many materials do not exhibit well-defined yield points, and the *yield strength* is defined as the stress at which the material exhibits a specified limiting permanent set. The specified set (or offset) most commonly used is 0.2%, which corresponds to a strain of 0.002 in./in. The yield strength is therefore the stress corresponding to the intersection of a line parallel to the straight-line portion of the stress–strain curve and the stress–strain curve.

The defined yield strength does not represent a physical property of the material, and its value is a function of the defined offset. Actually, this yield strength of a material is of importance in determining when a material is stressed beyond its ability to act elastically. For most applications, the onset of inelastic action is considered to be the point at which a member can no longer perform its structural function.

Ultimate Strength (Tensile Strength)

The *ultimate strength* of a material is defined as the stress obtained by dividing the maximum load reached before the specimen breaks by the initial cross-sectional area of the specimen. For the test data shown in Fig. 7.6, this corresponds to a stress of approximately 64 000 psi at point C. It will be noted from Fig. 7.6 that beyond the yield point, the stress continues to increase until it reaches the ultimate strength. This is true when the initial area of the test specimen is used to define the stress. The true stress–strain curve based upon the actual area of the specimen shows a continual increase in stress until the breaking point of the specimen is reached. The ultimate strength (often called the *tensile strength)* of the material is commonly used as a basis for establishing working stresses for a material.

Elongation

We earlier noted that *elongation* is a measure of the ability of a material to undergo deformation without rupture. Percentage elongation, defined by Eq. (7.1), is a measure of the ductility of a material. This property, ductility, is a desirable and necessary property, and a member must possess it to prevent failure due to local overstressing. From the data of Table 7.1, percent elongation is

$$\left(\frac{9.67 - 8.00}{8.00}\right) \times 100 = 20.9\% \text{ in 8 in.}$$

Breaking Strength (Rupture Strength, Fracture Strength)

The *breaking strength* of a material is the load on the material at the time of failure divided by the original cross-sectional area of the specimen. As we see from Fig. 7.6, it is the final ordinate on this stress–strain curve. For the data of Fig. 7.6 the breaking strength is very close to 50 000 psi. It should be reemphasized that this definition of the breaking strength is based upon the original area of the test specimen during the test. The breaking stress on this basis is less than the ultimate stress of the data of Fig. 7.6, while the true stress at failure is the maximum stress on the material. For the test interval between the ultimate stress and the breaking stress, the specimen continues to elongate even though the resisting stress based on the original area decreases.

Reduction of Area

As the load on the material undergoing testing is increased, the original cross-sectional area decreases until it is at a minimum at the instant of fracture. It is usual to express this reduction in area as the ratio of the change in area to the original specimen cross-sectional area, expressed as a percentage. As Fig. 7.9 shows, the failed specimen exhibits a local decrease in diameter, known as *necking down,* in the region where failure occurs. It is very difficult to determine the onset of necking down and to differentiate it from the uniform decrease in diameter of the specimen. The percent reduction in area is

$$\text{percent reduction in area} = \left(\frac{A_o - A_f}{A_o}\right) \times 100 \qquad (7.3)$$

where A_f is the final cross-sectional area at the point of fracture and A_o is the initial cross-sectional area of the specimen. The percent reduction in area is also a measure of the ductility of a material. Brittle materials exhibit almost no reduction in area, while ductile materials exhibit a high percent reduction in area.

Mechanical Properties of Materials

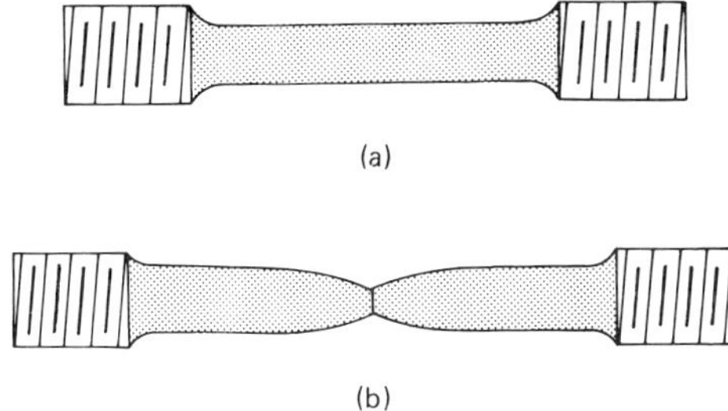

FIGURE 7.9   Necking down of a ductile specimen. (a) Initial specimen. (b) Failed specimen.

The reduction in area is less dependent on the dimensions of the test specimen than is percent elongation, and it is also independent of the gage length of the test specimen. For the data of Table 7.1, the percent reduction in area is

$$\left(\frac{0.1979 - 0.0881}{0.1979}\right) \times 100 = 55.5\%$$

Toughness

The area under the stress–strain curve of Fig. 7.6 is a measure of the work required to cause fracture to occur. The ability of a material to absorb energy up to fracture is also used by designers as a characteristic property of a material and is called *toughness*. This property of a material is a function of both the strength and the ductility of the material. The area under the stress–strain curve yields units of (in./in.) (lb/in.^{2}), or in.·lb/in.^{3}, that is, energy per unit volume of material. This quantity is also known as the *modulus of toughness*. A typical value of modulus of toughness for cast iron is 500 in.·lb/in.^{3}, while for low carbon steel it is approximately 17 000 in.·lb/in.^{3}.

7.4   OTHER TESTS

The tensile test is the most important single test that can be conducted to evaluate the properties of a material for design purposes. Other tests are conducted on materials in order to obtain data concerning properties of materials that are needed, but cannot be obtained from the simple static tension test. A direct correlation between the results of these tests and the desired material property is often not feasible, and in many cases only qualitative results are obtainable. We shall briefly indicate the more important of these tests.

Compression Tests

The *compression test* is used primarily to test brittle materials such as cast iron and concrete. The universal testing machine is used for this test, and data are taken in a manner similar to that discussed for the tension test. It is usual to make the test specimen in the form of a prism, whose height in the compression direction is several times larger than its lateral dimensions in order to minimize the effect of friction on the contact surfaces of the specimen. The results of the compression test are an elastic range, a proportional limit, and a yield strength. In the compression test the cross-sectional area of the specimen increases, and this produces a continuously rising engineering stress–strain curve.

High-Temperature Tests

The evaluation of the physical properties of materials at high temperatures is very important in modern technology. Most properties, such as the yield point and ultimate strength, are temperature dependent. In addition, when materials are kept at constant stress at high temperatures for considerable periods of time, a continuous deformation of the material, known as *creep,* occurs. An allied phenomenon, known as *relaxation,* occurs when an initial stress, such as in the bolts of a flanged joint, decreases with time at an elevated temperature, causing a decrease in the tightness of the bolted joint.

In general, the strength of metallic materials varies inversely with temperature. By definition, the *creep limit* of a material is taken to be the stress required to produce an elongation of 1% in 100 000 h. The creep limit decreases markedly as the temperature of the material increases. Since the creep limit decreases, the load-carrying capacity decreases as the temperature increases.

The creep test is a long-term, elaborate, expensive test. Preliminary data can be obtained by the *stress rupture test*. The stress rupture test is what the name implies. A specimen is subjected to a temperature, and a stress level to produce a failure in a given time is determined. When the data for, say, 10, 100, and 1000 h are plotted on a semilogarithmic plot of stress versus the logarithm of time, a straight-line relation usually results. Quite often both stress-rupture data and creep data are needed for evaluation of material properties at elevated temperatures. In order to better understand the creep phenomenon, let us refer to Fig. 7.10, which shows a typical plot of strain versus time for a metal.

For metals tested at high values of stress or temperature, three stages in the creep–time relation are observed. The initial stage, often called the *stage of primary*

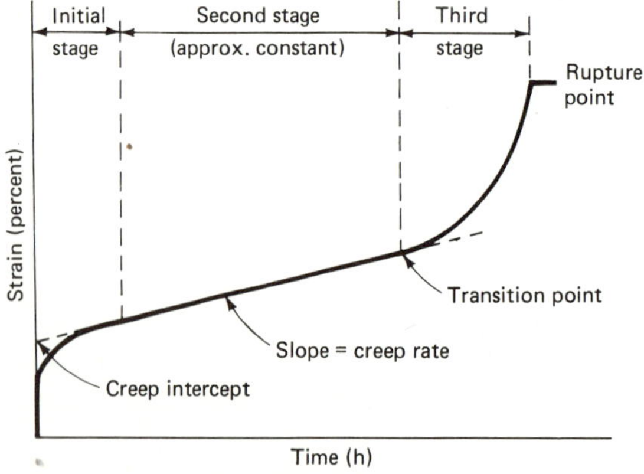

FIGURE 7.10 Typical creep-rupture curve.

                                    Mechanical Properties of Materials

creep, includes the elastic deformation and that region where the rate of creep deformation decreases rather rapidly with time. The second stage, often referred to as the *secondary creep stage*, represents a stage where the rate of strain has decreased to a constant value for a considerable time period, representing the period of minimum creep rate. The third stage, often called the *tertiary creep stage*, represents the period where the reduction in cross-sectional area leads to a higher stress, a greater creep rate, and finally rupture. The inflection point between the constant creep rate of the second stage and the increasing rate of the third stage is referred to as the *transition point*. Failure generally occurs shortly after the transition point. The minimum creep rate is that indicated in the second stage, where the creep rate is practically constant.

Cyclic Stress (Fatigue)

It is well known that a member subjected to repeated conditions of loading or unloading, or to repeated stress reversals, will fail at a stress considerably lower than the ultimate stress obtained in a simple tension test. Failures that occur as a result of this type of repeated loading are known as *fatigue failures*. One type of fatigue test is a *rotating beam test*. In this test a specimen is mounted in a test machine and loaded to produce a given stress. The test member is then rotated causing the stress to change sign every half-revolution (i.e., it is completely reversed from compression to tension and vice versa). The number of cycles of stress is equal to the number of revolutions of the machine. The test is run until the specimen ruptures, and the number of cycles to rupture is noted. The data are usually plotted as shown in Fig. 7.11, with stress as the ordinate and the logarithm of the number of cycles to failure as the abscissa. This curve is generally known as an *S–N* diagram. At the beginning, the stress decreases as the number of cycles increases. After several million cycles, the curve becomes a horizontal line whose stress value is known as the *endurance limit* of the material for complete stress reversal. Thus the endurance limit can be defined as the maximum completely reversed stress to which a material can be subjected without failure. To date no adequate theory has been developed to clearly explain the fatigue failure of materials. Fatigue failure appears to begin with a crack at a point of weakness in the material, with the crack progressing along crystal boundaries. A microscopic exami-

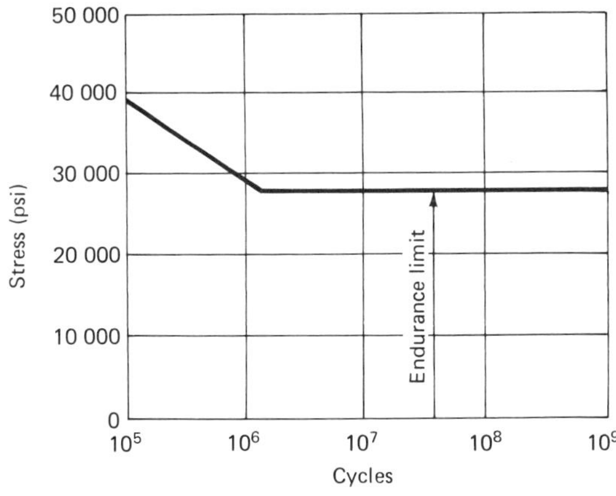

FIGURE 7.11 Typical S–N curve.

nation of metals indicates that there are many small cracks scattered throughout a material. Under the action of repeated stress, these small cracks open and close during the stress cycle. The cracks cause higher stress at the base of the crack as compared to the stress if there were no crack. Under this repeated concentration of stress, the cracks will gradually extend across the section of the member and finally cause complete failure of the member.

Although there appears to be a relation between the endurance limit and the tensile strength of a material, the influence of such items as the surface conditions of the test specimen, the speed of testing, the prior treatment of the material, and so on, is such as to make generalized statements of such a relation of limited value.

Hardness

The *hardness test* is used as an engineering test for a quick, inexpensive method of obtaining approximate values of some of the mechanical properties of materials to be used in design. The test usually measures the ability of a material to resist penetration. The hardness test is usually designated by the instruments used to determine the hardness. These most commonly are the Brinell, Rockwell, Vickers, and the Shore scleroscope. The basic test consists of applying a standard load to an indenter (a ball or pyramid) for a fixed time period and then measuring the size of the resulting indentation. Thus in the Brinell test, the hardened steel ball indenter is 10 mm in diameter, and it is pressed into the test piece with a force of 3000 kg for hard materials (or 500 kg for soft materials) for 30 s. The *Brinell hardness number* is defined as the load in kilograms divided by the area of the impression in square millimetres. A nonpenetrating test is the Shore scleroscope test, which consists of dropping a small diamond-tipped hammer from a fixed height and measuring the height of its rebound. The height of rebound corresponds to the hardness of the material. It should be noted that this test measures the resilience of a material rather than its resistance to penetration.

The results of hardness tests have been correlated to the ultimate strength of many materials. The ratio of the ultimate stress of steel to Brinell hardness is reasonably consistent and varies from approximately 470 to 530. In other words, a steel having a Brinell hardness number (BHN) of 280 will have an ultimate stress in the range of 132 000–148 000 psi. Although hardness and strength are interrelated, care should be exercised when comparing the hardness and strength of materials, especially when only limited data are available.

Impact Testing

When a load is applied rapidly to a part or structure, appreciable shock and vibration can occur. This type of loading, referred to as *impact loading,* therefore produces conditions in the structure that differ markedly from those produced by equal-magnitude static loads. A further effect of this manner of loading is that materials behave differently than they do under static loading. There are instances when a metal is ductile when loaded statically, but fractures in a brittle manner when subjected to impact loading.

There are in general two types of tests to determine the behavior of materials

                    Mechanical Properties of Materials

under impact loads. The usual impact test is referred to as the *notched bar test* and consists of subjecting notched specimens to axial, bending, and torsional loads by the Charpy or Izod impact testing machines. In both of these machines an impact load is applied to the specimen by swinging a weight W from a certain vertical height h to strike and rupture the notched specimen, and then the load stops at a vertical height h'. The energy expended in rupturing the specimen is then equal approximately to $(Wh - Wh')$. This type of test is primarily used for studying the influence of metallurgical variables.

Another type of impact testing is made on unnotched specimens, and the general purpose is to obtain a stress–strain diagram of materials under impact load, or a load–distortion diagram of a structural member as it is completely fractured under an impact load.

These tests are of little use in design, and the meaning of the results obtained is open to question. The tests are, at best, qualitative in nature, and the result obtained from a notched specimen is not directly applicable to design problems. The data from unnotched tests can be used directly in design problems. Tension and torsion impact testing has become the subject of increasing study in recent years. For a further discussion of this type of impact test the student is referred to general metallurgical and testing references.

7.5 CLOSURE

This chapter has as its purpose the presentation of a brief introduction to the mechanical properties of materials, and methods of obtaining data on these properties. The literature on this subject is vast and, at best, we have only touched on a small portion of it. The student will undoubtedly be exposed to more detail in materials and design courses. The importance of these studies cannot be overemphasized since the selection of the proper material for a given application is one of the principal functions of a designer.

REFERENCES

Arges, K. P., and A. E. Palmer, MECHANICS OF MATERIALS. New York: McGraw-Hill, 1963.

Bruhn, E. F., ANALYSIS AND DESIGN OF FLIGHT VEHICLE STRUCTURES. Cincinnati: Tri-State Offset Co.

Byars, E. F., and R. D. Snyder, ENGINEERING MECHANICS OF DEFORMABLE BODIES, 3rd ed. New York: Harper and Row, 1975.

Case, J., and A. H. Chilver, STRENGTH OF MATERIALS. London: Edward Arnold.

Clark, D. S., ENGINEERING MATERIALS AND PROCESSES. International Textbook Co.

Keyser, C. A., MATERIALS OF ENGINEERING. Englewood Cliffs, NJ: Prentice-Hall.

Lipson, C., and R. C. Juvinall, HANDBOOK OF STRESS AND STRENGTH. New York: Macmillan.

Olsen, G. A., STRENGTH OF MATERIALS. Englewood Cliffs, NJ: Prentice-Hall.

Rosenthal, E., and G. P. Bischof, ELEMENTS OF MACHINE DESIGN. New York: McGraw-Hill, 1955.

Shigley, J. E., APPLIED MECHANICS OF MATERIALS. New York: McGraw-Hill, 1976.

Slaymaker, R. R., MECHANICAL DESIGN AND ANALYSIS. New York: John Wiley.

Timoshenko, S., and D. H. Young, Elements of Strength of Materials, 5th ed. New York: D. Van Nostrand, 1968.

PROBLEMS

7.1   A rod has a uniform diameter of 0.50 in. An axial tensile force of 1000 lb causes the rod to elongate by 1% of its original length. Determine the modulus of elasticity of the material.

7.2   A rod has a uniform diameter of 50 mm. An axial tensile force of 1000 N causes the rod to elongate an amount equal to 1% of its original length. Determine the modulus of elasticity of the material.

7.3   A material is tested by applying a tensile load to a specimen that is ⅝ in. in diameter. If the load is 10 000 lb, and the elongation is 0.020 in. in an 8-in. gage length, determine the modulus of elasticity of the material.

7.4   If the diameter of the specimen in Problem 7.3 is 0.6245 in. under load, determine Poisson's ratio.

7.5   A tensile test machine applies a load of 20 kN to a test specimen that is 30 mm in diameter. If the elongation is 0.5 mm in a gage length of 200 mm, determine the modulus of elasticity of the material.

7.6   If the diameter of the specimen in Problem 7.5 is 29.97 mm under load, determine Poisson's ratio.

7.7   A test specimen is originally 0.505 in. in diameter and supports an ultimate load of 16 000 lb. The specimen was found to have a length at failure of 2.8 in. and an initial gage length of 2.000 in. What is its ultimate strength and percent elongation?

7.8   A test specimen is originally 12.5 mm in diameter and supports an ultimate load of 65 kN. At failure the length is found to be 70 mm. The initial gage length was 50 mm. What is the ultimate strength of the material and its percent elongation at failure?

7.9   A test specimen is originally 0.75 in. in diameter. A load of 12 000 lb causes the stress–strain curve to become nonlinear, and failure is found to occur at a load of 42 000 lb. Calculate the proportional limit of the material and the stress at failure.

7.10 A test specimen is 12.5 mm in diameter, and under load it is found that a load of 50 kN causes the stress–strain curve to deviate from linearity. Failure in the specimen is found to occur at 170 kN. What is the stress at failure, and what is the proportional limit of the material?

7.11 A ⅝-in. diameter rod is tested in tension. The force on the rod at the yield point is 14 000 lb, the ultimate load is 16 000 lb, and the rod breaks at a load of 15 000 lb. Calculate the breaking stress, the ultimate stress, and the yield stress.

7.12 If the rod in Problem 7.11 necks down to a diameter of ⅜ in. at fracture, determine the percent change in area and the true breaking stress.

7.13 A material is known to have a modulus of elasticity of 30×10^6 psi. If a rod is subjected to a 10 000-psi stress, what is its percent elongation?

7.14 A test specimen having an original diameter of 0.506 in. has a final diameter of 0.302 in. at failure. The original gage length was 2.000 in., and at failure it was found to be 2.75 in. Determine the percentage elongation and the percentage reduction in area of the specimen.

7.15 A test specimen has an original diameter of 12.5 mm. At failure its diameter is 8 mm.

                                              Mechanical Properties of Materials

The original gage length was 50 mm, and at failure it was found to be 70 mm. Determine the percent elongation and the percent reduction in area of the specimen.

7.16 The following data were obtained from a tension test:
(a) Original diameter of test specimen $= 0.505$ in.
(b) Minimum diameter after failure $= 0.312$ in.
(c) Final length at failure $= 2.79$ in.
(d) Initial gage length $= 2.000$ in.
Determine the percent reduction of area and the percent elongation.

7.17 A tensile load is applied to a $\frac{1}{2}$-in. diameter specimen. If the measured elongation of the specimen is 0.0052 in. over an 8-in. gage length, determine the unit deformation, the stress, and the applied load if E is 30×10^6 psi.

7.18 A tensile test specimen is 12.5 mm in diameter. The specimen elongates 0.0625 mm in a gage length of 200 mm. Determine the unit deformation, the stress, and the applied load if $E = 210$ GPa.

7.19 At the proportional limit it is found that a $\frac{3}{4}$-in. diameter steel bar has elongated by 0.0030 in. in a gage length of 8 in., while its diameter was reduced by 0.000084 in. If the axial load was 5000 lb, determine the modulus of elasticity and Poisson's ratio.

7.20 A bar 20 mm in diameter elongates 0.075 mm in a gage length of 200 mm at the proportional limit where the axial load is 20 kN. If its diameter is reduced by 0.002 mm, determine E and Poisson's ratio.

7.21 A stress–strain diagram for a material has a straight-line portion similar to that of steel. If a stress of 5000 psi produces a strain of 5×10^{-4} in./in. and a stress of 15 000 psi produces a strain of 15×10^{-4} in./in., determine the modulus of elasticity of the material.

7.22 A material is tested in tension, and it is found that a stress of 40 MPa produces a strain of 6×10^{-4} m/m and a stress of 110 MPa produces a strain of 20×10^{-4} m/m. If these data are taken within the proportional limit of the material, determine E.

7.23 During a tension test, a stress of 2500 psi is found to cause a strain of 8.3×10^{-5} in./in., and a stress of 10 000 psi caused a strain of 33.2×10^{-5} in./in. Assuming that the proportional limit of the material is 20 000 psi, determine the modulus of elasticity of the material.

7.24 A cylindrical test specimen has an initial diameter of 0.752 in. During the tension test of this test specimen it is found that a load of 15 000 lb is the maximum load that the specimen can carry and still maintain stress proportional to strain. Yielding occurs at a load of 20 000 lb, and failure occurs at 40 000 lb. The diameter at failure is 0.485 in. Determine the proportional limit, yield stress, breaking stress, true breaking stress, and percent change in area.

7.25 A cylindrical test specimen has an initial diameter of 20 mm. When tested in tension, it is found that the maximum load that it can carry while still maintaining stress proportional to strain is 70 kN. Yielding is found to occur at 180 kN. At failure the diameter is found to be 12.5 mm. Determine the proportional limit, yield stress, breaking stress, true breaking stress, and the percent change in area.

7.26 The following data were taken from the test of a specimen having a diameter of 0.505 in. initially and an initial gage length of 2.000 in. Plot the stress–strain curve and determine the proportional limit, the modulus of elasticity, the percent elongation, and the percent reduction of area. Also determine the ultimate strength and breaking strength based upon both the initial and the final areas of the piece. The final length of the piece is 2.79 in., and the final necked down diameter is 0.302 in.

Stress (psi)	Strain (in./in.)
0	0
5 000	0.000 167
10 000	0.000 33
20 000	0.000 66
30 000	0.000 99
35 000	0.001 20
40 000	0.002 5
50 000	0.060
60 000	0.14
46 000	0.40 fracture

7.27 A tensile test is conducted on a nonferrous alloy and the following data are obtained:

Stress (MPa)	Strain (m/m $\times$ 10^{-4})	Stress (MPa)	Strain (m/m $\times$ 10^{-4})
0	0	238	32.6
14	2	266	36.4
42	5.5	294	40.1
70	9.2	322	44.2
98	13.5	350	48.6
126	17.5	364	54.8
154	21.1	371	74.0
182	24.9	378	94.0
210	28.8		

Determine the proportional limit and the modulus of elasticity for this material by plotting the data.

7.28 A test is conducted on a specimen having an initial diameter of 12.5 mm and a gage length of 200 mm. Plot the data obtained as a stress–strain curve and determine the proportional limit and E. This test was stopped before failure occurred.

Load (MN)	Extensometer Reading (mm)	Load (MN)	Extensometer Reading (mm)
4.83	0.025	55.6	0.375
13.5	0.075	57.0	0.400
22.7	0.125	57.92	0.475
30.6	0.175	58.06	0.525
38.2	0.225	58.19	0.575
45.2	0.275	58.61	0.675
51.7	0.325		

Mechanical Properties of Materials

Center of Gravity and Moment of Inertia

8.1 INTRODUCTION

In subsequent chapters of our study it will become necessary to calculate two properties of an area, namely, the first and second moments of the area with respect to a specified axis. Because these terms bear a resemblance to similar expressions that are developed in mechanics for solid bodies, the terms *center of gravity* (or centroid) and *moment of inertia,* respectively, are commonly given them. Although an area cannot be said to have a center of gravity or inertia (since it has no mass), these terms are in such universal use that we will subsequently use them in the development of this chapter. As the expressions for center of gravity and moment of inertia are studied, the student should note that use is again made of principles developed in statics for parallel force systems.

8.2 CENTER OF GRAVITY

Recall that the weight of a body is the resultant force of attraction on each particle of the body with respect to the earth. Since each unit of mass of the body experiences the same force, and each force is parallel to the forces acting on other particles, the weight of the body is simply the resultant of a parallel force system. Weight, therefore, is a force and can be represented and treated as a vector. Like all vectors, it must have three characteristic attributes: magnitude, direction, and line of application. If the body is irregularly shaped and oriented differently in space, its weight will remain constant, and the direction of its resultant weight vector will be toward the center of the earth, but the line of action of the resultant will be in a different position with respect to the various parts of the body. Most important, however, is the fact that regardless of the

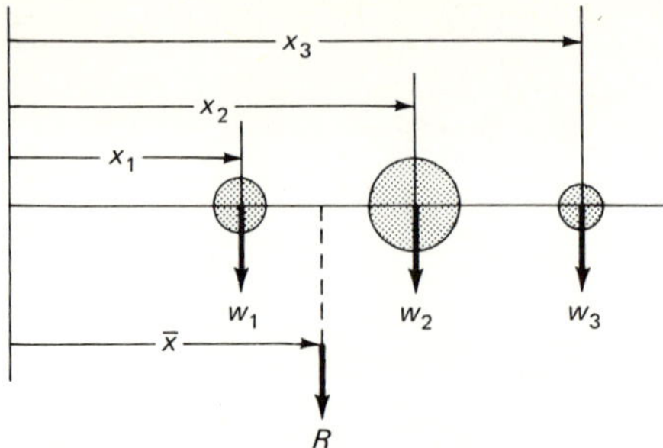

FIGURE 8.1   Center of gravity of discrete weights.

orientation of the body, the resultant weight vector (extended if necessary) will pass through a point in the body (or outside of the body) known as the *center of gravity* (or *centroid*) of the body. This point can be considered to be the point of application of the weight of the body.

Consider the system shown in Fig. 8.1, which consists of three weights w_1, w_2, and w_3, mounted on a rigid, weightless rod. We wish to determine the resultant of this system, which requires us to evaluate both its magnitude and its location along the rod. It will be noted that the arrangement shown in Fig. 8.1 consists of three parallel forces due to the effect of gravity on each of the bodies. From statics we may state that the magnitude of the resultant of a parallel force system is equal to the algebraic sum of the forces on the system. Thus the resultant R for the system shown in Fig. 8.1 can be written as

$$R = w_1 + w_2 + w_3 \tag{8.1}$$

The location of the resultant is found from the condition that the moment of the resultant about an arbitrary point equals the sum of the moments of the individual weights about this same point. For the situation being considered, a moment summation about the left end of the bar yields

$$R\bar{x} = w_1 x_1 + w_2 x_2 + w_3 x_3 \tag{8.2}$$

or

$$\bar{x} = \frac{w_1 x_1 + w_2 x_2 + w_3 x_3}{R} \tag{8.2a}$$

where $\bar{x}$ is the location of the resultant from the left end of the bar. Combining Eqs. (8.1) and (8.2) yields

$$\bar{x} = \frac{w_1 x_1 + w_2 x_2 + w_3 x_3}{w_1 + w_2 + w_3} \tag{8.2b}$$

The location of the resultant $\bar{x}$ is the location of the center of gravity of the arrangement

                                    Center of Gravity and Moment of Inertia

shown in Fig. 8.1. If we now introduce the Σ notation so that

$$\sum_{1}^{3} (wx) = w_1x_1 + w_2x_2 + w_3x_3 \tag{8.3}$$

and

$$\sum_{1}^{3} (w) = w_1 + w_2 + w_3 \tag{8.4}$$

where the symbol Σ indicates a summation of the wx terms and w terms, respectively, we can rewrite Eq. (8.2b):

$$\bar{x} = \frac{\sum_{1}^{3}(wx)}{\sum_{1}^{3}(w)} = \frac{\sum(wx)}{\sum(w)} \tag{8.5}$$

Similarly, if we consider the y axis of a body,

$$\bar{y} = \frac{\sum(wy)}{\sum(w)} \tag{8.5a}$$

If w_1 is 50 lb, w_2 is 100 lb, and w_3 is 25 lb with the corresponding distances $x_1 = 5$ in., $x_2 = 10$ in., and $x_3 = 14$ in. for the arrangement shown in Fig. 8.1, determine the location of the center of gravity.

SOLUTION

The magnitude of the resultant is equal to the algebraic sum of the weights, namely,

$$R = 50 + 100 + 25 = 175 \text{ lb}$$

The moment of each weight about the left edge of the bar is $w_1x_1 = 50(5)$, $w_2x_2 = 100(10)$, and $w_3x_3 = 25(14)$. The total moment is therefore $250 + 1000 + 350 = 1600$ in.·lb. The location of the center of gravity $\bar{x}$ is given by Eq. (8.5) as

$$\bar{x} = \frac{1600}{175} = 9.14 \text{ in.}$$

Now consider a thin body, like the one shown in Fig. 8.2, having a uniform thickness. If we freely suspend the body from point A, a plumb bob dropped from A will give us the line of action of the weight of the body passing through point A,

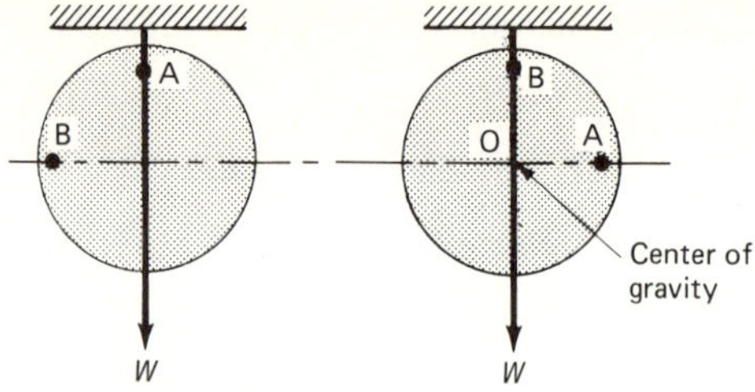

FIGURE 8.2 Center of gravity experimentally.

which we can scribe on the face of the plate. If we now suspend the plate from B and drop a plumb bob as before, we can again scribe the line of action of the weight, this time passing through point B. The intersection of these two lines gives us a point O on the face of the body, which locates the center of gravity in the face plane.

We can conclude that the first moment with respect to an axis through the centroid (center of gravity) must be zero. This is due to the fact that there must be equal positive and negative moments on either side of the centroid. If the area has an axis of symmetry, the centroid must lie on this axis (or axes).

ILLUSTRATIVE PROBLEM 8.2

A beam has the cross section shown in Fig. 8.3. Determine the location of the center of gravity of this area.

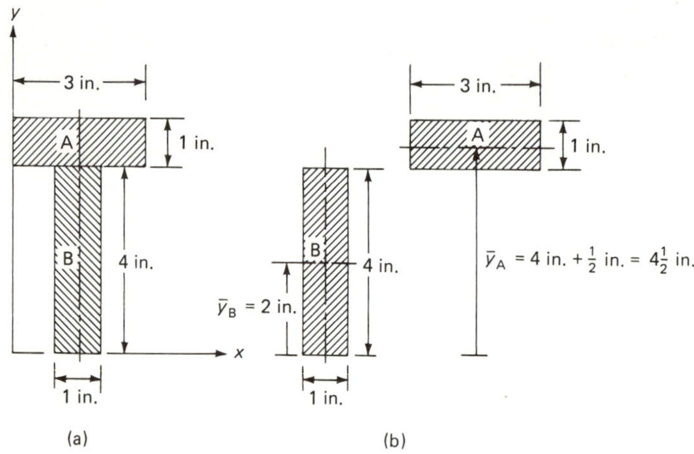

FIGURE 8.3 Illustrative Problem 8.2.

SOLUTION

Figure 8.3(a) shows the physical body divided into two rectangular parts A and B. Since the figure is symmetrical, $\bar{x}$ is located on the axis of symmetry. In Fig. 8.3(b) each part is shown separately, and the locations of the centers of gravity for both parts are shown as $\bar{y}_A$ and $\bar{y}_B$, respectively. In order to apply Eq.

         Center of Gravity and Moment of Inertia

(8.5a), we must calculate each area and its moment with respect to the y axis:

$$\text{area A} \quad A_A = 3 \times 1 = 3 \text{ in.}^2 \quad \bar{y}_A = 4 + \tfrac{1}{2} = 4\tfrac{1}{2} \text{ in.}$$

$$\text{area B} \quad A_B = 4 \times 1 = 4 \text{ in.}^2 \quad \bar{y}_B = 2 \text{ in.}$$

$$\sum A = 7 \text{ in.}^2$$

Applying Eq. (8.5a),

$$\bar{y} = \frac{A_A\bar{y}_A + A_B\bar{y}_B}{A_A + A_B} = \frac{3 \times 4\tfrac{1}{2} + 4 \times 2}{3 + 4} = 3.07 \text{ in.}$$

Determine the center of gravity of the angle shown in Fig. 8.4.

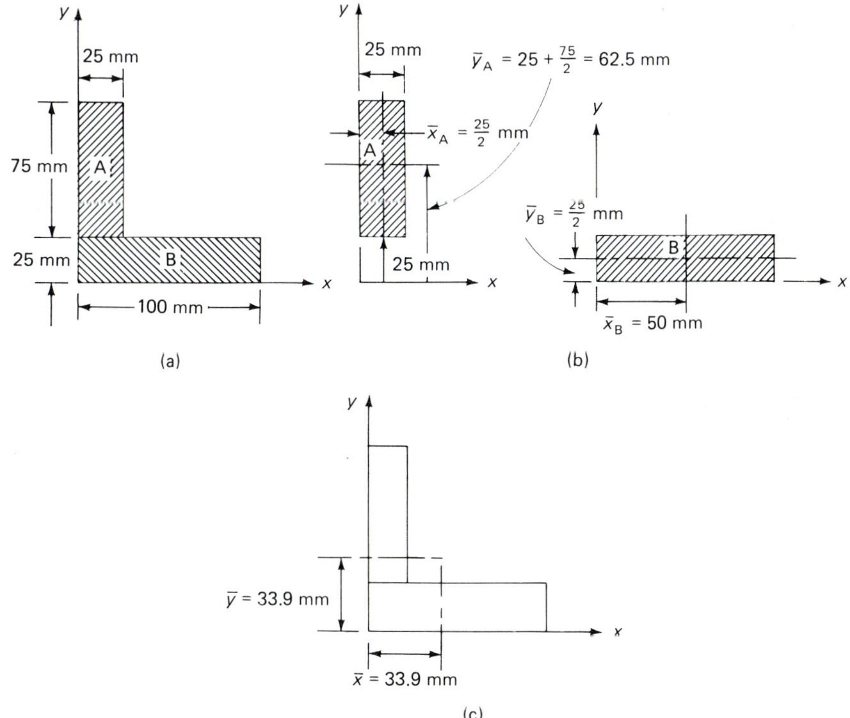

FIGURE 8.4   Illustrative Problem 8.3.

SOLUTION

Since there is no axis of symmetry, it will be necessary to determine both $\bar{x}$ and $\bar{y}$. For convenience we select the x and y reference axes as shown in

Fig. 8.4(a) to coincide with the left and bottom edges of the figure. Also, we subdivide the angle into two parts A and B, as shown. In Fig. 8.4(b) each part of the figure is shown separately, with the location of its center of gravity also shown, that is, $\bar{x}_A$, $\bar{y}_A$, $\bar{x}_B$, $\bar{y}_B$.

Let us proceed as was done in Illustrative Problem 8.2:

$$\text{area A} = A_A = 75 \times 25 = 1875 \text{ mm}^2 \qquad \bar{y}_A = 25 + \frac{75}{2} = 62.5 \text{ mm}$$

$$\text{area B} = A_B = 25 \times 100 = 2500 \text{ mm}^2 \qquad \bar{y}_B = \frac{25}{2} \text{ mm}$$

Applying Eq. (8.5a),

$$\bar{y} = \frac{A_A\bar{y}_A + A_B\bar{y}_B}{A_A + A_B} = \frac{1875 \times 62.5 + 2500 \times (25/2)}{1875 + 2500} = 33.9 \text{ mm}$$

Applying Eq. (8.5),

$$\bar{x} = \frac{A_A\bar{x}_A + A_B\bar{x}_B}{A_A + A_B} = \frac{1875 \times (25/2) + 2500 \times 50}{1875 + 2500} = 33.9 \text{ mm}$$

where $\bar{x}_A = 25/2$ mm and $\bar{x}_B = 50$ mm.

Thus far we have considered simple figures that could be built up as the sum of several rectangles. Let us now consider the case of an area with a portion cut out of it, as shown in Fig. 8.5. This figure can be thought of as consisting of the entire area A, minus the area B. For this case we can consider the area B to be negative and portray it as a vector in a direction opposite to that of the vector for the entire area. The moment of B, $A_B x_B$ is therefore the negative of the moment of A, $A_A x_A$. Thus the moment effect of the cutout is opposite to the moment effect of the solid area of the plate.

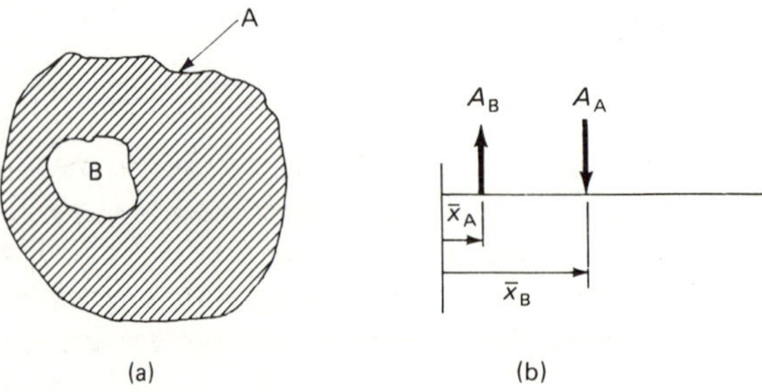

FIGURE 8.5   Area with cutout.

                          Center of Gravity and Moment of Inertia

Determine the center of gravity of the rectangle with the cutout shown in Fig. 8.6.

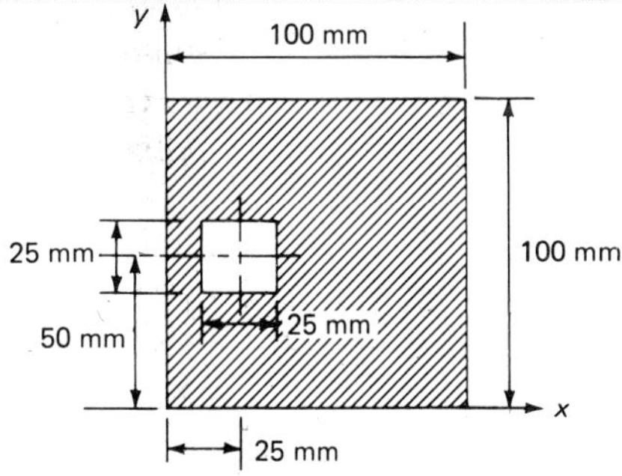

FIGURE 8.6   Illustrative Problem 8.4.

SOLUTION

Proceed as before, denoting the area of the entire plate as A_A and the area of the cutout as A_B:

$$A_A = 100 \times 100 = 10\ 000\ \text{mm}^2 \qquad \bar{x}_A = 50\ \text{mm} \qquad \bar{y}_A = 50\ \text{mm}$$

$$A_B = -(25 \times 25) = -625\ \text{mm}^2 \qquad \bar{x}_B = 25\ \text{mm} \qquad \bar{y}_B = 50\ \text{mm}$$

Applying Eqs. (8.5a) and (8.5), respectively,

$$\bar{y} = \frac{A_A\bar{y}_A + A_B\bar{y}_B}{A_A + A_B} = \frac{10\ 000 \times 50 + (-625) \times 50}{10\ 000 - 625} = 50\ \text{mm}$$

Note that the cutout is on the y axis of symmetry and that, consequently, $\bar{y}$ for the composite figure (50 mm) falls along this axis. Proceeding,

$$\bar{x} = \frac{A_A\bar{x}_A + A_B\bar{x}_B}{A_A + A_B} = \frac{10\ 000 \times 50 \times (-625) \times 25}{10\ 000 - 625} = 51.7\ \text{mm}$$

The effect of the cutout is to shift the center of gravity away from the cutout toward the uncut portion of the area.

Solve Illustrative Problem 8.2 using as reference axes the centroid of area B.

SOLUTION

Referring to Fig. 8.7, we have the following:

$$A_A = 1 \times 3 = 3\ \text{in.}^2 \qquad \bar{y}_A = 2 + \tfrac{1}{2} = 2\tfrac{1}{2} \qquad \bar{x}_A = 0$$

$$A_B = 1 \times 4 = 4\ \text{in.}^2 \qquad \bar{y}_B = 0 \qquad \bar{x}_B = 0$$

Center of Gravity

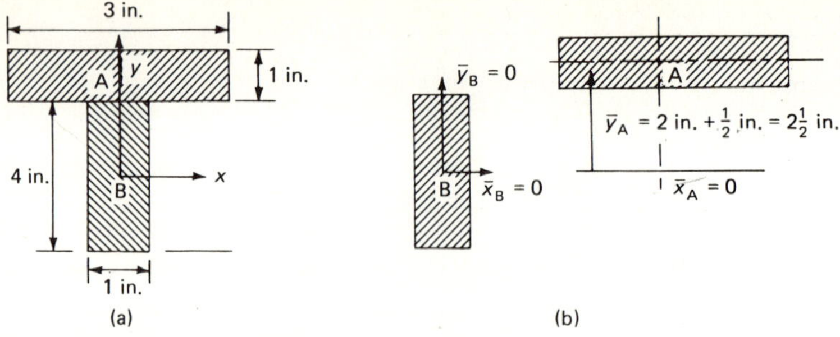

FIGURE 8.7 Illustrative Problem 8.5.

From Eq. (8.5a),

$$\bar{y} = \frac{A_A\bar{y}_A + A_B\bar{y}_B}{A_A + A_B} = \frac{3 \times 2\frac{1}{2} + 4 \times 0}{3 + 4} = 1.07 \text{ in.}$$

and for $\bar{x}$ we have $\bar{x} = 0$ by symmetry or from Eq. (8.5),

$$\bar{x} = \frac{A_A\bar{x}_A + A_B\bar{x}_B}{A_A + A_B} = \frac{3 \times 0 + 4 \times 0}{3 + 4} = 0$$

The center of gravity is therefore located 1.07 in. above the center of gravity of area B, or 2 + 1.07 = 3.07 in. from the bottom of the figure. This result is the same as that obtained from Illustrative Problem 8.2. Obviously the choice of axis does not change the result. It can, however, simplify the calculations for a particular problem if some preliminary thought is given to the selection of the reference axis before starting the problem.

Thus far we have treated simple rectangular shapes. In principle it is possible to divide the most complex figure into a number of small areas sufficient to obtain the center of gravity of the figure. This procedure is most readily carried out using the methods of calculus. Since most shapes can be considered to consist of simple areas consisting of rectangles, triangles, circles, and so on, the results for these figures are given in Table 8.1.

ILLUSTRATIVE PROBLEM 8.6

Determine the location of the centroid of a quarter circle, with a 6-in. radius with respect to the x and y axes, as shown in Fig. 8.8.

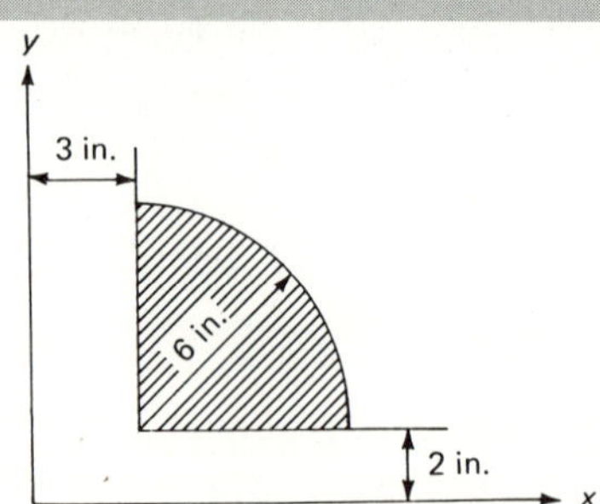

FIGURE 8.8 Illustrative Problem 8.6.

   Center of Gravity and Moment of Inertia

TABLE 8.1 Properties of Areas

Section	Area	$\bar{x}$	$\bar{y}$
Square	b_2	$\dfrac{b}{2}$	$\dfrac{b}{2}$
Rectangle	bh	$\dfrac{b}{2}$	$\dfrac{h}{2}$
Triangle	$\frac{1}{2}\,bh$	$\dfrac{b}{3}$	$\dfrac{h}{3}$
Semicircle	$\dfrac{\pi R^2}{2}$	0	$\dfrac{4R}{3\pi}$ or $0.4244R$
Quarter circle	$\dfrac{\pi R^2}{4}$	$\dfrac{4R}{3\pi}$ or $0.4244R$	$\dfrac{4R}{3\pi}$ or $0.4244R$
Quadrant of ellipse	$\dfrac{\pi ab}{4}$	$\dfrac{4a}{3\pi}$ or $0.4244a$	$\dfrac{4b}{3\pi}$ or $0.4244b$
Spandrel	$\dfrac{bh}{n+1}$	$\dfrac{b}{n+2}$	$\left\{\dfrac{n+1}{4n+2}\right\}h$

SOLUTION

From Table 8.1, $\bar{x} = \bar{y} = 0.4244R$ for the quarter circle. Thus for the pres-

Center of Gravity

ent problem, $\bar{x} = \bar{y} = 6(0.4244) = 2.55$ in. From the x reference axis, $\bar{y} = 2 + 2.55 = 4.55$ in., and from the y reference axis, $\bar{x} = 3 + 2.55 = 5.55$ in.

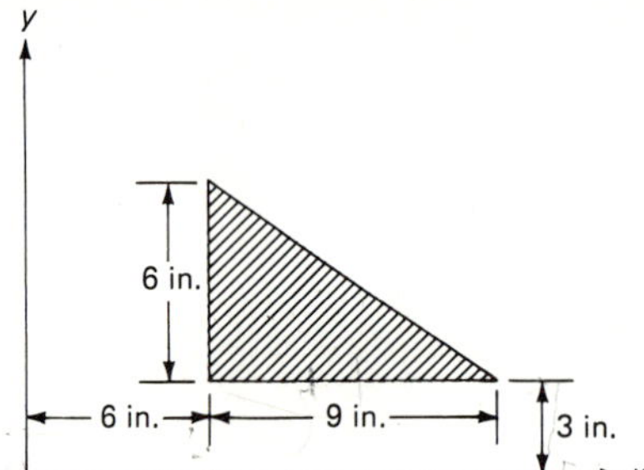

FIGURE 8.9   Illustrative Problem 8.7.

SOLUTION

From Table 8.1 we have $\bar{x} = b/3 = 9/3 = 3$ in. For $\bar{y}$ we have $\bar{y} = h/3 = 6/3 = 2$ in. With respect to the reference axes these become $\bar{x} = 6 + 3 = 9$ in. and $\bar{y} = 3 + 2 = 5$ in.

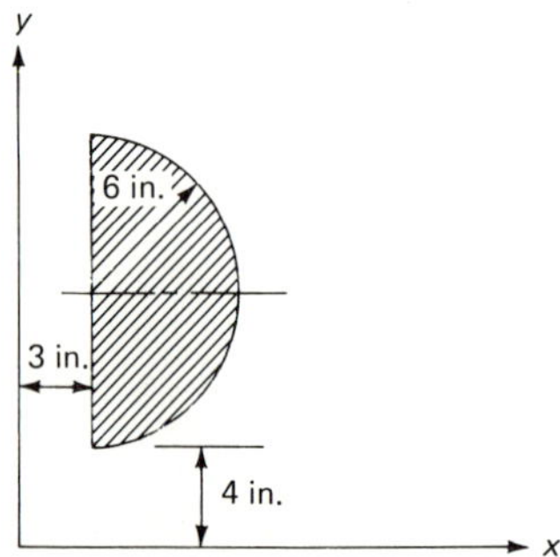

FIGURE 8.10   Illustrative Problem 8.8.

                    Center of Gravity and Moment of Inertia

SOLUTION

Using the data of Table 8.1, $\bar{x} = 0.4244R = 0.4244(6) = 2.55$ in. $\bar{y}$ is located on the axis of symmetry, 3 in. above the lowest point of the semicircle. With respect to the reference axes, $\bar{x} = 3 + 2.55 = 5.55$ in. and $\bar{y} = 3 + 4 = 7$ in.

8.3   CENTER OF GRAVITY OF COMPOSITE AREAS

Using the methods of the preceding section, we have seen how we can evaluate the center of gravity of several simple areas. The use of Table 8.1 enables us to extend these concepts to more complex plane areas. The procedure in such calculations consists of dividing the figure into simple areas, which are given in Table 8.1, and then selecting a convenient set of reference axes. Most calculations of this type are best handled in a tabular manner, as is shown in Illustrative Problem 8.9.

ILLUSTRATIVE PROBLEM 8.9

Locate the center of gravity of Fig. 8.11.

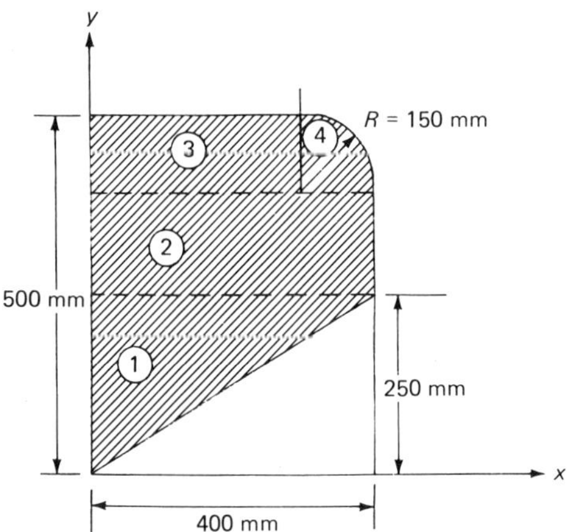

FIGURE 8.11   Illustrative Problem 8.9.

SOLUTION

Let us subdivide the figure into four parts, a triangle 400 mm × 250 mm, a rectangle 100 mm × 400 mm, a rectangle 150 mm × 250 mm, and a quarter circle with a 150-mm radius. We will also select the x and y axes as shown. For each subdivision (numbered ①, ②, ③, and ④) we will tabulate the areas $\bar{x}$, $\bar{y}$, $A\bar{x}$, and $A\bar{y}$, as is done in Table 8.2. Note that the area of the quarter circle is $\pi R^2/4$, and the center of gravity is $0.4244R$ plus the distance of the base of the figure from the axis. Also note the utility of using the tabular method as was done in this problem.

TABLE 8.2   Solution of Illustrative Problem 8.9

Section	Description	Area A	$\bar{x}$	$\bar{y}$	$\dfrac{A\bar{x}}{10^6}$	$\dfrac{A\bar{y}}{10^6}$
①		50 000	$\dfrac{400}{3}$	$\dfrac{2}{3}(250)$	6.67	8.33
②		40 000	200	300	8.00	12.00
③		37 500	125	425	4.69	15.94
④		17 671	313.66	413.66	5.54	7.31

$$\Sigma A = 145\ 171 \qquad \Sigma A\bar{x} = 24.9 \times 10^6 \qquad \Sigma A\bar{y} = 43.58 \times 10^6$$

$$\bar{x} = \frac{\Sigma A\bar{x}}{\Sigma A} = \frac{24.9 \times 10^6}{145\ 171} = 171.5 \text{ mm} \qquad \bar{y} = \frac{43.58 \times 10^6}{145\ 171} = 300.2 \text{ mm}$$

ILLUSTRATIVE PROBLEM 8.10

Locate the center of gravity of the figure used in Illustrative Problem 8.9 with the subdivisions shown in Fig. 8.12.

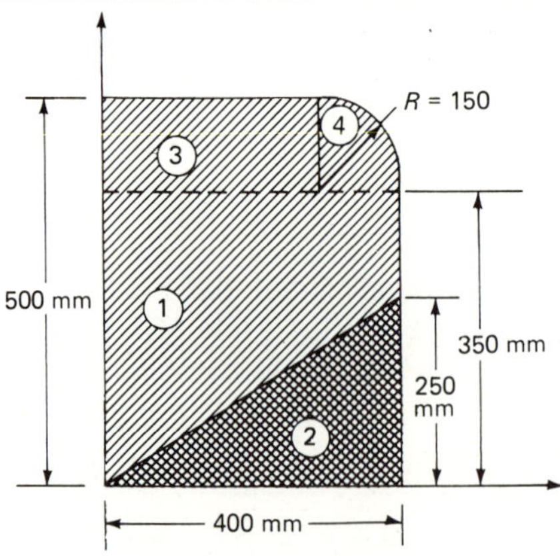

FIGURE 8.12   Illustrative Problem 8.10.

   Center of Gravity and Moment of Inertia

SOLUTION

In this alternative solution we have chosen the following four subdivisions: a rectangle 350 mm × 400 mm, a triangle cutout 250 mm × 400 mm, a rectangle 400 mm × 250 mm, and a quarter circle with a 150-mm radius. In this solution the cutout is treated as a negative area. As is to be expected, both approaches yield the same solutions for $\bar{x}$ and $\bar{y}$. The choice of method is up to the student and is based upon the procedure that is most comfortable to him or her.

TABLE 8.3 Solution of Illustrative Problem 8.10

Section	Description	Area A	$\bar{x}$	$\bar{y}$	$\dfrac{A\bar{x}}{10^6}$	$\dfrac{A\bar{y}}{10^6}$
①		140 000	200	175	28.0	24.5
②		$-50\ 000$	266.67	833	-13.33	-4.17
③		37 500	125	425	4.69	15.94
④		17 671	313.66	413.66	5.54	7.31

$$\Sigma A = 145\ 171 \qquad \Sigma A\bar{x} = 24.9 \times 10^6 \qquad \Sigma A\bar{y} = 43.58 \times 10^6$$

$$\bar{x} = \frac{\Sigma A\bar{x}}{\Sigma A} = \frac{24.9 \times 10^6}{145\ 171} = 171.5 \text{ mm} \qquad \bar{y} = \frac{\Sigma A\bar{y}}{\Sigma A} = \frac{43.58 \times 10^6}{145\ 171} = 300.2 \text{ mm}$$

8.4 THE SECOND MOMENT OF AREA—MOMENT OF INERTIA

When Newton's second law of motion is applied to the angular motion of rigid bodies, one of the terms that is developed is known as the *moment of inertia of the body*. As we will see later in our study of the stresses and deflections in beams and shafts, an analogous term appears for the property of an area that is also called the *moment of inertia of the area*. Since area has no mass, this term too is a misnomer; and since there is no physical analogy for this property of an area, we will simply resort to

The Second Moment of Area—Moment of Inertia

evaluating it based upon its mathematical definition. The moment of inertia or more properly, the second moment of area is defined as

$$I_x = \Sigma \, y^2 \Delta A \tag{8.6}$$

where I, the moment of inertia, has the physical units of (length)^4. In every case it is necessary to refer the moment of inertia to an axis. We can show this very clearly by referring to Fig. 8.13, where $I_x = y^2 \Delta A$ and $I_y = x^2 \Delta A$. In this case the axes in question are in the plane of the area (the page), and the moment of inertia as defined by Eq. (8.6) is known as the *rectangular moment of inertia*. Where the axis is perpendicular to the plane of the area, the resulting moment of inertia with respect to this axis is known as the *polar moment of inertia*. For example, let us consider an axis passing through O and perpendicular to the plane of the paper. (In mathematics this axis is the z axis.) The polar moment of inertia J_O is defined as

$$J_O = r^2 \Delta A \tag{8.7}$$

Note, however, that $r^2 = x^2 + y^2$. Therefore

$$J_O = x^2 \Delta A + y^2 \Delta A = I_x + I_y \tag{8.8}$$

We can express the result of Eq. (8.8) as follows. The polar moment of inertia of an area about an axis perpendicular to the plane of the area is equal to the sum of the rectangular moments of inertia of any two mutually perpendicular axes in the plane of the area whose intersection lies on the polar axis.

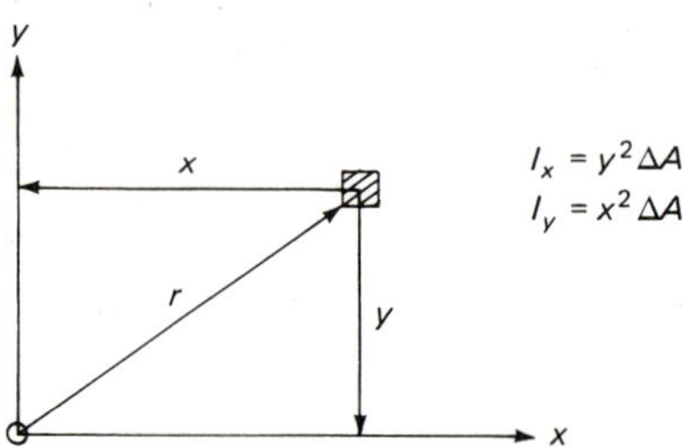

FIGURE 8.13   Definition of moment of inertia.

The rectangular moment of inertia defined by Eq. (8.6) will appear in those chapters of this book that deal with the stresses and deflections of beams, as well as in the chapters dealing with stresses in columns subjected to eccentric loads. In addition, the polar moment of inertia will be used when dealing with eccentrically loaded welds, and we shall also find it again when we study torsion.

Since the evaluation of the moment of inertia of even simple shapes involves the use of calculus, we will not attempt to derive any equations for the moment of inertia of these shapes. Table 8.4, however, gives the results of such analytical evaluations for most shapes of technical interest.

     Center of Gravity and Moment of Inertia

Shape	Moment of Inertia	Radius of Gyration
Rectangle	$\bar{I}_x = \dfrac{bh^3}{12}$ $I_x = \dfrac{bh^3}{3}$	$\bar{k}_x = \dfrac{h}{\sqrt{12}}$ $k_x = \dfrac{h}{\sqrt{3}}$
Any triangle	$\bar{I}_x = \dfrac{bh^3}{36}$ $I_x = \dfrac{bh^3}{12}$	$\bar{k}_x = \dfrac{h}{\sqrt{18}}$ $k_x = \dfrac{h}{\sqrt{6}}$
Circle	$\bar{I}_x = \dfrac{\pi r^4}{4}$ $J_o = \dfrac{\pi r^4}{2}$	$\bar{k}_x = \dfrac{r}{2}$ $\bar{k}_o = \dfrac{r}{\sqrt{2}}$
Semicircle	$I_x = \bar{I}_y = \dfrac{\pi r^4}{8}$ $\bar{I}_x = 0.11 r^4$	$k_x = \bar{k}_y = \dfrac{r}{2}$ $k_x = 0.264 r$
Quarter circle	$I_x - I_y - \dfrac{\pi r^4}{16}$ $\bar{I}_x = \bar{I}_y = 0.055 r^4$	$k_x = k_y = \dfrac{r}{2}$ $\bar{k}_x = \bar{k}_y = 0.264 r$
Ellipse	$\bar{I}_x = \dfrac{\pi a b^3}{4}$ $\bar{I}_y = \dfrac{\pi b a^3}{4}$	$\bar{k}_x = \dfrac{b}{2}$ $\bar{k}_y = \dfrac{a}{2}$

8.5 THE PARALLEL AXIS TRANSFER THEOREM

The most useful reference axis for the moment of inertia is an axis through the center of gravity of the area. The symbol $\bar{I}_x$ is used to denote the moment of inertia of the area about an axis passing through the center of gravity of the area and parallel to the x axis. It is possible to obtain the moment of inertia of an area about any axis parallel to an axis passing through the center of gravity by considering the following situation. Consider any shaped figure with the $\bar{x}$ axis as shown in Fig. 8.14. Let us assume that the moment of inertia about this axis is known, that is, $\bar{I}_x = \Sigma y^2 \Delta A$, and that we desire to find the moment of inertia about the axis x, parallel to the $\bar{x}$ axis, and a distance d from it. The expression we seek is

$$I_x = \Sigma (y + d)^2 \Delta A \tag{8.9}$$

Expanding the terms in Eq. (8.9),

$$I_x = \Sigma y^2 \Delta A + \Sigma 2 dy \Delta A + \Sigma d^2 \Delta A \tag{8.10}$$

In the second term we may remove d from the summation since a constant distance separates the two axes and we can also remove d^2 from the summation of the third term. Thus

$$I_x = \Sigma y^2 \Delta A + 2d \Sigma y \Delta A + d^2 \Sigma \Delta A \tag{8.11}$$

Let us now examine each term on the right-hand side of Eq. (8.11). The first term is the definition of $\bar{I}_x$; the second term contains the first moment of area about the center of gravity, and by definition it must be zero; the last summation simply yields the total area. Therefore we have

$$I_x = \bar{I}_x + Ad^2 \tag{8.12}$$

By analogous reasoning,

$$J_o = \bar{J}_o + Ad^2 \tag{8.12a}$$

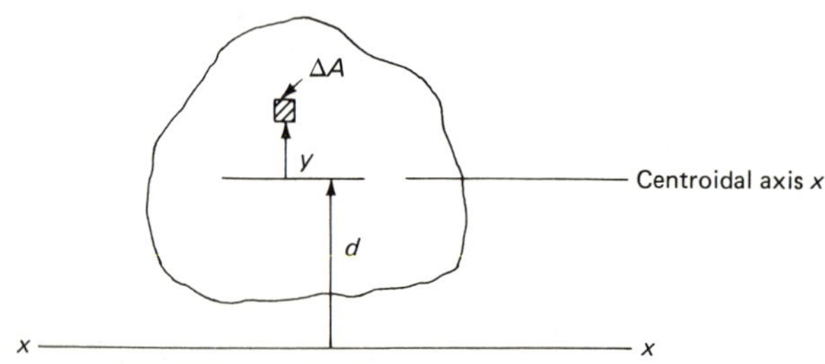

FIGURE 8.14   The parallel axis transfer theorem.

Equation (8.12) is known as the *parallel axis transfer theorem,* and it expresses the fact that the moment of inertia of an area with respect to any rectangular axis is equal to the moment of inertia of the area about a parallel centroidal axis plus the product of the area multiplied by the square of the distance separating the axes. The

                                    Center of Gravity and Moment of Inertia

student will note that $I_x > \bar{I}_x$, which indicates that the minimum moment of inertia of an area with respect to any axis in a given direction occurs when the axis passes through the center of gravity of the area.

SOLUTION

Table 8.4 gives us

$$\bar{I}_x = \frac{bh^3}{12}$$

Since the area of the rectangle is bh and we are moving the reference axis from the midpoint of h to the base, Eq. (8.12) gives us

$$I_x = \bar{I}_x + Ad^2 = \frac{bh^3}{12} + bh\left(\frac{h}{2}\right)^2$$

Simplifying,

$$I_x = \frac{bh^3}{3}$$

This result agrees with the value given in Table 8.4.

8.6 THE MOMENT OF INERTIA OF COMPOSITE AREAS

By the process of subdividing a body into sufficiently small areas, it is possible to calculate the moment of inertia of an area about any axis. However, most areas can be subdivided into squares, rectangles, circles, and so on, and for these shapes analytical expressions exist for the rectangular moment of inertia about axes through the center of gravity of the area. Table 8.4 gives the moment of inertia for various geometric shapes about both the centroidal axis and a convenient parallel axis.

In dealing with composite bodies, we find it convenient to use Table 8.4 with the caution that, before the moment of inertia is determined as the sum of the various shapes that comprise the figure, it is necessary that the moments of inertia of all shapes be found with respect to the same axis.

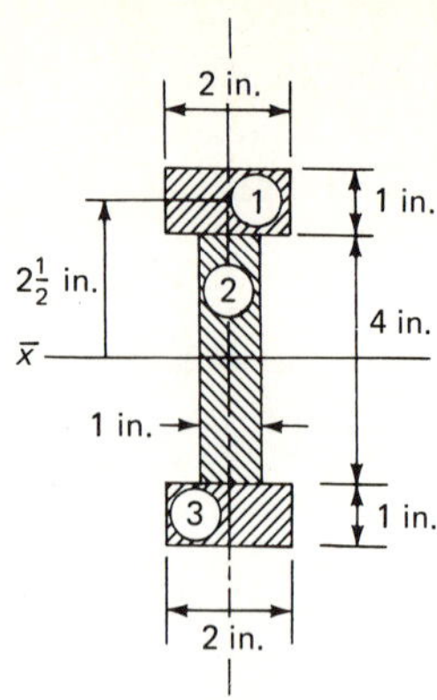

FIGURE 8.15   Illustrative Problem 8.12.

SOLUTION

Consider area ①; $\bar{I}_x = bh^3/12 = 2(1)^3/12 = \frac{1}{6}$. Transferred to the $\bar{x}$ axis of the figure,

$$I_{1\bar{x}} = \frac{1}{6} + Ad^2 = \frac{1}{6} + 2(2\frac{1}{2})^2 = \frac{1}{6} + 12.5 = 12.67 \text{ in.}^4$$

By symmetry, area ③ contributes an equal amount to the moment of inertia of the composite figure about the $\bar{x}$ axis, namely, 12.67 in.^{4}. For area ② its center of gravity coincides with the center of gravity of the composite figure. Thus its moment of inertia about this axis is $bh^3/12 = 1(4)^3/12 = 5.33$. Summing,

$$\bar{I}_x = 12.67 + 12.67 + 5.33 = 30.67 \text{ in.}^4$$

ILLUSTRATIVE PROBLEM 8.13

Determine the moment of inertia of the box section shown in Fig. 8.16 about its x centroidal axis.

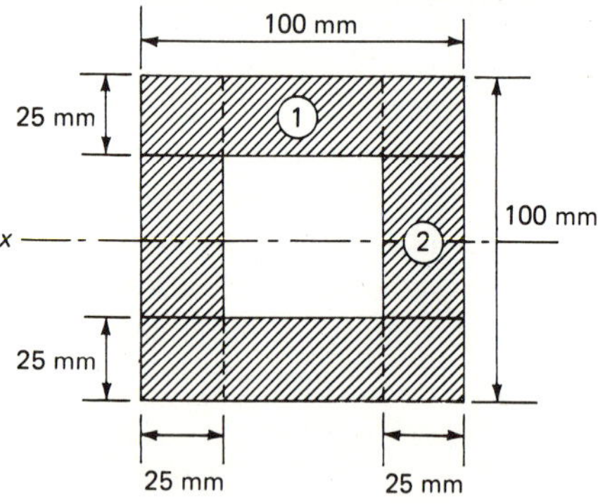

FIGURE 8.16   Illustrative Problem 8.13.

SOLUTION

Let us consider this problem to consist of a square with a square cutout. Thus for the 100-mm × 100-mm square,

   Center of Gravity and Moment of Inertia

$$\bar{I}_x = \frac{bh^3}{12} = \frac{100 \times 100^3}{12} = 8.333 \times 10^6 \text{ mm}^4$$

For the 50-mm × 50-mm cutout,

$$\bar{I}_x = \frac{-bh^3}{12} = \frac{-50 \times 50^3}{12} = -0.513 \times 10^6 \text{ mm}^4$$

where the negatives are due to the negative area. Then

$$\bar{I}_{x \text{ entire figure}} = 7.812 \times 10^6 \text{ mm}^4$$

As an alternate solution, let us calculate the moment of inertia as the sum of each of the four rectangular parts. For ①,

$$I = \frac{bh^3}{12} + Ad^2 = \frac{100 \times 25^3}{12} + 100 \times 25 \left(25 + \frac{25}{2}\right)^2$$

$$= 3.646 \times 10^6 \text{ mm}^4$$

For ②,

$$I = \frac{bh^3}{12} = \frac{25 \times 50^3}{12} = 0.260 \times 10^6 \text{ mm}^4$$

Since there are two vertical rectangles and two horizontal rectangles,

$$\bar{I}_x = 2(3.646 + 0.260) \times 10^6 = 7.812 \times 10^6 \text{ mm}^4$$

Determine the moment of inertia of the composite figure with a semicircular cutout with respect to the x axis as shown in Fig. 8.17.

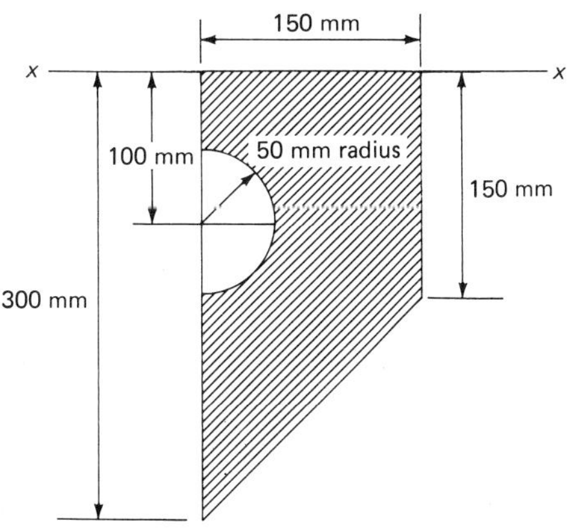

FIGURE 8.17   Illustrative Problem 8.14.

The Moment of Inertia of Composite Areas

SOLUTION

Let us divide the area into three shapes: a 150-mm $\times$ 150-mm square, a 150-mm $\times$ 150-mm triangle, and a semicircular cutout with a 50-mm radius. For the square,

$$I_x = \frac{bh^3}{3} = \frac{150 \times 150^3}{3} = 168.75 \times 10^6 \text{ mm}^4$$

For the triangle,

$$I_x = \bar{I}_x + Ad^2 = \frac{bh^3}{36} + \frac{1}{2}bh \times 200^2$$

$$= \frac{150 \times 150^3}{36} + \frac{1}{2}150 \times 150 \times 200^2$$

$$= 464.063 \times 10^6 \text{ mm}^4$$

For the semicircle,

$$I_x = \bar{I}_x + Ad^2 = \frac{\pi r^4}{8} + \frac{\pi r^2}{2} \times 100^2 = \frac{\pi \times 50^4}{8} + \frac{\pi \times 50^2}{2} 100^2$$

$$= -41.724 \times 10^6 \text{ mm}^4 \qquad \text{(negative)}$$

For the composite figure,

$$I_x = 168.75 \times 10^6 + 464.063 \times 10^6 - 41.724 \times 10^6$$

$$= 591.089 \times 10^6 \text{ mm}^4$$

8.7 THE RADIUS OF GYRATION

Later, in our study of columns, we shall use the term *radius of gyration*. The radius of gyration is defined as that length that, when squared and then multiplied by the area, will give us the moment of inertia of the area with respect to a given axis. Thus,

$$Ak^2 = I \tag{8.13}$$

$$Ak_x^2 = I_x \tag{8.13a}$$

$$Ak_y^2 = I_y \tag{8.13b}$$

$$A\bar{k}_x^2 = \bar{I}_x \tag{8.13c}$$

For a parallel axis we can write

$$I_x = \bar{I}_x + Ad^2 = A\bar{k}_x^2 + Ad^2 = Ak_x^2$$

Center of Gravity and Moment of Inertia

or

$$k_x^2 = \bar{k}_x^2 + d^2 \tag{8.14}$$

For a polar axis we can similarly obtain

$$J_o = Ak_o^2 = \bar{J}_o + Ad^2 = A\bar{k}_o^2 + Ad^2$$

and

$$k_o^2 = \bar{k}_o^2 + d^2 \tag{8.15}$$

ILLUSTRATIVE PROBLEM 8.15

Determine the radius of gyration about the x axis for Fig. 8.17 (Illustrative Problem 8.14).

SOLUTION

The area of the figure is

$$A = 150 \times 150 + \frac{150 \times 150}{2} - \frac{\pi}{2} \times 50^2 = 29\,823 \text{ mm}^2$$

Since $I_x = 591.089 \times 10^6$, $Ak_x^2 = 591.089 \times 10^6 \text{ mm}^4$

$$k_x = \sqrt{\frac{591.089 \times 10^6}{29\,823}} = 140.78 \text{ mm}$$

If we wish to give the radius of gyration a physical meaning, it may be useful for us to think of it as that distance from an axis at which all of an area is concentrated, resulting in the same moment of inertia about the axis as the actual area. With this in mind, the student should note that such a concentrated area will not have the same first moment about the given axis as does the actual area.

Equation (8.15) also leads us to the conclusion that the radius of gyration is always greater than the distance from the reference axis to the center of gravity of the area. Also, the symbol k has been used for radius of gyration rather than the AISC symbol r to avoid confusion with the common use of r for radius.

ILLUSTRATIVE PROBLEM 8.16

Determine the location of the center of gravity of the section in Fig. 8.17. Compare this result with that obtained for the radius of gyration in Illustrative Problem 8.15.

SOLUTION

Using the same three shapes and the x axis shown as the reference axis,

$$\bar{x} = \frac{150 \times 150 \times 75 + \tfrac{1}{2}150 \times 150 \times 200 - \tfrac{1}{2}\pi \times 50^2 \times 100}{150 \times 150 + \tfrac{1}{2}150 \times 150 - \tfrac{1}{2}\pi \times 50^2}$$

$$= 118.86 \text{ mm}$$

Closure

In Illustrative Problem 8.15 we found $k_x = 140.78$ mm. Comparison with the results above shows k_x to be greater than $\bar{x}$, as it must be.

8.8 CLOSURE

At this point in our study of "strength of materials," it may be difficult for the student to place a physical interpretation on the material covered in this chapter. If this material is thought of as a tool that will be required at a later time and background material that must be understood before further study can be undertaken, it may provide the student the rationale for mastering it. The mathematical definitions given in this chapter for the various properties discussed can be utilized without resorting to a physical model for these properties. However, where possible, an attempt has been made to relate the mathematical definitions to physical quantities. For the present, the student will have to accept the fact that we shall make use of all of these items at some later time in our studies, and the student should make every attempt to understand them and to be able to apply them when they are needed. In this chapter the student will note that the solution of a particular problem often involves a great deal of arithmetical computation. An orderly, systematic approach to each solution will help minimize errors in calculations.

REFERENCES

Bassin, M. E., S. M. Brodsky, and H. Wolkoff, STATICS AND STRENGTH OF MATERIALS, 3rd ed. New York: McGraw-Hill, 1979.

Higdon, A. E., and W. B. Stiles, ENGINEERING MECHANICS, 3rd ed. Englewood Cliffs, NJ: Prentice-Hall, 1968.

Higdon, A., E. E. Ohlsen, and W. B. Stiles, MECHANICS OF MATERIALS, 3rd ed. New York: John Wiley, 1976.

Levinson, I. J., INTRODUCTION TO MECHANICS, 2nd ed. Englewood Cliffs, NJ: Prentice-Hall, 1968.

Meriam, J. L., MECHANICS. New York: John Wiley, 1975.

Olsen, G. A., STRENGTH OF MATERIALS. Englewood Cliffs, NJ: Prentice-Hall.

Shames, I., ENGINEERING MECHANICS, 2 vols, 2nd ed. Englewood Cliffs, NJ: Prentice-Hall, 1966–67.

Singer, F. L., ENGINEERING MECHANICS, 2 vols, 3rd ed. New York: Harper and Row, 1975.

Timoshenko, S., and D. H. Young, ELEMENTS OF STRENGTH OF MATERIALS, 5th ed. New York: D. Van Nostrand, 1968.

PROBLEMS

8.1 Determine the location of the center of gravity with respect to the x axis shown.

 Center of Gravity and Moment of Inertia

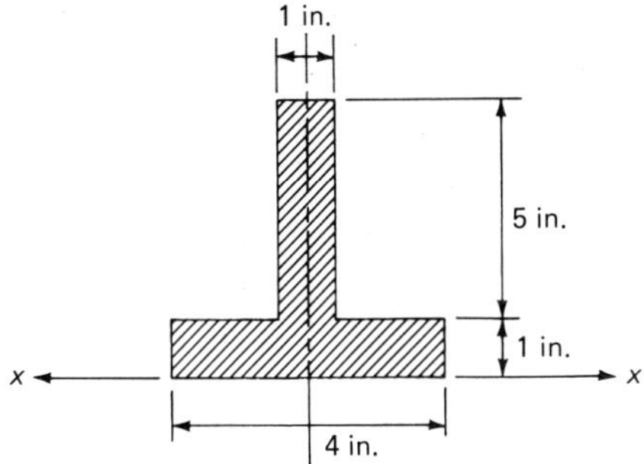

FIGURE P 8.1

8.2   Locate the center of gravity of the section shown.

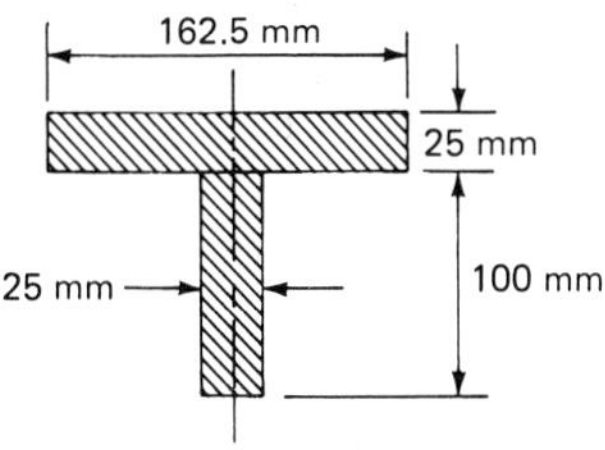

FIGURE P 8.2

8.3   If the section shown in Fig. P 8.2 is inverted, determine the location of its center of gravity.

8.4   Locate the center of gravity of the section shown.

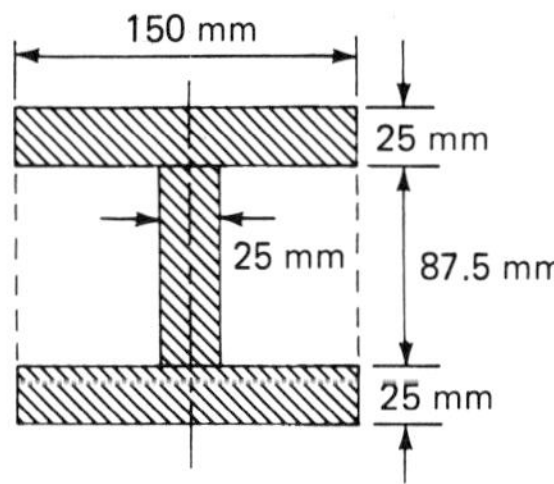

FIGURE P 8.4

8.5   Determine the location of the center of gravity with respect to the x and y axes for the sections shown.

Problems   211

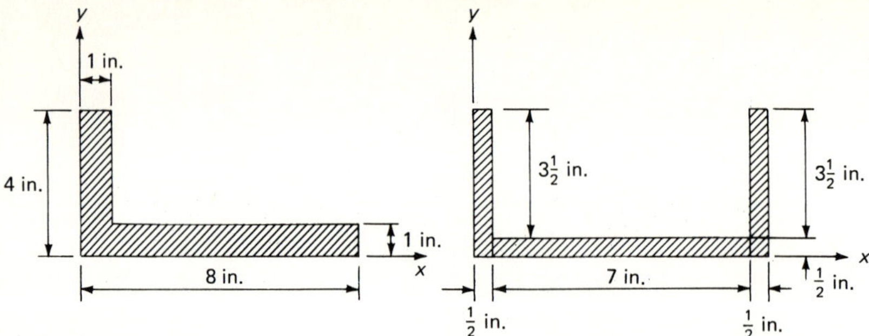

FIGURE P 8.5

8.6   Determine the location of the center of gravity with respect to both the x and the y axes shown.

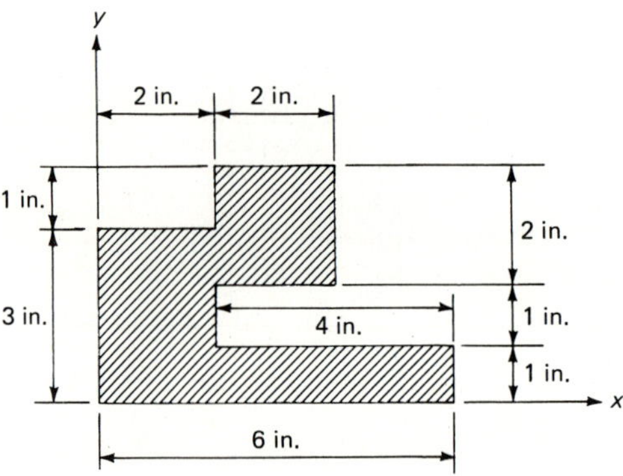

FIGURE P 8.6

8.7   Locate the center of gravity of the area shown with respect to an axis parallel to the base.

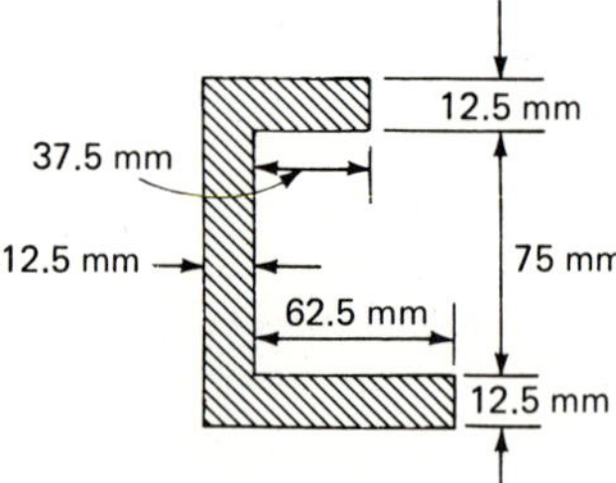

FIGURE P 8.7

8.8   Locate the center of gravity of Fig. P 8.7 with respect to an axis coinciding with the left-hand side of the figure.

8.9   Determine the center of gravity with respect to an axis coinciding with the base of the figure shown.

                    Center of Gravity and Moment of Inertia

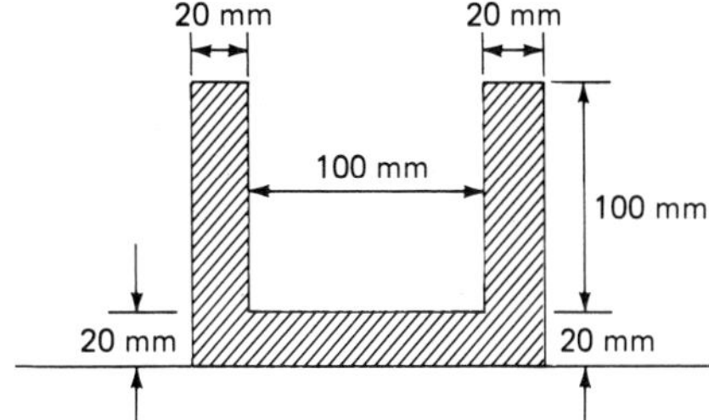

FIGURE P 8.9

8.10 Determine the location of the center of gravity with respect to both the x and the y axes shown.

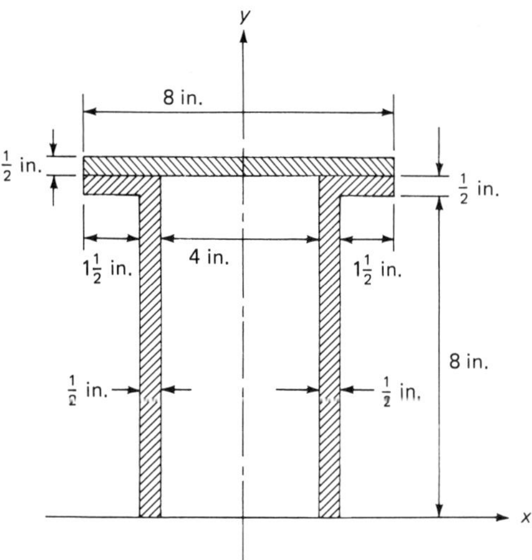

FIGURE P 8.10

8.11 Determine the location of the center of gravity with respect to both the x and the y axes shown.

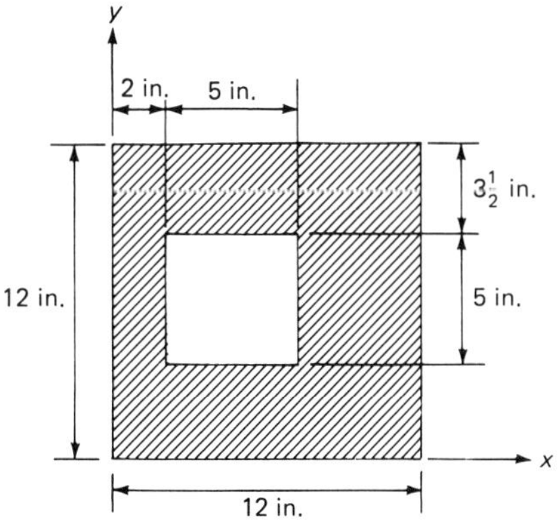

FIGURE P 8.11

Problems

8.12 Determine the location of the center of gravity with respect to both the *x* and the *y* axes shown.

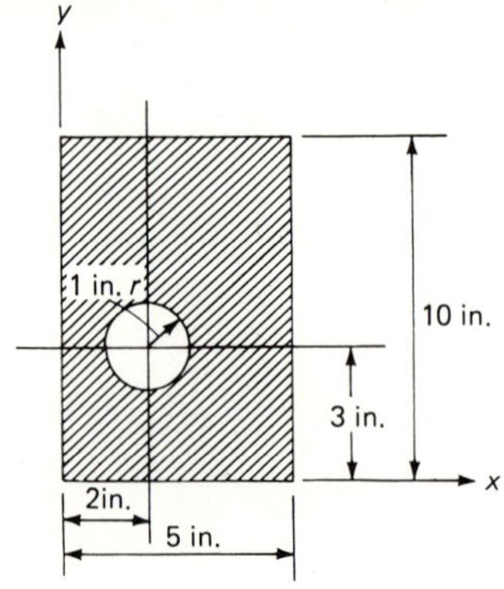

FIGURE P 8.12

8.13 Determine the location of the center of gravity with respect to both the *x* and the *y* axes shown.

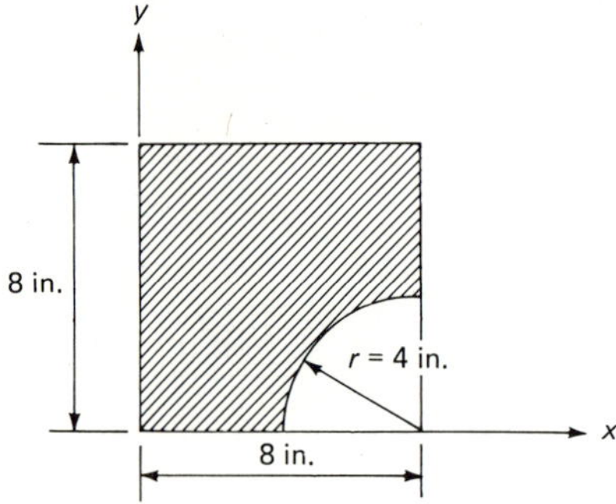

FIGURE P 8.13

8.14 Determine the location of the center of gravity with respect to the *x* axis shown.

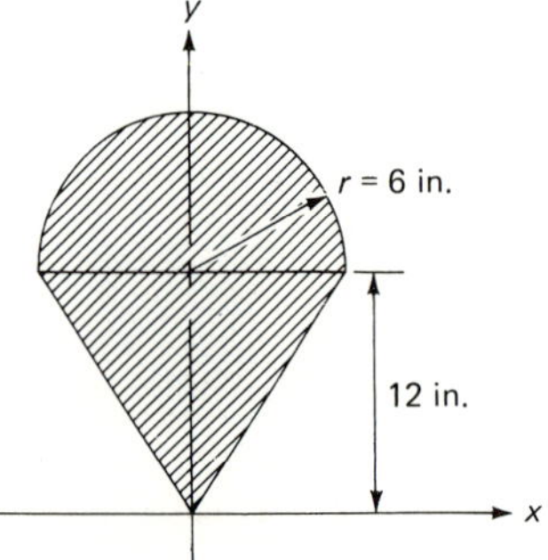

FIGURE P 8.14

        Center of Gravity and Moment of Inertia

8.15 Determine the location of the center of gravity with respect to both the x and the y axes shown.

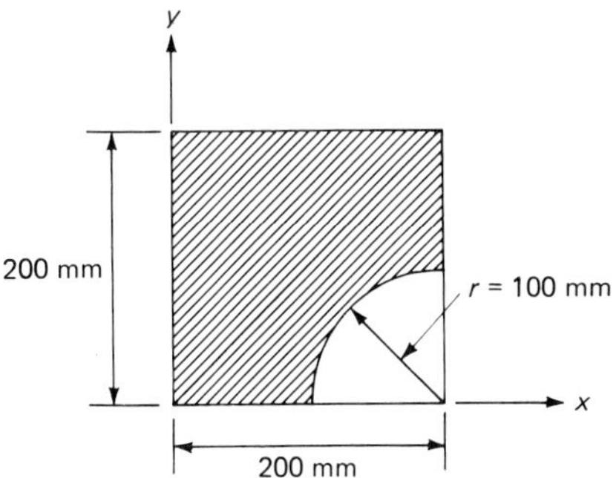

FIGURE P 8.15

8.16 Determine the radius of the semicircle that will cause the center of gravity of the section shown to lie on the y axis.

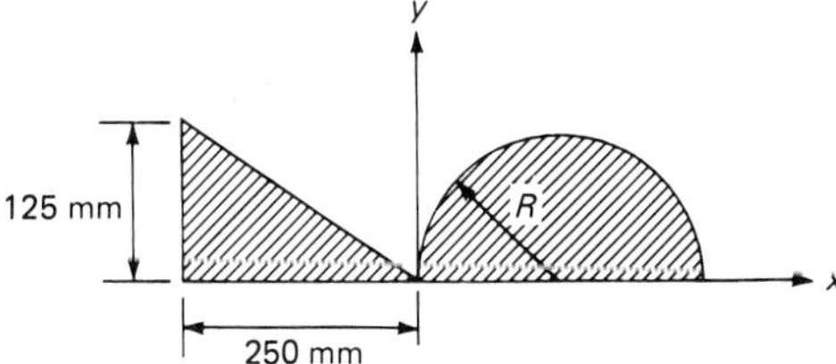

FIGURE P 8.16

8.17 Determine the moment of inertia for the section shown with respect to both the x and the y axes.

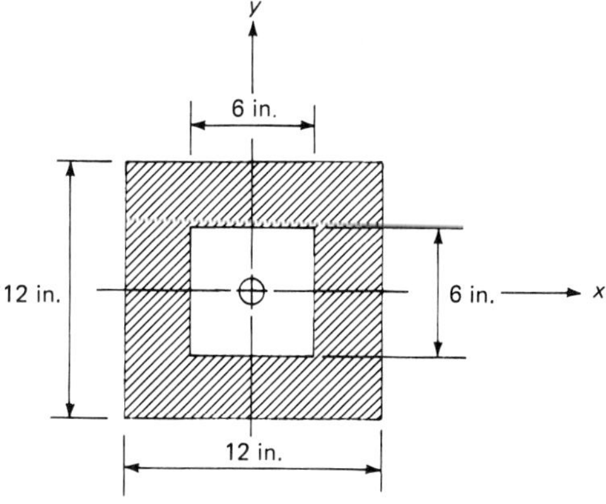

FIGURE P 8.17

8.18 Determine the location of the center of gravity with respect to the x axis shown.

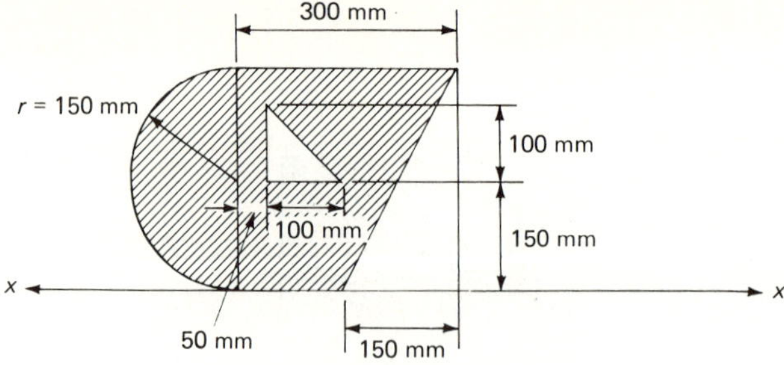

FIGURE P 8.18

8.19 A solid circular member is used as a beam. If it replaces a square 6-in. section and we desire to maintain the same moment of inertia, calculate the required diameter of the circular section.

8.20 Determine the moment of inertia of the T section shown with respect to its x centroidal axis.

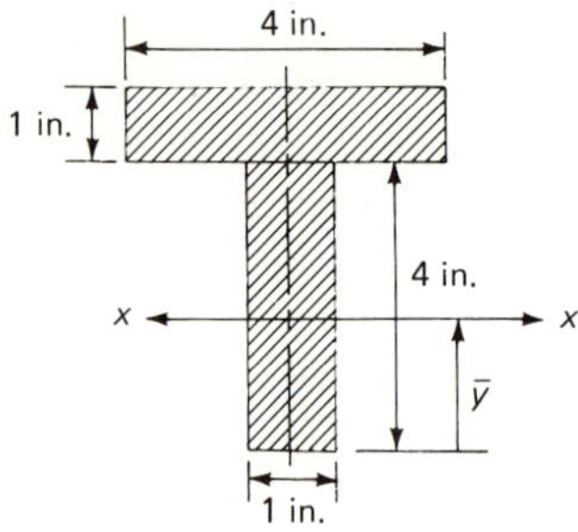

FIGURE P 8.20

8.21 Determine the moment of inertia and radius of gyration for the figure shown about the centroidal axes.

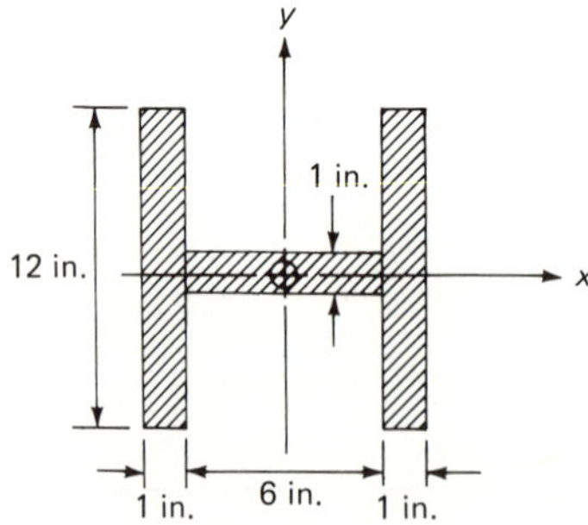

FIGURE P 8.21

8.22 Using the result of Problem 8.21, determine the polar moment of inertia about the O axis, perpendicular to the plane of the figure. Also determine the polar radius of gyration about this axis.

                    Center of Gravity and Moment of Inertia

8.23 Determine the polar moment of inertia with respect to an axis through O and perpendicular to the plane of the figure.

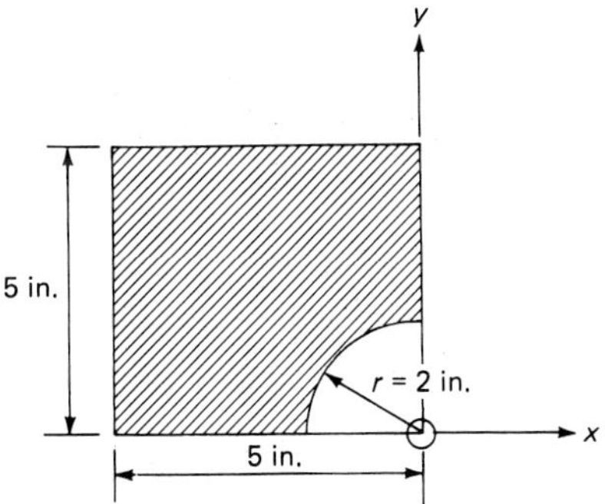

FIGURE P 8.23

8.24 Determine the polar radius of gyration for Fig. P 8.23 about the O axis.

8.25 Determine the moment of inertia and radius of gyration of the Z section shown with respect to both the x and the y centroidal axes.

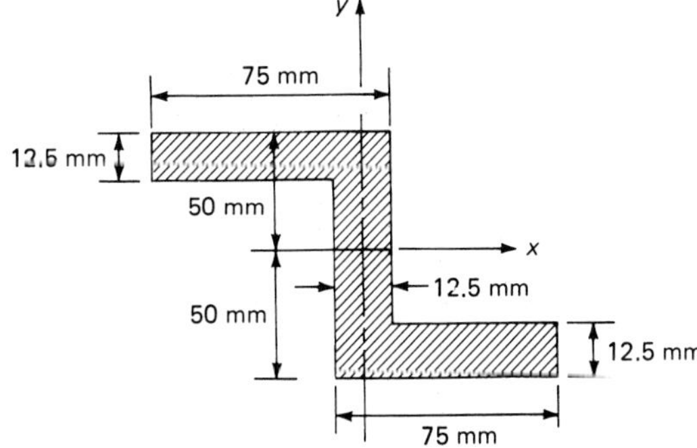

FIGURE P 8.25

8.26 Determine the moment of inertia for Fig. P 8.26 about its x centroidal axis if all thicknesses are ½ in.

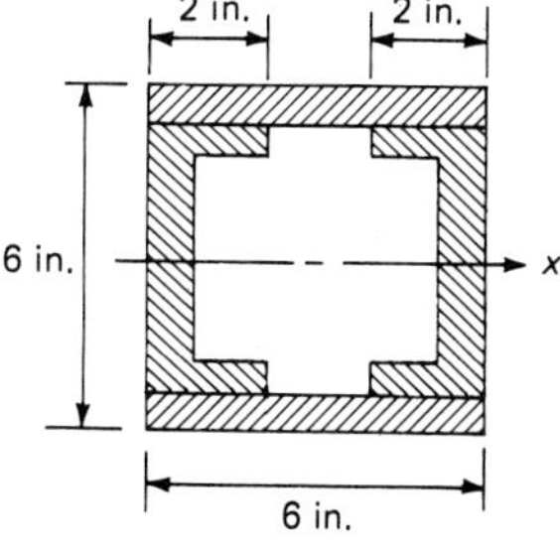

FIGURE P 8.26

Problems

8.27 Determine the radius of gyration about the x centroidal axis for Fig. P 8.26.

8.28 If it is desired that $\bar{I}_x$ and $\bar{I}_y$ be equal, determine the width of each rectangle shown.

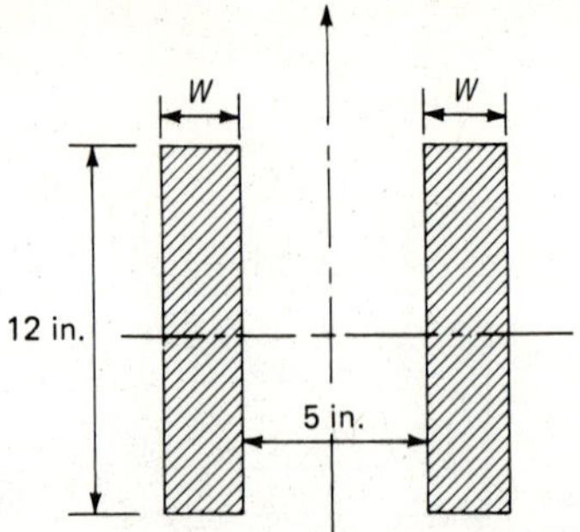

FIGURE P 8.28

8.29 Determine the moment of inertia about the x and y centroidal axes shown. Figure P 8.29 consists of two equal channels.

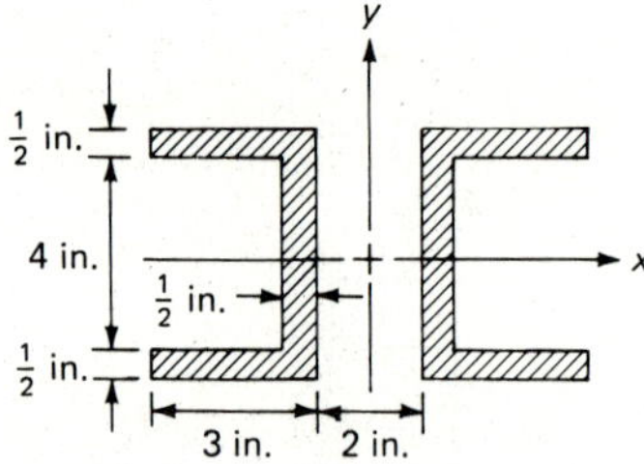

FIGURE P 8.29

8.30 An I beam is reinforced by a plate at the top as shown. Determine the moment of inertia of this composite beam with respect to its x centroidal axis.

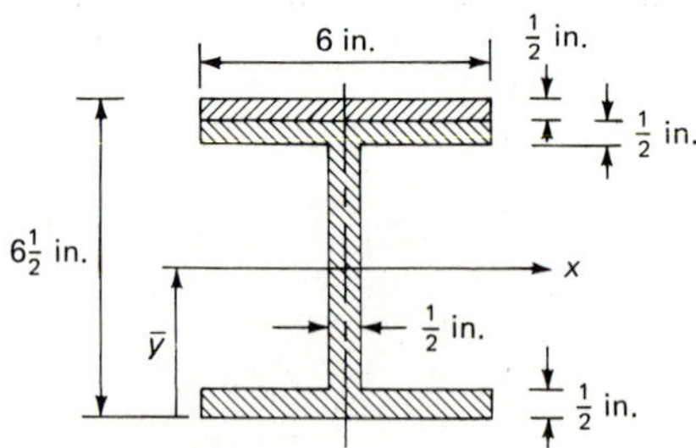

FIGURE P 8.30

8.31 Determine the radius of gyration for the x centroidal axis for Fig. P 8.30.

  Center of Gravity and Moment of Inertia

8.32 Two equal channels are welded together to form the section shown. Calculate the moment of inertia of the composite figure about its x centroidal axis.

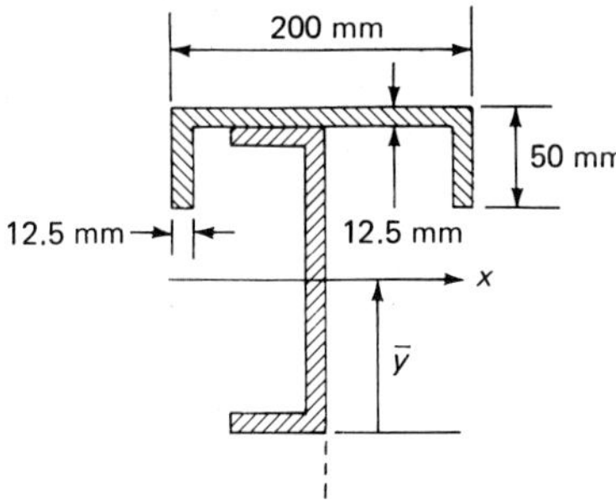

FIGURE P 8.32

8.33 If the upper channel of Fig. P 8.32 is inverted, determine the moment of inertia of the composite figure (Fig. P 8.33) about its x centroidal axis. Compare this with the results of Problem 8.32.

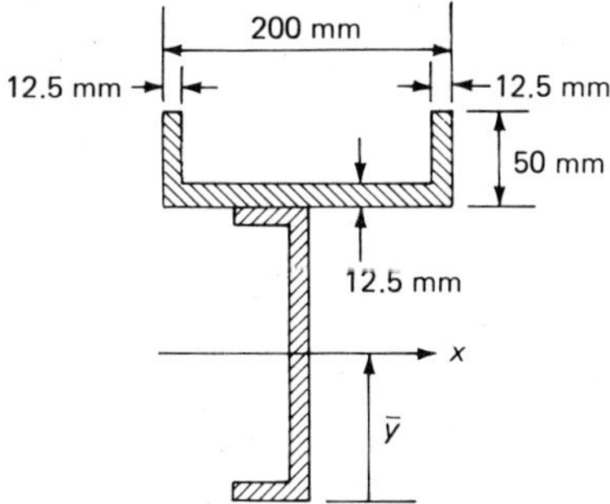

FIGURE P 8.33

8.34 Determine the moment of inertia and radius of gyration about the base of the figure shown.

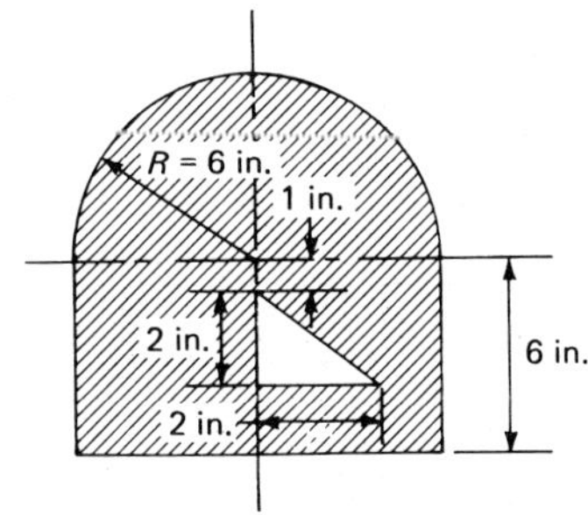

FIGURE P 8.34

Problems

219

8.35   Determine the moment of inertia of the figure shown about its center of gravity ($\bar{I}_x$ and $\bar{I}_y$). Also determine the moment of inertia about the x' and y' axes.

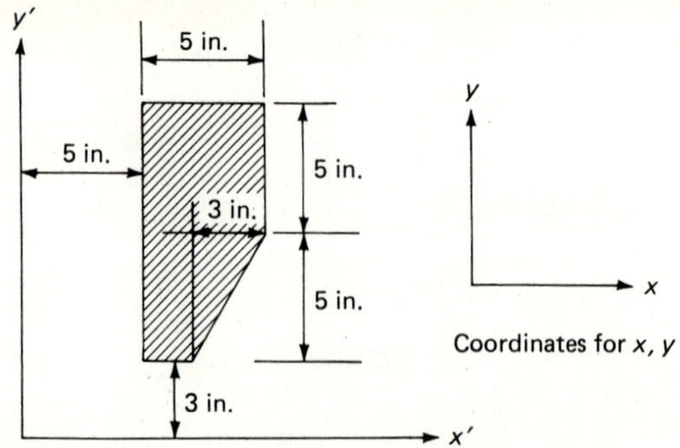

FIGURE   P 8.35

                          Center of Gravity and Moment of Inertia

9.1 INTRODUCTION

When a shaft transmits power, or a spring absorbs energy, or a torsion bar is used to provide a restoring force, the elastic properties of the member resist the applied twisting force. This twisting action is resisted by the material principally by internal shearing forces, and this twist is known as *torsion*; a member subjected to this action is said to be *in torsion*. The term *torque* will also be used, and it will denote a couple acting in a plane at right angles to the longitudinal axis of the member in question. Since the conditions for equilibrium require that the member have equal and opposite couples lying in parallel planes, we note that the portion of the member between these couples is in torsion. Torsion always produces rotation, and we shall be concerned with determining both the stresses and the strains caused by these rotations. The student should note that the theory developed in this chapter is primarily limited to cases of pure twisting in circular cross sections.

9.2 THE HORSEPOWER-TORQUE EQUATION

The transmission of power, the development of torque, and the performance of work at a specified rate are requirements imposed upon shafts by common industrial usage. It is therefore of interest to us to develop a relationship between horsepower, torque, and speed to be used in designing and rating shafts.

Consider the situation shown in Fig. 9.1, where a weight is being lifted at a constant rate as the cable is wrapped around the shaft due to the steady rotation of the shaft. As shown by the free-body diagram of Fig. 9.1(b), dynamic equilibrium can exist if it is considered that the shaft is supported by a force equal and opposite to the weight W (and directed through the center) to satisfy the condition that the sum of the forces in the vertical direction be zero.

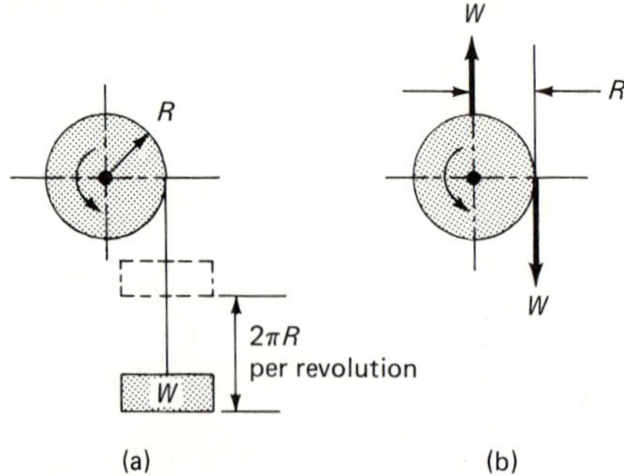

FIGURE 9.1   Diagram for horsepower-torque equation.

As a result of this condition, that is, that the sum of the moments must be zero, it is necessary to supply a moment (torque) T equal and opposite to the applied moment of WR, which yields the relation

$$T = WR \tag{9.1}$$

Now let us look at the weight W and its motion. If the weight is moving upward at a steady speed, we can evaluate the rate at which work is being done using the following reasoning. *Work* is defined as the product of a force multiplied by the displacement in the direction of the force. Each time the shaft completes one revolution, the weight is lifted by a distance equal to the circumference of the shaft, $2\pi R$ m, if R is expressed in metres. Thus the work done per shaft revolution is $W(2\pi R)$ N·m. However, the shaft is turning at the rate of N revolutions per minute (rev/min), so that per minute the work done is $2\pi RWN$ N·m/min.

At this point we note that *power* is defined as the rate at which work is being done per unit time, and that 1 horsepower (hp) is defined in common English units as a rate of work done at 33 000 ft·lb/min. In terms of watts, 1 hp is 746 W. Utilizing this definition with the rate of doing work by the shaft in raising the weight W and the results of Eq. (9.1), we have

$$\text{horsepower (hp)} = \frac{2\pi RWN}{60 \times 746} = \frac{2\pi TN}{60 \times 746} \tag{9.2}$$

or, rearranging,

$$T = \frac{7124 \times \text{hp}}{N} \ (\text{N·m}) \tag{9.3}$$

where T is in newton metres, and N is in revolutions per minute.

Torsion

In English units,

$$T = \frac{63\,000 \times \text{hp}}{N} \qquad (9.3a)$$

with T in inch pounds.

To obtain kW, note from Table I-6 that hp $\times$ 0.746 $=$ kW.

SOLUTION

(a) The torque being transmitted is obtained by direct substitution in
Eq. (9.3a). Thus

$$T = \frac{63\,000 \times \text{hp}}{N} = \frac{63\,000 \times 100}{150} = 42\,000 \text{ in.·lb}$$

or

$$T = 3500 \text{ ft·lb}$$

(b) 100 hp is a rate of doing work equal to $100 \times 33\,000$ ft·lb/min. Per
revolution,

$$\text{work per revolution} = \frac{\text{work/time}}{\text{revolution/time}} = \frac{\text{hp} \times 33\,000}{N} \text{ ft·lb/rev}$$

Therefore,

$$\text{work per revolution} = \frac{100 \times 33\,000}{150} = 22\,000 \text{ ft·lb/rev}$$

As an alternate solution we can consider that the torque arises from
a weight being lifted by a cable attached to the shaft. Since $T = WR$,

$$W = \frac{T}{R} = \frac{42\,000}{1} = 42\,000 \text{ lb}$$

Per revolution, the weight rises an amount equal to the circumference
of the shaft. Thus per revolution the lift is

$$\frac{\pi d}{12} = \frac{3.1416 \times 2}{12} = 0.524 \text{ ft}$$

and the work done is

$$42\,000 \times 0.524 = 22\,000 \text{ ft·lb/rev}$$

The Horsepower-Torque Equation

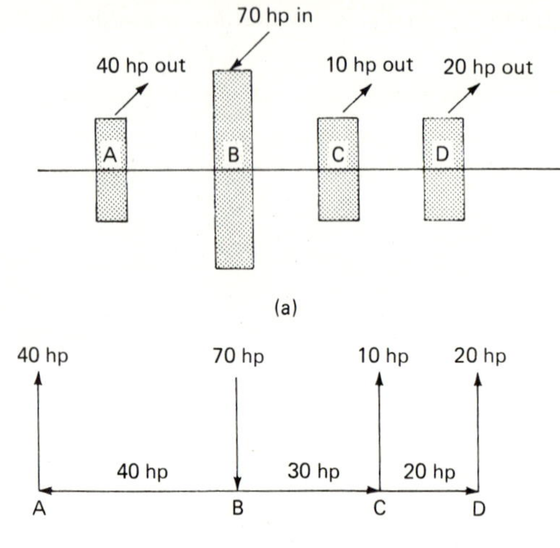

FIGURE 9.2   Illustrative Problem 9.2.

SOLUTION

We can consider the power flow from pulley B to the other pulleys just as we would consider the flow of a fluid from a reservoir. Thus 70 flow units enter at B and branch off toward A, C, and D. This is shown diagrammatically in Fig. 9.2(b). We then have

$$\text{hp}_{A\text{-}B} = 40, \qquad T = \frac{63\,000 \times \text{hp}}{N} = \frac{63\,000 \times 40}{150} = 16\,800 \text{ in.·lb}$$

$$\text{hp}_{B\text{-}C} = 30, \qquad T = \frac{63\,000 \times \text{hp}}{N} = \frac{63\,000 \times 30}{150} = 12\,600 \text{ in.·lb}$$

$$\text{hp}_{C\text{-}D} = 20, \qquad T = \frac{63\,000 \times \text{hp}}{N} = \frac{63\,000 \times 20}{150} = 8400 \text{ in.·lb}$$

9.3   TORSION OF CIRCULAR SHAFTS

In order to study the torsion of circular shafts, we shall have to make several assumptions concerning the nature of the imposed loads and the internal resistance of the shaft to these loads. Consider the situation shown in Fig. 9.3(a), where a circular shaft is fixed at its left end and is subjected to a torque at its right end by the applied couple $F \times d$. If the longitudinal line ab is scribed on the shaft prior to the application of this couple, it will be located as shown by the line ab' after the torque is applied. The radius originally shown as Ob will rotate through the angle θ and will be located at Ob' after equilibrium has been established. At this point we will make two initial assumptions:

1. Every diameter of any cross section through the shaft remains straight, and all rotate through the same angle. Thus a diameter before twisting remains a diameter after twisting.
2. As a consequence of assumption 1, all cross sections of the shaft remain plane and rotate as if they were absolutely rigid.

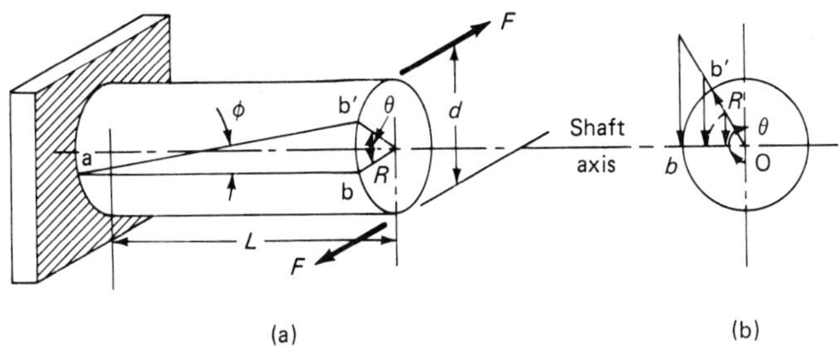

FIGURE 9.3 A shaft subjected to twisting only.

Figure 9.3(b) shows the consequence of assumption 2. The deformation of the shaft is proportional to the distance from the center of the shaft, and therefore the strain varies directly as the distance from the center of the shaft. Since we will assume a linear stress–strain relation for the material, the stress in the shaft [as shown superimposed in Fig. 9.3(b)] must vary linearly with the distance from the shaft center and is a pure shearing stress.

The third assumption we will make is a consequence of the first two and is self-explanatory.

3. The motion of any point in the shaft lies in a plane transverse to the shaft, and its direction is in a circle whose center is on the longitudinal axis of the shaft. Also, the resultant of the forces on any transverse plane must be a couple.

The final explicit assumption that we shall make concerns the line ab that we originally scribed on the outside of the shaft. This line will be assumed to be undergoing the following action.

4. A longitudinal straight line parallel to the axis of the shaft will be twisted to yield a helical line after twisting, having a constant rise (lead) per unit length of the shaft.

We can now consider any transverse cross section of the shaft as a free body and study the forces on it that must exist if the shaft is in equilibrium under the applied torque. We will use the assumptions discussed above, as necessary, to analyze this situation.

Torsion of Circular Shafts

Figure 9.4 shows a cross section of the shaft subjected to the couple $F \times d$. Consider the circular element of area shown, which we shall take to be equal to ΔA. The thickness of annulus t will be taken to be small enough to consider all the elements of the area to be located at a distance r from the center. Since all elements are located at the same distance from the center, the shear stress on this area will be constant, and the shear forces will be tangent to the annulus as shown. Summing up the internal moments and equating them to the external moment we have

1. The external moment (torque) is $F \times d = T$.
2. S_s' at any radius is related to the maximum shearing stress at the outside S_s linearly as the distance from the center. Therefore $S_s' = S_s(r/R)$.
3. The resisting moment of the area ΔA is $(S_s'\Delta A)r$ or $(S_s\Delta Ar/R)r$. Thus

$$F \times d = T = \sum \frac{S_s\Delta Ar^2}{R} = \frac{S_s}{R} \sum \Delta Ar^2 \qquad (9.4)$$

But we have already defined the summation of the ΔAr^2 terms to be the polar moment of inertia J_0. Using this definition, denoting $R = C$ (by convention and also by analogy to the terms that will be used when considering the bending of beams), we can rearrange Eq. (9.4) to yield

$$S_{s\ max} = \frac{TR}{J} = \frac{TC}{J} \qquad (9.5)$$

Equation (9.5) is known as the *torison equation,* and it has been developed for circular shafts. From Chapter 8 we have the polar moment of inertia of a solid circular shaft about a longitudinal central axis:

$$\bar{J}_o = \frac{\pi d^4}{32} \qquad (9.6)$$

which, when combined with Eq. (9.5), yields the torsion equation for a solid shaft, namely,

$$S_{s\ max} = \frac{16T}{\pi d^3} \qquad (9.7)$$

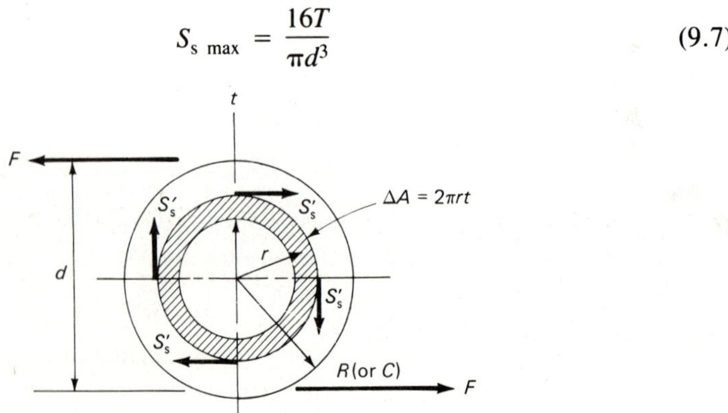

FIGURE 9.4 Development of the torsion equation.

Torsion

Since all of the assumptions that we have made concerning solid circular shafts can be applied to hollow concentric circular shafts, we can also evaluate the torsion equation for hollow circular shafts by noting $\bar{J}_o$ for this case to be

$$\bar{J}_o = \frac{\pi}{32}(d^4 - d_i^4) \tag{9.8}$$

where d_i is the inside diameter of the shaft. Therefore for a hollow circular shaft

$$S_{s\ max} = \frac{16Td}{\pi(d^4 - d_i^4)} \tag{9.9}$$

ILLUSTRATIVE PROBLEM 9.3

Compare the stresses in a solid shaft having a diameter d to those of a hollow shaft whose mean annulus diameter is d and whose weight per metre is equal to the weight per metre of the solid shaft.

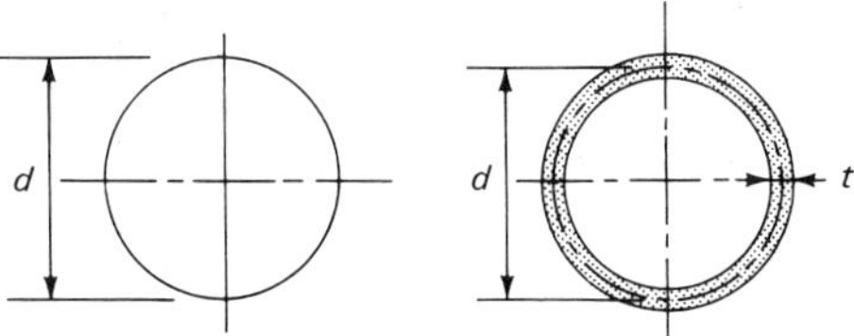

FIGURE 9.5   Illustrative Problem 9.3.

SOLUTION

If the weight per foot of each shaft is the same, then the cross sectional areas must be equal. Thus

$$\pi\frac{d^2}{4} = \pi dt$$

and

$$t = \frac{d}{4}$$

The outside diameter of the hollow shaft is $d + d/4$ or $\tfrac{5}{4}d$, and the inside diameter is $d - d/4 = \tfrac{3}{4}d$. Using these values and Eq. (9.9),

$$S_{s\ max} = \frac{16T\ \tfrac{5}{4}d}{\pi[(\tfrac{5}{4}d)^4 - (\tfrac{3}{4}d)^4]} = \frac{16T\ \tfrac{5}{4}d}{\pi d^4\left(\dfrac{625}{256} - \dfrac{81}{256}\right)}$$

and

$$S_{s\ max} = \frac{16T\ \tfrac{5}{4}d \times 256}{\pi d^4 \times 544} = \frac{16T}{\pi d^3} \times 0.588 = \frac{16T}{\pi d^3}\left(\frac{1}{1.7}\right)$$

But the stress in the solid shaft is $S_{s\,max} = 16T/\pi d^3$. Therefore

$$\frac{S_{s\,max}\ \text{solid}}{S_{s\,max}\ \text{hollow}} = \frac{1.7}{1}$$

and we can conclude that a hollow shaft of a given weight can carry a torque 70% greater than a solid shaft of the same weight whose outside diameter equals the mean diameter of the hollow shaft.

ILLUSTRATIVE PROBLEM 9.4

A hollow circular shaft has an outside diameter of 2 in. and an inside diameter of 1 in. If the maximum allowable shear stress is 20 000 psi, determine:
    (a) The torque that it can transmit.
    (b) The shear stress at the inside and at a radius of ¾ in.

SOLUTION

(a) Applying Eq. (9.9),

$$T = \frac{S_{s\,max}\,\pi(d^4 - d_i^4)}{16d} = \frac{20\,000(2^4 - 1^4)\pi}{16 \times 2} = 29\,450\ \text{in.·lb}$$

(b) Since the stress is proportional to the distance from the center of the shaft,

$$S_{s\,inside} = 20\,000 \times \frac{½}{1} = 10\,000\ \text{psi}$$

and

$$S_{s\,(r\,=\,3/4)} = 20\,000 \times \frac{¾}{1} = 15\,000\ \text{psi}$$

In addition to our being concerned with the stresses in the shaft, we are interested in the deflection (angular) that occurs when the shaft is subjected to torsion. Figure 9.6 shows point b, an initial location in the right face and on the outside of the shaft. Twisting of the shaft moves b to b'. The amount of this motion bb' can be determined from considering the motion both in the plane and also along the shaft. Thus

$$bb' = R\theta$$

But

$$bb' = L\phi$$

$$(9.10)$$

Therefore we can express θ in terms of ϕ,

$$\theta = \frac{L\phi}{R}$$

$$(9.11)$$

Torsion

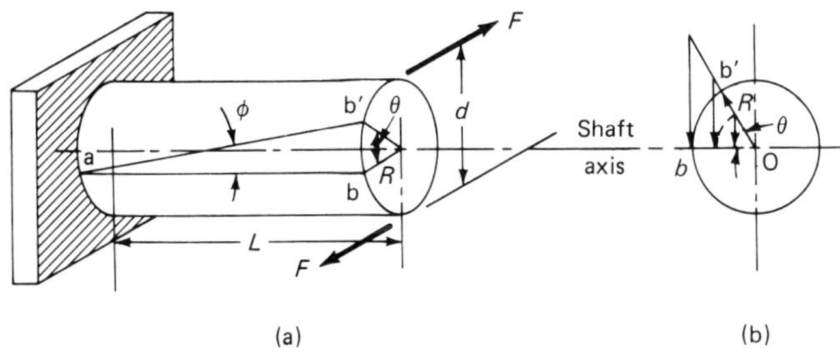

FIGURE 9.6   A shaft subjected to twisting only. (Repeated from Fig. 9.3.)

By definition G, the *shear modulus*, is defined as the shearing stress divided by the shearing strain, giving

$$G = \frac{S_s}{\phi} \tag{9.12}$$

Placing Eq. (9.12) into Eq. (9.11) yields

$$\theta = \frac{S_s L}{GR} \tag{9.13}$$

But

$$S_s = \frac{TR}{J} \tag{9.5}$$

Thus the angular twist of the shaft at a given cross section is

$$\theta = \frac{TL}{JG} \tag{9.14}$$

The student should note that θ is given in radians, where 1 rad is approximately $=$ 57.3 degrees, and that dimensionally this requires the right-hand side of Eq. (9.14) to be dimensionless. Thus if T is in newton metres and L is in metres, J must be in metres^4 and G must have the dimension of newtons per square metre, that is, pascal (Pa). Any self-consistent set of units will give the correct solution. In English units, typical values for T are in inch pounds, L is in inches, J must be in inches^4, and G must have the dimensions of psi.

ILLUSTRATIVE PROBLEM 9.5

A solid shaft is to transmit 30 hp at 1000 rev/min. If the allowable shear stress is 10 000 psi and $G = 12 \times 10^6$ psi, determine the shaft diameter required
    (a)   If it is based upon the stress only.
    (b)   If the allowable angular deflection is ½ degree per foot of shaft.

Torsion of Circular Shafts

SOLUTION

(a) From Eq. (9.3a)

$$T = \frac{63\,000 \times hp}{N} = \frac{63\,000 \times 30}{1000} = 1890 \text{ in.·lb}$$

Applying Eq. (9.7),

$$d^3 = \frac{16T}{\pi S_{s\ max}} = \frac{16 \times 1890}{\pi \times 10\,000} = 0.96$$

and $d = 0.987$ in., say, 1 in. in diameter.

(b) ½ deg/ft is ½/57.3 = 0.00873 rad/ft. Thus

$$\theta = \frac{TL}{JG} = 0.008\,73 = \frac{T(1 \times 12)}{(\pi d^4/32)(12 \times 10^6)} = \frac{T(1 \times 12 \times 32)}{\pi d^4(12 \times 10^6)}$$

But $T = 1890$ in.·lb. Therefore

$$d^4 = \frac{1890 \times 12 \times 32}{0.008\,73\ \pi(12 \times 10^6)} = 2.21$$

and

$$d = 1.219 \text{ in.}$$

For this example, the required shaft diameter is determined by the limiting angular deflection.

ILLUSTRATIVE PROBLEM 9.6

A hollow shaft having an outside diameter of 50 mm and an inside diameter of 25 mm is transmitting power at 2000 rev/min. If the allowable shear stress is 140 MPa, and the allowable angular deflection is 3 deg/m of shaft, determine the horsepower capacity of the shaft. Use $G = 84 \times 10^9$ Pa.

SOLUTION

For a hollow shaft

$$J = \frac{\pi(d^4 - d_i^4)}{32} = \frac{\pi}{32}(0.050^4 - 0.025^4) = 5.752 \times 10^{-7} \text{ m}^4$$

Since $S_s = \dfrac{TC}{J}$,

$$T = \frac{S_s J}{C} = \frac{140 \times 10^6 \times 5.752 \times 10^{-7}}{0.025} = 3221 \text{ N·m}$$

Therefore

$$hp = \frac{TN}{7124} = \frac{3221 \times 2000}{7124} = 904 \text{ hp}$$

Torsion

If the allowable angular deflection is 3 deg/m of shaft,

$$\theta = \frac{3}{57.3} = 5.236 \times 10^{-2} \text{ rad/m}$$

Thus

$$T = \frac{\theta J G}{L} = \frac{5.236 \times 10^{-2} \times 5.752 \times 10^{-7} \times 84 \times 10^{9}}{1} = 2530 \text{ N·m}$$

and

$$\text{hp} = \frac{TN}{7124} = \frac{2530 \times 2000}{7124} = 710 \text{ hp}$$

The shaft shown in Fig. 9.7 is rigidly supported at each end. If a torque of T in.·lb is applied at the intersection of the two shafts, determine the distribution of the torque T in each shaft.

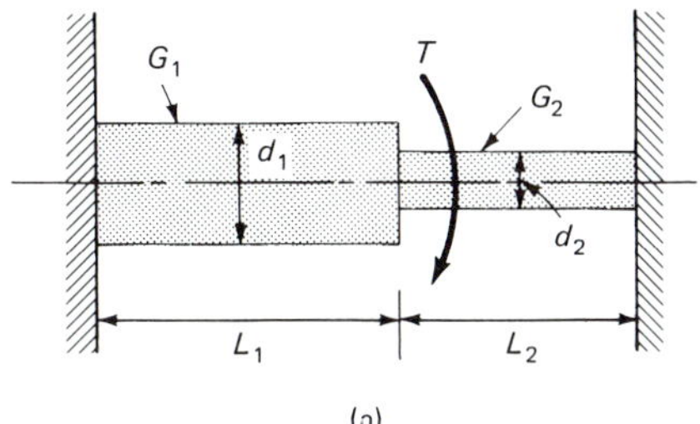

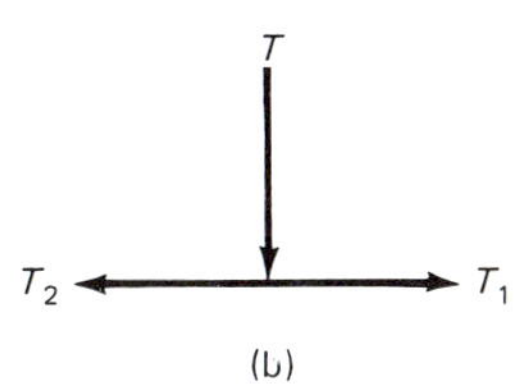

FIGURE 9.7　Illustrative Problem 9.7.

SOLUTION

Consider that the torque T distributes as shown in Fig. 9.6(b). Thus

$$T = T_1 + T_2$$

At the intersection of the two shafts $\theta_1 = \theta_2$, yielding the condition that

$$\frac{T_1 L_1}{J_1 G_1} = \frac{T_2 L_2}{J_2 G_2}$$

We can solve these two equations to obtain

$$T_1 = \frac{T}{1 + (L_1/L_2)(J_2/J_1)(G_2/G_1)}$$

and

$$T_2 = \frac{T}{1 + (L_2/L_1)(J_1/J_2)(G_1/G_2)}$$

9.4 SHAFT COUPLINGS

Shaft couplings are commonly used in machinery to provide for misalignments of shafts, to connect differing machines, to join lengths of shafts, and to absorb shock loads. Of the many types of shaft couplings that are used, we shall be concerned here with the *flange coupling*, an essentially rigid coupling used to connect shafts that are in relatively good alignment. A coupling of this type is shown diagrammatically in Fig. 9.8. The two parts of this coupling are held together by a single circle of bolts. In order to properly distribute the loads uniformly in the bolts, the bolts should be ground and fitted into close tolerance with reamed holes in the radial part of the flange.

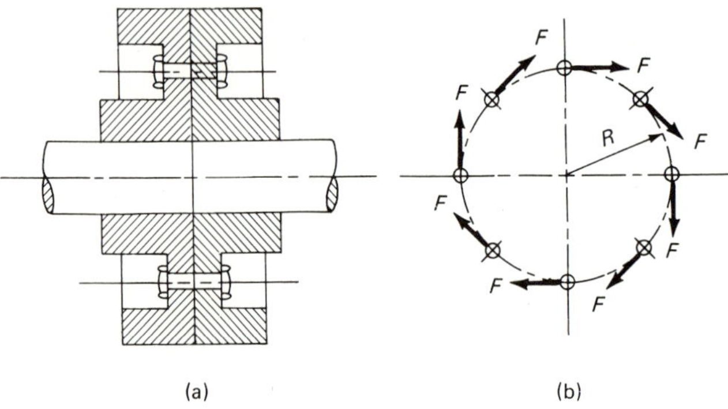

FIGURE 9.8   A flange coupling.

Let us assume that we have such a coupling with a single circle of n bolts, each having a diameter d and arranged in a circle of radius R. The shear area of each bolt A_s is $\frac{1}{4}\pi d^2$, the shear force is $\frac{1}{4}\pi d^2 S_s$, and the torque exerted by each bolt on the driven shaft is $F \times R = \frac{1}{4}\pi d^2 S_s R$. For n bolts the torque is

$$T = \frac{\pi d^2}{4} S_s R n \tag{9.15}$$

ILLUSTRATIVE PROBLEM 9.8

A 1-in. diameter shaft is used to transmit 600 hp at 1500 rev/min from a motor. Using a material with an allowable shear stress of 10 000 psi, find how many ½-in. diameter bolts would be needed if they were to be placed on a 3-in. diameter circle in a flange coupling.

SOLUTION

Calculating the torque, we have

$$T = \frac{63\,000 \times \text{hp}}{N} = \frac{63\,000 \times 600}{1500} = 25\,200 \text{ in.·lb}$$

Torsion

The shear area per bolt is $\frac{1}{4}\pi d^2 = \frac{1}{4}\pi(\frac{1}{2})^2 = 0.196$ in.^{2}. Therefore

$$n = \frac{T}{A_s S_s R} = \frac{25\ 200}{0.196 \times 10\ 000 \times \frac{3}{2}} = 8.57$$

We could specify 9 bolts, but for symmetry 10 bolts might be used.

In the event that more than one row of bolts is used, it is necessary to be able to find the load taken by each row. When we considered the deflections in a shaft, we found them to be proportional to their distance from the center of the shaft. For bolt circles at differing radii from the center of the shaft, the deflections of the bolts and, consequently, the loads on the bolts will be proportional to their location relative to the center. We may write this relation in the following form for the situation shown in Fig. 9.9:

$$\frac{F_1}{F_2} = \frac{R_1}{R_2} \tag{9.16}$$

and

$$T = T_1 + T_2 = F_1 R_1 n_1 + F_2 R_2 n_2 \tag{9.17}$$

Equations (9.16) and (9.17) can readily be modified to take into account three or more rows of bolts.

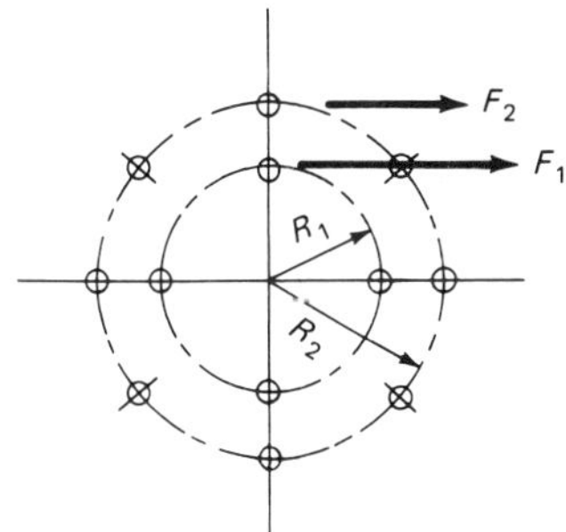

FIGURE 9.9   A coupling with two rows of bolts.

ILLUSTRATIVE PROBLEM 9.9

Twelve $\frac{1}{2}$-in. diameter bolts are arranged in two concentric bolt circles having diameters of 5 and 10 in., respectively. If there are four bolts on the inner bolt circle and eight bolts on the outer bolt circle, determine the torque they can transmit if the allowable shear stress is 10 000 psi.

SOLUTION

Assume that the outer bolts are stressed to their allowable stress of 10 000 psi. The shear force per bolt is

$$F_2 = \tfrac{1}{4}\pi(\tfrac{1}{2})^2 \times 10\ 000 = 1960\ \text{lb/bolt}$$

On the inner row

$$F_1 = \frac{R_1}{R_2} F_2 = \frac{5}{10} \times 1960 = 980 \text{ lb/bolt}$$

The torque is

$$T_1 = F_1 R_1 n_1 = 980 \times 2.5 \times 4 = 9800 \text{ in.·lb}$$

$$T_2 = F_2 R_2 n_2 = 1960 \times 5 \times 8 = 78\,400 \text{ in·lb}$$

and the total torque is

$$T = T_1 + T_2 = 9800 + 78\,400 = 88\,200 \text{ in.·lb}$$

9.5   CLOSURE

In most applications the ability of materials to resist an applied torque can yield a machine element that is useful. Examples of such elements have been studied in this chapter so that we could understand their principles of operation and be able to design them. It must be apparent to the student that it has not been possible to cover all facets of each topic comprehensively. We have limited ourselves to those basic considerations appropriate to the level of our study. The student desiring more information on the these topics will find that he or she will be led to texts on advanced mechanics of materials, theory of elasticity, and/or machine design. However, all roads, so to speak, lead from the material developed in this chapter, and the mastery of this material (in addition to present necessity) is a prerequisite for future professional studies. We shall use some of these concepts later in studying the shear in beams and in elements subjected to both shear and other direct stresses. The principles espoused herein are not difficult, and the effort required to understand and apply them will yield gratifying results.

REFERENCES

Arges, K. P., and A. E. Palmer, MECHANICS OF MATERIALS. New York: McGraw-Hill, 1963.

Black, P. H., and O. E. Adams, Jr., MACHINE DESIGN, 3rd ed. New York: McGraw-Hill, 1968.

Breneman, J. W., STRENGTH OF MATERIALS, 3rd ed. New York: McGraw-Hill, 1965.

Faires, V. M., DESIGN OF MACHINE ELEMENTS, 4th ed. New York: Macmillan, 1965.

Higdon, A., E. E. Ohlsen, and W. B. Stiles, MECHANICS OF MATERIALS, 3rd ed. New York: John Wiley, 1976.

Jensen, A., and H. H. Chenoweth, APPLIED STRENGTH OF MATERIALS, 3rd ed. New York: McGraw-Hill, 1971.

Levinson, I. J., MECHANICS OF MATERIALS, 2nd ed. Englewood Cliffs, NJ: Prentice-Hall, 1970.

Olsen, G. A., ELEMENTS OF MECHANICS OF MATERIALS, 3rd ed. Englewood Cliffs, NJ: Prentice-Hall, 1974.

Seely, F. B., and N. E. Ensign, Analytical Mechanics for Engineers. New York: John Wiley.

Seely, F. B., and J. O. Smith, Advanced Mechanics of Materials, 2nd ed. New York: John Wiley, 1952.

Singer, F. L., Strength of Materials, 2nd ed. New York: Harper and Row, 1962.

PROBLEMS

9.1 A weight of 100 lb is lifted at the rate of 60 mi/h. What horsepower is being used?

9.2 A weight of 400 N is lifted at the rate of 30 m/s. What horsepower is being used?

9.3 A motor runs at N rev/min and transmits hp horsepower. If the motor is lifting a weight, show that the weight can be expressed as $W = 12 \times 33\,000 \times \text{hp}/\pi DN$, where D is the shaft diameter in inches.

9.4 Determine the torque transmitted in each portion of the shaft shown in Fig. P 9.4. The shaft rotates at 150 rev/min.

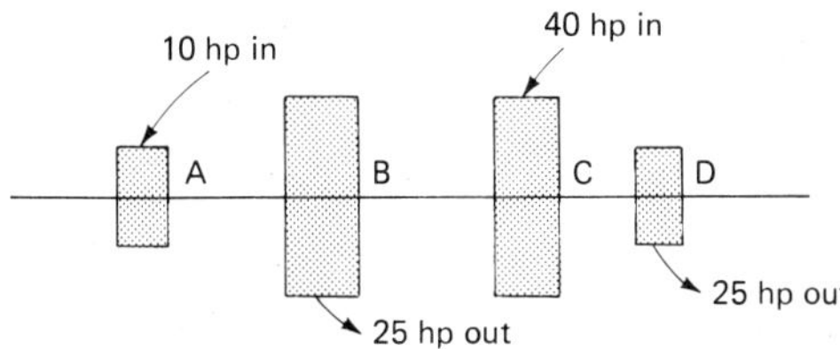

FIGURE P 9.4

9.5 A solid circular shaft is used to transmit 75 hp at 250 rev/min. If the allowable shear stress is 8000 psi, determine the diameter of the shaft.

9.6 A solid circular shaft is used to transmit 75 hp at 250 rev/min. If the allowable shear stress is 56 MPa, determine the diameter of the shaft.

9.7 A hollow circular shaft is stressed in shear to 9500 psi due to torsion. If the outside diameter of the shaft is 4.5 in. and the inside diameter is 2 in., determine the power transmitted when the shaft is operated at 125 rev/min.

9.8 It is desired to limit the angular deflection of a shaft to ¼ deg/ft of shaft. If $G = 11 \times 10^6$ psi and the applied torque is 7500 in.·lb, determine the diameter of the shaft.

9.9 Show that the stress in a hollow shaft of outside diameter d and inside diameter $\frac{1}{2}d$ is approximately 6% greater than a solid shaft of d, but the hollow shaft will weigh approximately 25% less than the solid shaft (Fig. P 9.9).

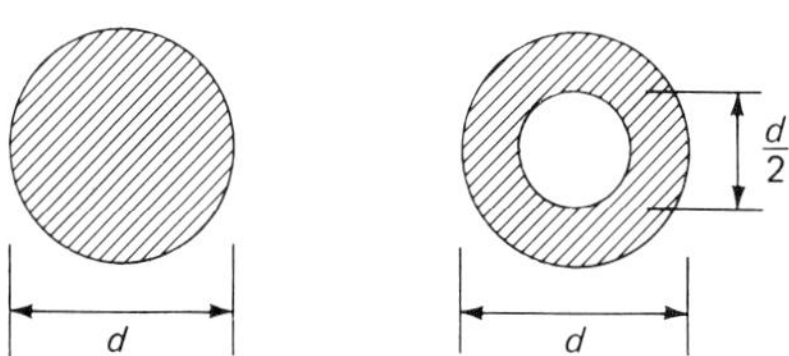

FIGURE P 9.9

Problems

9.10   A torsion-testing machine is used to test an unknown material. If a specimen ¾ in. in diameter and 1 ft in length is used, and it is experimentally found that a torque of 1800 in.·lb produces an angular deflection of 5 deg, determine G.

9.11   A solid circular shaft is 50 mm in diameter. What horsepower can it transmit at 200 rev/min if the allowable shear stress is 70 MPa?

9.12   A solid circular shaft is to transmit 100 hp at 110 rev/min. If the allowable shearing stress is 70 MPa, determine the shaft diameter. $G = 84$ GPa.

9.13   Determine the angular twist of the shaft in Problem 9.12 if the shaft is 1.5 m long.

9.14   A torque of 1000 in.·lb is applied to the shaft shown in Fig. P 9.14. Determine the total angle of twist of the free end. Take G to be 10×10^6 psi for both sections.

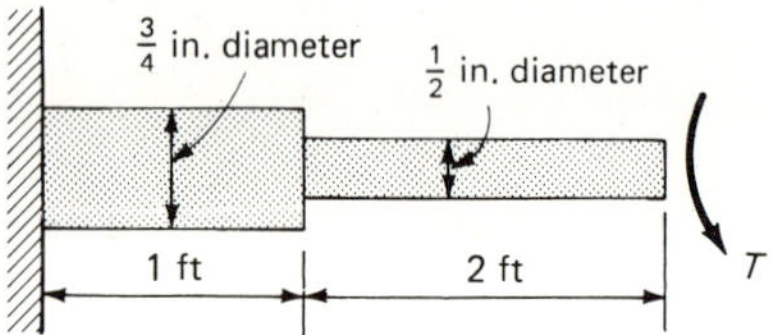

FIGURE P 9.14

9.15   The shafting shown in Fig. P 9.15 has a power input and takeoff as indicated. Determine the relative angle of twist between pulleys A and B.
(*Hint:* The angle of twist between A and B is the arithmetic sum of the angles between A and C and between C and B with due respect to the direction of twist.)

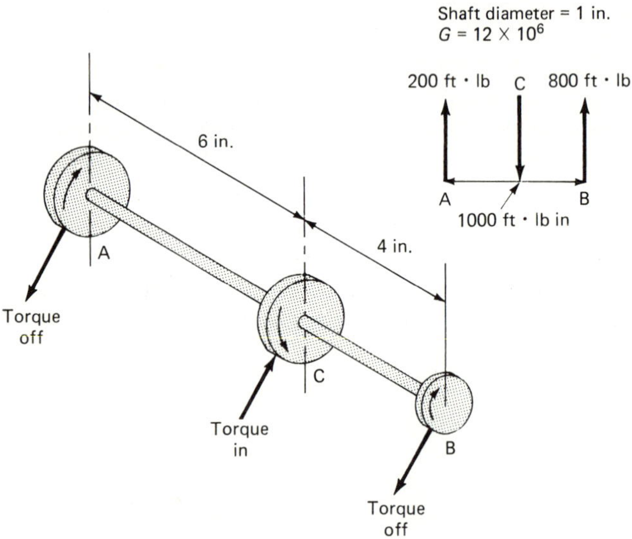

FIGURE P 9.15

9.16   Solve Problem 9.15 if the torque takeoff at B is in a direction opposite to that shown.

Torsion

9.17 A large marine propeller shaft, 50 ft long, is to be made of steel ($G = 12 \times 10^6$). If the outside diameter of the shaft is 20 in. and the inside diameter is 12 in., determine its angle of twist when subjected to a torque of 100 000 ft·lb.

9.18 A solid circular shaft is 1.25 m long and must transmit 50 hp at 200 rev/min. If the permissible shearing stress is 56 MPa and the permissible angular deflection is 1 deg for the shaft, determine the diameter of the shaft. $G = 84$ GPa.

9.19 The shaft shown in Fig. P 9.19 is subjected to a torque of 100 N·m clockwise at A. Determine the torque absorbed by each portion of the shaft and their angular deflection. $G = 77$ GPa.

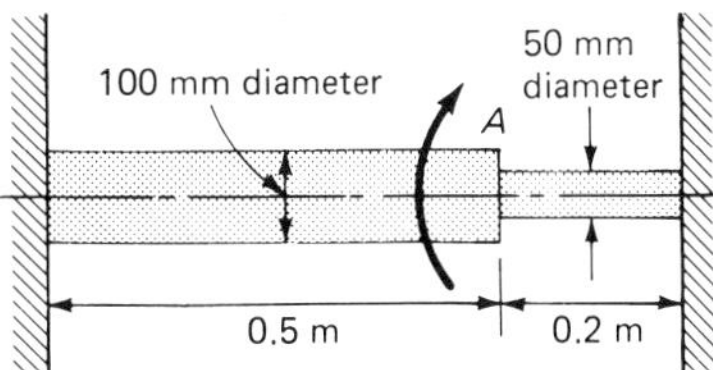

FIGURE P 9.19

9.20 A shaft coupling having a single circle of six ½-in. bolts is used to couple two shafts operating at 500 rev/min. If the bolt circle is 4 in. in diameter, what horsepower can this coupling transmit? The maximum bolt shear stress is 15 000 psi.

9.21 A coupling is required to transmit 100 hp at 100 rev/min. What bolt circle diameter is required if ten 10-mm diameter bolts are used having a design shear stress of 50 MPa?

9.22 A coupling uses two rows of bolts on bolt circles of 75-mm diameter and 200-mm diameter, respectively. If six 12.5-mm diameter bolts are used in each row, determine the torque capacity of the coupling. Take the maximum shear stress in the bolts to be 105 MPa.

9.23 A shaft coupling has two rows of bolts on bolt circle diameters of 50 mm and 100 mm, respectively. If four 15-mm diameter bolts are used in each row, determine the torque capacity of the coupling. Use a maximum allowable shear stress of 84 MPa.

9.24 Determine the bolt circle diameter for a coupling to transmit 250 hp at 100 rev/min if eight ½-in. diameter bolts having a design S_s of 12 000 psi are used.

9.25 A coupling uses two rows of bolts on bolt circles of 3-in. and 8-in. diameters, respectively. If six ½-in. diameter bolts are used in each row, determine the torque capacity of the coupling. Take the maximum shear stress in the bolts to be 15 000 psi.

10

Riveted and Welded
Structures

10.1   INTRODUCTION

The connection of the members of structures to carry the loads imposed upon them and the providing of the necessary structural support is usually accomplished by welding, riveting, or bolting the structures together. The principles which we shall discuss concerning riveted connections are directly applicable to bolted structures, and we will not treat bolted structures as a separate topic. A word of caution is in order at the beginning of our study. The techniques used by individual welders and riveters in making a specific joint can create situations that are not amenable to mathematical computations. A cold-worked rivet improperly set, a weld made with improper penetration of the weld material, the use of the improper weld rod, and so on, can all lead to unsatisfactory structural riveted or welded attachments. In order to preclude such bad joints or at least to limit the number of them that occur, various standard codes are in effect. Of these the *ASME* (American Society of Mechanical Engineers) *Boiler and Pressure Vessel Code* and the *AISC* (American Institute of Steel Construction) *Manual of Steel Construction* are the most comprehensive and most widely used in industry. Where desirable, we shall refer to them and the rules they contain or suggest.

10.2   RIVETED CONNECTIONS

A *riveted connection* is a mechanical connection made by upsetting (or distorting) the ends of a special pin, known as a *rivet*. Once a riveted connection is made, it is permanent and cannot be disassembled as can a bolted or screwed joint. As Fig. 10.1 shows, rivets are usually provided as a cylindrical shank on which a preformed head exists on one end. The rivet is inserted into holes drilled or punched in the plates that are to be fastened, and the rivet is "upset," that is, formed by mechanically deforming the rivet between appropriate dies. The forming process can be carried out either hot or cold. Steel rivets are usually worked hot (1600°F to 1900°F), while some nonferrous rivets are kept refrigerated before being used. The steps in setting up a rivet joint are shown schematically in Fig. 10.2. The formed part of the rivet is held in place by one die known as the backing bar, and the other die (the set) is positioned on the other end of the undeformed shank. The rivet is then upset by using a riveting machine (gun) operated pneumatically, mechanically, or by steam. The rivet can also be formed manually by using a hammer on the set.

Common rivets require access from both sides of the work and are installed by upsetting the extending shank to form a tail. They can be used for structural or nonstructural applications. Blind rivets require access from only one side of the work, and the tail is formed either chemically or mechanically. They can also be used for structural or nonstructural applications. Figure 10.3 gives the approximate dimensions used in selecting the length of a rivet which will yield a final rivet with the approximate tail dimensions shown. Commercially available steel structural rivets come in sizes from ⅜-in. shank diameter to 1¼-in. diameter, in increments of ⅛ in. SI sizes should be available in the near future.

Riveted joints are usually categorized by the type and complexity of the joint.

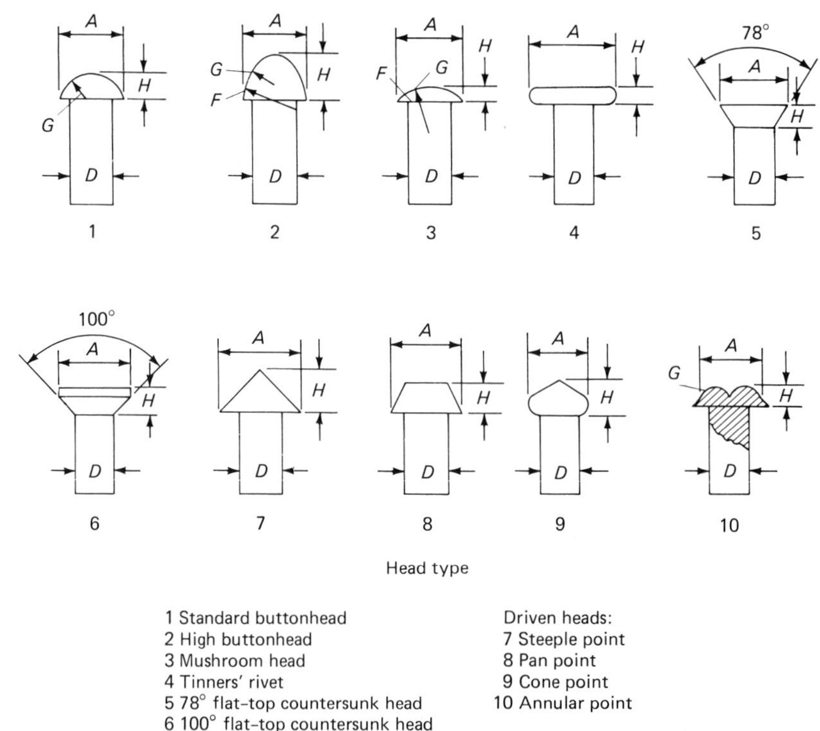

Head type

1 Standard buttonhead
2 High buttonhead
3 Mushroom head
4 Tinners' rivet
5 78° flat-top countersunk head
6 100° flat-top countersunk head

Driven heads:
7 Steeple point
8 Pan point
9 Cone point
10 Annular point

FIGURE 10.1 Common types of heads for rivets. (Reproduced with permission from H. A. Rothbart, *Mechanical Design and Systems Handbook.* New York: McGraw-Hill, Fig. 22-2, p. 22-3.)

                                                        Riveted and Welded Structures

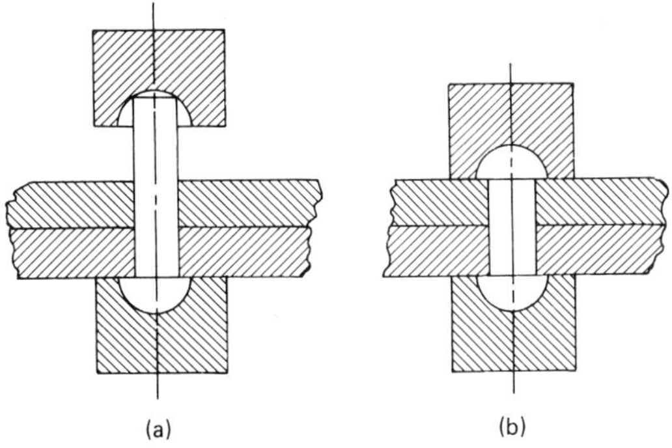

FIGURE 10.2   Setting-up of an ideal riveted joint.

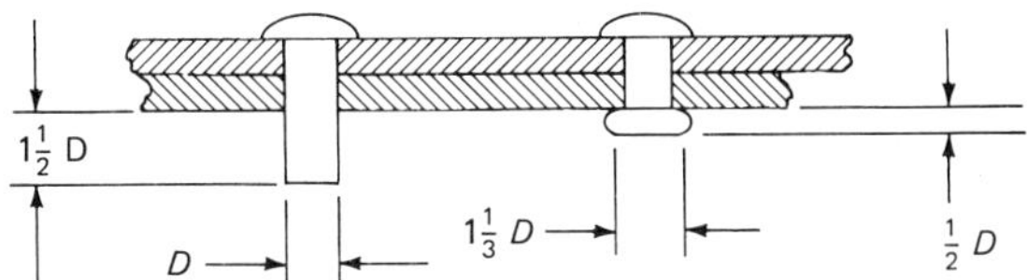

FIGURE 10.3   Approximate rivet sizes for structural joints.

The most commonly encountered riveted joints are the *lap joint* and the *butt joint*; both are shown schematically in Fig. 10.4. In the lap joint the two ends of the plates to be fastened are placed over one another and riveted together by placing rivets in holes made through both plates, as shown on Fig. 10.4(a). Figure 10.4(b) shows the second commonly used riveted joint, the butt joint, so named because the edges of the plates to be riveted are butted against each other. The joint is made by riveting a cover plate (or plates) to each of the main plates, as shown in Fig. 10.4(b).

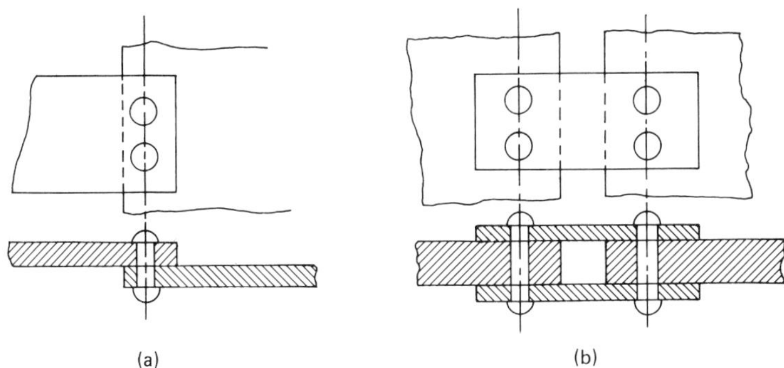

FIGURE 10.4   Types of riveted joints. (a) Lap joint. (b) Butt joint.

Riveted joints are also classified by the number of rows of rivets used, that is, a *double-riveted lap joint* is a lap joint having two rows of rivets. Thus Fig. 10.4(b) shows a *single-riveted butt joint*, since it consists of one row of rivets. The student should note that it is possible to have a butt joint with differing rows of rivets on either side of the joint. In this event it is necessary to identify the joint on each side.

TABLE 10.1   Allowable Stresses for the Design of Axially Stressed Riveted Joints*

	Shear Stress S_s		Tensile Stress S_t		Bearing Stress S_b	
	(psi)	(MPa)	(psi)	(MPa)	(psi)	(MPa)
ASME Boiler Code	8800	60.7	11 000†	75.8	19 000	131.0
A.R.E.A. Code for						
Railway Bridges						
Power-driven rivets	13 500	93.1	18 000	124.1	27 000	186.2
Hand-driven rivets	11 000	75.8	18 000	124.1	20 000	137.9
AISC Code for						
Buildings, prior to						
7th ed.	15 000	103.4	20 000	137.9	32 000,‡	220.6,
(see Table 10.2 for					single shear	
7th ed. values)					40 000,	275.8,
					double shear	

*Reproduced with permission from G. A. Olsen, *Elements of Mechanics of Materials*. Englewood Cliffs, NJ: Prentice-Hall, 1958. SI units added.

†Varies with steel specified.

‡A surface is in single- or double-shear bearing depending on whether one or two shear resisting surfaces bound the bearing area. If there is one, the area is in single-shear bearing. If there are two, it is in double-shear bearing.

10.3   DESIGN STRESSES FOR RIVETED CONNECTIONS

The rules for the design and fabrication of riveted joints are given in the *ASME Boiler and Pressure Vessel Code*, the *AISC Manual of Steel Construction*, and the *A.R.E.A.* (American Railway Engineering Association) *Code for Railway Bridges*. These codes differ due to the specific applications and methods of fabrication generally used. Table 10.1 gives a summary of the stresses allowed by these codes for shear, tension, and bearing. We shall not investigate further the rules given in these codes for the fabrication of riveted joints, such as punching of holes or minimum plate thicknesses, and the student is cautioned at this time that these codes should be carefully studied before undertaking the design of structural or pressure vessel joints.

The 7th edition of the *AISC Manual of Steel Construction* recognized recent advances in metallurgy which have produced high-strength alloy steels. The allowable tension and shear stresses on rivets, bolts, and threaded parts, as given in the 7th edition, are shown in Table 10.2. In this AISC edition the terms *friction-type* and *bearing-type* are used to categorize connections that transmit load by means of shear in their fasteners. Friction-type connections depend upon clamping forces sufficiently high to prevent slip of the connected parts. Bearing-type connections depend upon contact of the fasteners against the sides of their holes to transfer the load from one connected part to another.

For structural joints the values given in Table 10.2 should be used. Also, as is noted later, *the tensile stress in the connecting plate can be taken as $0.6S_y$, where S_y is the yield stress of the plate.*

Riveted and Welded Structures

10.4 FAILURE IN RIVETED JOINTS

Before we study the design of riveted joints and the assumptions made in conventional design procedures, let us consider the possible ways that failure can occur in such joints.

Rivet Failure

RIVET SHEAR

Figure 10.5 shows the condition of a joint failure due to the shear of a rivet. In Fig. 10.5(a) the mode of failure shown is caused by the shear of the rivet at the interface planes between the plates due to the relative motion of the upper and lower plates. Since Fig. 10.5(a) shows a single shear plane on which the rivet can fail, this

TABLE 10.2 Allowable Stresses (AISC Code, 7th ed.)*

Description of Fastener	Tension S_t		Shear S_s			
			Friction-Type Connections		Bearing-Type Connections	
	(MPa)	psi $\times$ 10^3	psi $\times$ 10^3	(MPa)	psi $\times$ 10^3	(MPa)
A502, grade 1, hot-driven rivets	(137.9)	20.0			15.0	(103.4)
A502, grade 2, hot-driven rivets	(186.2)	27.0			20.0	(137.9)
A307 bolts	(137.9)	20.0†			10.0	(69.0)
Threaded parts§ of steel (S_y = yield stress)	0.60S_y†				0.30S_y	
A325 and A449 bolts, threading is *not* excluded from shear planes	(275.8)	40.0‡	15.0	(103.4)	15.0	(103.4)
A325 and A449 bolts, threading is excluded from shear planes	(275.8)	40.0‡	15.0	(103.4)	22.0	(151.7)
A490 bolts, threading is *not* excluded from shear planes	(372.3)	54.0‡, ‖	20.0	(137.9)	22.5	(155.1)
A490 bolts, threading is excluded from shear planes	(372.3)	54.0‡, ‖	20.0	(137.9)	32.0	(220.6)

*Reproduced with permission from *AISC Manual of Steel Construction*, 7th ed. New York: American Institute of Steel Construction, 1970. SI units added.
†Applied to tensile stress area equal to $0.7854(D - 0.9743/n)^2$, where D is the major thread diameter and n is the number of threads per inch.
‡Applied to the nominal bolt area.
§Since the nominal area of an upset rod is less than the stress area, the former area will govern.
‖ Static loading only.

type of failure is called a *single shear* of the rivet. Figure 10.5(b) shows a failure that occurs due to the relative motion of the center plate with respect to both the upper and the lower plates, and this mode of failure is called *double shear* for obvious reasons.

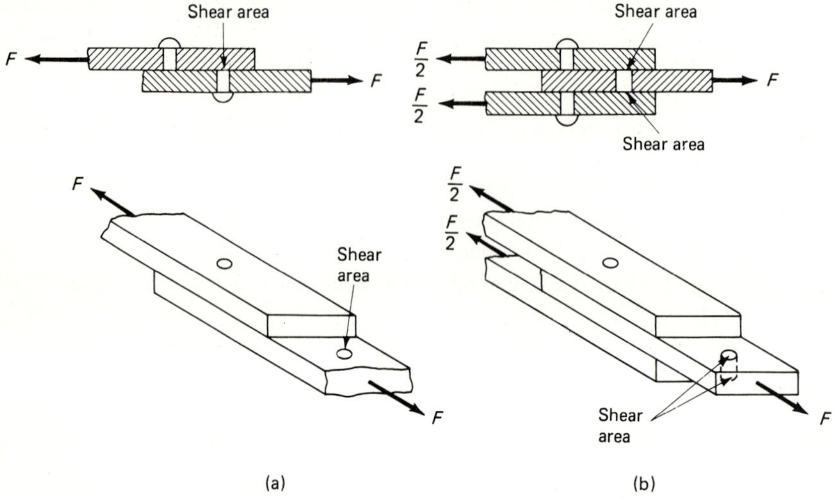

FIGURE 10.5   Failure by rivet shear. (a) Single shear. (b) Double shear.

RIVET BENDING

Consider the conditions shown in Fig. 10.6(a) and (b). In both instances the rivet is shown to be long compared to its diameter, and this condition causes the rivet to distort excessively. In the condition shown in Fig. 10.6(c) the axial load on each plate causes the moment Fd to exist on the plate, tending to bend it. This action places the rivet in tension and subjects it to bending as well. The use of a second row of rivets decreases the tendency of the plates to bend and alleviates greatly the conditions shown in Fig. 10.6(c).

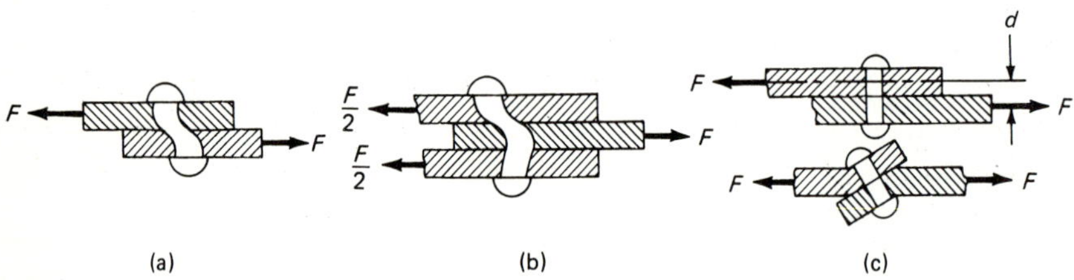

FIGURE 10.6   Rivets in bending.

RIVET CRUSHING (BEARING)

The bearing of the plate against a rivet can cause the rivet to crush. The failure that occurs in this case is a compressive failure which will cause the connection to loosen and deform excessively. Figure 10.7 shows this action in a joint subject to single shear. In addition to undergoing the action shown in Fig. 10.7, rivets in single

                                                       Riveted and Welded Structures

shear tend to rotate in the hole [see also Fig. 10.6(c)], causing a nonuniform distribution of bearing stresses between the plate and the rivet. When rivets are in double shear, as shown in Fig. 10.6(b), the joint is loaded symmetrically, reducing the tendency of the rivet to rotate and resulting in a uniform bearing stress distribution in the rivet. In order to account for these actions, the AISC decreased the allowable bearing stress for rivets in single shear to 80% of that allowed for rivets in double shear (as we can see in Table 10.1). Thus rivets in double shear are 25% stronger in bearing and 100% stronger in shear (since there are two shear areas) than rivets in single shear, as calculated from the values in Table 10.1.

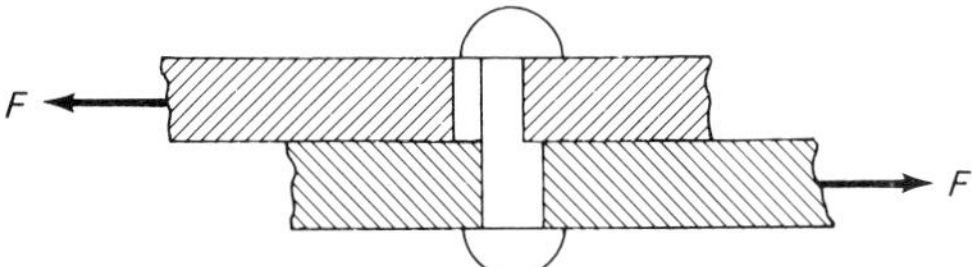

FIGURE 10.7   Rivet crushing.

Recent tests have shown that in bearing, the critical portion of a connection is not the fastener; rather the critical portion of a connection (for bearing) is the material of the connected parts. Recognizing this, the 1970 AISC manual permitted the fastener to be checked for bearing failure using the properties of the connected part. The allowable bearing stress on the projected area of bolts and rivets in bearing-type connections, as given by this code, is $S_b = 1.35S_y$, where S_y is the yield stress of the connected part.

RIVET SETUP

After a rivet is hot formed, it cools and in this process contracts, placing the plates in compression, which causes bearing under the heads of the rivet and sets up initial tensile stresses in the rivet equal to the yield point of the material of the rivet. Due to its diametrical contraction, the rivet does not fill the hole in the plates, allowing joint slippages to occur. All of these effects leave many unknowns in the actual stress distribution in the riveted joint.

Plate Failure

TENSION IN NET PLATE SECTION (TEARING)

A rivet decreases the area available to resist the direct load placed on a riveted joint. Figure 10.8 shows this type of plate failure. In addition to the simple joint shown, a more complex joint, such as a butt joint, could sustain this type of failure, which occurs in the plates being joined as well as in the connecting (or *strap*) plates.

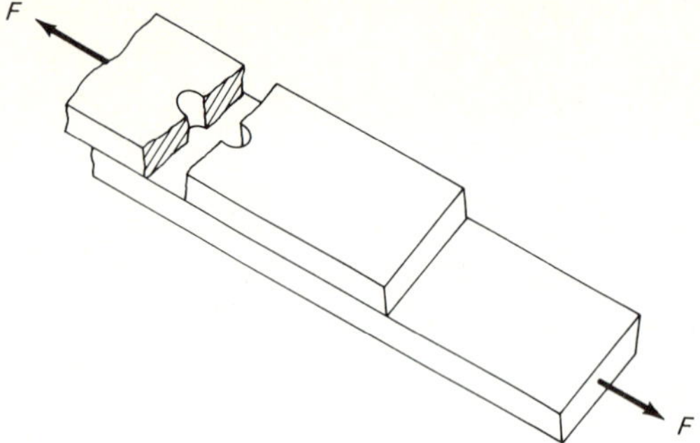

FIGURE 10.8   Plate tearing across net section.

BEARING FAILURE OF PLATE (CRUSHING)

In Fig. 10.7 we saw the material of a rivet being crushed by the plate. However, the plate against which the rivet bears is in compression, and the plate material can fail and wrinkle due to excessively high compressive stresses. Crushing of the plate will cause the joint to loosen due to the motion such action causes. In a butt joint the straps are also subject to this crushing action. However, by making the sum of the strap thicknesses greater than the plate thickness, it becomes unnecessary to investigate the straps in bearing.

PLATE SHEAR (EDGE FAILURE)

If a rivet is placed too near the edge of a plate, the plate may shear, as shown in Fig. 10.9. It is also possible for the joint to fail by the opening up or tearing of this plate in the area in back of the rivet. The failures so described can be eliminated by designing the joint for the proper edge distance, depending upon the rivet size.

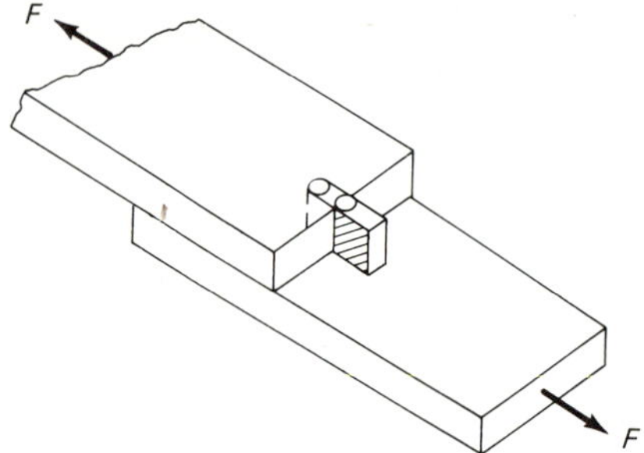

FIGURE 10.9   Plate edge shear failure.

PLATE TEARING BETWEEN ROWS (DIAGONAL FAILURE)

When a riveted connection has more than one row of rivet holes, failure can occur by a tearing of the plate between rows of holes. This type of failure (sometimes called *diagonal failure*) is due to the rivets in successive rows being too close. Figure 10.10 shows this type of failure, and it is noted that the proper proportioning of the dimensions of a riveted joint minimizes the possibility of this occurrence.

                                        Riveted and Welded Structures

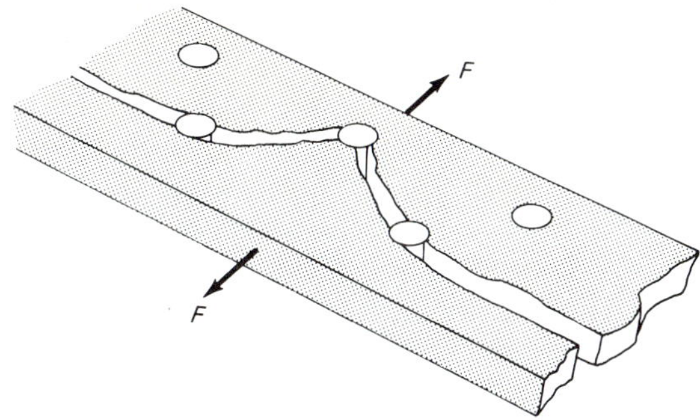

FIGURE 10.10   Diagonal failure in plate.

10.5   DESIGN OF RIVETED JOINTS

Due to the complex interactions that occur in a riveted connection, it is a practical impossibility to design such connections in a rigorous manner. Even attempting to account for the failure modes, as noted in the previous section, presents a task so fomidable as to be nearly impossible. Also, the various codes previously noted have design rules that differ from each other, depending upon the specific application. In order to rationally approach the design of riveted joints, we will make certain simplifying assumptions, which may not be applicable to an actual situation. Due to this, the student is once again cautioned to refer to the specific applicable code before attempting to design riveted structual connections.

With the foregoing in mind, the assumptions made are as follows.

1. Each rivet takes an equal share of the load, or the load is distributed among the rivets in proportion to the shear areas of the rivets. Actual conditions cause the outer row of a multirow joint to take a larger share than its proportional share of the load.
2. There is no failure mode due to edge tearing or shear. The design codes effectively eliminate this type of failure.
3. The effective shear area for a rivet in double shear is twice the effective shear area for a rivet in single shear. The nominal diameter of the rivet is used to calculate the shear area.
4. There are no bending, tension, or setup stresses in the rivets.
5. The crushing (bearing) is uniformly distributed over the projected area of the rivet and plate.
6. The tensile stresses in the plate are uniform over the net area of the plate.
7. Friction between the members of the connection is assumed to be zero and does not enter into the calculation of the strength of the riveted joint. The AISC manual should be consulted for friction-type connections.
8. The process of making the hole in the joint members does not damage the material. Actually, when holes are punched in a plate, the material adjacent to the hole is damaged, and the hole is ragged. Thus the AISC and A.R.E.A. codes both require that punched rivet holes be considered ⅛ in. (3.2 mm)

larger than the nominal rivet diameter when calculating the tension stresses on the net section across a row of rivets. Even where holes are drilled and reamed, they are made $\frac{1}{16}$ in. (1.6 mm) greater in diameter than the nominal diameter of the rivets. In calculations involving pressure vessels designed according to the ASME code, the nominal rivet diameter plus $\frac{1}{16}$ in. is to be used in all plate calculations. We shall use for our calculations a hole diameter $\frac{1}{8}$ in. (3.2 mm) larger than the nominal rivet diameter when calculating plate tension. For machined parts, the hole diameter can be taken equal to the fastener diameter.

9. After forming, the rivets completely fill the holes in the plate. For pressure vessel joints, the driven rivet is assumed to fill the hole completely; and for this application, the rivet diameter used in calculations is the diameter of the rivet hole in the plate. For structural joints, whether punched or drilled, the nominal rivet diameter will be used for rivet shear calculations and bearing calculations.

As a consequence of the previous discussions, we shall consider that a rivet joint can fail in one of three possible modes, and we express these failure modes mathematically based upon the assumptions made.

Failure by Rivet Shear

For single shear of a rivet subjected to the loading in Fig. 10.5(a),

$$S_s = \frac{F}{A_s} = \frac{F}{\pi d_s^2/4} \tag{10.1}$$

where S_s is the shear stress in psi, F is the shear force on the rivet, A_s is the effective shearing area of the rivet, and d_s is the nominal diameter of the rivet. For double shear [Fig. 10.5(b)],

$$S_s = \frac{F}{2A_s} = \frac{F}{2(\pi d_s^2/4)} \tag{10.2}$$

Crushing or Bearing of Plate and/or Rivet

Consider the condition shown in Fig. 10.11, where it is assumed that the rivet is bearing against the plate with a uniform radial pressure, which we shall denote by S_b, the bearing stress. If we draw a free-body diagram of the loading shown in Fig. 10.11(a), we obtain Fig. 10.11(b).

At the angle θ (both positive and negative) a small angle $\Delta\theta$ is shown. The force F on this segment is the stress S_b multiplied by the area of segment $t(r\Delta\theta)$. This force is directed normal to the arc, and F_h and F_v are the horizontal and vertical components of the force F. Reference to Fig. 10.11(b) shows us that the sum of all the vertical force components is zero, since there is a positive (up) and equal vertical force vector for each negative (down) vector. The horizontal vector (in each quadrant) is given by

$$F_h = S_b t(r\Delta\theta) \cos\theta \tag{10.3}$$

If we sum up all these horizontal vectors for angles θ from $+90°(\pi/2)$ to $-90°(-\pi/2)$, we obtain

$$F_h = S_b t(2r) = S_b td \qquad \text{or} \qquad S_b = \frac{F_h}{td} \tag{10.4}$$

Thus the horizontal force is equal to the product of the bearing stress and the projected area of the hole in the plate. This product is equivalent to a uniform stress of S_b acting over the projected area of the hole in the plate, as shown in Fig. 10.11. The rivet bearing is obtained in an identical manner and Eq. (10.4) is also applicable to the calculation of rivet bearing.

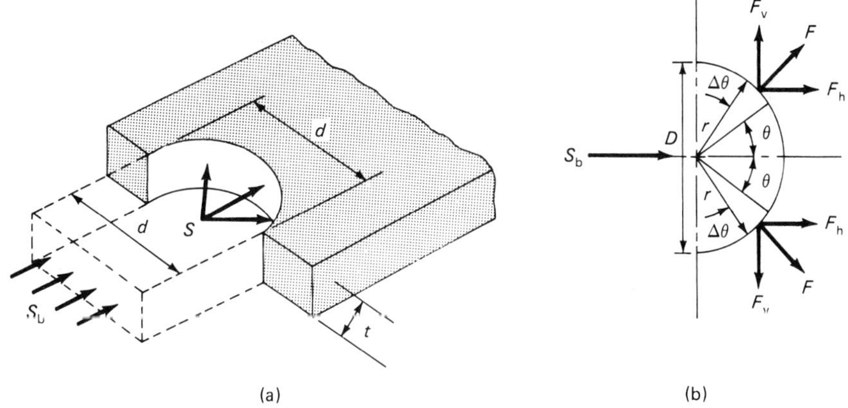

(a)   (b)

FIGURE 10.11   Bearing of plate.

Tension in the Plate

Figure 10.12 shows a plate in tension. As shown, we will consider a uniform tensile stress S_t to act across the net cross-sectional area of the plate. In terms of the dimensions shown in Fig. 10.12 and denoting the plate tensile area by A_t,

$$S_t A_t = F = S_t(b - d)t \tag{10.5}$$

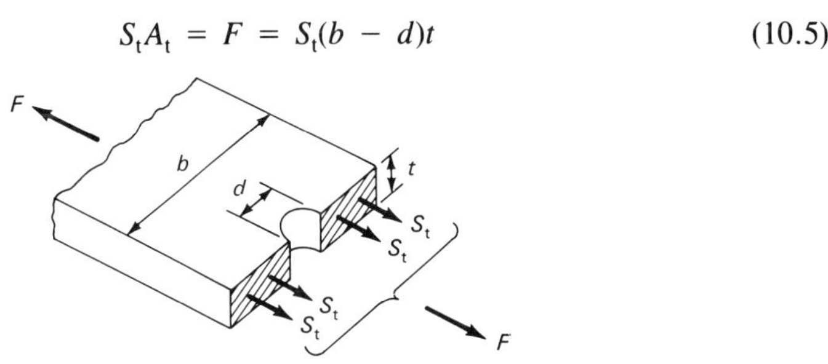

FIGURE 10.12   Plate in tension.

If a section of a riveted connection has n areas subjected to shear, Eq. (10.2) can be written for the rivet shear as

$$S_s = \frac{F}{A_s n} = \frac{F}{n(\pi d_s^2/4)} \qquad (10.6)$$

for single shear. For double shear,

$$S_s = \frac{F}{2A_s n} = \frac{F}{2n(\pi d_s^2/4)} \qquad (10.7)$$

The bearing stress of n bearing areas is

$$S_b = \frac{F}{ntd} \qquad (10.8)$$

Finally, for tension across the net section of the plate with n rivets in a given row, the tensile stress is

$$S_t = \frac{F}{(b - nd)t} \qquad \text{or} \qquad n = \frac{b}{d} - \frac{F}{dtS_t} \qquad (10.9)$$

The 8th edition of the AISC manual gives as an allowable stress in tension a value equal to $0.6S_y$, where S_y is the yield stress of the material.

ILLUSTRATIVE PROBLEM 10.1

A riveted lap joint has to carry a load of 30 000 lb.
  (a) Using the AISC allowable stresses from Table 10.1 (S_t = 20 000 psi, S_s = 15 000 psi, and S_b = 32 000 psi for single-shear bearing), design the joint. Assume each plate to be $\frac{3}{8}$ in. thick. Rivets $\frac{1}{2}$ in. in diameter are to be used, and the plate is 9 in. wide.
  (b) Compare this design with a connection designed after the 1970 AISC manual, using A502 grade 1 rivets in the connection, which has a plate yield stress of 50 000 psi.

SOLUTION

(a) For this problem we do not know how many rivets will be needed. The hole diameter in the plate will be taken to be $\frac{1}{8}$ in. greater than the rivet diameter. The hole is therefore $\frac{1}{2}$ in. + $\frac{1}{8}$ in. = $\frac{5}{8}$ in. in diameter.
  1. Rivet shear:

$$S_s = \frac{F}{A_s n} \qquad n = \frac{F}{S_s A_s} = \frac{30\,000}{15\,000(\pi/4)(\frac{1}{2})^2} = 10.2 \text{ rivets}$$

  2. Rivet (plate) bearing:

$$S_b = \frac{F}{nA_b} \qquad n = \frac{F}{S_b A_b} = \frac{30\,000}{32\,000(\frac{3}{8} \times \frac{1}{2})} = 5.0 \text{ rivets}$$

3. Plate tension:

$$S_t = \frac{F}{(b - nd)t} \qquad n = \frac{b}{d} - \frac{F}{dtS_t}$$

$$= \frac{9}{\tfrac{5}{8}} - \frac{30\,000}{\tfrac{5}{8} \times \tfrac{3}{8} \times 20\,000} = 8.0 \text{ rivets (in one row)}$$

Notice that at least 10.2 (say 10) rivets are needed to provide the required shear area, that at least 5.0 rivets must be provided to yield the necessary bearing area, and that the greatest number of rivets that can be used before tearing the plate is 8.0. Therefore this joint cannot be used. It is necessary to either thicken the plate or increase the rivet diameter, or both.

(b) The allowable tensile stress is $0.6S_y = 0.6 \times 50\,000 = 30\,000$ psi, the allowable bearing stress is $1.35S_y = 1.35 \times 50\,000 = 67\,500$ psi, and the allowable shearing stress for A502 grade 1 material is 15 000 psi. Proceeding as before,

1. Rivet shear:

$$n = \frac{F}{S_s A_s} = \frac{30\,000}{15\,000(\pi/4)(\tfrac{1}{2})^2} = 10.2 \text{ rivets}$$

2. Rivet (plate) bearing:

$$n = \frac{F}{S_b A_b} = \frac{30\,000}{67\,500(\tfrac{3}{8} \times \tfrac{1}{2})} = 2.4 \text{ rivets}$$

3. Plate tension:

$$n = \frac{b}{d} - \frac{F}{dtS_t} = \frac{9}{\tfrac{5}{8}} - \frac{30\,000}{\tfrac{5}{8} \times \tfrac{3}{8} \times 30\,000} = 10.1 \text{ rivets (in one row)}$$

For this case at least 10.2 rivets are needed for shear, at least 2.4 (say 3) rivets are needed for bearing, and the greatest number of rivets that can be used before tearing the plate is 10.1 (say 10). Therefore 10 rivets would be required using this design criterion.

The student will note the widely differing results obtained using these different design bases. It is for this reason that the student is cautioned to use the latest code methods.

ILLUSTRATIVE PROBLEM 10.2

A butt joint is used to connect two 12.5-mm thick plates. Assume the joint to be 75 mm wide. Assume also that there is a single rivet in double shear [see Fig. 10.5(b)], that each upper and lower strap plate is 9.5 mm thick, and that a 12.5-mm diameter rivet is used, as shown in Fig. 10.13. The hole diameter is 12.5 + 3.2 = 15.7 mm. What safe load can this joint (or section of a connection)

Design of Riveted Joints

carry? Use A502 grade 1 rivets and assume that the plate material has a yield stress of 275.8 MPa.

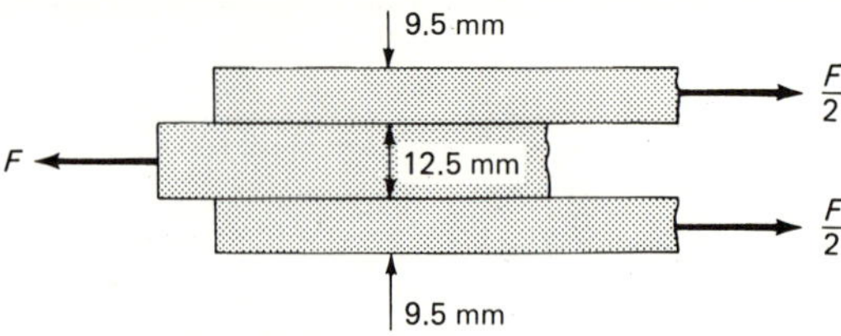

FIGURE 10.13   Illustrative Problem 10.2.

SOLUTION

(a) Rivet shear: $F_s = A_s S_s$ and $A_s = \frac{1}{4}\pi \times 0.0125^2 = 1.23 \times 10^{-4}$ m². For double shear we have twice the area. Also $S_s = 103.4$ MPa. Thus

$$F_s = 2(1.23 \times 10^{-4})(103.4 \times 10^6) = 25\ 486 \text{ N}$$

(b) Rivet (plate) bearing: This is a case of double-shear bearing. We also note that the sum of the thicknesses of the upper and lower straps exceeds the thickness of the main plates. Referring to Fig. 10.13 we can see that half of the total force is taken in each strap, but the thickness of each strap exceeds half of the plate thickness. Therefore it is not necessary to calculate the bearing and tensile stresses in the straps since they will not be the limiting stresses for this joint. Using $S_b = 1.35 S_y$, we have for the main plate

$$F_b = S_b A_b = 275.8 \times 10^6 \times 1.35 \times 0.0125^2 = 58\ 177 \text{ N}$$

(c) Plate tension: $S_t = 0.6 S_y = 0.6(275.8 \times 10^6) = 165.5 \times 10^6$ Pa. Thus

$$F_t = S_t A_t = 165.5 \times 10^6(0.075 - 0.0153) \times 0.0125 = 123\ 504 \text{ N}$$

The safe load is limited to the rivet shear load of 25 486 N.

A term commonly used to evaluate riveted joints is *joint efficiency*. The efficiency of a riveted joint is defined as the ratio of the load-carrying capability of the joint to the tensile load that the main unperforated plate can carry. Ideally it would be desirable to have the strength of a joint equal in bearing, shear, and tension. In practice this may not be possible, and the least load capacity is the design load for a joint. If we consider Illustrative Problem 10.2, we can determine the joint efficiency η as

$$\eta = \left(\frac{25\ 486}{0.075 \times 0.0125 \times 165.5 \times 10^6}\right) \times 100 = 16.4\%$$

which is quite low when compared to the values given in Table 10.3. For this example

Riveted and Welded Structures

TABLE 10.3   Typical Riveted Joint Efficiencies

Type of Joint	Efficiency (%)
Lap	
Single row	53
Double row	67
Triple row	76
Butt	
Single row	57
Double row	77
Triple row	85
Quadruple row	90

we could increase the efficiency of the joint by increasing the diameter of the rivet or by adding another rivet.

ILLUSTRATIVE PROBLEM 10.3

Using A502 grade 1 rivets and a plate yield stress of 40 000 psi, determine the safe load of the butt joint shown in Fig. 10.14. What is its efficiency?

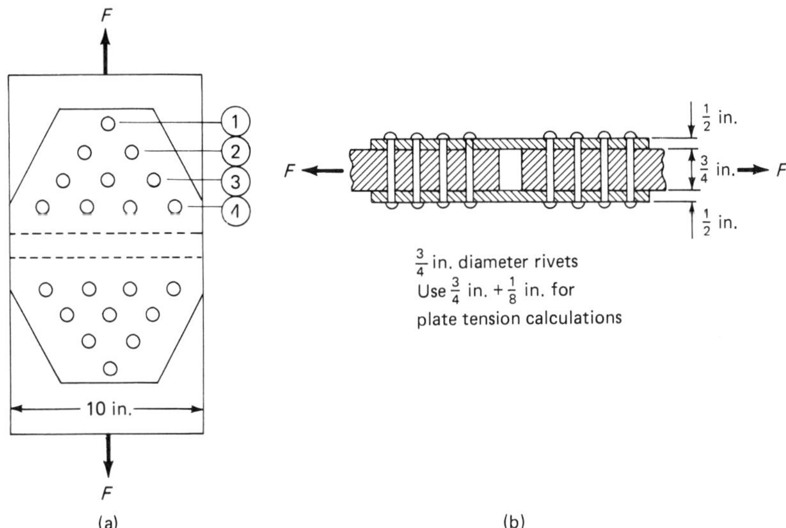

FIGURE 10.14   Illustrative Problem 10.3.

SOLUTION

The load will be assumed to be uniformly distributed over all of the rivets. The resisting load flow is shown in Fig. 10.15, where the branching flows denote the load into the straps. Note that each one of the 10 rivets is assumed to take

$\frac{1}{10}$ of the applied load. For this problem we need not investigate the straps (except to assure ourselves that they are wide enough at any section) since their combined thickness exceeds that of the main plates.

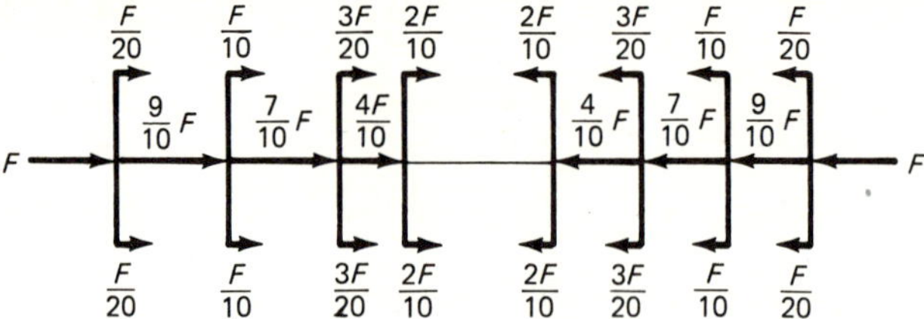

FIGURE 10.15   Load flow for Illustrative Problem 10.3.

(a) Rivet shear: We have 10 double-shear rivets, giving an allowable shear load, for $S_s = 15\,000$ psi, of

$$F_s = n\,2AS_s$$

$$F_s = 10 \times 2 \times \frac{\pi}{4}\,(\tfrac{3}{4})^2 \times 15\,000 = 132\,500 \text{ lb}$$

(b) Bearing: For (double-shear) bearing $S_b = 40\,000 \times 1.35 = 54\,000$ psi. Therefore

$$F_b = S_b\,(dt)n$$

$$F_b = 54\,000 \times \tfrac{3}{4} \times \tfrac{3}{4} \times 10 = 304\,000 \text{ lb}$$

(c) Tension: We shall have to investigate each section using the information in Fig. 10.15.
1.   Section 1 must carry the full load F. Therefore we have for $S_t = 0.6S_y = 24\,000$ psi, and 1 rivet hole:

$$F_t = S_t\,t[w - n(d + c)]$$

where c is clearance. Then

$$F_t = 24\,000 \times \tfrac{3}{4}\,(10 - \tfrac{7}{8}) = 164\,000 \text{ lb}$$

2.   Section 2 carries $\tfrac{9}{10}F$, and there are 2 rivet holes:

$$\tfrac{9}{10}F_t = 24\,000 \times \tfrac{3}{4} \times [(10 - 2(\tfrac{7}{8})] = 148\,000 \qquad F_t = 165\,000 \text{ lb}$$

3.   Section 3 carries $\tfrac{7}{10}F$, and there are 3 rivet holes:

$$\tfrac{7}{10}F_t = 24\,000 \times \tfrac{3}{4}\,[(10 - 3(\tfrac{7}{8})] = 133\,000 \qquad F_t = 190\,000 \text{ lb}$$

4.   Section 4 carries $\tfrac{4}{10}F$, and there are 4 rivet holes:

$$\tfrac{4}{10}F_t = 24\,000 \times \tfrac{3}{4}\,[(10 - 4(\tfrac{7}{8})] = 117\,000 \text{ lb} \qquad F_t = 293\,000 \text{ lb}$$

Riveted and Welded Structures

5. At the inner row each strap must carry the sum of the strap load:

$$\frac{F}{20} + \frac{F}{10} + \frac{3F}{20} + \frac{2F}{10} = \frac{F}{2}$$

(This is a check since by inspection of Fig. 10.5(b) the inner force is obviously $F/2$.) The strap has four holes at the inner row and can carry in tension

$$F_{t/2} = \tfrac{1}{2}[(10 - 4(\tfrac{7}{8})]24\,000 = 78\,000 \qquad F_t = 156\,000 \text{ lb}$$

The joint can therefore carry only 132 500 lb, the limiting factor being rivet shear. The joint efficiency is

$$\eta = \frac{132\,500}{10 \times \tfrac{3}{4} \times 24\,000} = 73.6\%$$

which compares well with the efficiency for the quadruple row butt joint given in Table 10.5.

10.6 ECCENTRICALLY LOADED RIVETED JOINTS

Thus far we have considered each rivet to be loaded equally. In order for this condition to exist, it is necessary for the resultant transmitted force to pass through the center of gravity of the rivet group. Figure 10.16(a) shows the resultant force passing through the center of gravity of the rivet group. In this case all of the resisting forces are equal, parallel to, and opposed to the applied force F. The situation shown in Fig. 10.16(b) differs from the case shown in Fig. 10.16(a) in that the former joint is being twisted by the eccentric loading. If the student visualizes a cantilever beam that is riveted to a column, it will be noted that this is just one case of the situation shown in Fig. 10.16(b). We can consider this joint to be loaded by the force F through the center of gravity of the rivet group and the couple Fe, which is shown. The direct shear of each of the rivets due to the load F through the center of gravity of the rivet group is simply the load divided by the total rivet shearing area. In addition, the couple rotates the bracket about the center of gravity of the rivet group. Denoting the distances of the rivets from the center of gravity as r_1, r_2, r_3, and r_4, respectively, we will assume each rivet to be distorted (and therefore stressed) an amount proportional to its distance from the center of gravity. Therefore we write as a general equation

$$\frac{F_1'}{F_n'} = \frac{r_1}{r_n} \qquad \text{or} \qquad F_n' = \left(\frac{F_1'}{r_1}\right)r_n \qquad (10.10)$$

where F_n' denotes the force on rivet n due to twisting. The moment of this force about the center of gravity of the rivet group is

$$F_n' r_n = \left(\frac{F_1'}{r_1}\right)r_n^2 \qquad (10.11)$$

But the sum of all of the moments on the rivets about the center of gravity of the rivet group must equal the applied moment Fe. Denoting this summation for n rivets as $\sum_{1}^{n}$, we have

$$Fe = \frac{F'_1}{r_1} \sum_{1}^{n} r_n^2 \tag{10.12}$$

Solving for F'_1,

$$F'_1 = r_1 \left(\frac{Fe}{\sum_{1}^{n} r_n^2} \right) \tag{10.13}$$

If we once again refer to Fig. 10.16(b), we can write r_n^2 as

$$r_n^2 = x_n^2 + y_n^2$$

x_n and y_n are coordinate distances from the center of gravity. Using this and rearranging Eq. (10.13), we arrive at the twisting force on any rivet F'_n

$$F'_n = r_n \left(\frac{Fe}{\sum_{1}^{n} x_n^2 + \sum_{1}^{n} y_n^2} \right) \tag{10.14}$$

This force must be added vectorially to the direct force to obtain the total force on the rivet. For rivet 3 in Fig. 10.16(c) we have the direct resisting force F/n and the force due to twisting F'_3 acting as shown with the ensuing resultant F_{3R}. The foregoing is best illustrated by a specific example.

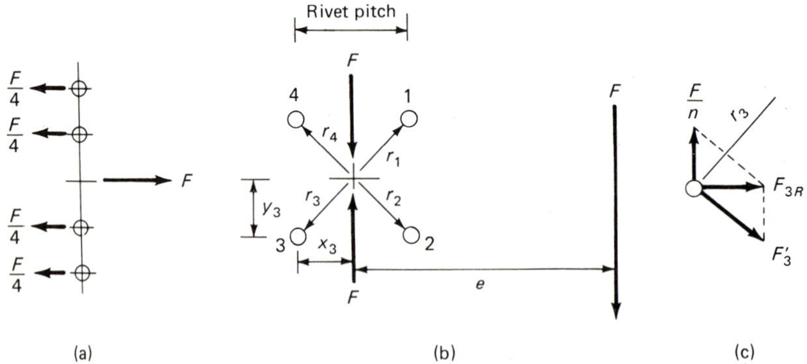

FIGURE 10.16   (a) Centric loading. (b) Eccentric loading. (c) Rivet 3.

ILLUSTRATIVE PROBLEM 10.4

If the rivet group shown in Fig. 10.16(b) has a square pattern in which the rivet

                                              Riveted and Welded Structures

SOLUTION

The direct load on each rivet is $F/4 = 45\,000/4 = 11\,250$ N. The twisting moment is $45\,000 \times 0.250 = 11\,250$ N·m. For each rivet, $x = y = 50$ mm, and $r = \sqrt{0.050^2 + 0.050^2} = 0.0707$ m. The sum

$$\sum_1^4 x_n^2 = 0.050^2 + 0.050^2 + 0.050^2 + 0.050^2 = 0.01$$

and

$$\sum_1^4 y_n^2 = 0.050^2 + 0.050^2 + 0.050^2 + 0.050^2 = 0.01$$

Applying Eq. (10.14),

$$F_1' = \frac{0.0707 \times 11\,250}{0.01 + 0.01} = 39\,768 \text{ N}$$

A vector diagram of the forces on rivet 1 is shown in Fig. 10.17. The resultant of the two vectors shown will have its x and y components equal to the vector sum of the x and y components of these vectors. For the 39 768-N force, the x component is $39\,768 \cos 45° = 28\,120$ N, and the y component is $39\,768 \sin 45° = 28\,120$ N. Thus the resultant will have a left-directed x component of 28 120 N and an upward-directed component of $28\,120 + 11\,250 = 39\,370$ N. The resultant will therefore be $R_1 = \sqrt{(28\,190)^2 + (39\,370)^2} = 48\,422$ N directed as shown.

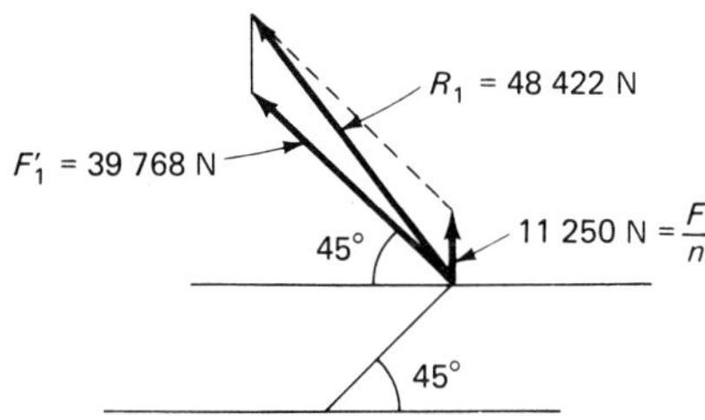

FIGURE 10.17 Illustrative Problem 10.4.

10.7 WELDED STRUCTURES—GENERAL

Welding is a metal joining process characterized by the melting of the parts being joined or by the placing of these parts at a temperature approaching their melting point.

Welded Structures—General

TABLE 10.4 Welding Process Comparison*

Process	Advantages	Disadvantages
Resistance welding	Very flexible Economical Joins three or more pieces with one weld Properties of joint independent of process Can join any combination of metals that alloy together	Poor in fatigue Requires expensive equipment Must be able to reach both sides of joint
Gas welding	Good control of weld metal Heat input anneals weld metal and heat-affected zone	High cost Low production rate Higher heat input; steep temperature gradient More warpage, distortion, etc. Material thicknesses must be nearly equal
Arc welding	High production rate Minimum distortion Low annealing of base metal	Stress concentrations at edge of welds, undercutting, overlaps, etc., likely

*Reproduced with permission from H. A. Rothbart, *Mechanical Design and Systems Handbook*. New York: McGraw-Hill, 1964.

The welding process, when properly executed, offers the advantage of being more economical than riveted or bolted connections, does not require the drilling or punching of holes in the plates which the latter process does, yields a lighter joint, and yields a more uniform load distribution than do riveted joints.

The three common types of welding are *resistance welding*, *gas welding*, and *arc welding*. Resistance welding obtains the required heat from the electrical resistance at the interface of the parts being joined. Gas welding provides the required heat from the combustion of the gas (acetylene, etc.) with air. Arc welding generates heat from the resistance of an electric current across a gas gap (filler metal may or may not be added to the joint), and an inert gas may be used to envelope the area being welded. Table 10.4 compares the advantages and disadvantages of these various welding processes.

As in any technology a certain amount of empirical background has evolved, as well as a nomenclature specific to the topic. Figure 10.18 shows the nomenclature used for certain basic weld joints. Once again the student should be aware that a considerable body of technology has evolved concerning welding, and that the information presented in this and subsequent sections of the present chapter is introductory material. Also standard codes, such as the AISC code, govern in many cases and should be consulted for specific applications.

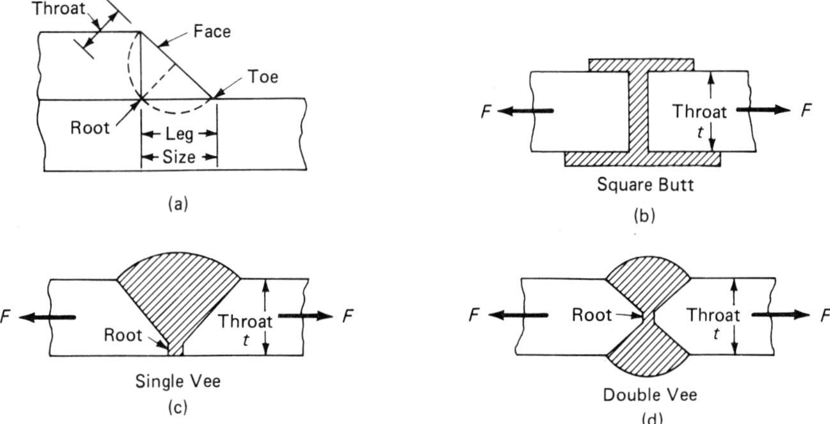

FIGURE 10.18   Basic weld nomenclature.

10.8   WELD DESIGN

The *butt joint* and the *fillet joint* are the two basic joint types in welding. In the design
of butt joints, the cross-sectional area used for tension is that of the thinner plate being
joined away from the weld. Thus for the butt welds shown in Fig. 10.18(b), (c), and
(d), the minimum throat dimension is used. Denoting the minimum throat dimension
as t,

$$F_t = S_t t L \tag{10.15}$$

where L is the length of the weld and S_t is the allowable tensile stress (see Table 10.5).
Due to the possibility of heat damage in the parent metal adjacent to the weld,
Eq. (10.15) is sometimes modified by introducing an additional efficiency term on the
right-hand side to decrease the design load-carrying capacity of the joint. We will
assume this joint efficiency to be 100% and use Eq. (10.15) as it stands since code
design stresses already include this factor.

Fillet welds are used to make joints having surfaces that are usually at right angles
to each other. The basic assumptions made in the design of a fillet weld are that failure
will occur due to shearing of the weld across the minimum section of the weld [the
throat in Fig. 10.18(a)] and that the load is taken up uniformly over the length of the
weld. For the weld shown in Fig. 10.19, the side welds, if they are the only ones
present, will resist the shearing force F throughout their length, and the failure section
is assumed to be the throat of the weld. For an equal-sided triangle, $t =
(\sqrt{2}/2)a = 0.707a$. The joint design criteria therefore yield the following equation
for each side weld:

$$F_1 = 0.707 a_1 L_1 S_s \qquad F_3 = 0.707 a_3 L_3 S_s \tag{10.16}$$

Table 10.5 gives the values of permissible stress for welds from the 1970 edition of

TABLE 10.5 Permissible Weld Stresses*

Kind of Stress	Permissible Stress	Required Electrode	"Matching" Base Metal
Tension and compression parallel to axis of any complete penetration groove weld	Same as for base metal		
Tension normal to effective throat of complete-penetration groove weld	Same as allowable tensile stress for base metal		
Compression normal to effective throat of complete- or partial-penetration groove weld	Same as allowable compressive stress for base metal		
Shear on effective throat of complete- and partial-penetration groove welds	Same as allowable shear stress for base metal		
Shear stress of effective throat of fillet weld regardless of direction of application of load; tension normal to axis on effective throat of a partial-penetration groove weld; and shear stress on effective area of a plug or slot weld. The given stresses shall also apply to such welds made with the specified electrode on steel having a yield stress greater than that of the "matching" base metal. The permissible stress, regardless of electrode classification used, shall not exceed that given in the table for the weaker "matching" base metal being joined.	18 000 psi (124.1 MPa) 21 000 psi (144.8 MPa)	AWS A5.1, E60XX electrodes AWS A5.17, F6X-EXXX fluxelectrode combination AWS A5.20, E60T-X electrodes AWS A5.1 or A5.5, E70XX electrodes AWS A5.17, F7X-EXXX fluxelectrode combination AWS A5.18, E70S-X or E70U-1 electrodes AWS A5.20, E70T-X electrodes	A500 grade A A570 grade D A36 A53 grade B A242 A375 A441 A500 grade B A501 A529 A570 grade E A572 grades 42 to 60 A588

24 000 psi (165.5 MPa)	AWS A5.5, E80XX electrodes Grade 80 submerged arc, gas metal-arc, or flux cored arc weld metal	A572 grade 65
27 000 psi (186.2 MPa)	AWS A5.5, E90XX electrodes Grade 90 submerged arc, gas metal-arc, or flux cored arc weld metal	A514 over 2½ in. thick (63.5 mm)
30 000 psi (206.7 MPa)	AWS A5.5, E100XX electrodes Grade 100 submerged arc, gas metal-arc, or flux cored arc weld metal	A514 over 2½ in. thick (63.5 mm)
33 000 psi (227.5 MPa)	AWS A5.5, E110XX electrodes Grade 110 submerged arc, gas metal-arc, or flux cored arc weld metal	A514, 2½ in. and less in thickness (63.5 mm)

*Reprinted with permission from *AISC Manual of Steel Construction*. New York: American Institute of Steel Construction, 1970. SI units added.

the *AISC Manual of Steel Construction*. It will be noted that for tension or compression, the allowable stress is that of the base metal (as is also the case for shear on groove welds). For fillet welds the allowable shear stress is determined by the combination of electrode and base metal. The allowable shear stress in fillet welds varies from 124.1 to 227.5 MPa.

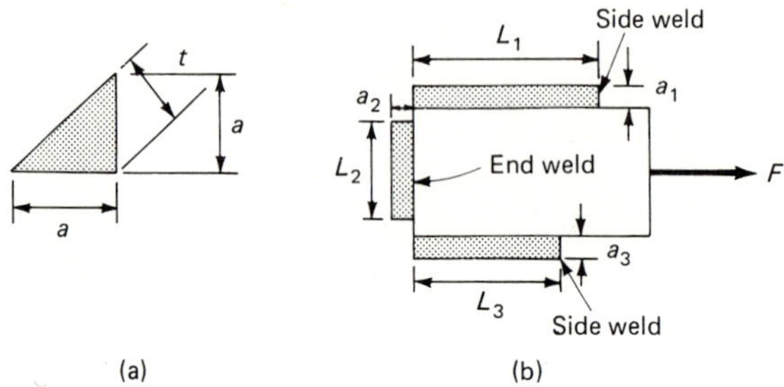

FIGURE 10.19   Fillet weld analysis. (a) Weld cross section. (b) Fillet welds.

If we use the lower value of S_s = 127.3 MPa, Eq. (10.16) becomes

$$F = 90.0 \times 10^6 a_1 L_1 \text{ MPa} \tag{10.17}$$

or

$$F = 13\,000 a_1 L_1 \text{ lb} \tag{10.17a}$$

End fillet welds are probably stronger than side fillet welds, but this is not usually considered in design calculations. An end weld is considered to behave like a side weld, and the design is based on the throat section of the weld. Thus for only the end fillet weld [Fig. 10.19(b)]

$$F = 0.707 a_2 L_2 S_s \tag{10.18}$$

and if S_s = 127.3 MPa,

$$F = 90 \times 10^6 a_2 L_2 \text{ MPa} \tag{10.19}$$

or

$$F = 13\,000 a_2 L_2 \text{ lb} \tag{10.19a}$$

If both end and side fillets are used,

$$F = 90 \times 10^6 (a_1 L_1 + a_2 L_2 + a_3 L_3) \text{ MPa} \tag{10.20}$$

or

$$F = 13\,000 (a_1 L_1 + a_2 L_2 + a_3 L_3) \text{ lb} \tag{10.20a}$$

Riveted and Welded Structures

SOLUTION

Applying Eq. (10.20) and noting $a_1 = a_2 = 0.0125$ m, we have $225\,000 = 90 \times 10^6 \times 0.0125(2 \times 0.050 + L_2)$ and $L_2 = 0.1$ m $= 100$ mm.

In the foregoing discussion we had assumed the welds to be subjected to shear only. This type of shear stress, produced in the direction of the applied load, is known as *primary shear*. In a manner analogous to that considered for eccentrically loaded riveted joints, we find that nonconcentric loads on welds tend to rotate the welded joint, giving rise to secondary shear stresses. Consider the condition shown in Fig. 10.20, where two fillet welds of equal length are subjected to a twisting moment Fe about their center of gravity. In addition to the primary direct shear load on the welds, the structure will tend to rotate about the center of gravity of the welds, inducing a secondary shear stress. We could use the same methods used to calculate the secondary shear stresses in rivets, but we would find the calculations become cumbersome. Instead, we will apply the torsion formula of Chapter 9 to calculate the secondary shear stresses in welds. Using S_s' to denote the secondary shear stress and the notation of Fig. 10.20, the secondary shear stress is,

$$S_s' = \frac{Fer}{J_o} \tag{10.21}$$

where J_o is known as the *polar moment of inertia* of the weld group. Table 10.6 gives the polar moment of inertia of common weld configurations. The total stress in the weld at a given point will be the vector sum of the direct and secondary stresses.

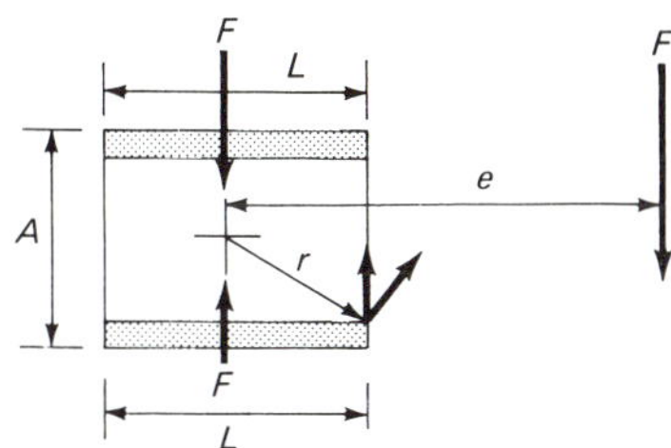

FIGURE 10.20   An eccentrically loaded welded joint.

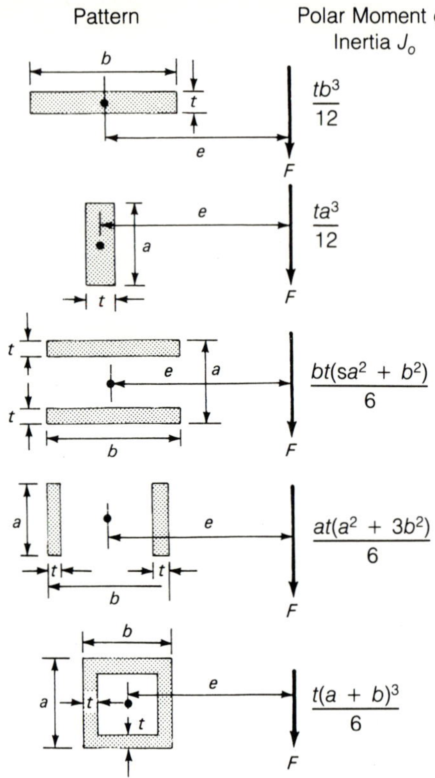

SOLUTION

The direct stress is the force divided by the shear area of the weld. Therefore

$$S = \frac{10\ 000}{0.707 \times (\tfrac{1}{2} \times 5) \times 2} = 2830 \text{ psi}$$

From Table 10.6

$$J_o = \frac{bt(3a^2 + b^2)}{6} = \frac{5(\tfrac{1}{2})(3 \times 4^2 + 5^2)}{6} = 30.4$$

and

$$S' = \frac{Fer}{J} = \frac{10\ 000 \times 4 \times \sqrt{2^2 + 2.5^2}}{30.4} = 4210 \text{ psi}$$

The vector diagram at the point in question is shown in Fig. 10.21. The x and y components of the 4210 psi are: $x = 2.0/3.2 \times 4210 = 2630$ and $y = 2.5/3.2 \times 4210 = 3290$ psi. The resultant stress $R = \sqrt{2630^2 + (3290 + 2830)^2} = 6660$ psi.

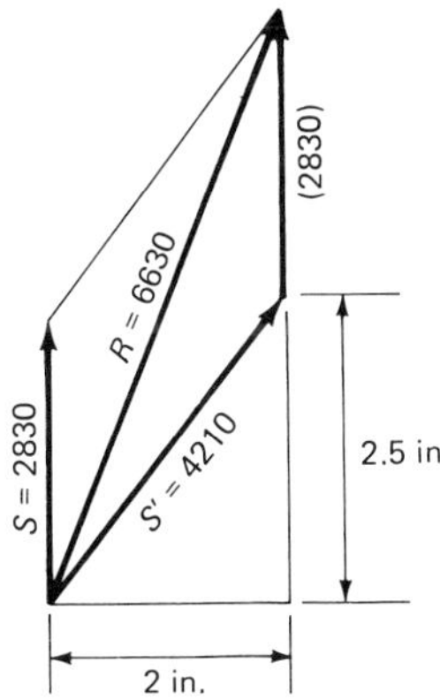

FIGURE 10.21   Illustrative Problem 10.6.

10.9 CLOSURE

The methods of material joining discussed in this chapter, riveting and welding, are widely accepted and used throughout industry. The designs of these connections, however, are quite complex and, to an extent that will not be found elsewhere in this book, it has been necessary to make a multitude of simplifications in our analysis in order to obtain solutions. Also, the connections in question are the subject of extensive empirical studies, which have been incorporated into specific, detailed design codes. While the author is quite aware of the frustration of the student when cautioned to consult other works before doing the design of a particular structural connection, it cannot be helped. Quite often the rationale for a given rule or design criterion is not evident—the best that can be said for these instances is that *these rules work*. These codes are continuously reviewed by cognizant code committees to ensure that their adequacy and applicability are in line with current state-of-the-art knowledge and practices. Revisions of these design rules are incorporated into these codes continuously, and even the most experienced structural designer must spend time with the code to ensure that the data are current. No other satisfactory approach has yet evolved, and the student should heed all of the precautionary notes scattered throughout this chapter.

REFERENCES

American Institute of Steel Construction, THE MANUAL OF STEEL CONSTRUCTION, 7th ed. 1970.

American Society of Mechanical Engineers, THE BOILER AND PRESSURE VESSEL CODE.

Arges, K. P., and A. E. Palmer, MECHANICS OF MATERIALS. New York: McGraw-Hill, 1963.

Black, P. H., and O. E. Adams, Jr., MACHINE DESIGN, 3rd ed. New York: McGraw-Hill, 1968.

Doughtie, V. L., and A. Vallance, DESIGN OF MACHINE MEMBERS, 4th ed. New York: McGraw-Hill, 1964.

Jensen, A., and H. H. Chenoweth, APPLIED STRENGTH OF MATERIALS, 3rd ed. New York: McGraw-Hill, 1971.

Olsen, G. A., ELEMENTS OF MECHANICS OF MATERIALS, 3rd ed. Englewood Cliffs, NJ: Prentice-Hall, 1974.

Rosenthal, E., and G. P. Bischof, ELEMENTS OF MACHINE DESIGN. New York: McGraw-Hill, 1955.

Rothbart, H. A., MECHANICAL DESIGN AND SYSTEMS HANDBOOK. New York: McGraw-Hill, 1965.

Sheiry, E. S., ELEMENTS OF STRUCTURAL ENGINEERING. International Textbook Co.

Timoshenko, S., and D. H. Young, ELEMENTS OF STRENGTH OF MATERIALS, 5th ed. New York: D. Van Nostrand, 1968.

PROBLEMS

In all of the following problems, except as specifically noted, use $S_t = 137.9$ MPa, $S_s = 103.4$ MPa, $S_b = 465.4$ MPa. (These values correspond to A502 grade 1 rivets.) Use 127.3 MPa as the allowable shear stress in fillet welds. Use $0.6S_y$ (plate) for S_t in riveted plates, in accordance with the latest AISC code, with S_y (plate) = 344.8 MPa. In English units, $S_t = 20\,000$ psi, $S_s = 15\,000$ psi, and $S_b = 67\,500$ psi. Use 18 400 psi as the allowable shear stress in fillet welds. Use $0.6S_y$ (plate) for S_t in riveted plates, corresponding to $0.6 \times 50\,000 = 30\,000$ psi.

10.1   A ¾-in. diameter rivet is used to fasten two plates as shown in Fig. P 10.1. Determine the load the joint can sustain.

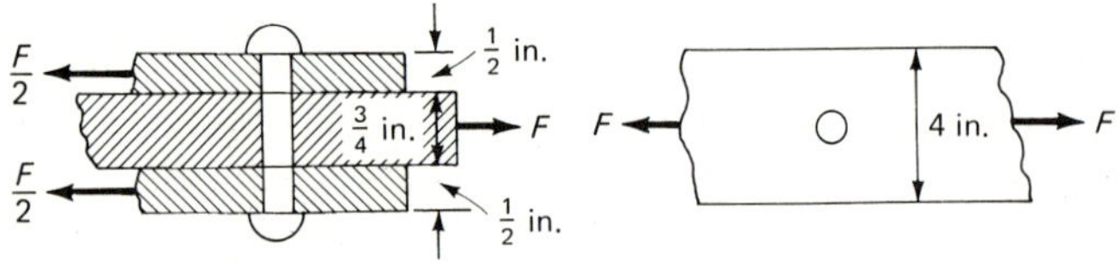

FIGURE P 10.1

10.2   A double-row riveted lap joint has an efficiency of 65%. Determine the size of rivet required if there are two rivets in each row and the plate is ½ in. thick and 4 in. wide.

10.3   If three 16-mm diameter rivets are used to carry the load in the joint shown in Fig. P 10.3, determine the shear stress in the rivets.

FIGURE P 10.3

10.4   Four 12.5-mm rivets carry the load equally in the arrangement shown in Fig. P 10.4. Determine the shear stress in each of the rivets if each rivet is in double shear.

                                            Riveted and Welded Structures

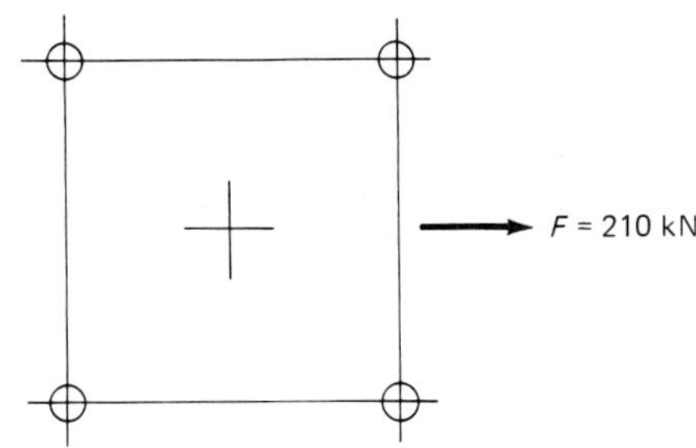

FIGURE P 10.4

10.5   Two 12.5-mm diameter rivets are used to carry the load in a lap joint as shown in Fig. P 10.5. Determine the maximum load the joint can carry.

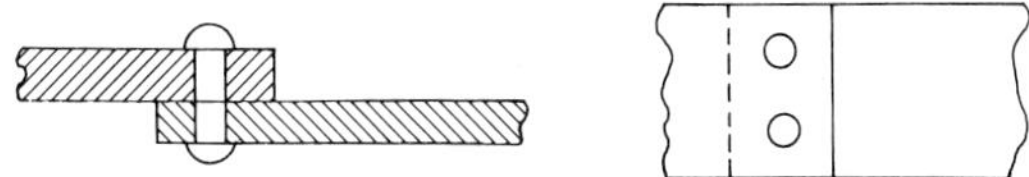

FIGURE P 10.5

10.6   A 19-mm diameter rivet is used to fasten two plates as shown in Fig. P 10.6. Determine the load the joint can sustain.

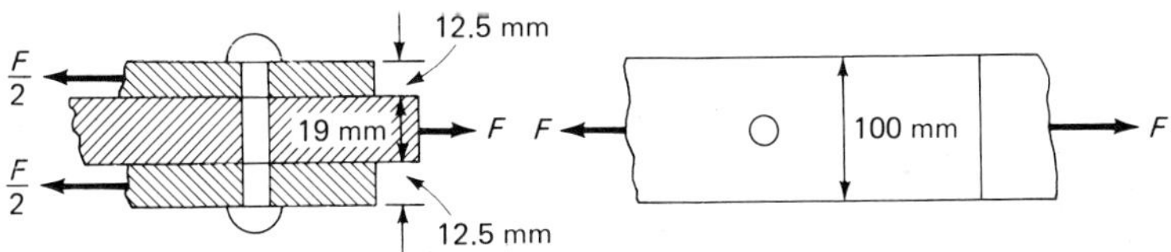

FIGURE P 10.6

10.7   A lap joint is made as shown in Fig. P 10.7. Determine the maximum load the joint can sustain. What is its efficiency?

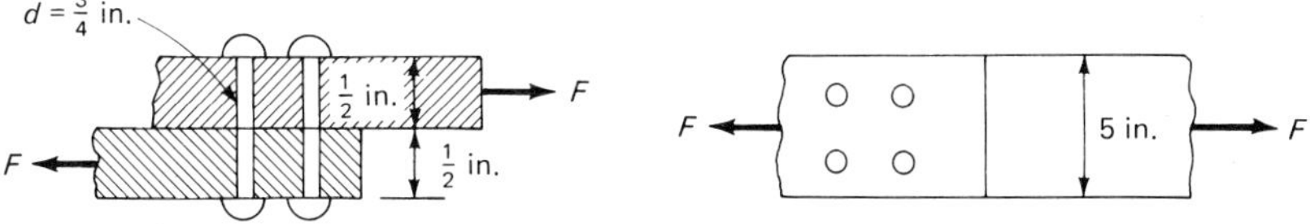

FIGURE P 10.7

10.8   If the lower plate in Problem 10.7 is made ⅜ in. thick, determine the maximum load the joint can sustain and its efficiency.

10.9   A butt joint has two rows of rivets as shown in Fig. P 10.9. If the nominal rivet diameter is ½ in., determine the allowable strength of the joint.

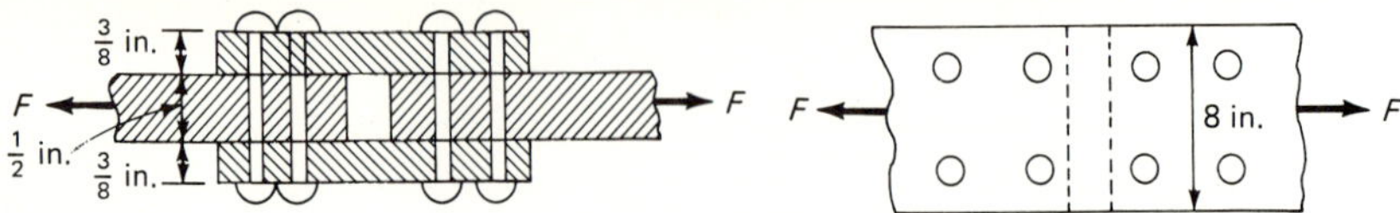

FIGURE P 10.9

10.10   For the triple-riveted butt joint shown in Fig. P 10.10, determine the load that can be carried.

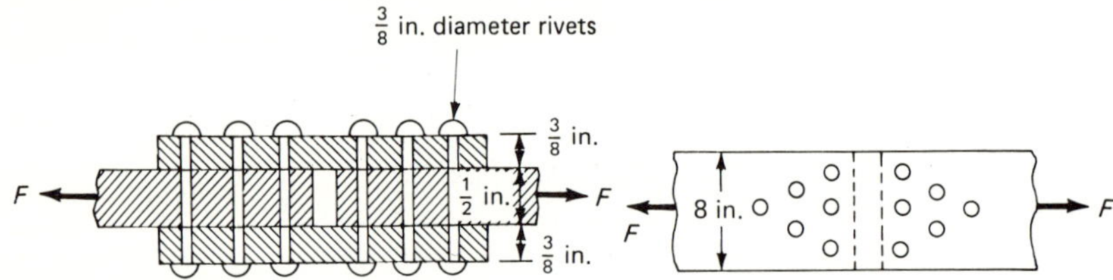

FIGURE P 10.10

10.11   Determine the maximum shear force in the joint shown in Fig. P 10.11.

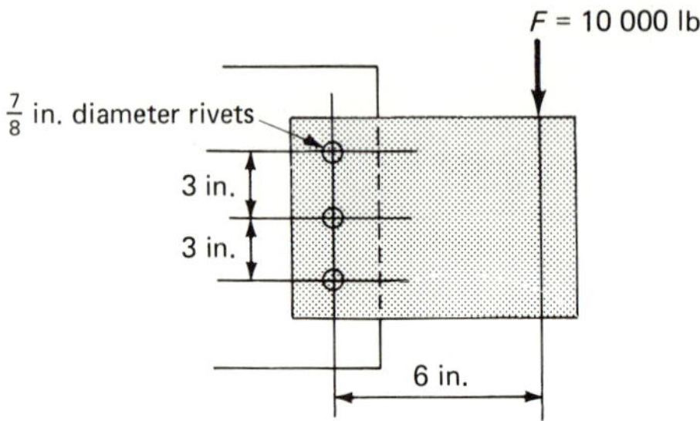

FIGURE P 10.11

10.12   Determine the maximum shear force in the joint shown in Fig. P 10.12.

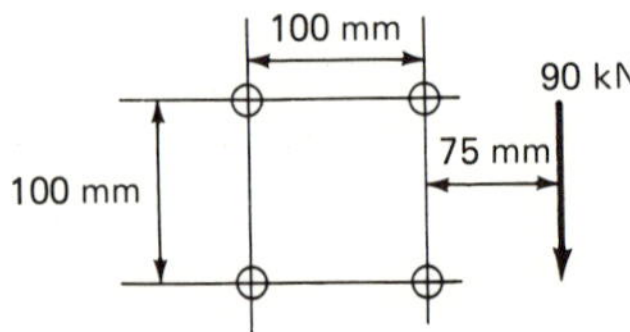

FIGURE P 10.12

                                    Riveted and Welded Structures

10.13   What is the maximum shearing force in the connection shown in Fig. P 10.13?

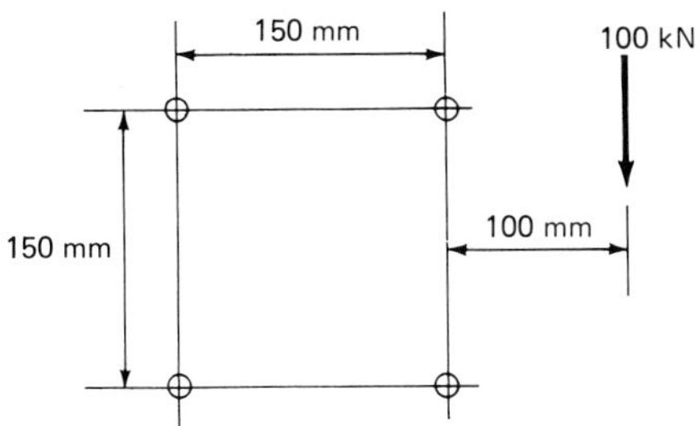

FIGURE P 10.13

10.14   Determine the maximum shear force in the joint shown in Fig. P 10.14.

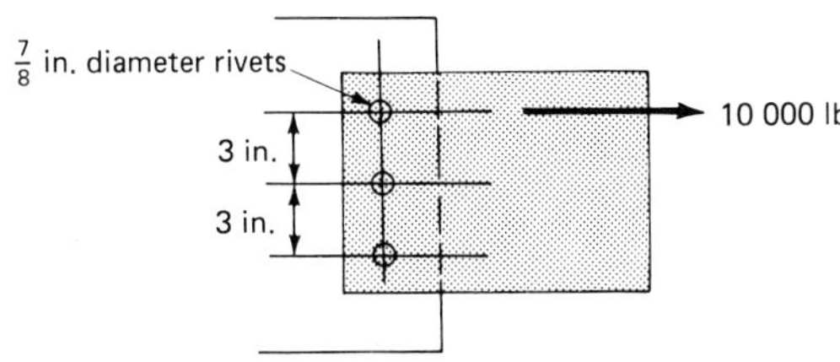

FIGURE P 10.14

10.15   Which of the joints shown in Fig. P 10.15 is stronger?

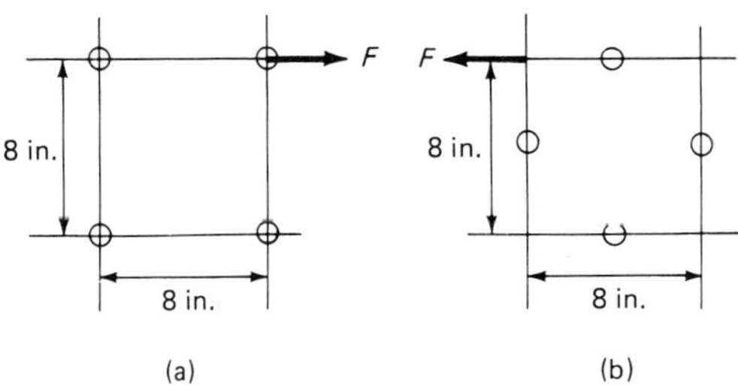

FIGURE P 10.15

10.16   Which of the two arrangements shown in Fig. P 10.16 yields a lower shear stress in the welds?

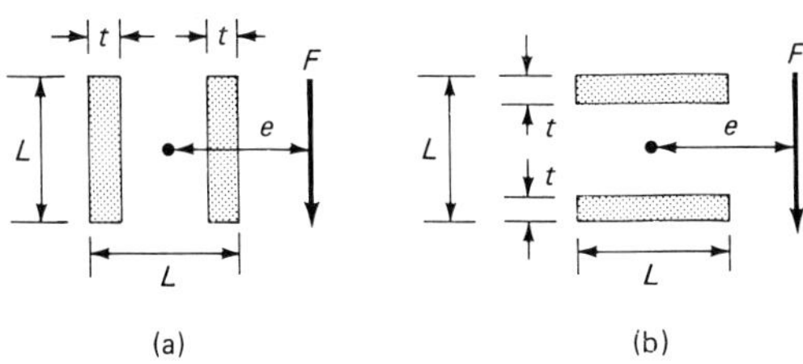

FIGURE P 10.16

Problems   269

10.17 Determine the lengths of weld required in the joint shown in Fig. P 10.17. Assume that the side and end fillet welds are equally long. $S_s = 18\,400$ psi allowable.

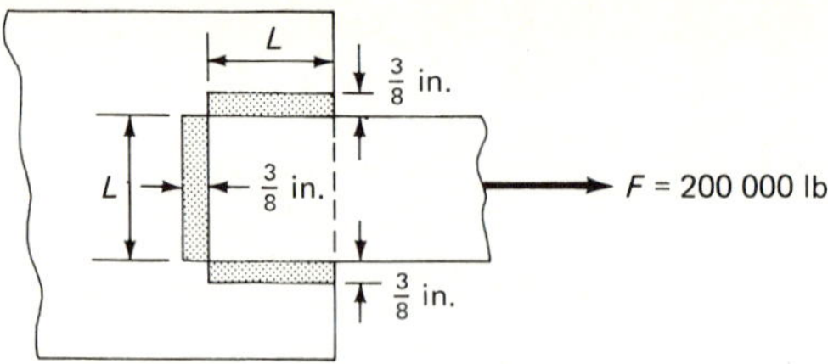

FIGURE P 10.17

10.18 In order to increase the strength of a welded joint, a designer suggests that the design shown in Fig. P 10.18 be used. Comment on it.

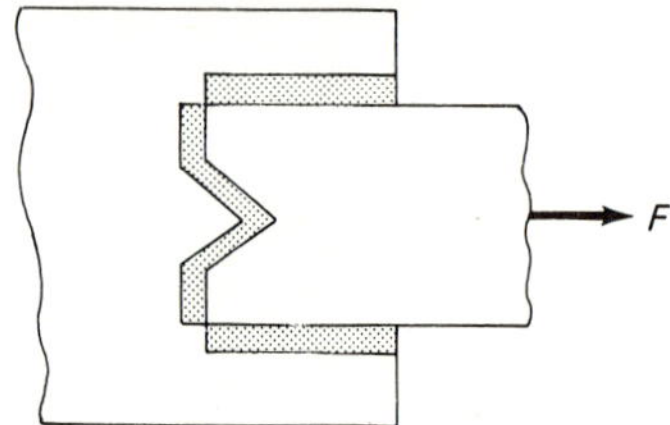

FIGURE P 10.18

10.19 Determine the condition for the placement of F if each length of each side weld is to be subjected to the same stress (Fig. P 10.19). Let L equal the total length, $L_A + L_B$. The joint is not to be subject to twisting.

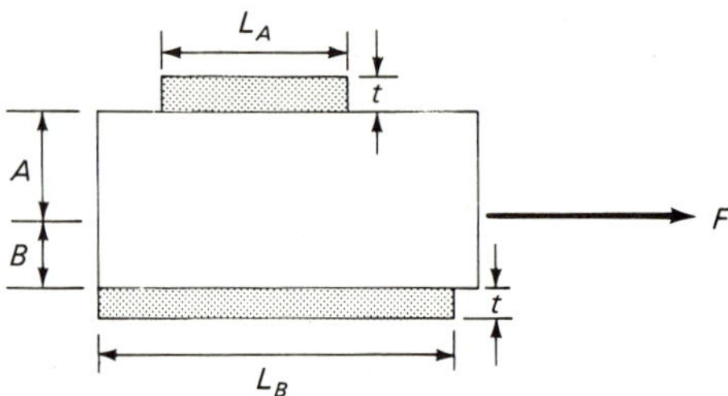

FIGURE P 10.19

10.20 Which corner of the square shown in Fig. P 10.20 will have the highest stress if the periphery of the square is welded completely around?

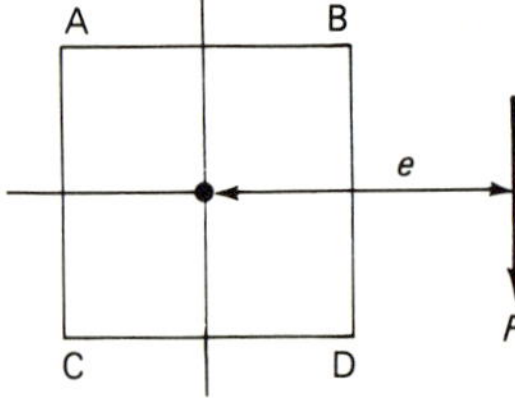

FIGURE P 10.20

Riveted and Welded Structures

10.21 A welded joint is loaded as shown in Fig. P 10.21. Determine the maximum shear stress.

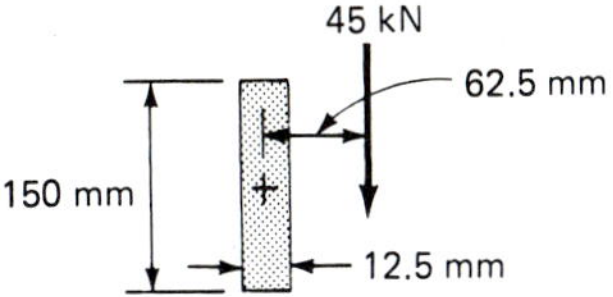

FIGURE P 10.21

11

Shear and Bending Moments in Beams

11.1 INTRODUCTION

In the previous chapters of this book we were primarily concerned with the behavior of structural components subjected to axial forces or torques, not transverse forces. When a member, such as a rivet, was subjected to a transverse force, it was assumed that the member was short and that bending could be neglected. This chapter and the following two chapters concern themselves with a different type of structural member, the *beam*. A beam is a structural member that resists transverse forces normal to its longitudinal axis. These transverse forces cause the beam to bend in the plane of the applied loads, and internal stresses are set up in the material as its resists these loads. Chapters 12 and 13 are concerned with the evaluation of the stresses and deflections in beams and are based upon the work of this chapter. None of the material in this chapter should be novel to the student who has understood statics, and thus it will be presented in a manner that will make our subsequent studies easier. Also, certain conventions are adopted in this chapter that will be retained throughout the rest of this text.

11.2 BEAM SUPPORTS AND BEAM LOADINGS

Supports

The beams can be classified, according to the types of support they have, in terms of the following conventional designations.

SIMPLE BEAMS

A *simple beam* is one that is freely supported (rests on supports) at each end. It is possible to have a roller support at one end and also, at most, one hinged end. Figure 11.1 shows several types of simple beams subject to a single concentrated load. In Fig. 11.1(a) and (b) the support reactions are assumed to be vertical, while in Fig. 11.1(c) the left, hinged support resists the horizontal component of the oblique load and has a horizontal component. In every case we can apply the conditions for equilibrium, $\Sigma F_x = 0$, $\Sigma F_y = 0$, $\Sigma M = 0$, and fully determine the support reactions. For this reason a simple beam is said to be *statically determinate*.

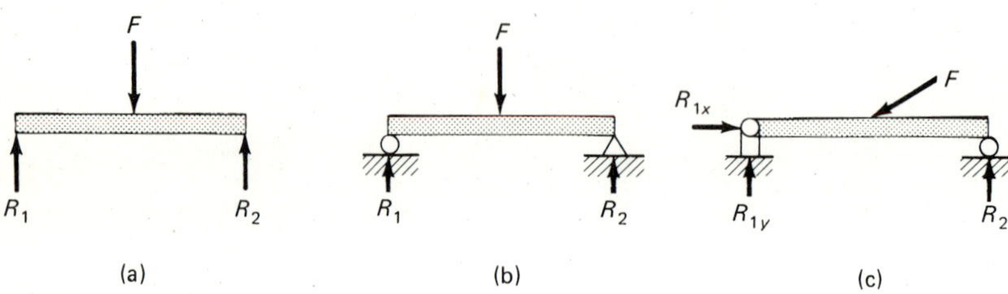

FIGURE 11.1   Simple beams.

CANTILEVER BEAMS

A *cantilever beam* is a beam that is unsupported at one end and is built in at the other end so as to be able to support moments and rotation. Figure 11.2(a) shows a cantilever beam with a concentrated load at its free end. The conditions that exist at the fixed end are complex, but we can, in principle at least, replace the force system at the fixed end either by a moment and a force, as shown in Fig. 11.2(b), or by the support system shown in Fig. 11.2(c). Figure 11.2(b) represents a free-body diagram of the beam, and we can readily see that the three conditions for equilibrium suffice to fully determine the support reactions—this beam is statically determinate.

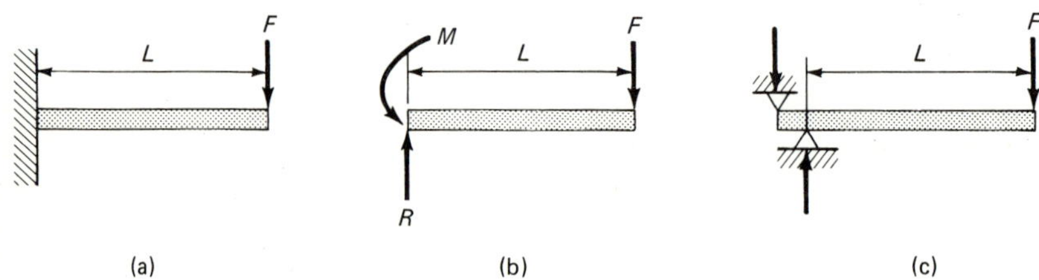

FIGURE 11.2   Cantilever beams.

OVERHANGING BEAMS

An *overhanging beam* is a simple beam with one or both of its ends extending beyond the supports, as shown in Fig. 11.3. All support reactions are assumed to be vertical and can be evaluated from the conditions for static equilibrium. This type of overhanging beam is also statically determinate.

                                              Shear and Bending Moments in Beams

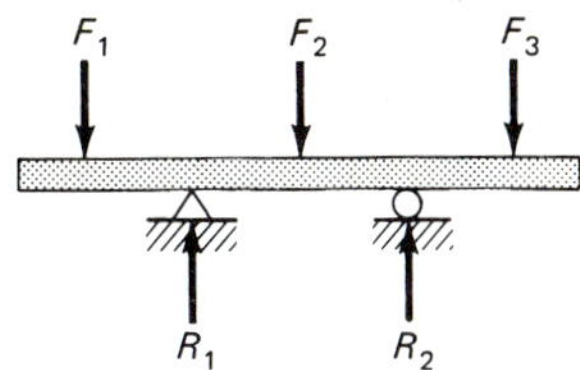

FIGURE 11.3   Overhanging beams.

The foregoing three types of beams are illustrative of those beams whose reactions can be found from the conditions of static equilibrium, and thus we designate them as *statically determinate beams*. There are situations in which the conditions for static equilibrium are insufficient to determine all of the reactions at the supports of a beam. The following three types of beams are illustrative of conditions in which there are either more supports or conditions at the supports than can be evaluated from the three conditions for equilibrium. These beams are called *statically indeterminate beams* and can only be solved by adding additional conditions evaluated from the elastic deformations (deflections) under load. We shall discuss this further in Chapter 13. Let us continue with our classification of beam types.

BUILT-IN BEAMS

A *built-in beam* has both ends built into masonry, with each end fully restrained, as shown in Fig. 11.4(a). It can be considered (schematically) to be a double-ended cantilever beam with the end conditions illustrated in Fig. 11.4(b) and (c).

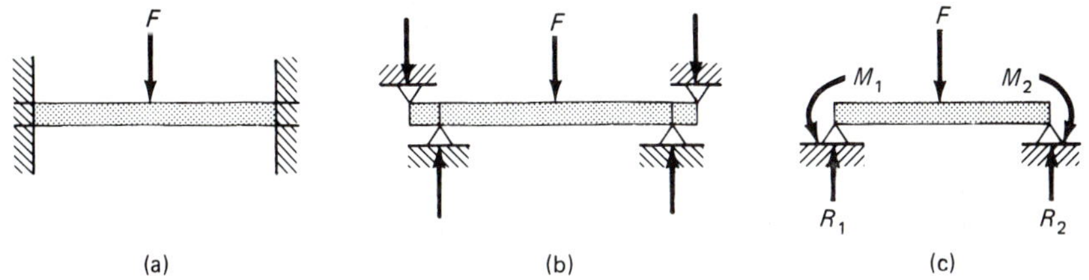

FIGURE 11.4   Built-in beams.

PROPPED BEAMS

The *propped beam* is a cantilever beam with a support under the free end, as shown in Fig. 11.5(a). The equivalent beams, from a force and moment viewpoint, are shown in Fig. 11.5(b) and (c).

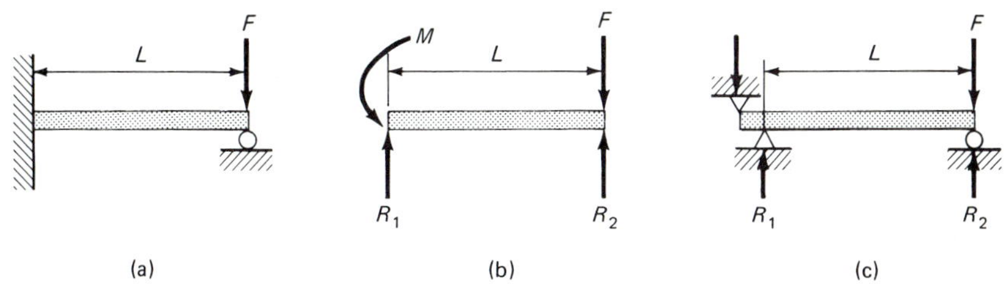

FIGURE 11.5   Propped beams.

Beam Supports and Beam Loadings

The *continuous beam,* a beam having multiple supports, is our last example of statically indeterminate beams. In this case there are more supports than are necessary (from a consideration of statics only) and conditions insufficient to explicitly evaluate all of the reactions. This beam is shown in Fig. 11.6.

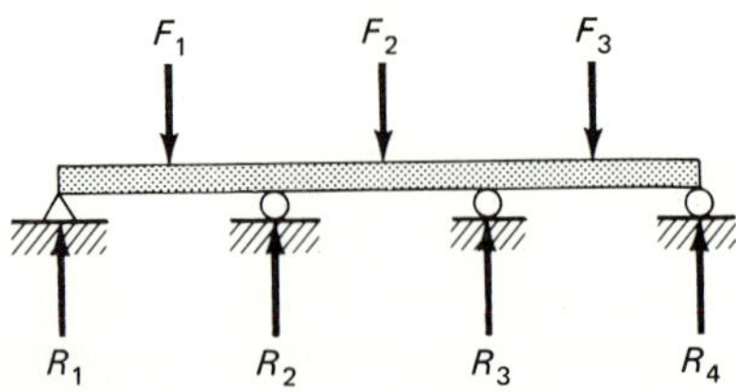

FIGURE 11.6   Continuous beams.

Loads

In the previous discussion we have used the concept of a concentrated load without defining it. A weight hung from a wire is an example of a concentrated load; that is, a *concentrated load* is a load that is being supported by an area sufficiently small so that the load can be considered as being carried by a single point on the beam, as shown in Fig. 11.7. In addition to the concentrated load, we shall deal with other types of beam loads.

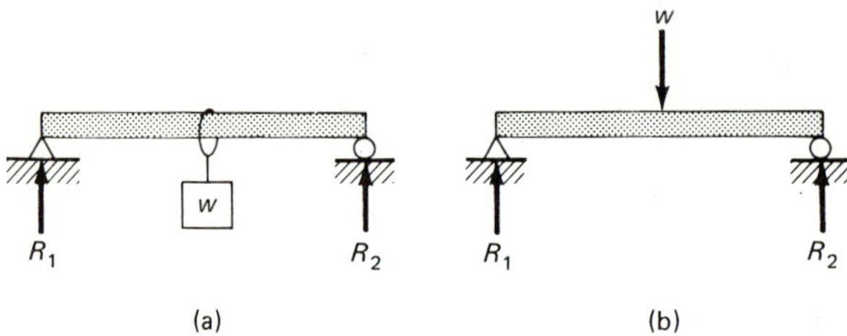

FIGURE 11.7   A concentrated load.

A load that is uniformly distributed over a portion of a beam is called a *uniformly distributed load.* Examples of this type of loading occur on roofs, warehouse floors, trucks loaded with sand, the base of water tanks, and so on. Diagrammatically this type of loading is conventionally shown as in Fig. 11.8(a), and it is usually expressed in terms of pounds per linear foot of beam or newtons per metre, w. The resultant of a uniform load is equal to the load per unit length (or newtons per metre) multiplied by the length over which the load acts. Since this type of loading is a series of equal parallel forces, the resultant can be taken to act at the midpoint of the loaded length when determining external moments. Figure 11.8 illustrates these concepts.

                    Shear and Bending Moments in Beams

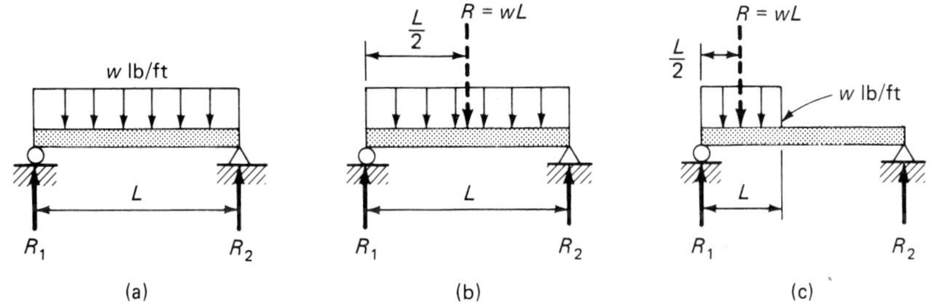

FIGURE 11.8   Uniformly distributed beam loading.

Beams may also carry distributed loads that are nonuniform in character. One example of a well-defined, *nonuniform load* is the load carried by a vertical wall (cantilever) used as a dam. In this case the loading varies linearly from zero at the top of the wall to a maximum value at the bottom of the wall. This loading is shown diagrammatically in Fig. 11.9.

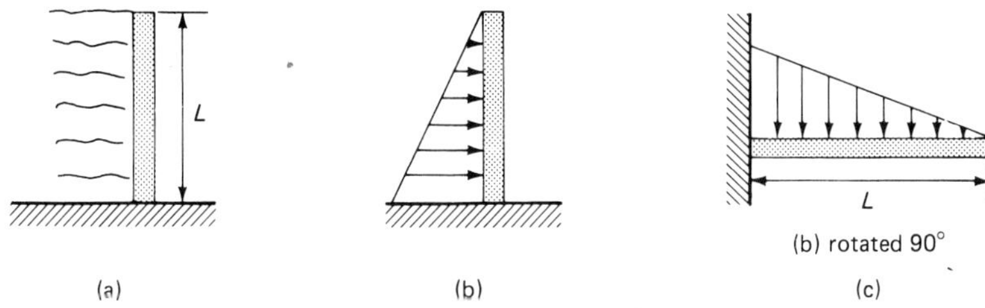

FIGURE 11.9   Nonuniform distributed load (linear).

The final type of loading which we shall consider is the *moment*. The term *moment*, as used here, is synonymous with the term *couple* used in Chapter 9 and is presumably understood from our earlier study of statics. An example of this type of loading is shown in Fig. 11.10, where the applied forces constitute a couple whose magnitude is $F \times d$.

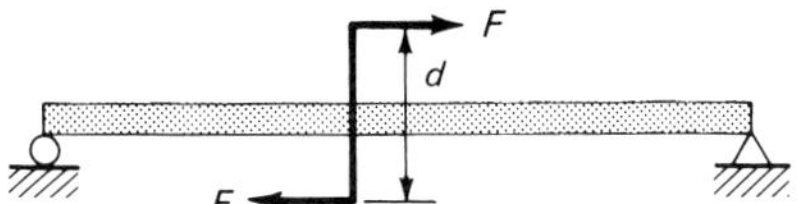

FIGURE 11.10   Beam subject to pure bending.

11.3   SHEAR AND BENDING MOMENT

When a beam is subjected to any of the loadings previously discussed, either singly or in any arbitrary combination, the beam must resist these loads and remain in equilibrium. In order for the beam to remain in equilibrium, an internal force system must exist within the beam to resist the applied forces and moments. In Chapter 12 we shall concern ourselves with the internal stresses in beams arising from these

Shear and Bending Moment   277

resisting forces; at present our concern is to define the external shear forces and moments applied to beams.

For the present let us consider the beam shown in Fig. 11.11 to be rigid and inelastic. When subjected to the concentrated load shown in Fig. 11.11(a), a section of the beam will tend to shear out of the beam, as illustrated in Fig. 11.11(b). If we draw a free-body diagram of the section that is being sheared (we are neglecting the effect of bending at this point), we obtain the diagram in Fig. 11.11(c), where the resultant of the internal shear forces at each surface being sheared is designated as V. Let us now consider the free-body diagram (still neglecting bending) from the left support to section A of our hypothetical beam [Fig. 11.11(d)]. For equilibrium, $V = R_1$ since $\Sigma F_y = 0$. If in place of the force F, the forces F_1 and F_2 [dotted in Fig. 11.11(d)] were placed on the beam, our argument would be the same, and the conditions for equilibrium yield

$$V = R_1 - F_1 - F_2 \tag{11.1}$$

In other words, the shearing force on any section is equal to the summation of the vertical forces on one side of the section (we have denoted this force to be V). It is important to emphasize that the shearing force V is the resultant external transverse shear force acting on the beam at a given section of the beam. This shearing force is equal in magnitude, but opposite in direction, to the resultant of the internal shearing forces in the beam. By convention and for consistency, we shall adopt a standard designation for positive and negative shear. For this purpose we shall consider the portion of the beam to the *left* of the section in question and state that shear is positive when its effect is to push the portion of the beam to the left upward with respect to the portion on the right. Figure 11.12 illustrates this convention. An alternate (but equivalent) sign convention for vertical shear can be simply obtained by considering upward acting forces to be positive and downward acting forces to be negative. Therefore the sign of vertical shear (external as distinguished from internal) is the algebraic sum of the forces to the left of the section in question. Either definition is satisfactory, and both should be fully understood.

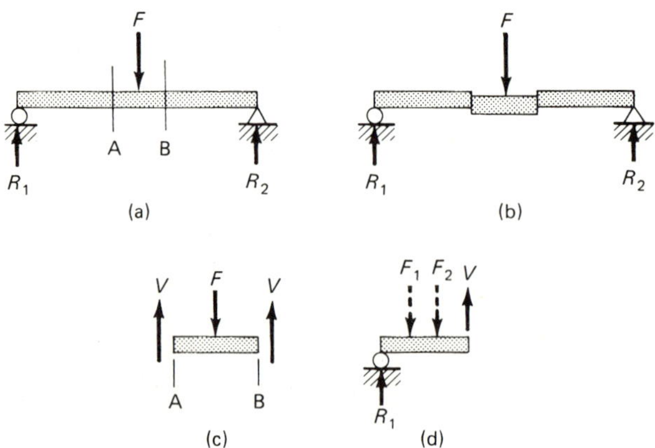

FIGURE 11.11   Shear of a beam.

                                        Shear and Bending Moments in Beams

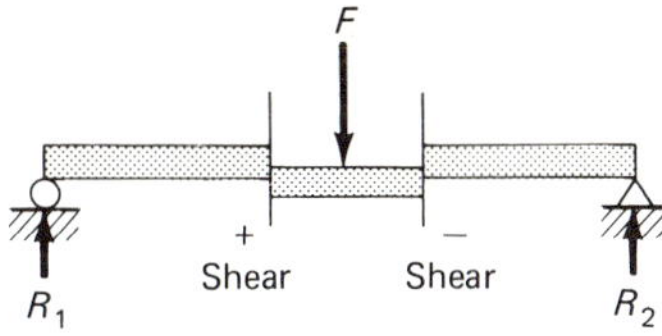

FIGURE 11.12   Shear convention.

In the foregoing discussion, we neglected the bending of the beam in the interest of simplicity. Let us now ignore the shearing forces and turn our attention to bending of the beam. In order to study the action of the beam in bending, we must consider its elastic action under load. The force F shown in Fig. 11.13(a) causes the beam to bend. The free-body diagram of Fig. 11.13(b) shows that the beam must provide a moment M at the cut end in order to satisfy the condition that $\Sigma M = 0$. The elastic deformation of the beam places the upper portion of the beam in compression and the lower portion of the beam in tension. As a result of this action we have the resultant internal force vectors C and T shown in Fig. 11.13(c). These forces constitute a couple M, which resists the rotation of the beam. In order for equilibrium to exist, this internal couple must equal the external moment acting on the beam at this section. The external moment at section A is

$$M = R_1 L \tag{11.2}$$

which must equal the internal couple Ta or Ca. At this point we shall define the term *bending moment* to be the algebraic sum of the moments acting on either side of the section in question. Consequently the internal moment will have the same magnitude but will be of the opposite sense to the bending moment.

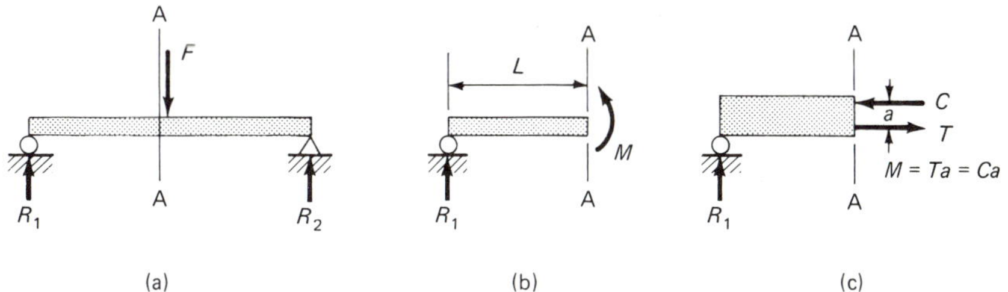

FIGURE 11.13   Bending of a beam.

Consistency of notation requires us to define now positive and negative bending moments. Several equally correct definitions of positive and negative bending moments are as follows.

1. The bending moment is positive when the curvature of the member causes compression in the fibers of the beam on the loaded side and tension on the side opposite the loads.
2. The bending moment at any section of a beam is positive when the algebraic sum of the moments *to the left* of the section is directed clockwise.

Shear and Bending Moment

3. The bending moment is positive when downward forces cause the beam to bend so that it will hold water—the opposite curvature indicates a negative bending moment.

The three definitions are illustrated in Fig. 11.14.

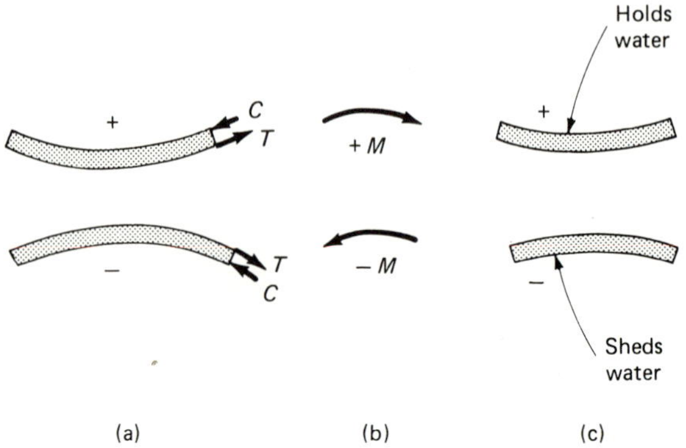

FIGURE 11.14   Bending moment convention.

ILLUSTRATIVE PROBLEM 11.1

Determine the shear and the bending moment at sections A–A, B–B, and C–C for the beam shown in Fig. 11.15.

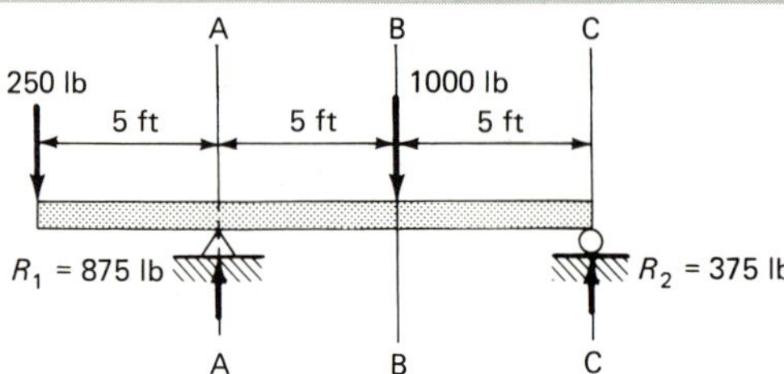

FIGURE 11.15   Illustrative Problem 11.1.

SOLUTION

The first step in the solution is to determine the reactions R_1 and R_2. Taking moments about R_2, we have for $\Sigma M = 0$,

$$-250 \times 15 + R_1 \times 10 - 1000 \times 5 = 0$$

and consequently, $R_1 = 875$ lb. Since $\Sigma F_y = 0$, $R_1 + R_2 = 250 + 1000$ and $R_2 = 375$ lb. Considering the loads to the left of section A–A we have 250 lb acting down and causing a negative curvature. Thus the shear must be −250 lb. The bending moment at this section is also negative due to the curvature and is found to be $-250 \times 5 = -1250$ ft·lb. To the left of section B–B, we find 250 lb acting down and 875 lb acting up. The shear at this section is therefore 875 − 250 = 625 lb acting up (+625). The bending moment at section B–B is −250

                                    Shear and Bending Moments in Beams

× 10 + 875 × 5 = +1875 ft·lb. The final section C–C has a shear of −250 + 875 − 1000, or −375 ft·lb. The bending moment is thus −250 × 15 + 875 × 10 − 1000 × 5 = 0. To summarize:

Section	Shear (lb)	Bending Moment (ft·lb)
A–A	−250	−1250
B–B	+625	+1875
C–C	−375	0

11.4 SHEAR AND BENDING MOMENT IN CANTILEVER BEAMS

In order to continuously portray the shear and bending moment at any section of a beam, it is convenient to plot curves of these quantities along the beam. A *shear diagram* is a curve (plotted at every section of the beam) whose ordinate is the shear acting on the beam at any given section in both magnitude and sense. A *bending moment diagram* is a curve (plotted at every section of the beam) whose ordinate is the bending moment acting on the beam at any given section in both magnitude and sense. These two definitions are quite simple, and the interpretations of the resulting shear and bending moment diagrams by engineers are the same universally.

It will be found to be most advantageous to begin our study of shear and bending moment diagrams by examining the cantilever beam under various types of loadings and subsequently to generalize our study to any type of beam subject to arbitrary loadings. The four types of loadings which we had defined earlier in this chapter and which must be considered on cantilever beams are *concentrated load, constant moment, uniformly distributed load,* and *uniformly varying (triangular) load.*

The cantilever beam illustrated in Fig. 11.16 is shown subjected to a concentrated load at its free end. The shear at any section x has a constant value, equal to $-F$. At the free end the shear goes discontinuously from 0 to $-F$, stays constant along the beam, and at the built-in end goes discontinuously from $-F$ to 0. This is depicted in Fig. 11.16(b). From these facts we can generalize the following.

1. At a concentrated load there is an abrupt change in the shear diagram, which is portrayed as a vertical discontinuity at the section of the beam where the force is applied.
2. The shear diagram starts at zero on the left and ends at zero at the right. We shall find this to be always true, regardless of the loading and the manner in which the beam is supported.
3. If there are no loads on a section of a beam, the shear remains constant on this section of the beam.

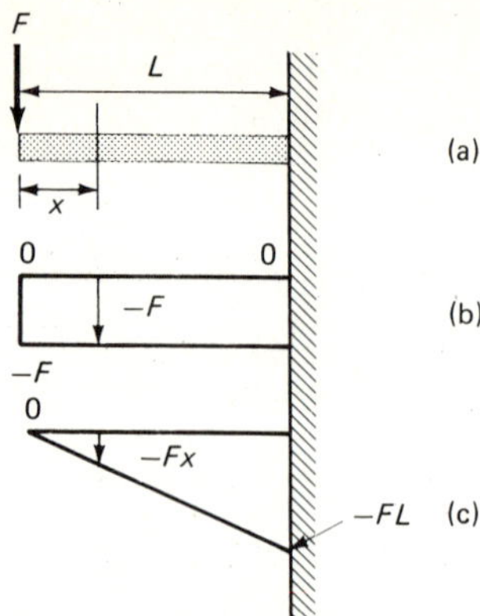

FIGURE 11.16   Cantilever with concentrated load.

The moment at any section of the beam is $-Fx$, increasing (in absolute magnitude) to a value of $-FL$. The moment diagram is therefore a straight line with constant slope, starting at zero on the left and increasing (in absolute magnitude) to a maximum value of $-FL$ at the right.

The bending moment curve is shown in Fig. 11.16(c). Let us summarize the bending moment curve.

1. A concentrated load yields a bending moment with a constant slope, which increases linearly from zero at the point of application of the load.
2. The moment of a given concentrated force F is simply Fx, where x is the distance from the point of application of the load to the section in question. It is imperative that the proper attention be given to the sign convention that we have adopted, namely, a positively directed force (upward) yields a positive bending moment, and a negatively directed force (downward) yields a negative bending moment in the cantilever beam.

ILLUSTRATIVE PROBLEM 11.2

A cantilever beam is loaded by a downward acting concentrated load of 1000 N at its free end. If the beam is 10 m long, draw the shear and moment diagrams, and determine the maximum shear and the maximum bending moment.

SOLUTION

Figure 11.16 is repeated here as Fig. 11.17 with the numerical data for this problem included. As previously discussed, the shear diagram is as shown in Fig. 11.17(b) with the value of the shear constant over the length of the beam and equal to -1000 N. The free-end and the fixed-end shears undergo a discontinuous change from 0 to -1000 N and from -1000 N to 0, respectively. The moment diagram varies linearly from zero at the free end of the beam to a value of $-1000 \times 10 = -10\,000$ N·m at the fixed end of the beam. At any section x, the bending moment is $-1000x$ N·m.

   Shear and Bending Moments in Beams

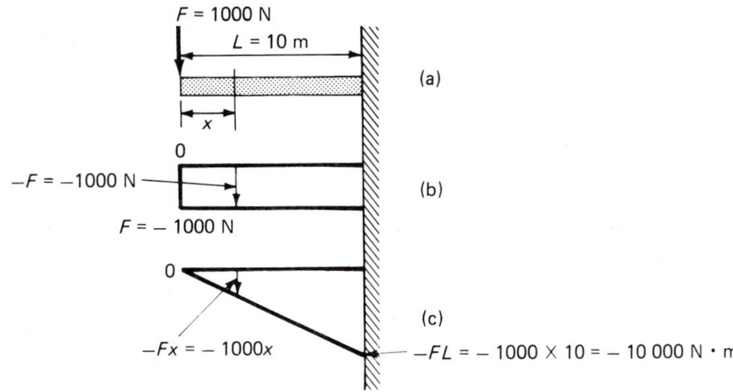

FIGURE 11.17   Illustrative Problem 11.2. (a) Load. (b) Shear (constant = −1000 N). (c) Bending moment.

The second type of loading which we shall consider on a cantilever beam is that of a pure moment at its free end. This loading is shown in Fig. 11.18(a) as a moment M or a couple Fd at the free end. At any section of the beam x there is no net force acting transverse to the beam. Therefore the shear curve is a horizontal straight line whose magnitude is always zero. The bending moment everywhere on the beam is constant and equal to M, the applied moment. Thus the bending moment curve discontinuously rises to M at the free end and stays constant along the length of the beam.

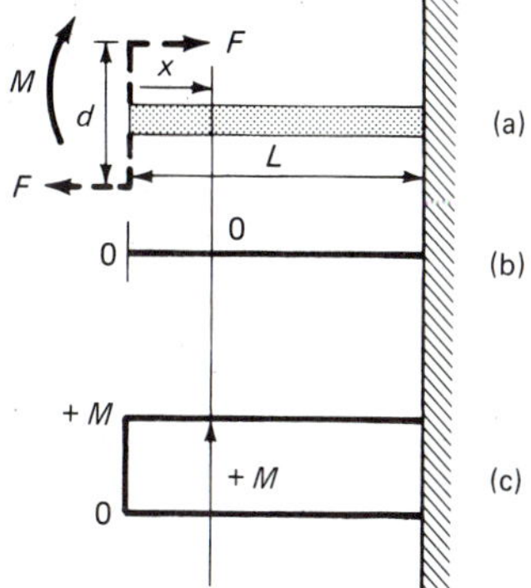

FIGURE 11.18   Cantilever beam with constant moment. (a) Load. (b) Shear. (c) Bending moment.

The third type of cantilever loading which we shall consider is a uniformly distributed load which is denoted as being equal to w lb/ft or N/m. Roof loads and floor loads are common examples of this type of loading. Figure 11.19(a) shows the loading on the beam. At any section x from the left end of the beam we find that the total downward force equals the sum of the individual constant loads of w lb/ft acting for the distance x. Thus the shear at section x is $-wx$, and it varies linearly from zero at the left end to a maximum value (in the absolute sense) of $-wL$ at the right built-in end. This is shown in Fig. 11.19(b), where the shear is denoted as negative in accordance with our sign convention. The moment curve can be established by considering the section of the beam to the left of section x, as shown in Fig. 11.19(d). The resultant of all of the uniform forces is wx acting downward, with its point of application at $x/2$ [Fig. 11.19(d)]. Thus the moment of the uniformly distributed load is $-wx(x/2)$ or $-wx^2/2$, and it varies from zero at the left end to a maximum value of $-wL^2/2$ at the right end of the beam. The equation ax^n is the equation of a parabola where the exponent n denotes the degree or order of the parabola, that is, ax^3 is a cubic parabola, and so on. Note that a uniform load always yields a linear shear variation and a parabolic moment variation.

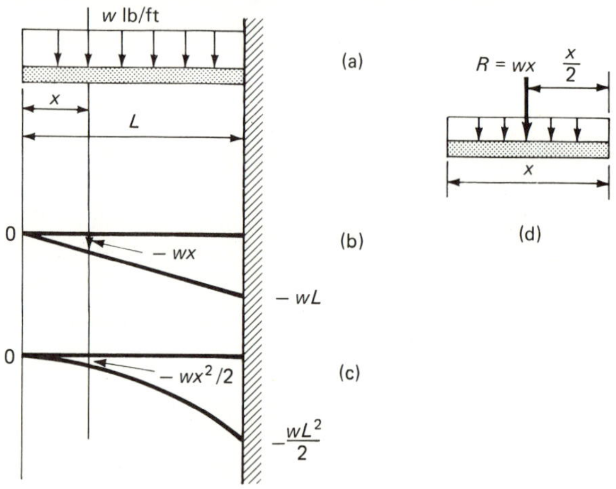

FIGURE 11.19 Cantilever with uniformly distributed load. (a) Load. (b) Shear.
(c) Bending moment. (d) Resultant.

ILLUSTRATIVE PROBLEM 11.3

A cantilever beam is loaded with a uniform load of 500 lb/ft over 5 ft of its length, as shown in Fig. 11.20, which is a modification of Fig. 11.19. Plot the shear and moment diagrams, and determine the maximum shear and the maximum bending moment.

SOLUTION

As is shown in Fig. 11.20, the shear diagram is zero up to the point of the beginning of the load. From this point to the end of the load the shear diagram is a linearly varying curve from 0 to $-500 \times 5 = -2500$ ft·lb. From the end of the uniform load to the end of the beam there are no other loads, and the shear remains constant at its maximum numerical value of -2500 lb. The moment diagram is zero up to the beginning of the load. For the length of the load the bending moment varies parabolically to a maximum numerical value of -500

$\times\ 5^2/2\ =\ -6250$ ft·lb. From the end of the uniform load to the end of the beam the moment increases linearly to its maximum numerical value of $-18\,750$ ft·lb. [From $x\ =\ 10$ ft to $x\ =\ 15$ ft the *increase* in moment is $(x\ -\ 10)(-2500)$ft·lb.]

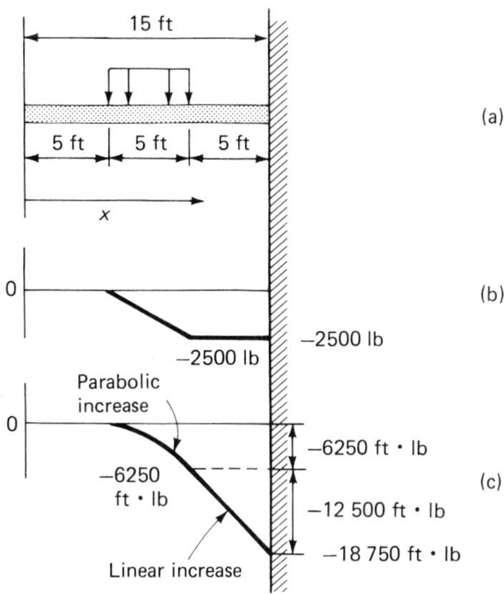

FIGURE 11.20   Illustrative Problem 11.3. (a) Load. (b) Shear. (c) Bending moment.

As the fourth and last case of cantilever beam loading, let us consider a load that varies linearly from zero at the left (free) end of the beam to a maximum value of w lb/ft or N/m at the built-in end of the beam. This type of loading is also called a triangular load, and it is conventionally portrayed as shown in Fig. 11.21(a). The shear at any section x can be obtained by reference to Fig. 11.21(d). The loading diagram is a triangle, and the total downward acting load (shear) is the area of the load triangle having a base x and an altitude wx. The shear is therefore $(-wx/L)(x/2)\ =\ wx^2/2L$, varying from zero at the free end to a maximum of $-wL/2$ at the built-in end. The complete shear diagram is shown in Fig. 11.21(b). We can also construct the moment diagram by referring once again to Fig. 11.21(d). The total loading acting to the left of section x is $-wx^2/2L$, and since the loading can be taken as acting at the location of the center of gravity of the triangular area $(x/3)$, the moment is $(-wx^2/2L)(x/3)$ $=\ -wx^3/6L$, varying from zero at the left end of the beam to a maximum value of $-wL^2/6$ at the built-in end. The triangular loading therefore gives us a parabolic shear variation and a parabolic bending moment variation. One word of caution must be advanced at this time. For the triangular loading, w is the maximum load per foot or metre, and it occurs at the built-in end.

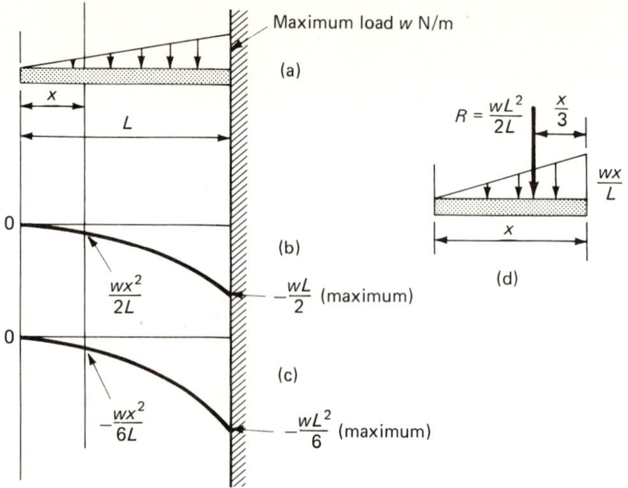

FIGURE 11.21   Cantilever beam with triangular loading. (a) Load. (b) Shear.
(c) Bending moment. (d) Resultant.

Table 11.1 summarizes our findings concerning cantilever beams subjected to the four types of loading that we have studied. In addition, we have had to use the properties of several curves in the previous discussion. Since we shall have occasion to use the properties of the area bounded by these curves, their properties have been tabulated in Table 11.2. Tables 8.1 and 11.2 will provide us with all of the properties of areas that we shall require.

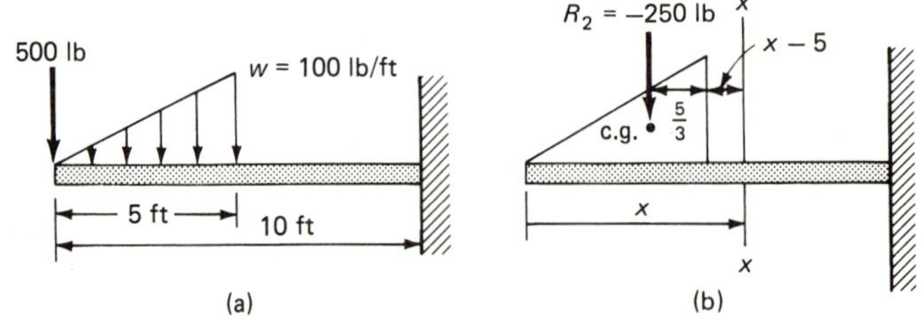

FIGURE 11.22   Illustrative Problem 11.4.

SOLUTION

One method of solving shear and bending moment problems is to consider

          Shear and Bending Moments in Beams

TABLE 11.1 Cantilever Beams

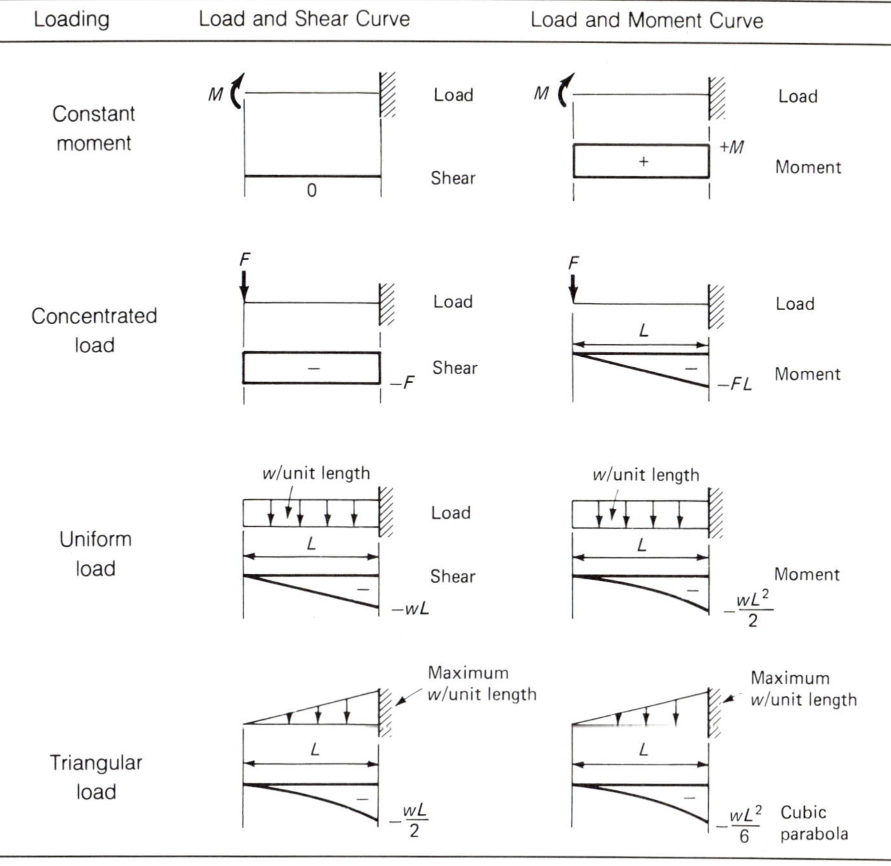

the effect of each item of loading separately and then to add them together. Consider the 500-lb load. Its effect at the end of the beam is to produce a shear and bending moment curve such as was obtained in Fig. 11.16. The maximum moment due to this load is -500×10 or -5000 ft·lb at the built-in end. The shear will be constant and equal to -500 lb due to the concentrated load. The triangular load will give us shear and bending moment curves such as the ones obtained in Fig. 11.21, with the maximum value of shear equal to $-100 \times 5/2 = -250$ lb and a moment at the 5-ft section from the left equal to $-wL^2/6$ or $-100 \times 5^2/6 = -2500/6$ ft·lb. In the portion of the beam where there is no load, the shear will remain constant, but the bending moment will not. The center of gravity (centroid) of the triangular loading is $5/3$ ft from the section that is 5 ft from the left end of the beam, and the resultant load is -250 lb. As shown in Fig. 11.22(b), the moment arm will be located at a distance of $x - 5 + 5/3$ ft when a section to the right of the 5-ft location is considered. Therefore the moment will be $-250(x - 5 + 5/3)$, and it will vary linearly to a maximum (absolute) value of $-5000/3$ ft·lb. Figure 11.23 has been plotted to show the individual effects as well as the composite shear and bending moment curves.

Shear and Bending Moment in Cantilever Beams

Shape and Center of Gravity (c.g.)	Area
Rectangle 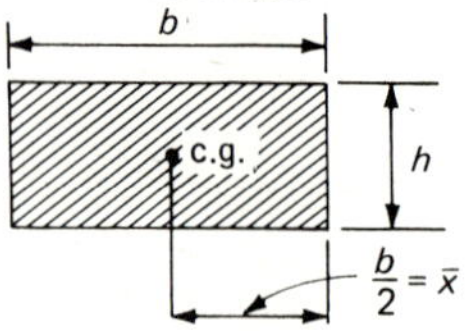 $\dfrac{b}{2} = \bar{x}$	bh
Triangle $\dfrac{b}{3} = \bar{x}$ 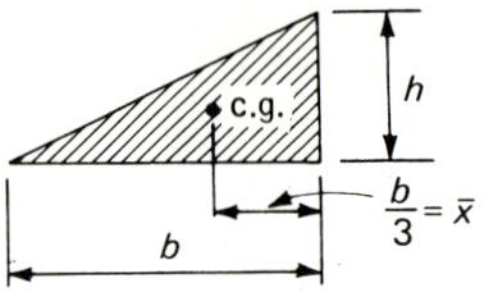	$\dfrac{bh}{2}$
Second-order parabola 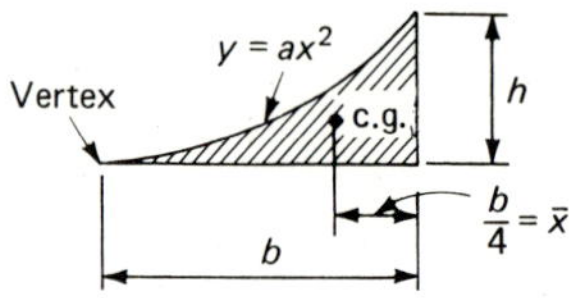 $y = ax^2$ Vertex $\dfrac{b}{4} = \bar{x}$	$\dfrac{bh}{3}$
Third-order parabola 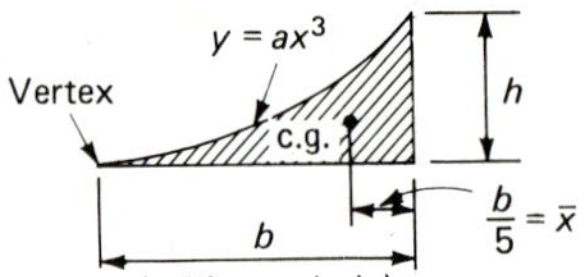 $y = ax^3$ Vertex $\dfrac{b}{5} = \bar{x}$ (cubic parabola)	$\dfrac{bh}{4}$
mth order parabola 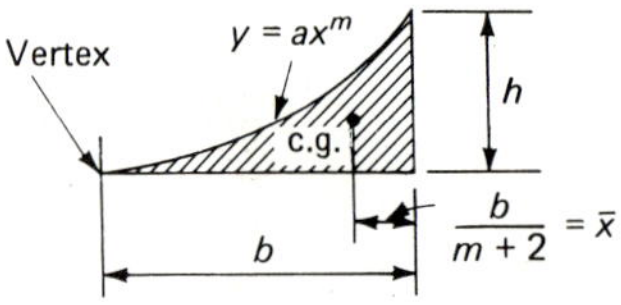 $y = ax^m$ Vertex $\dfrac{b}{m + 2} = \bar{x}$	$\dfrac{bh}{m + 1}$
mth order parabola 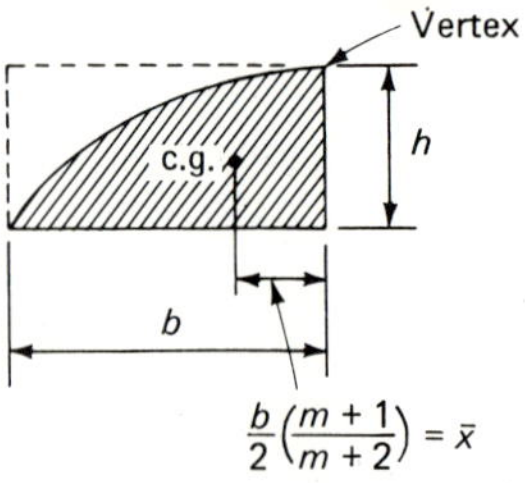 Vertex $\dfrac{b}{2}\left(\dfrac{m + 1}{m + 2}\right) = \bar{x}$	$bh\left(\dfrac{m}{m + 1}\right)$

Shear and Bending Moments in Beams

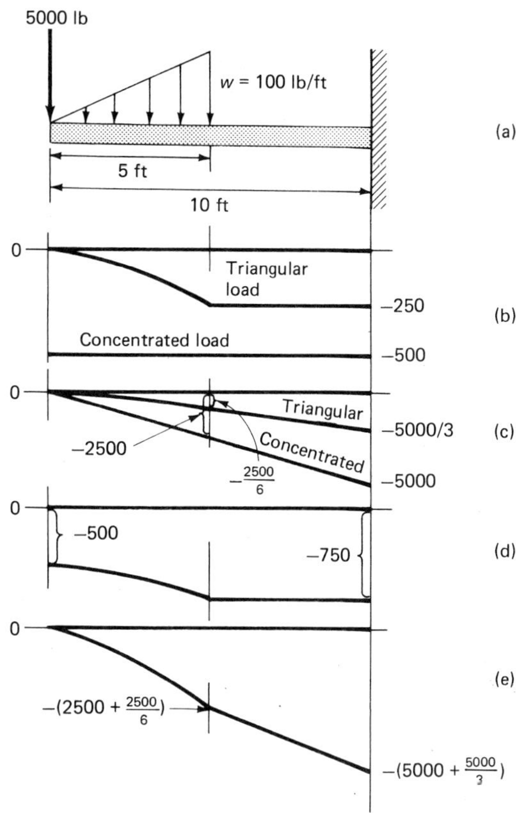

FIGURE 11.23 Illustrative Problem 11.4. (a) Load. (b) Shear. (c) Bending moment. (d) Shear (combined). (e) Bending moment (combined).

11.5 GENERAL RELATION BETWEEN SHEAR AND BENDING MOMENT—APPLICATIONS

The cantilever beam studied in the preceding section represents one of the simplest beams for which shear and bending moment curves can be constructed. More complicated loadings on beams with several different types of supports would require a time-consuming effort if we used the method of step-by-step calculation of these curves. It is possible, however, to derive certain general mathematical relations which we shall find most useful in the construction of these diagrams. In addition, these general expressions will give us an insight into the location of those sections of the beam where the maximum bending moment occurs. For this purpose let us take a beam subject to an arbitrary loading and consider a free-body diagram of a short section of the beam of length Δx. If the section is taken to be sufficiently short, the loading can be effectively considered to be uniform. As shown in Fig. 11.24, there is a shear V and a moment M on the left portion of the element. In the interval Δx there is a uniformly distributed vertical load causing the shear on the right to increase to $V +$

ΔV and the moment to increase to $M + \Delta M$. Since the element is in equilibrium, $\Sigma F_y = 0$, which yields

$$V - w\Delta x - (V + \Delta V) = 0 \tag{11.3}$$

and therefore

$$\frac{\Delta V}{\Delta x} = w \tag{11.4}$$

Since $\Delta V/\Delta x$ is the slope of the shear diagram, we can generalize and state that *the slope of the shear curve is equal to the intensity (magnitude) of the loading at that point*. If we now rewrite Eq. (11.4), we obtain

$$V_2 - V_1 = \Sigma\Delta V = \Sigma w\Delta x \tag{11.5}$$

Equation (11.5) can be interpreted to give us the fact that *the change in shear between any two sections of a beam is equal to the area of the load diagram between these sections*.

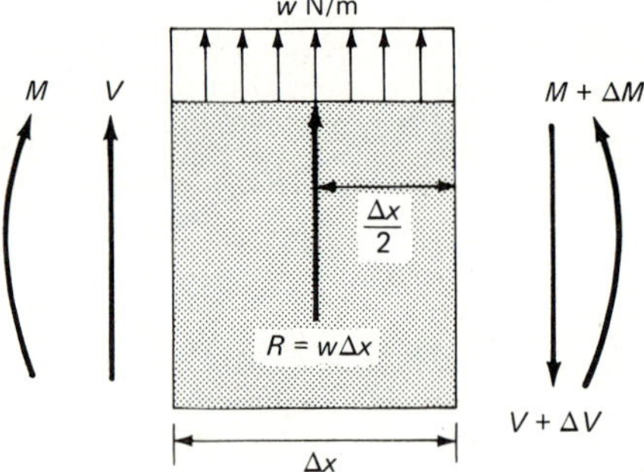

FIGURE 11.24   General shear and bending.

The conditions for equilibrium also require that $\Sigma M = 0$, which, for Fig. 11.24, yields Eq. (11.6) when taken from the right edge of the section:

$$M + V\Delta x + \frac{w\Delta x(\Delta x)}{2} - (M + \Delta M) = 0 \tag{11.6}$$

If we neglect the resulting term $w(\Delta x)^2/2$, since $(\Delta x)^2$ is very small compared to Δx (which was initially taken to be small), we have

$$\frac{\Delta M}{\Delta x} = V \tag{11.7}$$

and

$$M_2 - M_1 = \Sigma\Delta M = \Sigma V\Delta x \tag{11.8}$$

   Shear and Bending Moments in Beams

Equations (11.7) and (11.8) yield the following significant conclusions.

1. *The slope of the moment diagram at a point $\Delta M/\Delta x$ must equal the shear at the point.*
2. *The change in moment between any two sections of a beam is equal to the area of the shear diagram between these two sections.*
3. *The maximum and minimum moments will occur at the section where the shear is zero.*

We have developed all of the information that will be required to draw any shear or bending moment curve. The following summary is a recapitulation of these points, which should be studied and understood before proceeding further in this chapter.

1. Shear is positive when the resultant of all of the loads to the left of the section acts upward.
2. The bending moment is positive when the algebraic sum of the moments to the left of the section is directed clockwise.
3. The shear curve for a beam that supports only concentrated loads is a series of horizontal and vertical straight lines.
4. The change in shear between two sections of a beam is equal to the sum of the loads between the sections.
5. Where the shear diagram is horizontal, the moment diagram will be a straight line. The only exception that will occur is the case in which there is a couple on a beam, causing an abrupt change in the moment diagram, which is not indicated by the shear diagram.
6. Discontinuities occur in shear curves at sections where concentrated loads or reactions act.
7. An abrupt change in shear (as at a concentrated load) indicates a discontinuity of the slope of the moment diagram.
8. Under a uniformly distributed load the shear curve is a sloping straight line.
9. Where the shear diagram is a sloping straight line, the moment diagram will be a segment of a parabola.
10. For concentrated loads the moment curve is a series of connected straight lines. The slope of the curve changes at each concentrated load or reaction.
11. The ordinate of the moment curve at any section of the beam equals the area under the shear curve between that section and a section at which the moment is zero.
12. The slope of the moment curve at any section of the beam is equal to the shear at that section.
13. The slope of the moment curve is greatest where the shear is greatest.
14. The maximum moment always occurs at a section of the beam at which the shear is zero.
15. The change in moment between any two sections of a beam is equal to the area of the shear diagram between these two sections.
16. The slope of the shear curve is equal to the intensity of the loading at that point.

To illustrate the preceding discussion, we shall consider a number of problems. As an aid in doing these and other beam problems, the following procedure will be utilized. It is suggested that the student employ it in doing problems.

1. Calculate the reactions. Where symmetry exists, use it to minimize the arithmetical computations.
2. Determine the shear at sections at which the load changes.
3. Based upon the type of loading, the principles enumerated in the preceding paragraphs, and Tables 11.1 and 11.2, sketch the shear diagram. Accurately determine the sections of zero shear.
4. Compute the bending moment at sections at which the load changes and at sections having zero shear.
5. Using the principles enumerated in the preceding paragraphs and Tables 11.1 and 11.2, sketch the bending moment curve.

Draw the shear and bending moment curves for a beam, simply supported, and carrying a uniform load of w lb/ft or N/m.

SOLUTION

The total load on the beam is wL. By symmetry, the load will be equally divided and the support reactions will each equal $wL/2$. For a uniform load, the slope of the shear curve will be a straight line (item 8) connecting each end reaction. Zero shear (and by item 14, maximum moment) occurs at the center of the beam. We evaluate the maximum moment at the center by taking moments to the left of the center, which gives us $(wL/2)(L/2) - w(L/2)(L/4) = wL^2/8$. By item 9, the resultant bending moment curve will be a parabola having its vertical axis in the middle of the beam.

The shear and bending moment curves for the simply supported beam with

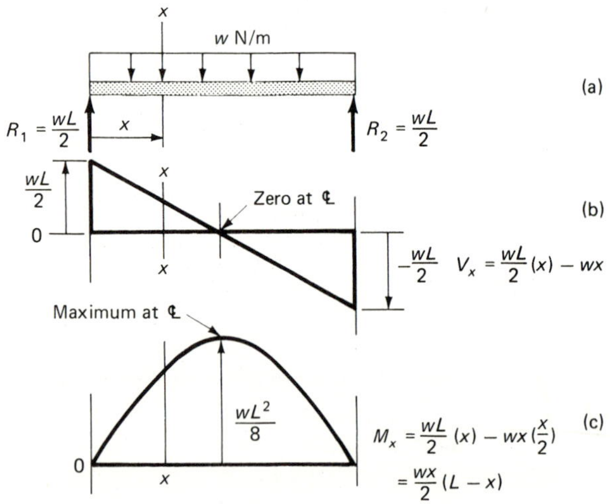

FIGURE 11.25   Illustrative Problem 11.5.

Shear and Bending Moments in Beams

a uniform load are shown in Fig. 11.25. The general expressions for the shear and bending moment at any section x are also given in Fig. 11.25.

ILLUSTRATIVE PROBLEM 11.6

Construct the shear and bending moment curves for the beam of Illustrative Problem 11.1.

SOLUTION

In the solution of Illustrative Problem 11.1 we determined that $R_1 = 875$ lb and $R_2 = 375$ lb. Using these values and noting that there are only concentrated loads on this beam, we can draw a shear curve consisting of a series of horizontal and vertical straight lines. The shear will pass through zero at the left support and at the section where the 1000-lb load is applied. We must investigate both of these sections to determine the location of the maximum moment. The moment at the left support is $-250 \times 5 = -1250$ ft·lb. The moment goes from zero to this value linearly (item 10). At the support there is a shear change (item 6), which causes a change in the slope of the moment curve (item 7). We can calculate the moment at the 1000-lb load by taking moments to the left of the load. Thus $-250 \times 10 + 875 \times 5 = 1875$ ft·lb. We can also obtain the moment at this section by summing the area under the shear curve between the left end (zero moment) and the section (item 11). Thus $-250 \times 5 + (875 - 250) \times 5 = 1875$ ft·lb. The moment curve is therefore a straight line between the value of -1250 ft·lb at the left support and 1875 ft·lb under the 1000-lb load. The value of $+1875$ ft·lb is therefore the maximum moment. Since the shear curve between the 1000-lb load and the right support is a horizontal line having a negative value, the moment curve will decrease linearly from $+1875$ to 0 ft·lb at the right support. The complete shear and bending moment curves are shown in Fig. 11.26.

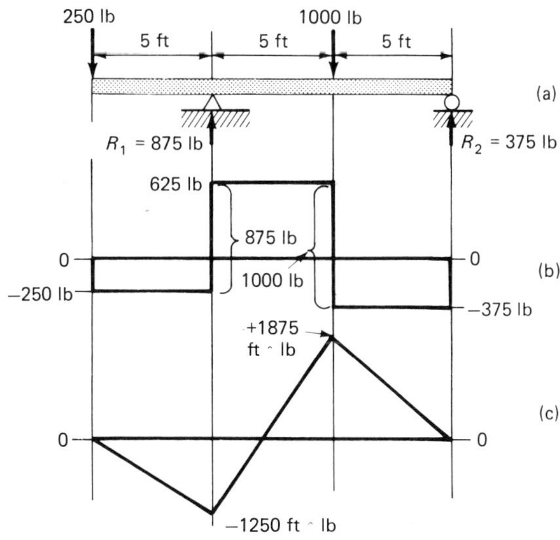

FIGURE 11.26   Illustrative Problem 11.6.

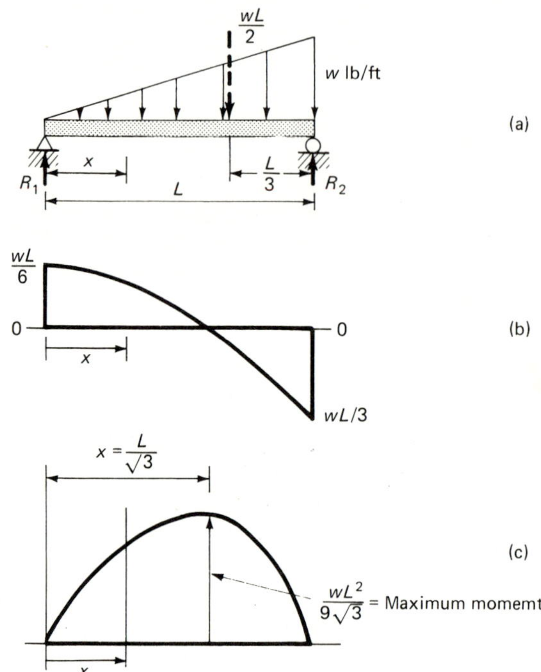

FIGURE 11.27   Illustrative Problem 11.7.

SOLUTION

Our first step in the solution of this problem is to determine the forces at the supports. Since the total load is the area $wL/2$, under the load curve,

$$R_1 + R_2 = \frac{wL}{2} \tag{11.9}$$

In order for the beam to be in equilibrium, $\Sigma M = 0$. The center of gravity of the triangular load from the right-hand side of the beam is found from Table 11.2 to be located at $L/3$. Taking moments about the right-hand support, we obtain

$$R_1(L) - \frac{wL}{2}\frac{L}{3} = 0 \tag{11.10}$$

from which $R_1 = wL/6$. Substituting this value in Eq. (11.9), $R_2 = wL/3$. The shear at any section x from the left end can be found by subtracting from the concentrated end load $wL/6$ the effect of the triangular load. The effect due to the triangular load is obtained from our discussion of cantilever beams (see Fig. 11.21) as $wx^2/2L$. The shear is therefore

$$V_x = \frac{wL}{6} - \frac{wx^2}{2L} = \frac{w}{6L}(L^2 - 3x^2) \qquad (11.11)$$

We can obtain the location of the maximum moment by determining the section at which the shear is zero. Solving Eq. (11.11) for $V_x = 0$ yields

$$x = \frac{L}{\sqrt{3}}$$

We may find the moment at any section x in an analogous manner, that is, by subtracting the moment of the triangular load from the moment due to the concentrated load at the support. Figure 11.21 gives us the effect of the triangular load as $wx^3/6L$. The moment at x is therefore

$$M_x = \frac{wL}{6}x - \frac{wx^3}{6L} \qquad (11.12)$$

The maximum moment is obtained by placing $x = L/\sqrt{3}$ in Eq. (11.12), which yields a maximum moment of $wL^2/9\sqrt{3}$.

It should be noted that each load on a beam can be considered independently and that the final result can be obtained by summing these individual effects. This procedure is known as the *superposition method*. This method can often be used to simplify a specific problem and, if properly executed, will always lead to the correct complete shear and bending moment diagrams for any loading on any beam.

11.6 MOVING LOADS

A unique situation occurs when we consider a vehicle such as a train, a truck, or other moving load as it passes over a beam. The problem of finding the maximum moment is complicated by the fact that as a load moves along a beam, the moment caused by this load changes with its position. The same situation is true of all of the loads that move across the span as a group. Additionally, it is necessary to evaluate the maximum shear, which varies with the position of the load group on the span. Let us examine the system of loads shown in Fig. 11.28. The loads F_1 and F_2 move across the span as a fixed unit with the distance between them kept constant. The bending moment is maximum under one of the loads since the shear must go through zero under one of these concentrated loads. On Fig. 11.28 the resultant of these loads is indicated as being a variable distance x from the left end. The magnitude of R is $F_1 + F_2$, and

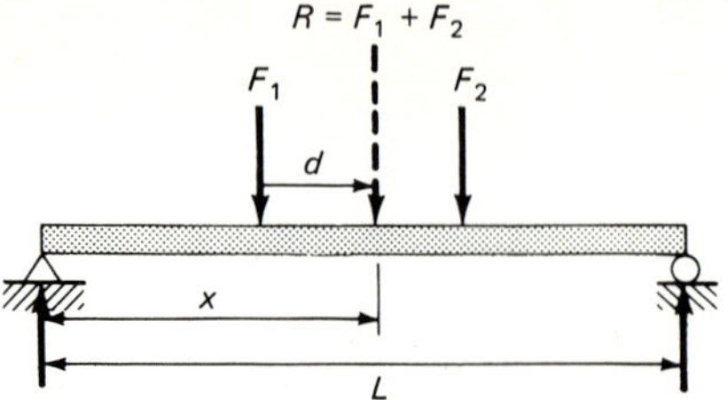

FIGURE 11.28 Moving load on a beam.

its location is determined by taking moments about F_1, that is,

$$F_2 a = Rd = (F_1 + F_2)d \tag{11.13}$$

and

$$d = \frac{F_2 a}{F_1 + F_2} \tag{11.14}$$

If we now consider that the only load on the beam is R, we can obtain the reactions readily. The left reaction is

$$R_1 = \left(\frac{L - x}{L}\right)R \tag{11.15}$$

The moment under the load F_1 is M_1, which is found to be

$$M_1 = R_1(x - d) = \left(\frac{L - x}{L}\right)(x - d)R \tag{11.16}$$

If we expand the terms on the right-hand side of Eq. (11.16),

$$M_1 = -\frac{R}{L}\left[x^2 - (L + d)x + Ld\right] \tag{11.17}$$

Since we wish to determine when M_1 is maximum, it is necessary to determine when the quadratic expression in brackets is maximum. It will be recalled that the solution of a quadratic equation is

$$x = \frac{-b}{2a} \pm \frac{\sqrt{b^2 - 4ac}}{2a} \tag{11.18}$$

when the equation is of the form

$$ax^2 + bx + c = 0 \tag{11.19}$$

The only time that x has a single real value is when $x = -b/2a$. This value of x

                    Shear and Bending Moments in Beams

yields the maximum value of M_1 when substituted into Eq. (11.17). Therefore the maximum moment is

$$x = \frac{L}{2} + \frac{d}{2} \qquad (11.20)$$

We may generalize this result to state that the maximum bending moment under a particular load occurs when the bisector of the distance between that load and the resultant of all of the loads then on the span coincides with the center of the span. The maximum shear occurs at one of the supports and is equal to the maximum support force. This occurs at the left support when the leftmost load is over it or at the right support when the rightmost load is over the right support. Of these two possibilities, the greater reaction will occur at that support nearest the resultant of the group of loads. If one or more of the loads falls off the span, it is necessary to investigate this condition separately.

ILLUSTRATIVE PROBLEM 11.8

A truck having a wheelbase of 15 m carries a load of 12 000 N distributed between the front and back wheels, as shown in Fig. 11.29. Determine the maximum moment and the maximum shear as the truck moves over the span.

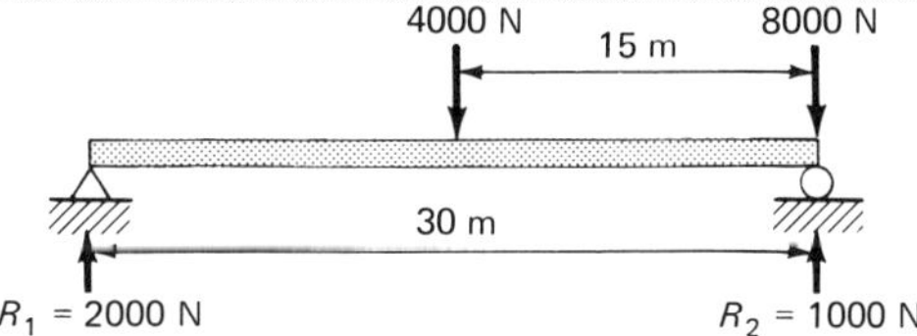

FIGURE 11.29   Illustrative Problem 11.8.

SOLUTION

The resultant of the two axle loads is 12 000 N, which is located at $d =$ (8000/12 000) $\times$ 15 = 10 m from the left load. The position of the wheels necessary for the maximum moment to occur under the left load is as shown in Fig. 11.30. Taking moments about R_1

$$20 \times 12\ 000 - 30R_2 = 0$$

which gives us R_2 equal to 8000 N and R_1 equal to 4000 N. The moment about the 4000-N load is therefore 4000 $\times$ 10 = 40 000 N·m, the maximum value for this load.

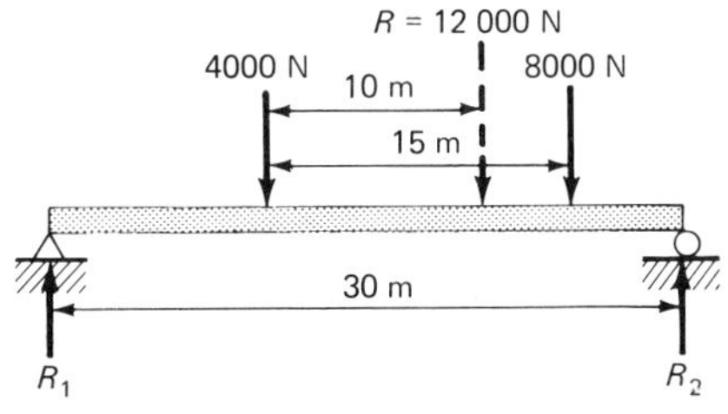

FIGURE 11.30   Illustrative Problem 11.8.

Moving Loads

Now consider the 8000-N load. For the maximum moment due to this load the truck must have moved to the position shown in Fig. 11.31. Again taking moments about R_1,

$$12\,000 \times 12.5 - 30R_2 = 0$$

which gives us $R_2 = 5000$ N and $R_1 = 7000$ N. The moment about the 8000-N load is $17.5 \times 7000 - 4000 \times 15 = 62\,500$ N·m. The maximum moment due to the 8000-N load is therefore the maximum moment that this moving load group imposes on the beam.

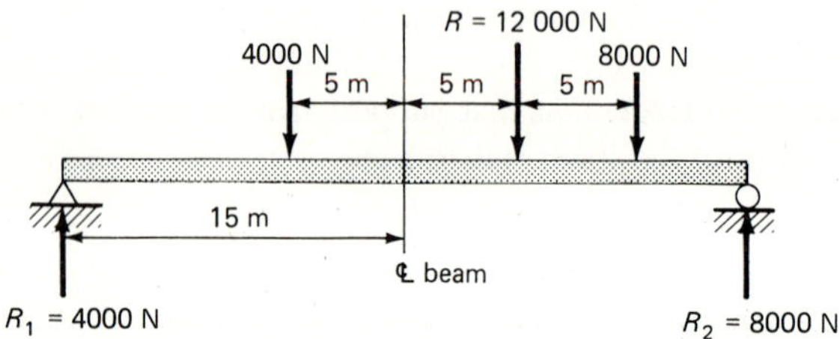

FIGURE 11.31   Illustrative Problem 11.8.

Since the resultant of the load group is nearest the 8000-N load, the maximum shear will occur at the right support when the 8000-N load is over the support. This configuration is shown in Fig. 11.32. Once again taking moments about R_1,

$$4000 \times 15 + 8000 \times 30 - 30R_2 = 0$$

and R_2 is found to equal 10 000 N. This is the maximum shear on the beam due to the truck moving over the beam.

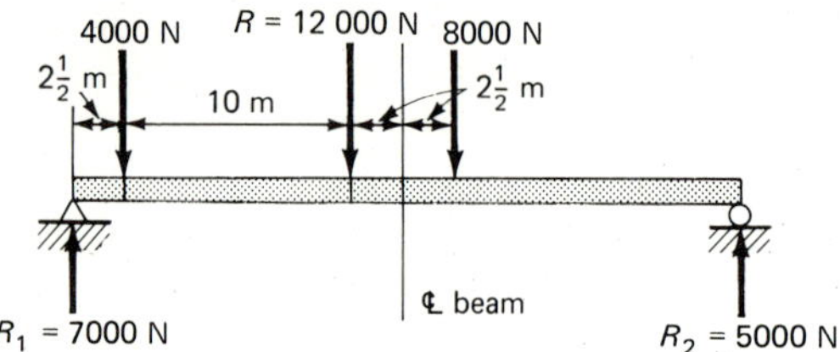

FIGURE 11.32   Illustrative Problem 11.8.

11.7   CLOSURE

Having defined the terms shear and bending moment, we have been able to determine these quantities at any section of a beam using the principles of statics. In addition, general relations have been obtained relating load, shear, and bending moment. The end result of these studies has been the development of shear and bending moment

Shear and Bending Moments in Beams

diagrams. These diagrams permit one to readily identify the critical sections of a beam and to evaluate the shear and bending moment at these sections.

In the introduction to this chapter it was noted that the information developed herein would be used in later sections of this book. This point cannot be stressed strongly enough—*this material must be thoroughly understood before proceeding*. Merely doing problems by rote will not give the student the mastery required for further studies of the stresses and deflections in beams.

REFERENCES

Arges, K. P., and A. E. Palmer, MECHANICS OF MATERIALS. New York: McGraw-Hill, 1963.

Bassin, M. E., S. M. Brodsky, and H. Wolkoff, STATICS AND STRENGTH OF MATERIALS, 3rd ed. New York: McGraw-Hill, 1979.

Borg, S. F., FUNDAMENTALS OF ENGINEERING ELASTICITY. D. Van Nostrand, 1973.

Breneman, J. W., STRENGTH OF MATERIALS, 3rd ed. New York: McGraw-Hill, 1965.

Conway, H. D., MECHANICS OF MATERIALS. Englewood Cliffs, NJ: Prentice-Hall.

Higdon, A., E. E. Ohlsen, and W. B. Stiles, MECHANICS OF MATERIALS, 3rd ed. New York: John Wiley, 1976.

Jensen, A., and H. H. Chenoweth, APPLIED STRENGTH OF MATERIALS, 3rd ed. New York: McGraw-Hill, 1971.

Levinson, I. J., MECHANICS OF MATERIALS, 2nd ed. Englewood Cliffs, NJ: Prentice-Hall, 1970.

Olsen, G. A., ELEMENTS OF MECHANICS OF MATERIALS, 3rd ed. Englewood Cliffs, NJ: Prentice-Hall, 1974.

Sheiry, E. S., ELEMENTS OF STRUCTURAL ENGINEERING. International Textbook Co.

Singer, F. L., STRENGTH OF MATERIALS, 2nd ed. New York: Harper and Row, 1962.

PROBLEMS

11.1   A concentrated load is placed on a simple beam as shown in Fig. P 11.1. Determine the reactions R_1 and R_2. Also determine the shear and the bending moment at section X–X if this section is to the left of the force, as shown in the figure.

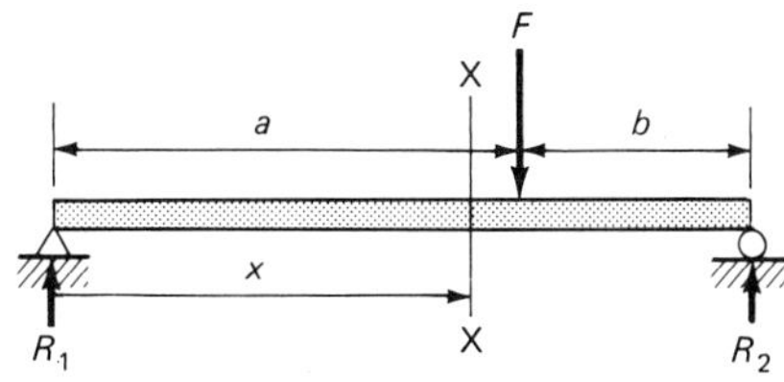

FIGURE P 11.1

11.2 Determine the shear and the bending moment at section X–X if it is to the right of the load, as shown in Fig. P 11.2.

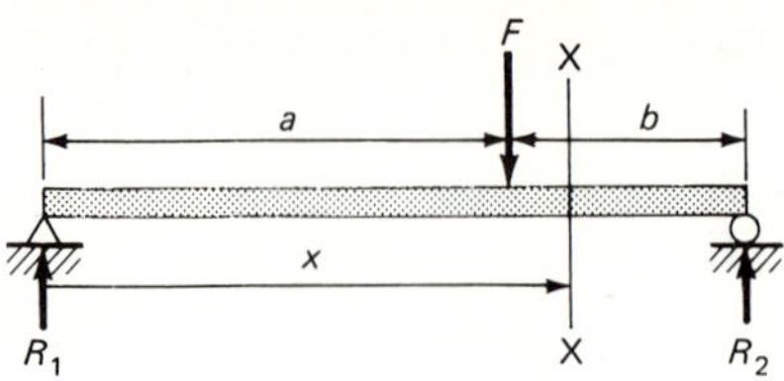

FIGURE P 11.2

11.3 A cantilever beam carries the load shown in Fig. P 11.3. Determine the shear and the bending moment at the built-in end.

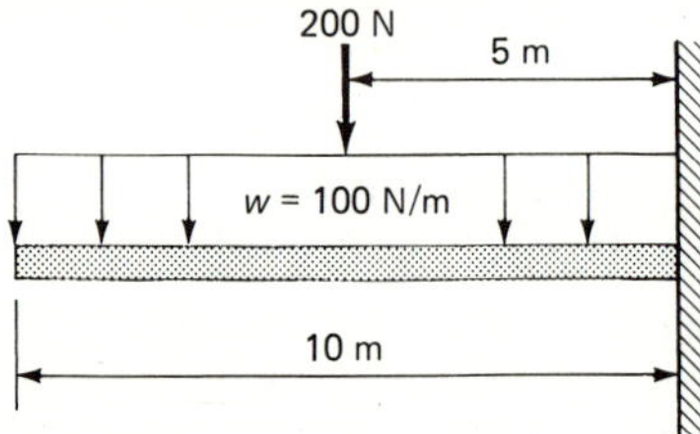

FIGURE P 11.3

11.4 A cantilever beam carries the load shown in Fig. P 11.4. Determine the shear and bending moment at the built-in end.

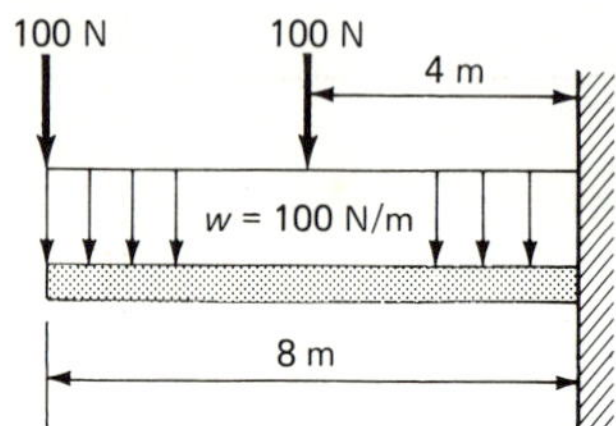

FIGURE P 11.4

11.5 A cantilever beam carries the load shown in Fig. P 11.5. Calculate the shear and the bending moment at the built-in end.

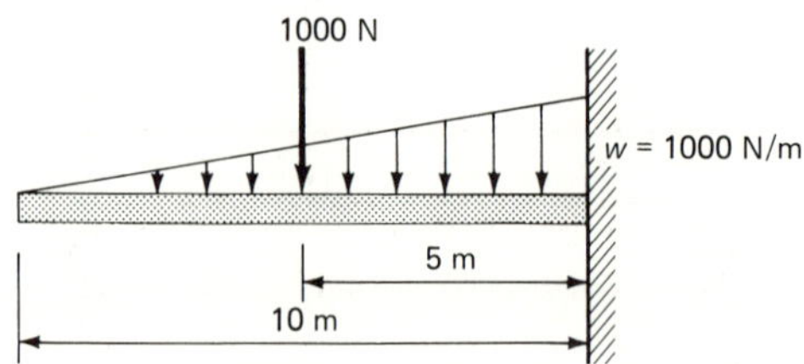

FIGURE P 11.5

                                   Shear and Bending Moments in Beams

11.6   For the cantilever beam shown in Fig. P 11.6, calculate the shear and the bending moment at the built-in end.

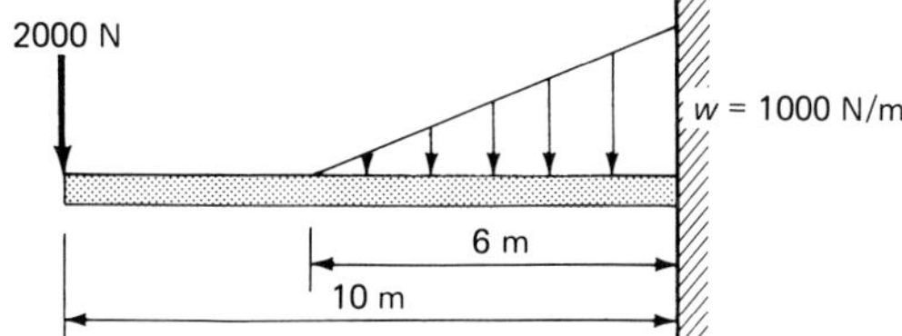

FIGURE P 11.6

11.7   A cantilever beam carries a tapezoidal load. Determine the shear and the bending moment at the built-in end (Fig. P 11.7).

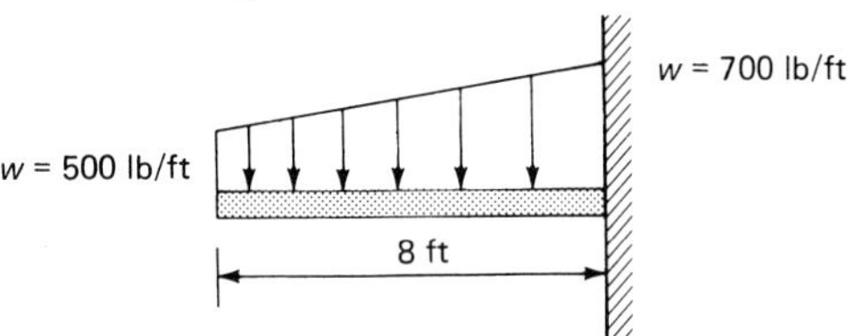

FIGURE P 11.7

11.8   A cantilever beam has a concentrated load at its free end and a distributed load over a portion of its length. Determine the shear and the bending moment at the built-in end (Fig. P 11.8).

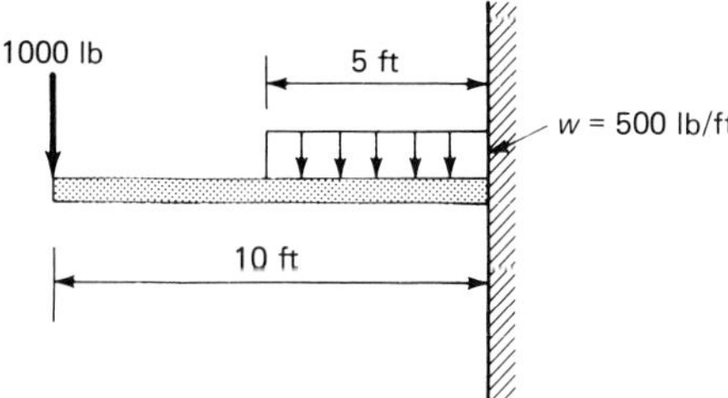

FIGURE P 11.8

11.9   Draw the shear and bending moment curves for the cantilever beam of Problem 11.8.

11.10   Determine R_1 and R_2, and plot the shear and the bending moment curves for the beam shown in Fig. P 11.10.

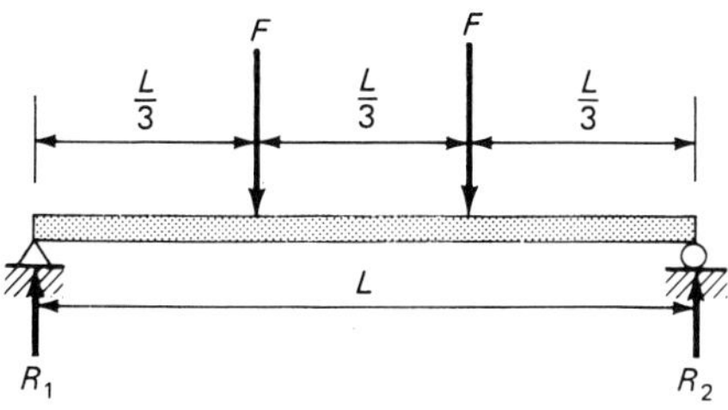

FIGURE P 11.10

Problems

11.11   A simple beam carries a trapezoidal loading as shown in Fig. P 11.11. Calculate the support reactions R_1 and R_2. Determine the location of the section of zero shear.

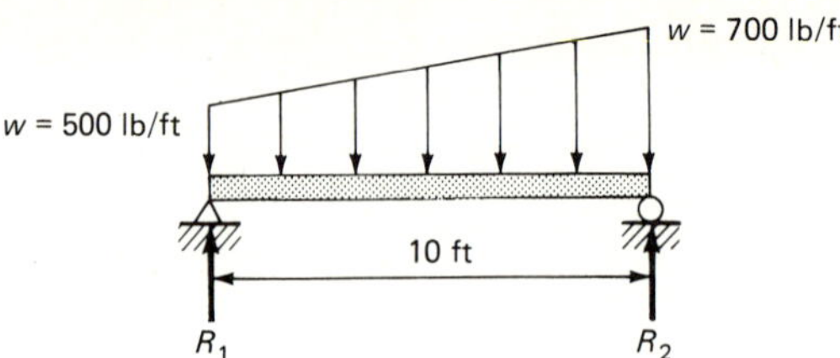

FIGURE P 11.11

11.12   Determine the reactions and draw the shear and bending moment diagrams for the beam shown in Fig. P 11.12.

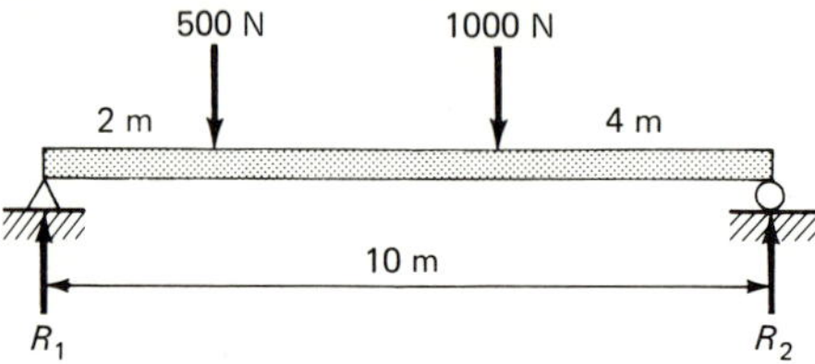

FIGURE P 11.12

11.13   Determine the reactions and draw the shear and bending moment diagrams for the beam shown in Fig. P 11.13.

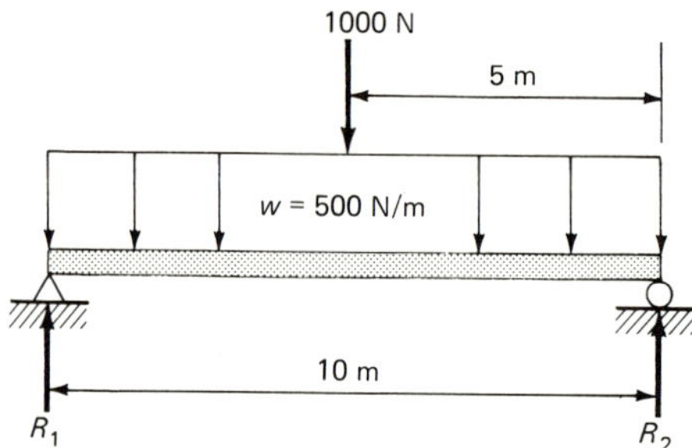

FIGURE P 11.13

11.14   Determine the reactions and draw the shear and bending moment diagrams for the beam shown in Fig. P 11.14.

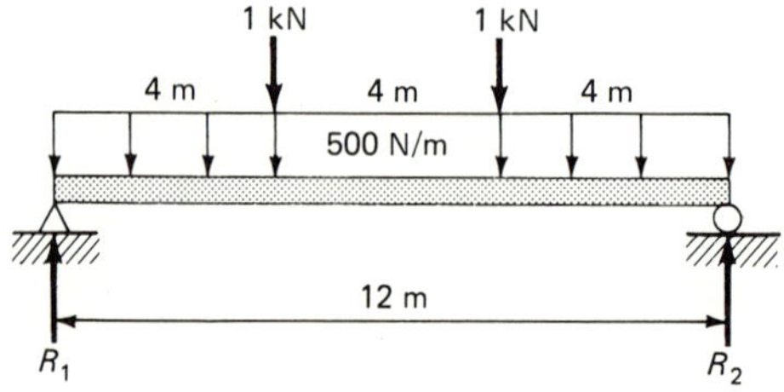

FIGURE P 11.14

                                             Shear and Bending Moments in Beams

11.15 Determine the reactions, and draw the shear and bending moment diagrams for the beam shown in Fig. P 11.15.

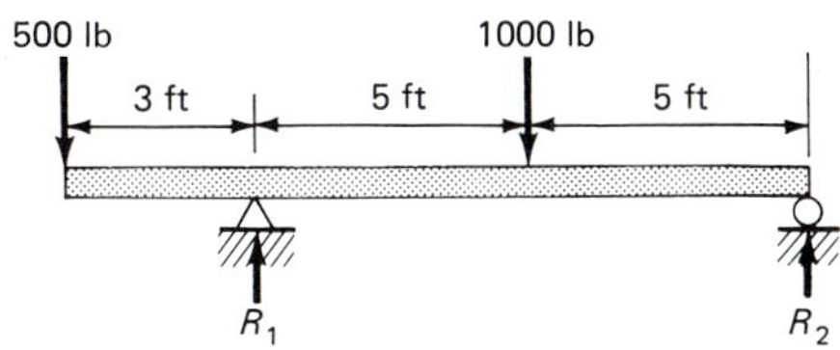

FIGURE P 11.15

11.16 Determine the reactions, and draw the shear and bending moment curves for the beam shown in Fig. P 11.16.

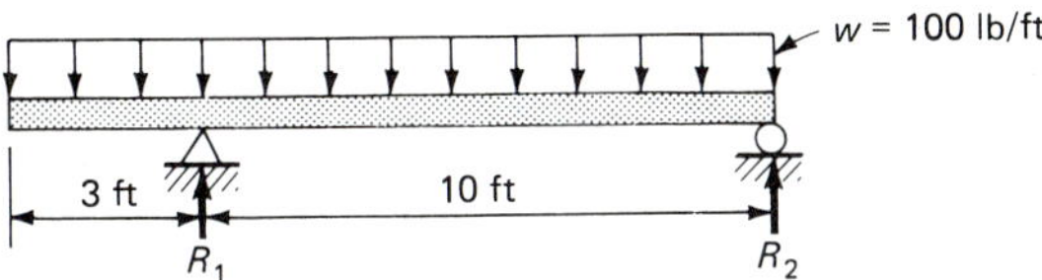

FIGURE P 11.16

11.17 Determine the reactions, and draw the shear and bending moment curves for the beam shown in Fig. P 11.17.

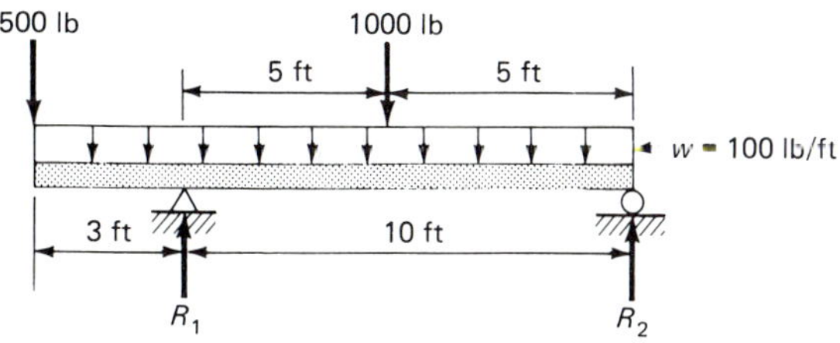

FIGURE P 11.17

11.18 Determine the reactions, and draw the shear and bending moment diagrams for the beam shown in Fig. P 11.18 if the couple is placed at the center of the beam.

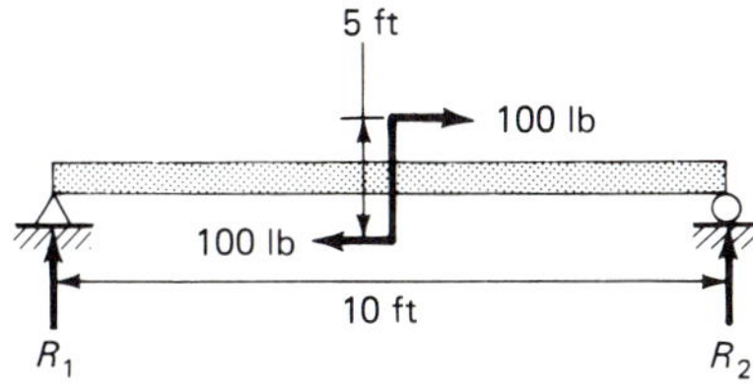

FIGURE P 11.18

Problems

303

11.19   Draw the shear and bending moment curves for the beam shown in Fig. P 11.19.

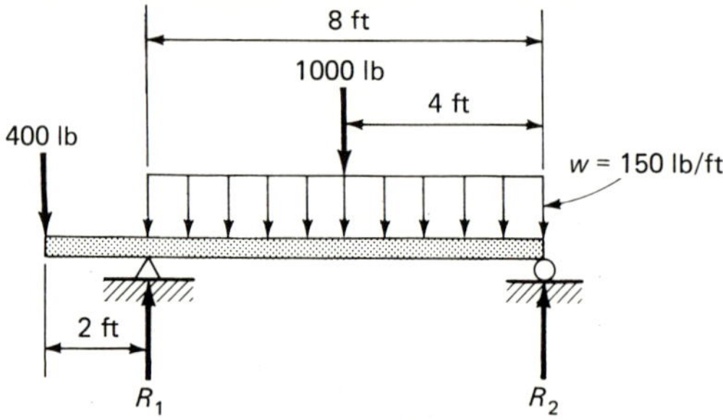

FIGURE P 11.19

11.20   Determine the maximum bending moment and its location for the beam shown in Fig. P 11.19.

11.21   Draw the shear and bending moment curves for the beam shown in Fig. P 11.21.

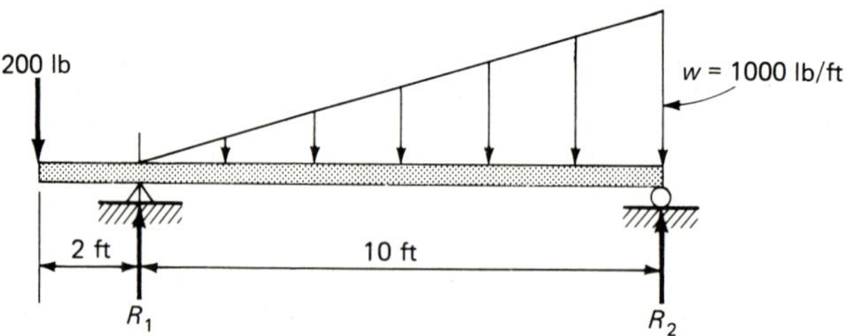

FIGURE P 11.21

11.22   Determine the maximum bending moment and its location for the beam shown in Fig. P 11.21.

11.23   Determine the maximum bending moment in the beam shown in Fig. P 11.23.

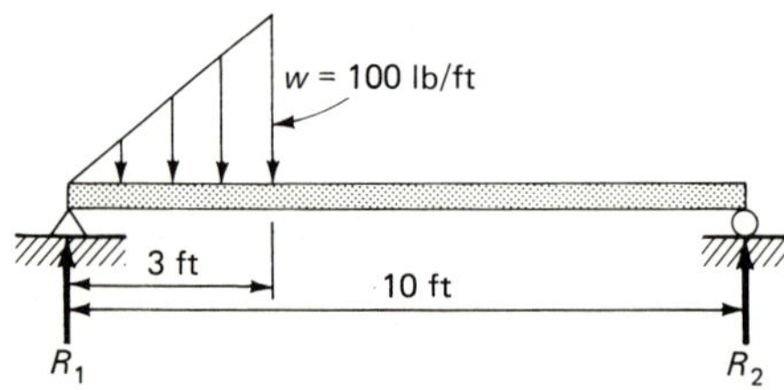

FIGURE P 11.23

                                        Shear and Bending Moments in Beams

11.24   The shear diagram in Fig. P 11.24 is shown for a cantilever beam. Determine the load
diagram and the bending moment diagram for this beam.

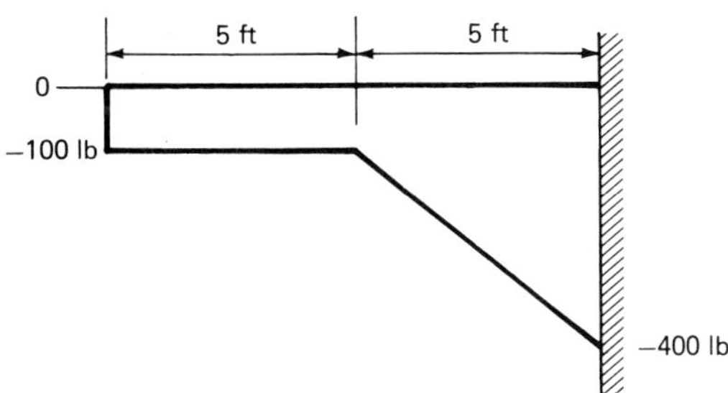

FIGURE P 11.24

11.25   Determine the load and bending moment curves for the beam having the shear curve
shown in Fig. P 11.25.

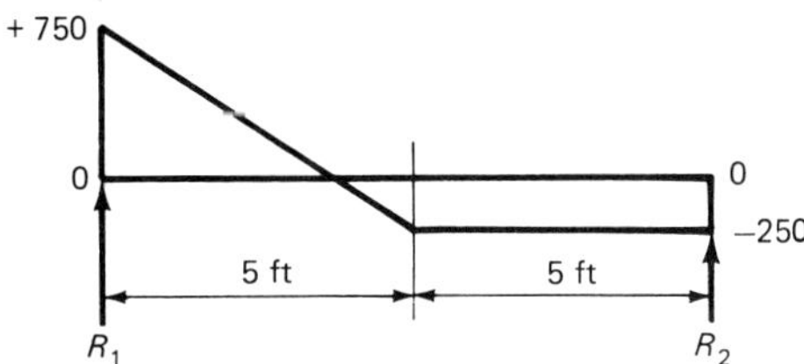

FIGURE P 11.25

11.26   Determine the shear and load diagrams for the beam having the bending moment diagram
shown in Fig. P 11.26.

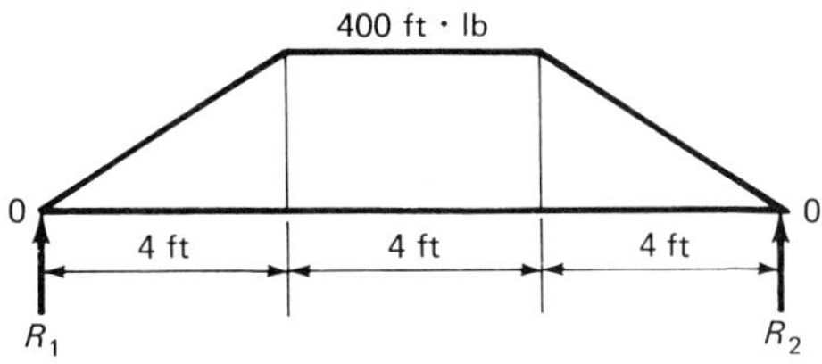

FIGURE P 11.26

Problems   **305**

11.27 Determine the load and bending moment curves for the shear curve shown in Fig. P 11.27. Also calculate the maximum moment.

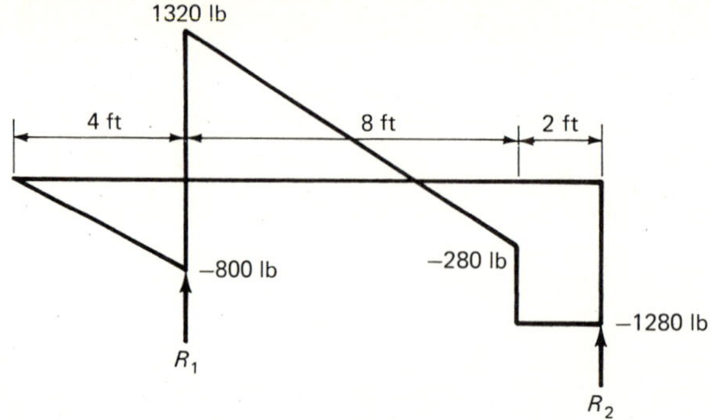

FIGURE P 11.27

11.28 Determine the load diagram, the bending moment diagram, and the maximum bending moment for the beam having the shear curve shown in Fig. P 11.28.

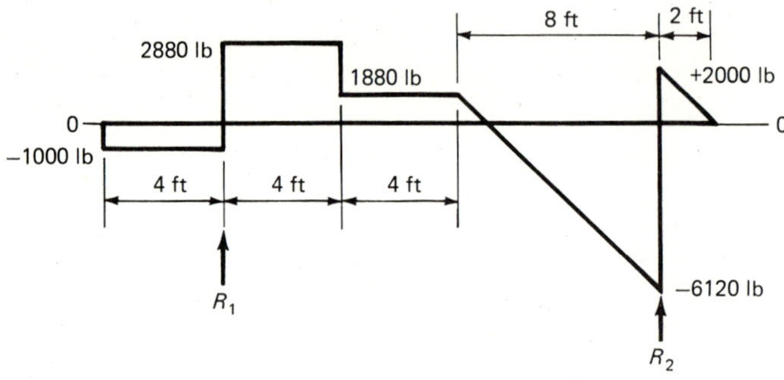

FIGURE P 11.28

11.29 An automobile having a wheelbase of 10 ft moves across a bridge whose span is 30 ft. If the car plus its occupants weighs 5000 lb and 40% of the load is on the rear wheels, calculate the maximum bending moment and the maximum shear as the car travels over the bridge.

11.30 A truck carries a load of 15 000 lb including its own weight. Assuming that the load is distributed to the front and rear wheels at a ratio of 40% and 60%, respectively, determine the maximum bending moment and the maximum shear as the truck crosses a bridge which has a span of 50 ft. The wheelbase of the truck is 18 ft.

11.31 A trailer having a front load of 4000 N separated from a rear load of 8000 N by 12 m passes over a simply supported span which has 22 m between supports. Determine the maximum shear and the maximum bending moment that occur as the trailer passes over the span.

                    Shear and Bending Moments in Beams

12

Stresses in Beams

12.1 INTRODUCTION

Using the principles of statics, we have seen that it is possible to determine the shear and the moment acting on any section of a beam in terms of the applied loads. Our next concern is to determine the beam's ability to resist these loads. In order to evaluate the resistive strength of the beam, it will be necessary to address ourselves to the following problems:

1. Stresses due to bending only
2. Stresses due to shear loads
3. Stresses in beams of several materials

The stresses of concern arise internally in a beam as it deforms and accommodates itself to the loading. The exact distribution of the stresses at any section of a beam is complex, and its calculation in detail is more properly relegated to studies in the theory of elasticity. It is possible, however, to develop mathematical expressions which are suitable for engineering calculations of the bending and shear stresses in beams and which are based upon simplifying, yet quite accurate, assumptions. The results of these calculations have been tested against experiments and found to be in good accord with the test data.

12.2 PURE BENDING—THE FLEXURE FORMULA

Consider a portion of a beam that is subjected to pure bending only by couples (designated as M) at each end, as shown in Fig. 12.1(a). Since the beam is in equilibrium, the moments at each end will be numerically equal, but of opposite sense. Due to these couples, the beam is bent from its original straight position (dashed) to the (exaggerated) curved shape indicated. Due to this bending action, the length of the upper parts of the beam decrease, while the bottom parts of the beam undergo a lengthening. This action has the effect of placing the upper portion of the beam in compression and the lower portion of the beam in tension. In the following discussion we will make certain assumptions concerning the section of beam subjected to pure bending.

1. The beam is straight before bending and has a uniform cross section throughout its length.
2. The material of the beam is homogeneous and its stress–strain curve is linear; that is, stress is proportional to strain. It obeys Hooke's law.
3. The beam material has the same modulus of elasticity in compression as in tension.
4. A transverse plane remains a plane after bending. This latter assumption can best be illustrated by reference to Fig. 12.1(b). Planes 1–1 and 2–2 have rotated to new positions $1'-1'$ and $2'-2'$, respectively, but are still planes.

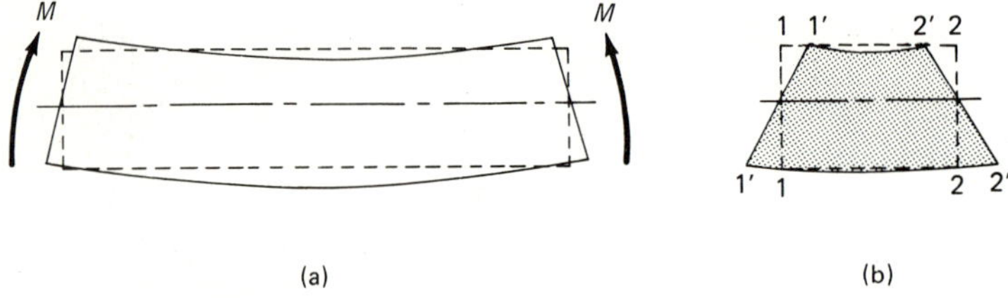

FIGURE 12.1   Beam subject to bending.

If we consider the beam to be a series of thin rods or fibers, we note that the action shown in Fig. 12.1(b) leads to the following conclusions.

1. The upper fibers are shortened and are in compression.
2. The lower fibers are lengthened and are in tension.
3. At some depth in the beam there is a fiber that is neither lengthened nor shortened. Since this fiber does not undergo any strain, it is free of bending stresses. This fiber is called the *neutral fiber,* and the plane of all such fibers is called the *neutral plane*.

Figure 12.2 shows the stress distribution in a beam of rectangular cross section subjected to the indicated end couples. Since plane sections remain plane after bending, a fiber located a distance y above the neutral plane will be compressed an amount equal to the extension of a fiber located a distance y below the neutral plane. Since it is assumed that the material obeys Hooke's law, we can state that the stress on a fiber at a distance y above the neutral plane is equal to the stress on a fiber at a distance y below the neutral plane. We can also conclude that the stress varies linearly from the neutral plane to the outermost fiber in either compression or tension. The portion

                                               Stresses in Beams

of the beam shown in Fig. 12.2 must be in equilibrium. It is therefore necessary that:

1. The sum of the longitudinal forces acting on the end planes be zero.
2. The resisting moment of the beam equal the applied moment M

The condition of zero longitudinal force can be evaluated by reference to Fig. 12.3. The compressive force F_c must equal the tensile force F_t for equilibrium.

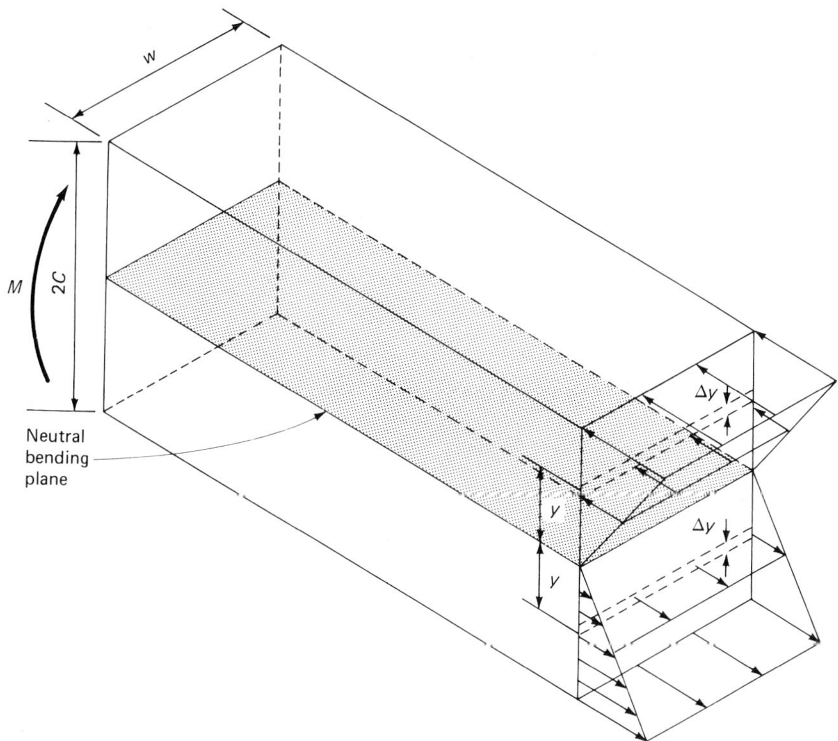

FIGURE 12.2   Stresses in a rectangular beam subject to pure bending.

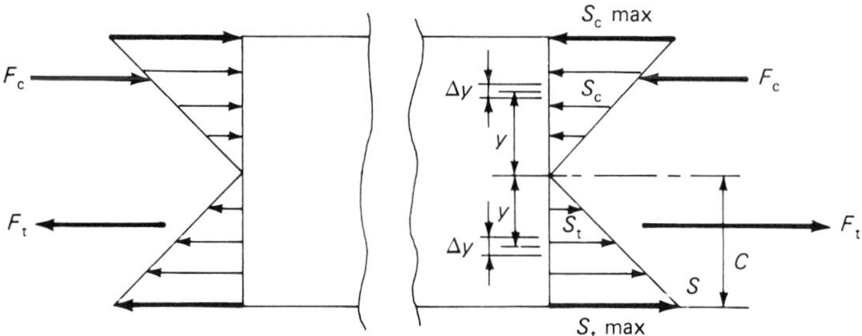

FIGURE 12.3   Forces on a rectangular beam subject to pure bending.

Pure Bending—The Flexure Formula

If we denote the element of area at any distance from the neutral axis y as ΔA and the stress on it as S, the requirement for equilibrium of forces yields

$$\Sigma F = 0 = \Sigma S \Delta A \tag{12.1}$$

However, the stress at any distance y from the neutral axis is proportional to y. Thus

$$S = ky \tag{12.2}$$

where k is a proportionality constant. Placing Eq. (12.2) in Eq. (12.1),

$$\Sigma S \Delta A = \Sigma ky \Delta A = k \Sigma y \Delta A = 0 \tag{12.3}$$

where k has been taken out of the summation since it is a constant. However, we note that this summation term is the first moment of area as we defined it in Chapter 8. Denoting $\bar{y}$ as the distance from the neutral axis to the centroid (center of gravity) of the area,

$$\bar{y}A = \Sigma y \Delta A \tag{12.4}$$

and

$$\bar{y} = \frac{\Sigma y \Delta A}{A} \tag{12.5}$$

From Eq. (12.3), $\Sigma y \Delta A$ must equal zero since k is not zero. Thus the numerator of Eq. (12.5) is zero; but the denominator A is not zero, which leads to

$$\bar{y} = 0 \tag{12.6}$$

Thus the neutral axis for bending coincides with the centroid of the section.

The second condition for equilibrium requires that the resisting moment of the beam be equal to the applied moment. Again let us refer to Figs. 12.2 and 12.3. The force at any distance y above the neutral axis is $S \Delta A$. The moment at this same distance is the moment arm multiplied by this force. The requirement that the sum of all of these moments equal the applied moment M gives

$$M = \Sigma y S \Delta A \tag{12.7}$$

We have already noted that the stress is proportional to the distance of the fiber from the neutral axis, that is, $S = ky$. Inserting this in Eq. (12.7),

$$M = \Sigma(ky)y \Delta A = k \Sigma y^2 \Delta A \tag{12.8}$$

If we denote the stress at the outer fiber as S_{max} and its distance from the neutral axis as C (as shown in Fig. 12.3), we have

$$S_{max} = kC \tag{12.9}$$

Stresses in Beams

and

$$k = \frac{S_{max}}{C} \qquad (12.10)$$

Placing Eq. (12.10) in Eq. (12.8),

$$M = \frac{S_{max}}{C} \Sigma y^2 \Delta A \qquad (12.11)$$

However, we defined the second moment of area (moment of inertia I) in Chapter 8 to be equal to $\Sigma y^2 \Delta A$. Therefore

$$M = \frac{S_{max}}{C} I \qquad (12.12)$$

or in terms of the maximum stress,

$$S_{max} = \frac{MC}{I} \qquad (12.13)$$

Equation (12.13) is known as the *flexure formula*, and it relates the maximum stress in the beam to the applied moment and the properties of the cross-sectional area of the beam. In Eq. (12.13) S is the normal stress in pounds per square inch, M is the applied moment in inch pounds, C is the distance to the outermost fiber from the neutral axis in inches, and I is the moment of inertia of the cross-sectional area with respect to the neutral axis in inches[4], all in English units. In SI units, S is in pascals, M is in newton metres, C is in metres, and I is in metres[4].

ILLUSTRATIVE PROBLEM 12.1

A steel beam having an allowable maximum stress of 20 000 psi in tension and compression is to have a rectangular cross section whose height is twice its width. Determine the beam's dimensions if it is to resist a bending moment of 250 000 in.·lb.

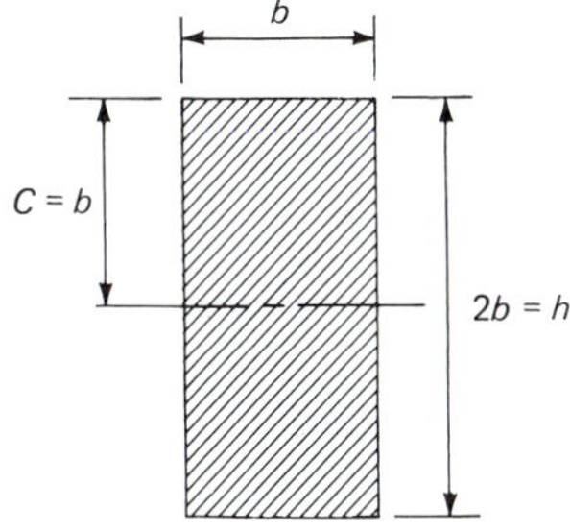

FIGURE 12.4   Illustrative Problem 12.1.

SOLUTION

Designating the width of the beam as b, the height becomes $2b$. From Chapter 8, $I = bh^3/12$ and $C = h/2$. However, $h = 2b$ in this problem. Therefore

$$S = \frac{MC}{I} = 20\,000 = \frac{250\,000(2b/2)}{\tfrac{8}{12}b^4} = \frac{250\,000b}{\tfrac{2}{3}b^4}$$

and

$$b^3 = \frac{250\,000}{20\,000}\left(\frac{3}{2}\right)$$

$$b = 2.66 \text{ in., say } 2\tfrac{11}{16} \text{ in.}$$

$$h = 2b = 5\tfrac{3}{8} \text{ in.}$$

When a beam has equal strength in both tension and compression, as was the case in Illustrative Problem 12.1, the most economical section is one that is symmetrical about the neutral axis since this would cause the beam to be stressed equally in both tension and compression. Thus for materials such as steel, which has very much the same properties in tension and compression, the beam sections are usually symmetrical to the neutral axis. However, for a material such as cast iron, which is strong in compression and weak in tension, the section should not be symmetrical about the neutral axis, but should be proportioned to yield C distances in the same ratio as the ratios of the compressive and tensile strengths.

In the flexure formula [Eq. (12.13)] the geometric properties of the cross-sectional area of the beam enter the equation as the distance from the neutral axis to the outermost fiber C and the moment of inertia I about the neutral axis. The ratio of I/C is known as the *section modulus* of the beam and is most commonly denoted by Z. Thus

$$Z = \frac{I}{C} \tag{12.14}$$

and

$$S = \frac{M}{Z} \tag{12.15}$$

The AISC uses the symbol S for section modulus. Because of the possible confusion with the same symbol for stress, Z is used for section modulus in this text and in many other texts on the strength of materials. Values of Z, the section modulus, are tabulated for standard beam shapes and will be found in Appendix B.

ILLUSTRATIVE PROBLEM 12.2

Select a W10 beam suitable to support a bending moment of 500 000 in.·lb if the allowable stress for steel in both tension and compression is 20 000 psi.

Stresses in Beams

First evaluate Z,

$$Z = \frac{M}{S} = \frac{500\ 000}{20\ 000} = 25 \text{ in.}^3$$

From Appendix B, a W10 $\times$ 25 beam having a $Z = 26.5$ in.^{3} is satisfactory.

A note of caution must be voiced at this point. Throughout this chapter we shall ignore the weight of the beam in our calculations. This is a common practice since the weight of the beam usually represents a small fraction of the maximum moment on the beam. For example, assume that the beam in Illustrative Problem 12.2 is 20 ft long. The maximum moment due to the beam's own weight $= wL^2/8 = (25 \times 20^2/8) \times 12 = 15\ 000$ in.·lb or 3% of the moment of 500 000 in.·lb for which the beam was selected. This is less than the difference between the minimum section modulus and the one selected, that is, $[(26.5 - 25.0)/25.0]100 = 6\%$. There will be cases in which this procedure might lead to difficulty, and the student is cautioned to remember that we have neglected the weight of the beam only in the interest of simplicity.

If a beam is symmetrical about the neutral axis (as noted earlier, this is usually the case for steel), the calculations are not difficult. Where a beam is not symmetrical about the neutral axis (as, say, for cast iron), it is desirable (if possible) to have the maximum stress in tension and compression occur simultaneously. This can be accomplished in some cases, but in others it is necessary to investigate both the maximum compressive stress and the maximum tensile stress.

12.3 BEAMS OF SEVERAL MATERIALS

The use of beams of two or more materials is a common industrial practice, and it is assumed that the student has probably seen such cases as wooden beams reinforced with steel plates or concrete reinforced with steel rods. It is entirely possible to analyze these cases by the methods already presented in this chapter. However, it is convenient and easier to apply a method known as the *method of the transformed section*, which we will now develop. Later on we shall apply this same method to reinforced concrete beams.

Consider the rectangular wooden beam reinforced with steel plates on top and bottom as shown in Fig. 12.5(a). If we assume that the steel and wood are firmly fastened to each other and are physically constrained to move as a single unit, then each must undergo the same deflections at their interfaces. In other words, the strain of the steel and wood at their junctions is the same. Assume that both materials are not stressed above their proportional limits and Hooke's law is applicable to both,

$$S_w = \epsilon E_w \quad \text{and} \quad S_{st} = \epsilon E_{st} \tag{12.16}$$

where S_w is the stress in the wood, S_{st} is the stress in steel, E_w and E_{st} are the moduli

Beams of Several Materials

of elasticity of the wood and steel, respectively, and ϵ is the common strain at the interface. Thus

$$\frac{S_{st}}{S_w} = \frac{E_{st}}{E_w} = n \tag{12.17}$$

Even though the strains are equal, the stress in the steel is n times greater than the stress in the wood at the same distance from the neutral axis of the beam. Thus even though the strain is linear throughout the beam, as shown in Fig. 12.5(b), the stress curve is as shown in Fig. 12.5(c).

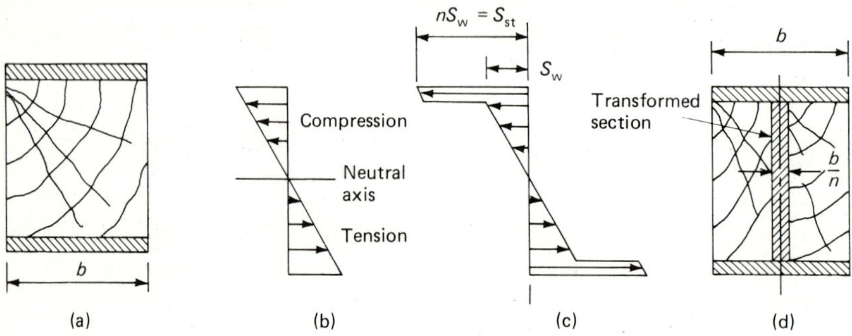

FIGURE 12.5   Stresses and strains in composite beams. (a) Reinforced beam. (b) Strain. (c) Stress. (d) Transformed section.

If the wooden portion of the beam were replaced with a steel section such that the width of the steel replacement was equal to the width of the wood divided by n, we would have the situation shown in Fig. 12.5(d). This transformed beam would have the same characteristics of stress and strain that the actual beam had. In other words, we may treat the transformed beam exactly as we have treated beams of a single material. For the case of a wooden beam reinforced only at the top, the beam would be transformed as noted, and the flexure formula would be applied to the transformed section. It is necessary in this case to find the neutral axis of the transformed section and the moment of inertia of the transformed section about this neutral axis.

After the transformed section has been solved using the flexure formula, it is necessary to calculate the stresses in the actual beam using Eq. (12.17). The following problem illustrates these procedures.

ILLUSTRATIVE PROBLEM 12.3

A 4-in. × 8-in. wooden beam is reinforced by ½-in. steel plates on top and bottom and is subjected to a moment of 500 000 in.·lb, as shown in Fig. 12.6. If $E_w = 1 \times 10^6$ psi, determine the maximum stress in each material. $E_{st} = 30 \times 10^6$.

Stresses in Beams

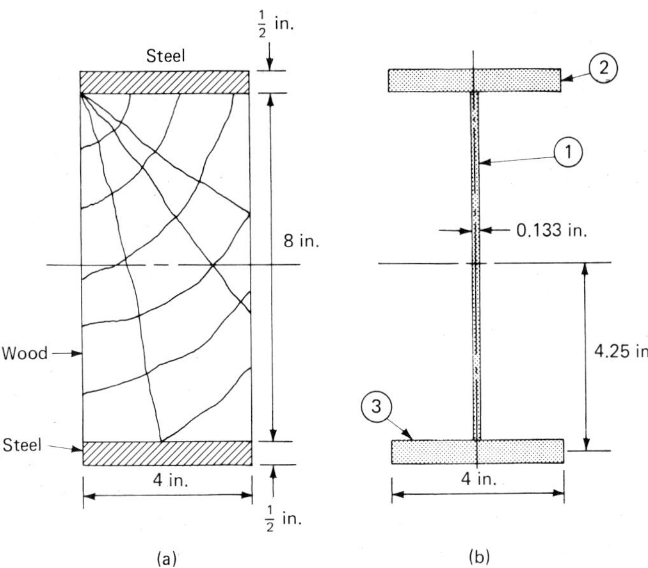

FIGURE 12.6   Illustrative Problem 12.3.

SOLUTION

For ease of solution, let us transform the wood to steel. Therefore the width of the transformed section is $4 \times (1 \times 10^6/30 \times 10^6) = 0.133$ in. The moment of inertia of the transformed beam is as follows. For ①,

$$I = \frac{bh^3}{12} = \frac{0.133 \times 8^3}{12} = 5.67 \text{ in.}^4$$

For ②,

$$I = \frac{bh^3}{12} + Ad^2 = \frac{4 \times (\frac{1}{2})^3}{12} + 4(\frac{1}{2}) \times 4.25^2 = 36.16 \text{ in.}^4$$

For ③,

$$I = 36.16 \text{ in.}^4$$

The total $I = 77.99$ in.^{4}. At the outer fiber of the steel

$$S = \frac{MC}{I} = \frac{500\,000 \times 4\frac{1}{2}}{77.99} = 28\,850 \text{ psi}$$

At the wood–steel interface the stress in the steel is proportional to the distance between the neutral axis and the interface. Therefore $S_{\text{interface}} = 28\,850 \times 4/4\frac{1}{2} = 25\,640$ psi. This figure of 25 640 psi is the stress in the steel—not the stress in the wood. The stress in the wood is $1/n$ the stress in the steel. Thus the maximum stress in the wood is $(1 \times 10^6/30 \times 10^6) \times 25\,640 = 25\,640/30 = 855$ psi.

Beam of Several Materials

12.4 REINFORCED CONCRETE BEAMS

Concrete is widely used as a material of construction since it possesses such desirable characteristics as ease of handling, fireproof qualities, and so on. There is one characteristic of concrete that is highly undesirable—it possesses almost no tensile strength. Because of this, concrete alone is not suitable as a beam material, and it must be reinforced on the tensile side with a structural material that will give it the required tensile strength. In addition, the reinforcing material must be properly bonded to the concrete so that the beam acts as a single unit. At this time we shall only consider the bending stresses in reinforced concrete beams and limit ourselves to *simple reinforced rectangular beams*, that is, beams having reinforcement only on the tension side of the neutral axis.

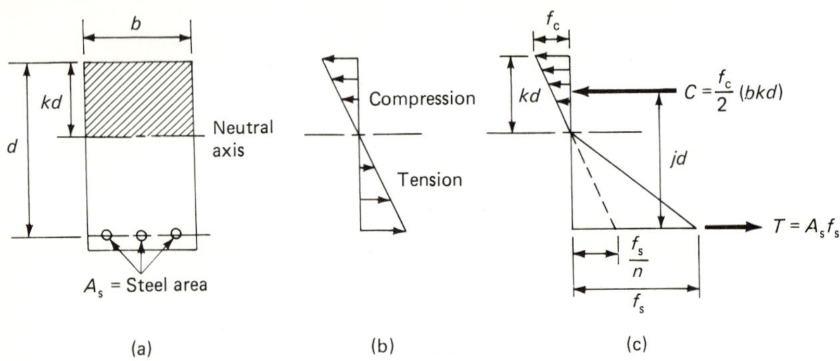

FIGURE 12.7 Rectangular reinforced concrete beam. (a) Cross section. (b) Deformation. (c) Stress.

Figure 12.7 shows a cross section of a rectangular reinforced concrete beam with steel reinforcing on the tension side. The deformation [Fig. 12.7(b)] is assumed to be linear, with zero deformation at the neutral axis of the beam. The steel portion of the beam is assumed to have deflected the same amount as it would have if it were concrete, but the stresses in the steel will exceed the stresses in the concrete by a factor n, where $n = E_{\text{steel}}/E_{\text{concrete}} = E_s/E_c$. The commonly used nomenclature for reinforced concrete differs from that which we have been using. We shall use the following nomenclature.

E_c = modulus of elasticity of concrete in compression
E_s = modulus of elasticity of steel
n = E_s/E_c
A_s = area of tension steel
A_t = transformed steel area, = nA_s
b = width of beam
C = total compressive force in concrete
d = depth of beam from top to center of steel reinforcement
f_c = maximum compressive stress in concrete
f_s = maximum tensile stress in steel
j = ratio of moment arm of C–T couple to depth d
k = ratio of location of neutral axis to depth d

                                             Stresses in Beams

$$p = A_s/bd$$
$$T = \text{total tensile force in concrete}$$

The first step in designing a reinforced concrete beam is to transform the steel area to equivalent concrete area, as shown in Fig. 12.8, which gives us

$$A_t = nA_s \qquad (12.18)$$

This area is assumed to be located at the centerline of the steel and to be so narrow as to have a uniform stress distribution. Therefore the total tensile force T is given by

$$T = f_s A_s = nA_s \frac{f_s}{n} = A_t f_t \qquad (12.19)$$

The total compression force in the beam (using only the portion above the neutral axis) is given by the average compressive stress multiplied by the area in compression

$$C = \frac{f_c}{2} bkd \qquad (12.20)$$

However, the condition for equilibrium requires that $C = T$, and we therefore have a couple with a moment arm of jd.

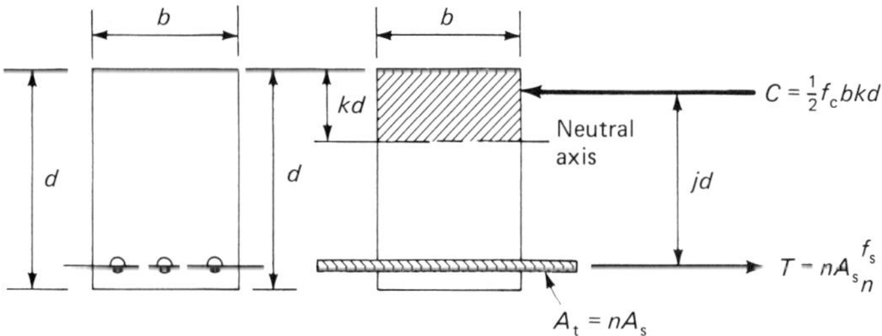

FIGURE 12.8   Rectangular reinforced concrete beam.

At this point we will make the assumption (which will be further discussed later in this section) that both the steel and the concrete reach their maximum stresses simultaneously. This condition is known as *balanced reinforcement*. The moment of the concrete force is

$$C(jd) = \frac{f_c}{2} (bkd)jd \qquad (12.21)$$

and the moment of the steel is

$$T(jd) = f_s A_s (jd) \qquad (12.22)$$

For balanced reinforcement the resisting moment M must equal either, or both, of the moments given by Eqs. (12.21) and (12.22). Thus

$$M = \frac{f_c}{2}\,(bkjd^2) = f_s A_s (jd) \qquad (12.23)$$

From Fig. 12.7(c) we can write

$$\frac{kd}{d} = \frac{f_c}{f_c + \dfrac{f_s}{n}} \qquad k = \frac{1}{1 + \dfrac{f_s}{nf_c}} \qquad (12.24)$$

and

$$jd = d - \frac{kd}{3} \qquad j = 1 - \frac{k}{3} \qquad (12.25)$$

Equations (12.23), (12.24), and (12.25) are sufficient for the *design* of a beam for bending when balanced reinforcement is assumed. We can obtain two useful relations from Eq. (12.23) for this case,

$$d = \sqrt{\frac{M}{(f_c/2)\,bkj}} \qquad (12.26)$$

and

$$A_s = \frac{M}{f_s jd} \qquad (12.27)$$

For initial design purposes k is often taken to be ⅜, giving a preliminary value for j of ⅞.

ILLUSTRATIVE PROBLEM 12.4

Design a reinforced concrete beam with balanced reinforcement to carry a bending moment of 100 000 N·m. The beam is to be 0.6 m wide, with $n = 15$, $f_s = 140$ MPa, and $f_c = 5.25$ MPa.

SOLUTION

The first step in the solution is to obtain k from Eq. (12.24). Thus

$$k = \frac{1}{1 + f_s/nf_c} = \frac{1}{1 + 140 \times 10^6/(15 \times 5.25 \times 10^6)} = 0.36$$

$$j = 1 - \frac{k}{3} = 1 - \frac{0.36}{3} = 0.88$$

From Eq. (12.26),

$$d = \sqrt{\frac{100\,000}{(5.25 \times 10^6/2)0.36 \times 0.88 \times 0.6}} = 0.448 \text{ m}$$

                                                    Stresses in Beams

The required steel area is obtained from Eq. (12.27),

$$A_s = \frac{M}{f_s j d} = \frac{100\,000}{140 \times 10^6 \times 0.88 \times 0.448} = 1.81 \times 10^{-3} \text{ m}^2$$

Equations (12.23) through (12.27) are design equations for the balanced reinforcement required when the external moments and allowable internal stresses are given. In those cases in which the reinforcement is not balanced, or in which all of the physical dimensions are known and we desire to calculate the maximum stresses (or the maximum allowable moment when the allowable stresses are given), we cannot use some of the foregoing equations. For example, we cannot use Eq. (12.24) since the *actual* values of f_c and f_s are not known. We can, however, solve for the location of the neutral axis of a given beam using the concept of the transformed section of the previous section. Referring to Fig. 12.8, we locate the neutral axis from the requirement that the first moment of the area above the neutral axis equal the first moment of the area below the neutral axis. Thus

$$(bkd)\,\frac{kd}{2} = nA_s\,(d - kd) \tag{12.28}$$

If we denote

$$p = \frac{A_s}{bd} \tag{12.29}$$

we can solve Eq. (12.28) for k,

$$k = \sqrt{2pn + (pn)^2} - pn \tag{12.30}$$

Once k is known, we can obtain j from Eq. (12.25), and the *actual* values of f_s and f_c by applying Eq. (12.23) twice.

ILLUSTRATIVE PROBLEM 12.5

Determine the maximum allowable bending moment in a rectangular reinforced concrete beam if $n = 15$ and the allowable maximum stresses in concrete and steel are $f_c = 5.25$ MPa and $f_s = 140$ MPa. Assume that $A_s = 0.0016$ m² and the beam dimensions are $b = 250$ mm and $d = 500$ mm.

SOLUTION

This is an investigation problem, and it is first necessary to find the location of the neutral axis. From Eq. (12.29),

$$p = \frac{A_s}{bd} = \frac{0.0016}{0.250 \times 0.500} = 0.0128$$

Reinforced Concrete Beams

From Eq. (12.30),

$$k = \sqrt{2 \times 15 \times 0.0128 + (0.0128 \times 15)^2} - 0.0128 \times 15 = 0.457$$

and

$$j = 1 - \frac{k}{3} = 1 - \frac{0.457}{3} = 0.848$$

For concrete,

$$M = \frac{f_c}{2}b(kd)(jd) = \frac{5.25 \times 10^6}{2} \times 0.250 \times 0.457 \times 0.500 \times 0.848 \times 0.500$$

$$= 63\,580 \text{ N·m}$$

For steel,

$$M = A_s f_s jd = 0.0016 \times 140 \times 10^6 \times 0.848 \times 0.500 = 94\,976 \text{ N·m}$$

This beam is obviously not balanced, and the stress in the concrete governs, giving a maximum allowable moment of 63 580 N·m.

The design and investigation of a reinforced concrete beam is not totally a process of examining the flexural stresses in the beam. It is equally important that shear, bond, diagonal tension, and anchorage be investigated.

The design of reinforced beams of shapes other than rectangular can be analyzed by the method we have used for rectangular sections. However, this becomes cumbersome, and the use of the transformed section method of the previous section is recommended for these cases. These topics are beyond the scope of this text, and the interested student is referred to specialized texts on structural engineering and reinforced concrete.

12.5 SHEAR STRESSES IN BEAMS

Our first concern is to establish the conditions in a beam that give rise to shearing stresses. When a transverse load is placed on a beam, the load tends to displace a portion of the beam, as shown in Fig. 12.9. We have already discussed this action in Chapter 11 when we reviewed the bending of beams. The action indicated is a pure shearing action and occurs even where there is no bending of the beam. However, beams do bend, and when they bend, fibers on one side of the neutral axis are placed in compression and those on the other side are placed in tension. In effect, the fibers on either side of the neutral plane tend to slip in directions opposite to one another. If we consider a beam that is composed of two planks that are placed one atop the other, and which are free to move longitudinally, the planks would bend as shown in Fig. 12.10. Each will move independently of the other so that the lower fibers of the upper plank move with respect to the upper fibers of the lower plank. If the member were solid, internal resistance to this sliding would act to prevent this sliding from occurring.

Stresses in Beams

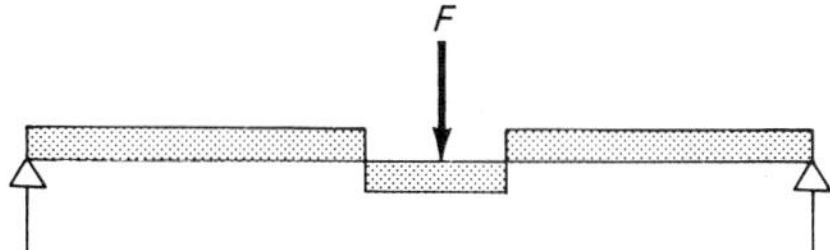

FIGURE 12.9   Transverse shear of a beam.

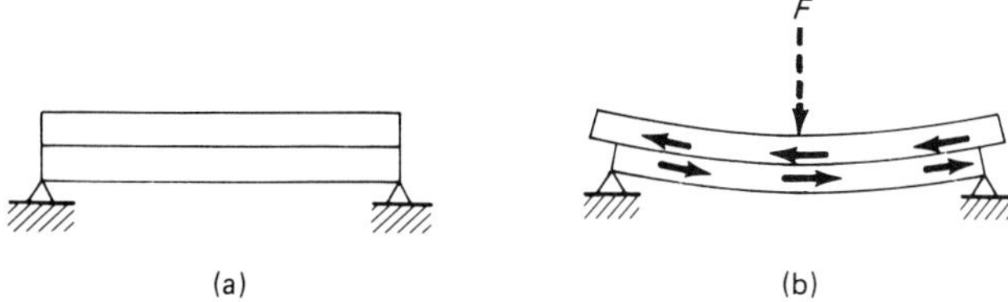

FIGURE 12.10   Shear due to bending.

Before considering this resistance to sliding, let us first look at the shearing stresses on an element of volume. Figure 12.11 shows an elemental cube with shearing stresses S_s on the yz faces. Due to these stresses, the shearing stresses S_s' are induced on the xz faces. Since the area of a yz face is $(\Delta y)(\Delta z)$, the force on each such face is $S_s(\Delta y)(\Delta z)$. For the xz faces the area is $(\Delta x)(\Delta z)$, and the force on each of these faces is $S_s'(\Delta x)(\Delta z)$. Taking moments about 0 for these force systems yields

$$S_s(\Delta y)(\Delta z)(\Delta x) = S_s'(\Delta x)(\Delta z)(\Delta y) \tag{12.31}$$

from which we obtain

$$S_s = S_s'$$

From the above we conclude that a shearing stress cannot occur alone; it must always induce a numerically equal shearing stress on a perpendicular plane.

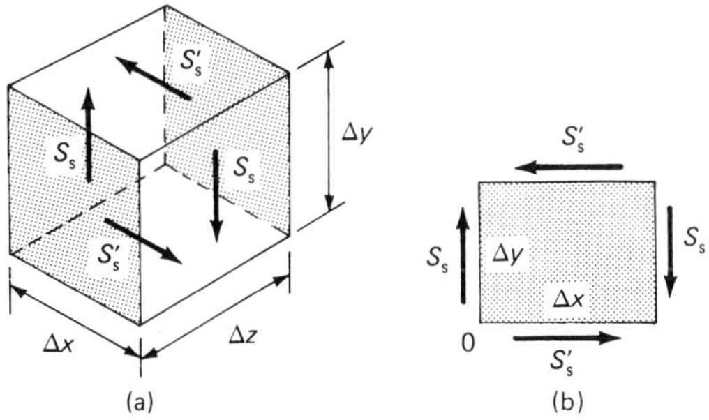

FIGURE 12.11   Stresses on an element.

Figure 12.12 shows the transverse face of a beam that is subjected to a vertical shearing force. Due to the vertical shear force, there are shearing stresses on the face at some location above the neutral axis, stresses that are essentially parallel to the

Shear Stresses in Beams

vertical shearing force. These vertical shearing stresses must therefore have associated with them a horizontal (or longitudinal) shear stress as shown. A good assumption is that these longitudinal shearing stresses are uniform across the width of the beam. Thus at any plane, $S_V = S_L$.

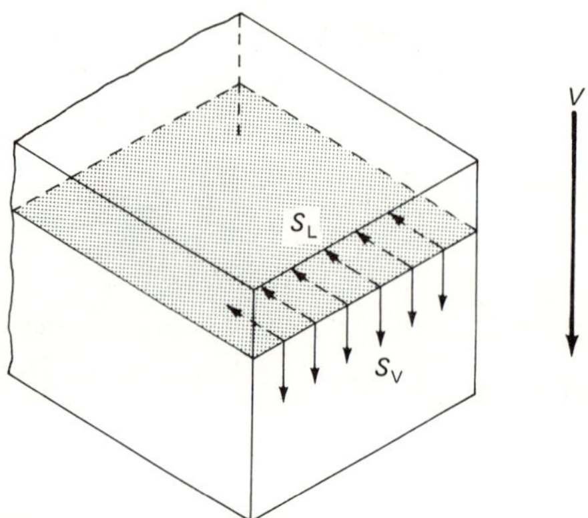

FIGURE 12.12   Longitudinal shear in beams.

It will be recalled from Chapter 11 that the difference in the moment at two positions along a beam can be obtained as the area under the shear diagram between these positions. Keeping this in mind, let us now consider a beam that is subject to any type of transverse loading. From this beam (Fig. 12.13) we will cut out a portion and consider it to be a free-body diagram in equilibrium. The left face of the element will, in general, be subject to a bending moment M. Due to this moment, there must be a horizontal force F, acting to the right as shown. At the right face, a distance Δx away, we shall assume that the moment increases from that on the left face by an amount ΔM and is thus equal to $M + \Delta M$. Due to this moment $M + \Delta M$, there must be a larger horizontal force F_2 acting in a direction opposite to that of F_1. As Fig. 12.13 shows, a shear force F_s must be present on the horizontal face [cross hatched in Fig. 12.13(a)] such that

$$F_1 + F_s = F_2 \tag{12.32}$$

Force F_1 can be evaluated using the flexure formula. Thus

$$F_1 = \sum \frac{M_1 y}{I} A \tag{12.33}$$

and similarly

$$F_2 = \sum \frac{M_2 y}{I} A \tag{12.34}$$

where Σ denotes the summation of all such terms over the area A, and where A is the

Stresses in Beams

area of each face [$b\,(C - y_1)$]. The resisting shearing force F_s is

$$F_s = S_s b(\Delta x) \tag{12.35}$$

If Eqs. (12.33), (12.34), and (12.35) are substituted in Eq. (12.32),

$$\sum \frac{M_1 y}{I} A + S_s b(\Delta x) = \sum \frac{M_2 y}{I} A \tag{12.36}$$

Rearranging,

$$S_s = \left(\frac{M_2 - M_1}{\Delta x}\right) \sum \frac{yA}{Ib} = \left(\frac{\Delta M}{\Delta x}\right) \frac{\sum yA}{Ib} \tag{12.37}$$

However, the vertical shear $V = \Delta M/\Delta x$. Therefore

$$S_s = \frac{V \sum yA}{Ib} \tag{12.38}$$

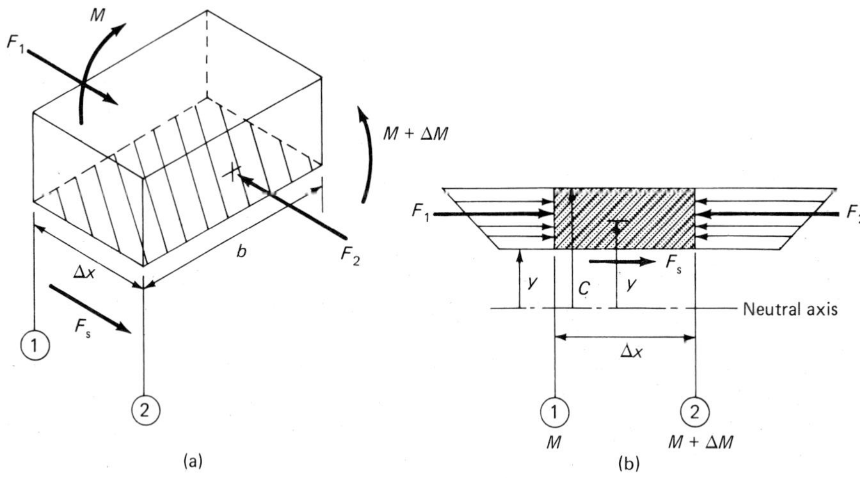

FIGURE 12.13 Shear in a beam.

Equation (12.38) can be further simplified by noting that $\sum yA$ is the first moment of area about the neutral axis, that is, $\bar{y}A$. Denoting

$$Q = \bar{y}A \tag{12.39}$$

the horizontal shearing stress in terms of the vertical shear can be expressed as

$$S_s = \frac{VQ}{Ib} \tag{12.40}$$

Shear Stresses in Beams

The student should note that Q is the first moment about the neutral axis of that part of the cross-sectional area of the beam above the horizontal plane at which the shearing stress is desired, and b is the width of the beam at the section in question. For shapes such as a rectangle, circle, I, T, and others, Eq. (12.40) yields the fact that the maximum shearing stress occurs at the neutral axis. While this is usually the case, it is not always true.

For the rectangular section shown in Fig. 12.14 we can derive both the maximum shear stress and the shear distribution quite readily. From Fig. 12.14(a) we can calculate the shear stress at the neutral axis.

For this section

$$Q = \left(\frac{bh}{2}\right)\frac{h}{4} \qquad I = \frac{by^3}{12} \qquad b = b$$

Therefore

$$S_{s\ max} = \frac{VQ}{Ib} = \frac{V\,(bh/2)\,\times\,(h/4)}{(bh^3/12)b} = \frac{3V}{2bh} \tag{12.41}$$

In other words, the maximum shear stress in a rectangular beam is 50% greater than the average shear stress across the section. The distribution of the stress can be obtained by considering Fig. 12.14(b). In this case consider b, V, and I to be constants.

Then $Q = b(h/2 - y)\,[y + \tfrac{1}{2}\,(h/2 - y)]$ which, when simplified, yields

$$S_s = \frac{V}{2I}\left(\frac{h^2}{4} - y^2\right) \tag{12.42}$$

Equation (12.42) is the equation of a parabola that has a maximum value at the neutral axis.

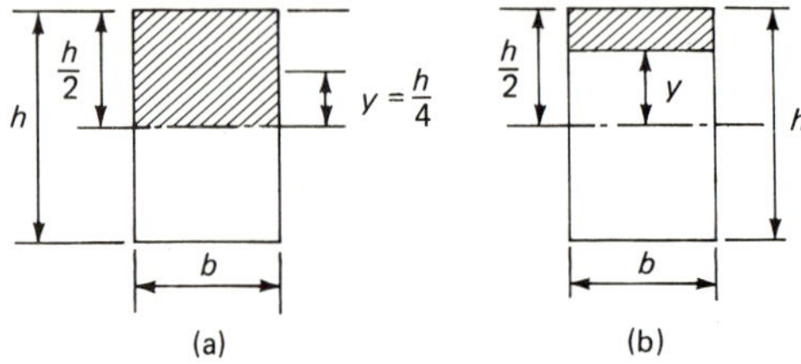

FIGURE 12.14   Rectangular beam.

For a circular cross section the maximum shearing stress also occurs at the neutral axis and is given by

$$S_{s\ max} = \frac{4V}{3\pi r^2} \tag{12.43}$$

ILLUSTRATIVE PROBLEM 12.6

A rectangular beam L metres long and having a concentrated load of F newtons

                                   Stresses in Beams

at its center is to be designed so that the maximum allowable flexure stress and the maximum allowable shear stress occur simultaneously. Determine the maximum length of the beam in terms of the allowable shear stress S_s, the allowable flexure stress S_t, and the dimensions of the beam, b and h.

SOLUTION

The maximum moment in the beam is at the center and equal to $FL/4$. Therefore the maximum flexural stress is

$$S_t = \frac{MC}{I} = \frac{(FL/4) \times (h/2)}{bh^3/12} = \frac{4FL}{3bh^2}$$

For maximum shear,

$$S_s = \frac{3V}{2bh} = \frac{3}{2}\left(\frac{F}{2}\right)\frac{1}{bh} = \frac{3F}{4bh}$$

From flexure,

$$F = \frac{S_t bh^2}{L}\left(\frac{4}{3}\right)$$

From the shear stress,

$$F = \frac{S_s 4bh}{3}$$

Equating these terms,

$$\left(\frac{4}{3}\right)\frac{S_t bh^2}{L} = \frac{S_s 4bh}{3}$$

and

$$L = \left(\frac{S_t}{S_s}\right)h$$

The variation of shearing stresses in beams of different cross sections is shown in Fig. 12.15. As noted earlier, the distribution of shear stresses in a rectangular beam gives rise to a parabolic curve with a maximum value at the center. For the I and T sections it is interesting to note that most of the shear is taken by the web of the beam; the flanges are not effective in resisting the vertical shear. Therefore as a good approximation to the maximum shear stress in an I beam, a channel, or a wide flange section, we can use the shear stress obtained by dividing the shearing force V by the web area. Thus

$$S_{s\ max} = \frac{V}{bt} \tag{12.44}$$

Shear Stresses in Beams

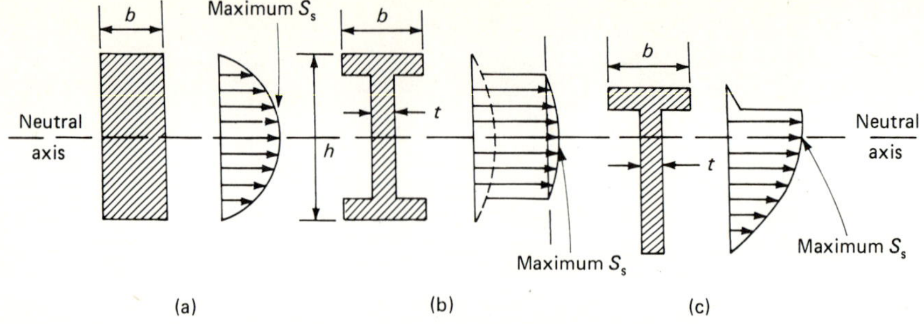

FIGURE 12.15　Variation of shearing stress in beams. (Reprinted with permission from A. Jensen and H. H. Chenoweth, *Applied Strength of Materials.* New York: McGraw-Hill, 1967.)

ILLUSTRATIVE PROBLEM 12.7

Determine the maximum shear stress in the I beam shown in Fig. 12.16. Compare the results with Eq. (12.44).

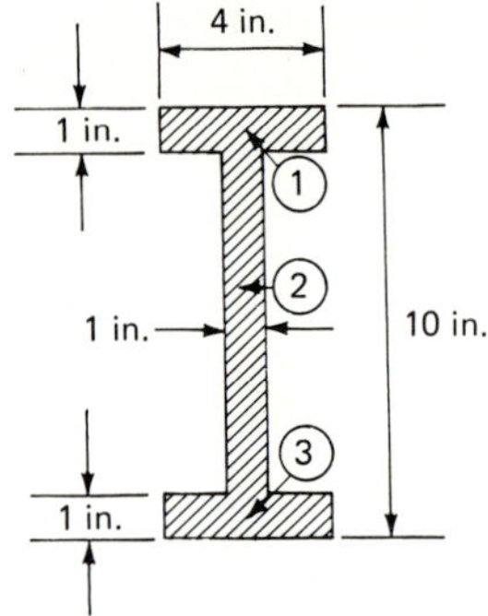

FIGURE 12.16　Illustrative Problem 12.7.

SOLUTION

The moment of inertia of the beam is for ①,

$$I = \frac{4 \times 1^3}{12} + (4 \times 4.5^2) = 81.33 \text{ in.}^4$$

For ②,

$$I = \frac{1 \times 8^3}{12} = 42.66 \text{ in.}^4$$

For ③,

$$I = \frac{4 \times 1^3}{12} + (4 \times 4.5^2) = 81.33 \text{ in.}^4$$

Then

$$I_{total} = 81.33 + 42.66 + 81.33 = 205.32 \text{ in.}^4$$

　　　　Stresses in Beams

Since

$$Q = (4 \times 1 \times 4.5) + (4 \times 1 \times 2) = 26$$

We have

$$S_{s\,max} = \frac{VQ}{Ib} = \frac{V \times 26}{205.32 \times 1} = 0.126V$$

Equation (12.44) yields $S_{s\,max} = V/bh = V/10 \times 1 = 0.1V$. Had we used the web area, not including the flanges, we would have obtained $S_{s\,max} = V/8 \times 1 = 0.125V$, which agrees quite well with the more exact value of 0.126. In any case, the procedure given by the applicable code should be used.

As a final application of shearing action in beams, let us consider the case of a beam composed of elements that are either riveted or bolted together. For the case shown in Fig. 12.17 it is necessary for the bolt to resist the shearing force tending to slide the two portions of the beam relative to each other. If the bolts are assumed to take the shearing force, then each bolt must take the shearing stress multiplied by the shear area between bolts

$$F_B = S_s \times e \times b \tag{12.45}$$

However, we can use Eq. (12.40) for S_s. Thus

$$F_B = \frac{VQ}{Ib} \times e \times b = \frac{VeQ}{I} = A_B S_{sB} \tag{12.46}$$

Equation (12.46) is not restricted to rectangular beams and can be used for other sections. For the beam shown in Fig. 12.18 the value of Q is the moment of the shaded cross section with respect to the neutral axis and I is the moment of inertia of the entire cross section of the beam without any deduction for rivet holes. If the vertical shear varies in the interval e, it is usual to take the maximum value in the interval, although the average value is more nearly correct. The use of the maximum value introduces an error on the safe side.

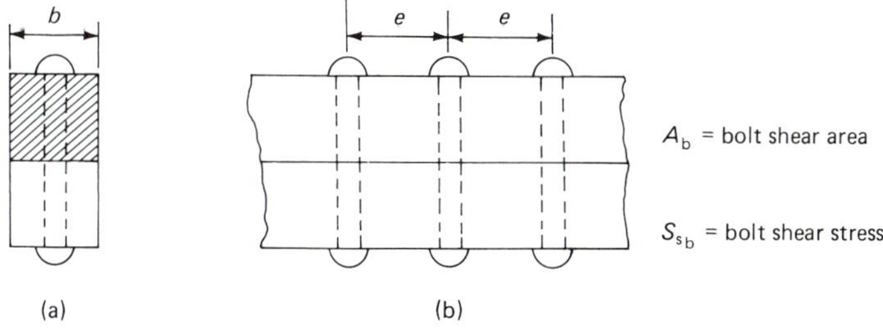

FIGURE 12.17   Composite beam.

Shear Stresses in Beams

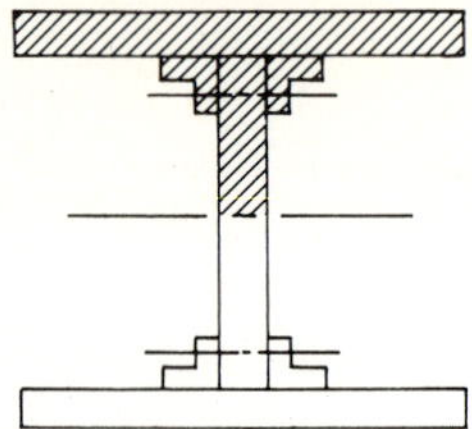

FIGURE 12.18 Built-up beam.

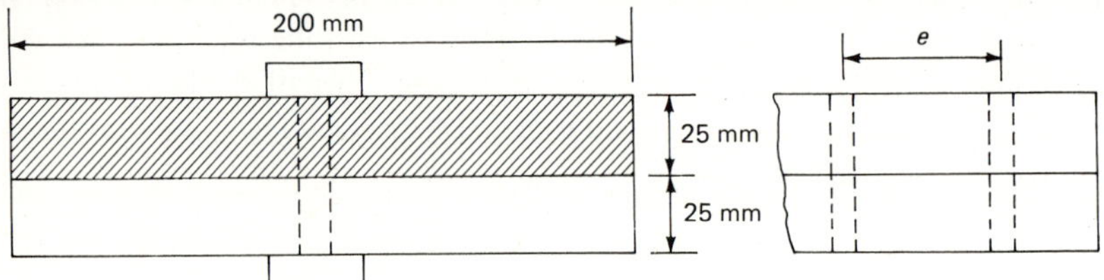

FIGURE 12.19   Illustrative Problem 12.8.

SOLUTION

Q above the neutral axis is

$$Q = 0.200 \times 0.025 \left(\frac{0.025}{2}\right) = 6.25 \times 10^{-5} \text{ m}^3$$

$$I = \frac{bh^3}{12} = \frac{0.200 \times 0.050^3}{12} = 2.083 \times 10^{-6} \text{ m}^4$$

$$\frac{VeQ}{I} = A_B S_{sB}$$

Thus

$$e = \frac{A_B S_{sB}}{VQ} = \frac{\tfrac{1}{4}\pi \times 0.025^2 \times 56 \times 10^6 \times 2.083 \times 10^{-6}}{4500 \times 6.25 \times 10^{-5}} = 0.204 \text{ m}$$

12.6 CLOSURE

The study of stresses in beams leads to formulas such as the flexure formula, the shear formula, formulas for reinforced concrete beams, and so on. Recognizing that students have limited time to devote to a given subject, that they must ''do'' a given number

Stresses in Beams

of homework problems, and that on tests they must satisfactorily apply this material under conditions of duress, one cannot blame most students for adopting as an operating procedure the memorization of formulas. However, this is not really a learning process and all too often leads to the misapplication of formulas to situations that are beyond their scope. It is of primary importance (and the author cannot overstress this concept) that the assumptions made, the free-body diagrams used, and the limitations imposed on each situation are of greater importance than the end formula. One very good example of this exists in the study of stresses in rectangular reinforced concrete beams— the formulation used for designing beams is not applicable to the investigation of these beams.

Study each topic, study the derivations, and do not become dependent upon the memorization of formulas. The dividends reaped from this approach will exceed the investment by many orders of magnitude. It will also be invaluable for other professional applications (such as structural design) and for more advanced studies in this field.

REFERENCES

American Institute of Steel Construction, MANUAL OF STEEL CONSTRUCTION, 8th ed. 1980.

Arges, K. P., and A. E. Palmer, MECHANICS OF MATERIALS. New York: McGraw-Hill, 1963.

Higdon, A., E. E. Ohlsen, and W. B. Stiles, MECHANICS OF MATERIALS, 3rd ed. New York: John Wiley, 1976.

Jensen, A., and H. H. Chenoweth, APPLIED STRENGTH OF MATERIALS, 3rd ed. New York: McGraw-Hill, 1971.

Levinson, I. J., MECHANICS OF MATERIALS, 2nd ed. Englewood Cliffs, NJ: Pentice-Hall, 1970.

Olsen, G. A., ELEMENTS OF MECHANICS OF MATERIALS, 3rd ed. Englewood Cliffs, NJ: Prentice-Hall, 1974.

Sheiry, E. S., ELEMENTS OF STRUCTURAL ENGINEERING. International Textbook Co.

Singer, F. L., STRENGTH OF MATERIALS, 2nd ed. New York: Harper and Row, 1962.

Timoshenko, S., and D. H. Young, ELEMENTS OF STRENGTH OF MATERIALS, 5th ed. New York: D. Van Nostrand, 1968.

PROBLEMS

The weight of the beam is to be neglected in all problems in this chapter.

12.1   A simply supported rectangular beam is 10 ft long and has cross-sectional dimensions of 3 in. × 6 in. Determine the maximum concentrated load that can be placed on the center of the beam if the maximum allowable stress is 20 000 psi.

12.2   If an S12 × 35 beam is stressed to 20 000 psi, determine the moment at the section of the beam where this occurs.

12.3   A simply supported steel beam, 10 ft long and having a cross-sectional area corresponding to an S5 × 10 beam is uniformly loaded. Determine the loading per foot of beam if the maximum bending stress is 22 000 psi.

12.4 A cantilever beam carries a concentrated load as shown in Fig. P 12.4. Determine the maximum bending stress in the beam.

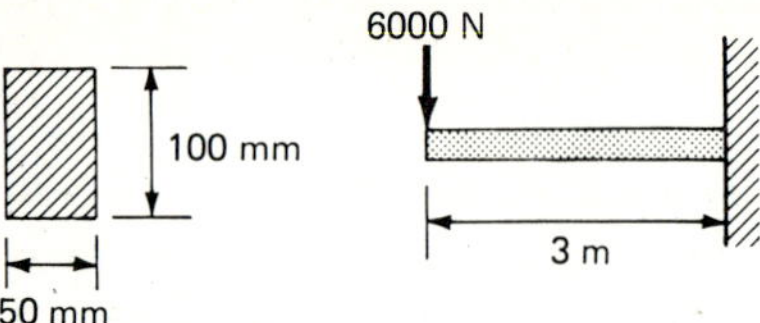

FIGURE P 12.4

12.5 A 4-in. Schedule 40 steel pipe is used to support a load as shown in Fig. P 12.5. If the maximum bending stress is 22 000 psi, determine the maximum load.

FIGURE P 12.5

12.6 A simply supported hollow square beam has a uniform load distributed over its length. If the maximum bending stress is 18 000 psi, determine the permissible load per foot of beam for a beam 10 ft long (Fig. P 12.6).

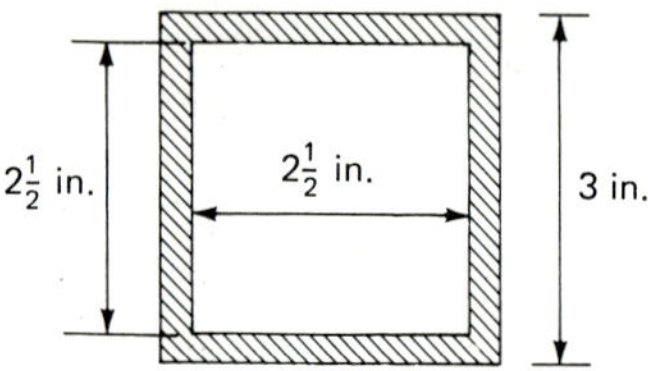

FIGURE P 12.6

12.7 Select a W10 beam to carry the loading shown in Fig. P 12.7 if the maximum allowable flexural stress is 24 000 psi.

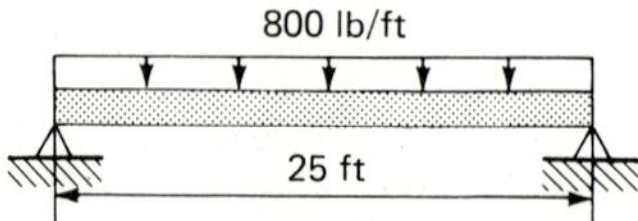

FIGURE P 12.7

                                             Stresses in Beams

12.8   A cantilever beam carries a uniform load over its entire span as shown in Fig. P 12.8. Determine the maximum bending stress in the beam.

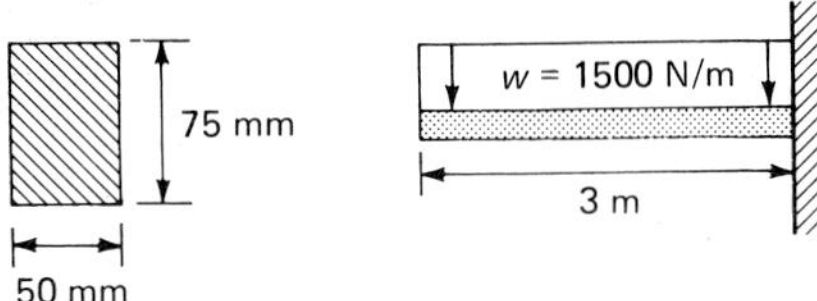

FIGURE P 12.8

12.9   A cantilever beam carries the loads shown in Fig. P 12.9. Calculate the maximum bending stress in the beam.

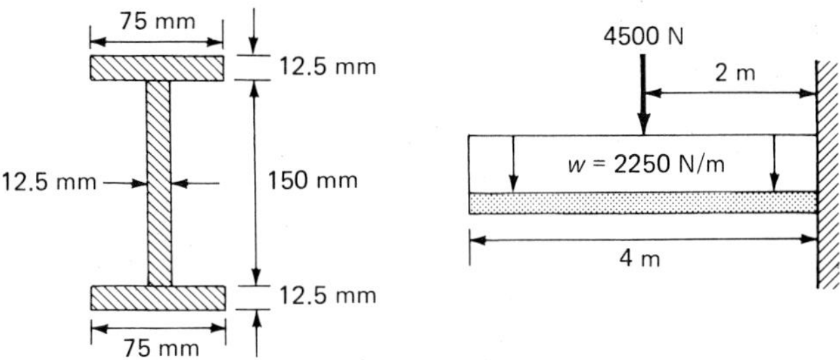

FIGURE P 12.9

12.10   A simply supported beam is loaded as shown in Fig. P 12.10. Determine the maximum bending stress in the beam.

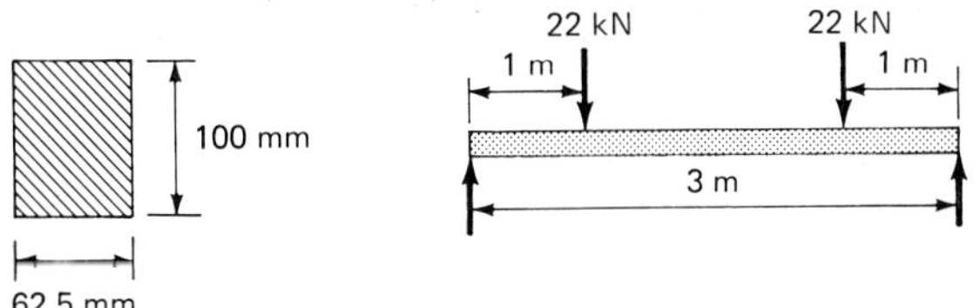

FIGURE P 12.10

12.11   If the beam in Problem 12.10 has an additional load of 10 kN placed at its center, what will the maximum bending stress be in the beam?

12.12   Select an S8 beam to carry the loading shown in Fig. P 12.12 if the maximum allowable bending stress is 140 MPa.

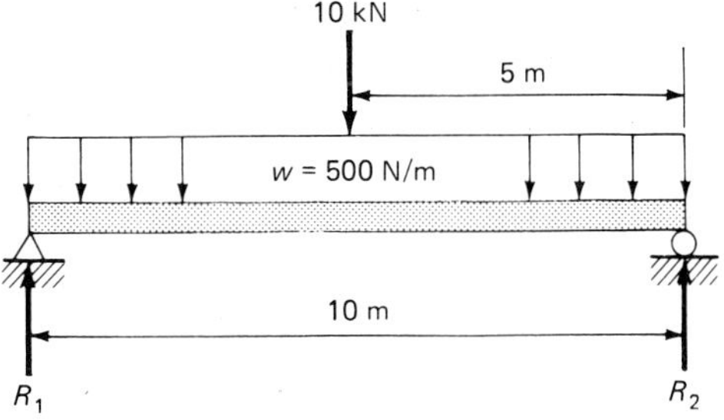

FIGURE P 12.12

12.13   Two channels are welded together to form a beam as shown in Fig. P 12.13. What is the maximum moment that this beam can support if the maximum allowable flexural stress is 20 000 psi?

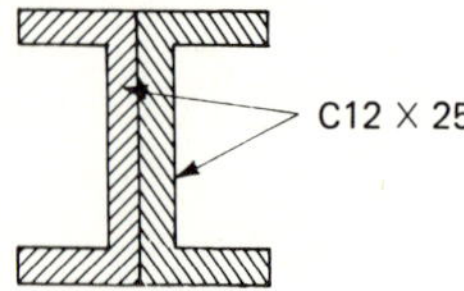

FIGURE 12.13

12.14   If 1-in. cover plates are welded to the beam of Problem 12.13 as shown in Fig. P 12.14, what increase in maximum moment will occur?

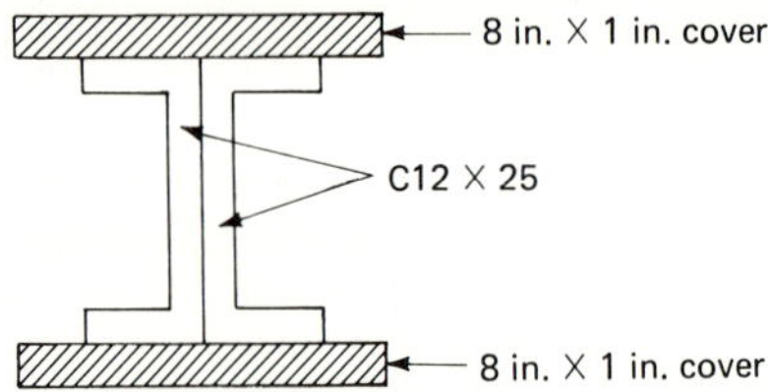

FIGURE P 12.14

12.15   Two equal loads are placed on a beam 20 ft long. If the maximum allowable bending stress is 22 000 psi, determine the maximum value of these loads (Fig. P 12.15).

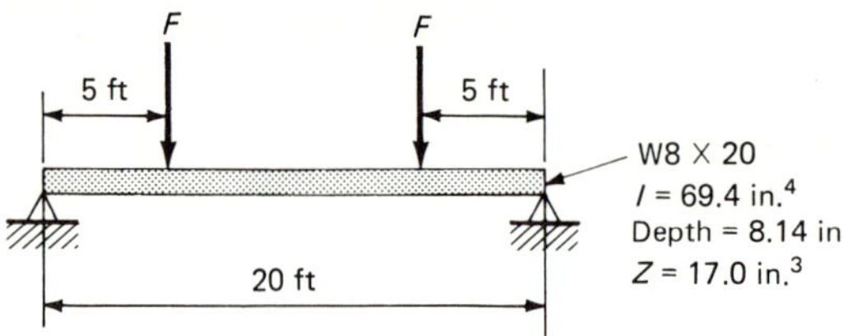

FIGURE P 12.15

12.16   A beam is made of a full 8-in. × 8-in. wooden beam, reinforced on top and bottom by ¾-in. steel plates as shown in Fig. P 12.6. If it is to support a maximum moment of 500 000 in.·lb, with $E_w = 1 \times 10^6$ psi and $E_s = 30 \times 10^6$ psi, determine the maximum stress in each material.

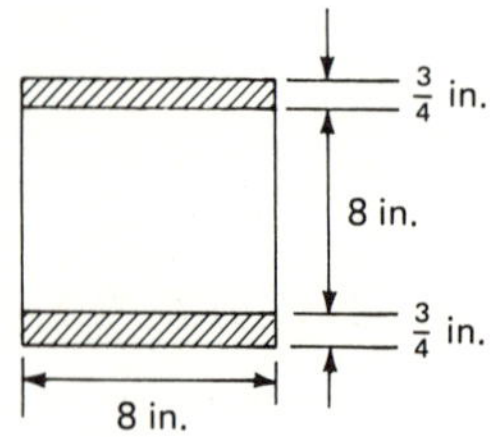

FIGURE P 12.16

                                        Stresses in Beams

12.17   If the reinforcing plates of Problem 12.16 are rotated through 90 deg, determine the maximum stress in each material (Fig. P 12.17).

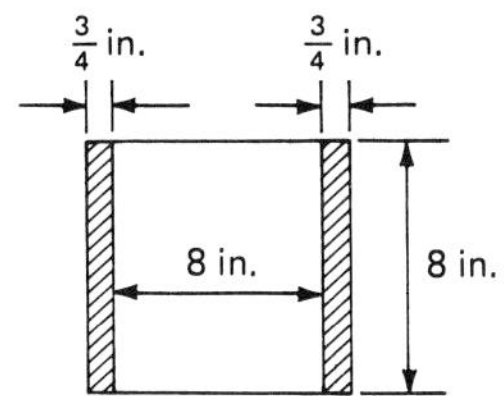

FIGURE P 12.17

12.18   Design a reinforced concrete beam with balanced reinforcement to carry a bending moment of 800 000 in.·lb. The beam is to be 20 in. wide, $n = 15$, $f_s = 18\ 000$ psi, $f_c = 600$ psi.

12.19   Determine the maximum allowable bending moment in the reinforced concrete beam shown in Fig. P 12.19.

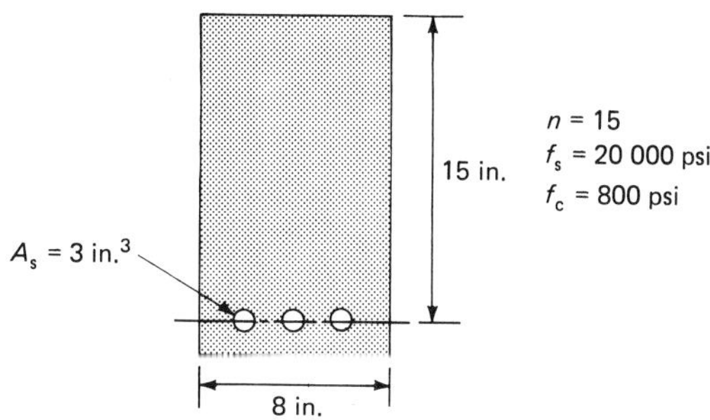

FIGURE P 12.19

12.20   Design a rectangular reinforced concrete beam to carry a bending moment of 750 000 in.·lb if $d = 2b$. Assume balanced reinforcement and $n = 12$, $f_s = 18\ 000$ psi, $f_c = 700$ psi.

12.21   Determine the area of steel required for balanced reinforcement of a rectangular reinforced concrete beam if $b = 12$ in., $d = 20$ in., $n = 15$, $f_s = 20\ 000$ psi, and $f_c = 800$ psi.

12.22   Determine the maximum allowable bending moment that the reinforced concrete beam shown can support. Take $n = 14$, $f_c = 700$ psi, and $f_s = 18\ 000$ psi (Fig. P 12.22).

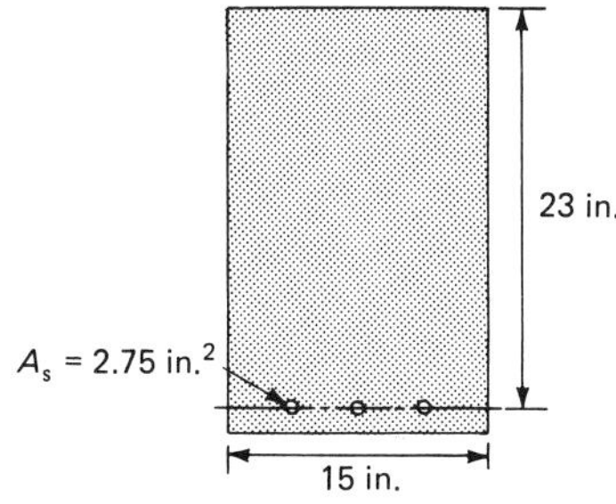

FIGURE P 12.22

Problems

12.23 Investigate the reinforced concrete beam shown in Fig. P 12.23 to determine the maximum allowable moment if $n = 20$, $f_c = 500$ psi, $f_s = 15\ 000$ psi, $b = 14$ in., $d = 25$ in., and $A_s = 3.5$ in.^{2}.

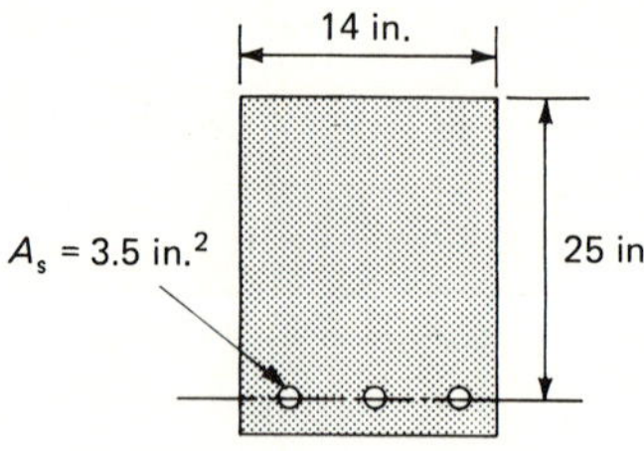

FIGURE P 12.23

12.24 Determine the maximum value of F for the wooden beam shown in Fig. P 12.24.

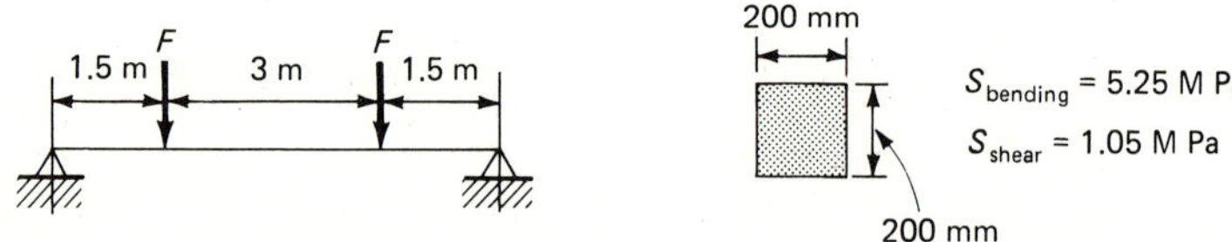

FIGURE P 12.24

12.25 A W10 × 33 beam has an allowable bending stress of 140 MPa and an allowable shear stress of 56 MPa. Determine the maximum value of F (Fig. P 12.25).

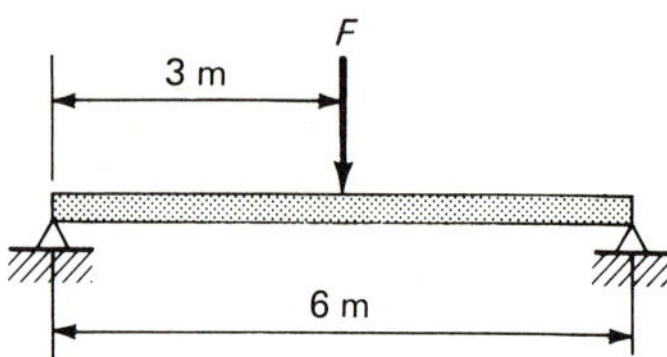

FIGURE P 12.25

12.26 A W10 × 33 beam has an allowable bending stress of 20 000 psi and an allowable shear stress of 8000 psi. Determine the maximum value of F (Fig. P 12.26).

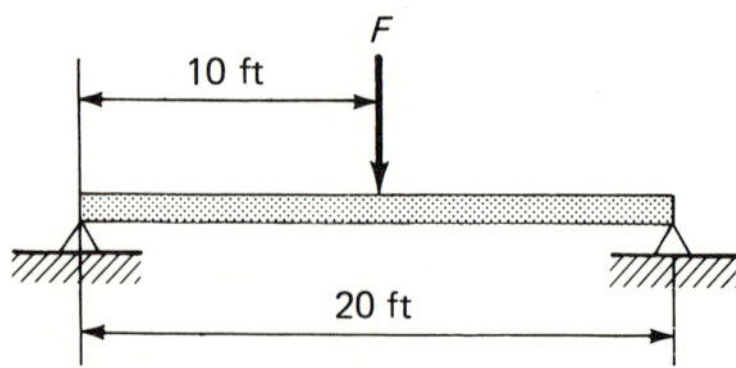

FIGURE P 12.26

Stresses in Beams

12.27 Determine the maximum allowable uniform load per foot of beam for an S4 × 9.5 beam loaded as shown in Fig. P 12.27. Use an allowable tensile stress of 22 000 psi and an allowable shear stress of 12 000 psi.

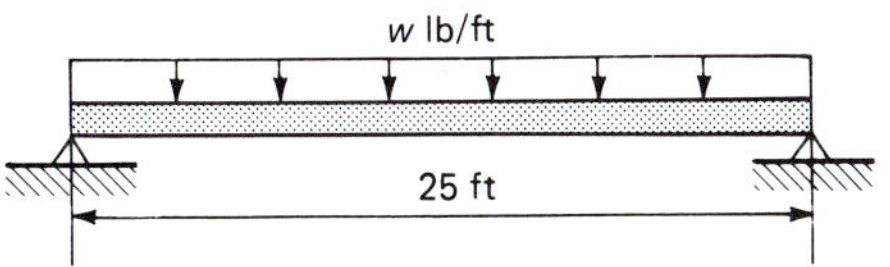

FIGURE P 12.27

12.28 If two 4-in. × 4-in. bars are bolted together to form a beam as shown, assuming a shear load of 5000 lb, bolts of ¾-in. diameter, and an allowable bolt shear stress of 8000 psi, determine the necessary bolt spacing.

12.29 Three 2-in. × 4-in. full planks are bolted together as shown in Fig. P 12.29. Determine the maximum shear force if the bolts are ¾ in. in diameter and are spaced 8 in. apart. The allowable shear stress in the bolts can be taken as 10 000 psi.

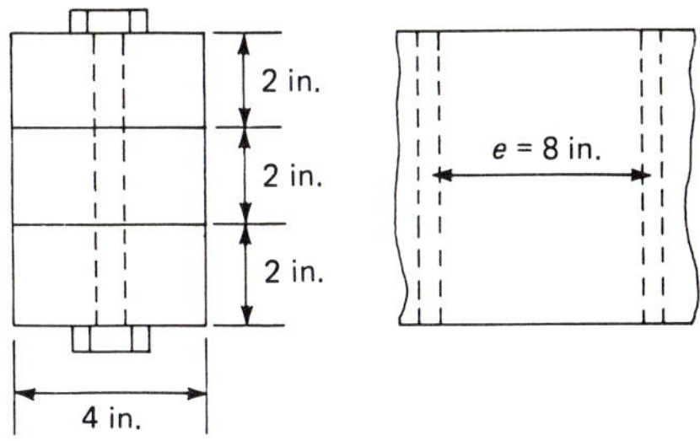

FIGURE P 12.29

12.30 Two S3 × 7.5 beams are fastened together as shown in Fig. P 12.30 by pairs of 1-in. rivets spaced at 6-in. intervals along the beam. If the allowable rivet shear stress is 32 000 psi, determine the maximum shear force the rivets can carry.

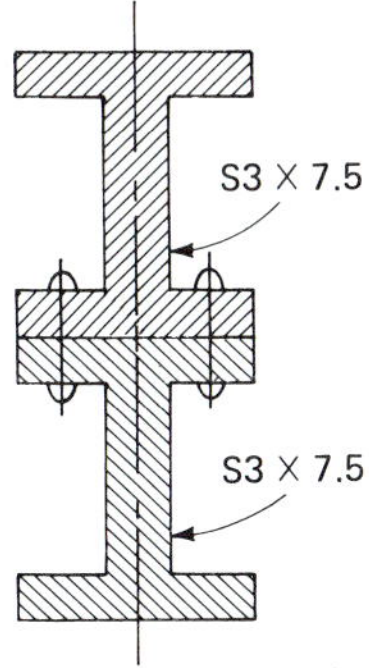

FIGURE P 12.30

Problems

13

The Deflection of Beams

13.1 INTRODUCTION

We have seen that when an elastic member is subjected to a load it will deflect and set up internal stresses as it resists the applied loading. In Chapter 12 our concern was to calculate these stresses; in this chapter we shall concern ourselves with the deflections of beams as they deform under the applied loading. Quite frequently a restriction on the maximum allowable deflection of a beam is the governing design factor and limits the load for which the beam can be designed. One typical example is the limitation that the maximum deflection of a ceiling beam shall be less than 1/360 of its span to prevent the cracking of plaster ceilings. The deformation of shafts under loads is also of considerable importance. The calculation of beam deflections is often approached from a mathematical viewpoint that requires the solution of a second-order differential equation subject to the loading and the type of end supports of the beam. While this approach is mathematically straightforward, it presents formidable problems associated with the evaluation of the proper boundary conditions as well as in the algebra required to obtain the solution. Fortunately an equivalent procedure can be carried out by relatively simple arithmetic computations, based upon general rules derived from the conditions for equilibrium of the deformed beam. There are available several such arithmetic procedures, and in subsequent portions of this chapter we shall consider the *moment area method,* the procedure most generally in use. In each case we shall derive the necessary principles and then apply them to specific problems. It should be noted at this time that the deflection of a beam causes its ends to rotate and that this angular rotation must be considered in the design of structures. Frequently the angular deflection of the ends of a beam and/or its deflection will yield the conditions necessary for the solution of the category of beams that we have previously denoted as being statically indeterminate.

13.2 THE ELASTIC CURVE

In Chapter 12 we studied the action of a beam as it flexed under the action of the applied loading and subsequently derived the flexure formula. The assumptions made concerning the beam action are equally applicable to the present discussion and are thus repeated here.

1. The beam is straight before bending and has a uniform cross section throughout its length.
2. The material of the beam is homogeneous and its stress–strain curve is linear, that is, stress is proportional to strain. It obeys Hooke's law.
3. The beam has the same modulus of elasticity in compression as in tension.
4. A transverse plane remains plane after bending.
5. The neutral axis is curved due to the bending, and the curved neutral plane is known as the *elastic curve* of the beam.

With the foregoing assumptions in mind, let us now consider the section of a beam shown in Fig. 13.1. On the left side of this section [Fig. 13.1(b)] a counterclockwise couple M is shown; on the right side there is a clockwise couple shown. Thus two equal but opposite couples M act on the section in question. Originally the segment, Δx long, is straight, as shown in Fig. 13.1(a). It bends as shown in Fig. 13.1(b). Due to the bending action, the upper fibers of the beam elongate by an amount e, and the lower fibers contract by an amount e'. This action causes B to move to B' and C to move to C', while the neutral axis (the neutral plane is also the elastic curve of the beam) does not change in length, although it does become curved. Due to the curvature, the plane BC rotates through the angle $\Delta\theta$ to the position B'C', and it remains plane. The center of curvature of the beam can be found as the intersection of AD and B'C' extended to meet at point O. The radius of curvature is denoted by R.

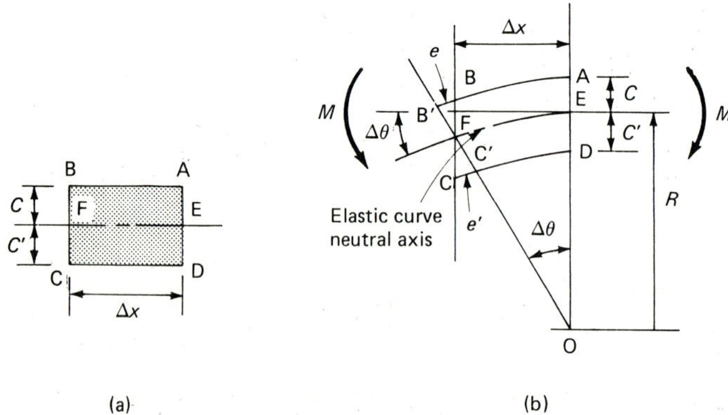

(a) (b)

FIGURE 13.1 The bending of a beam. (a) Before bending. (b) After bending.

  The Deflection of Beams

Let us now consider the sectors OEF and FBB′, which we can see are geometrically similar. The following relationship must therefore hold:

$$\frac{EF}{EO} = \frac{\Delta x}{R} = \frac{BB'}{FB} = \frac{e}{C} \qquad (13.1)$$

Therefore

$$\frac{\Delta x}{R} = \frac{e}{C} \qquad (13.2)$$

The symbol e has been used to denote the total deformation of the top fiber. The unit strain ϵ is $e/\Delta x$. Thus

$$\epsilon = \frac{C}{R} \qquad (13.3)$$

From Hooke's law we know that stress is proportional to strain, that is, $S = E\epsilon$ or $\epsilon = S/E$. Inserting this into Eq. (13.3) yields

$$\frac{S}{E} = \frac{C}{R} \qquad \text{or} \qquad S = \frac{EC}{R} \qquad (13.4)$$

At this point we invoke the flexure formula, $S = MC/I$, and substitute for S in Eq. (13.4) to obtain

$$\frac{MC}{EI} = \frac{C}{R} \qquad (13.5)$$

Simplifying,

$$\frac{1}{R} = \frac{M}{EI} \qquad (13.6)$$

or

$$M = \frac{EI}{R} \qquad (13.6a)$$

Equations (13.6) and (13.6a) are the basic equations for the elastic curve of the bent beam. It is interesting to note that these equations were derived for the case of a beam in pure bending, but since they express the relation between bending moment and radius of curvature, they are applicable to all bent members.

One other relation will be derived at this time since it will be used in the following sections of this chapter—an expression for the *angle of curvature* (also known as *deflection angle*). The tangent of the angle $\Delta\theta$ is $\Delta x/R$. However, since $\Delta\theta$ is a small angle, the tangent can be replaced by the angle $\Delta\theta$ in radians. Thus

$$\Delta\theta = \frac{\Delta x}{R} \qquad (13.7)$$

Substituting for R from Eq. (13.6),

$$\Delta\theta = \frac{M}{EI}(\Delta x) \tag{13.8}$$

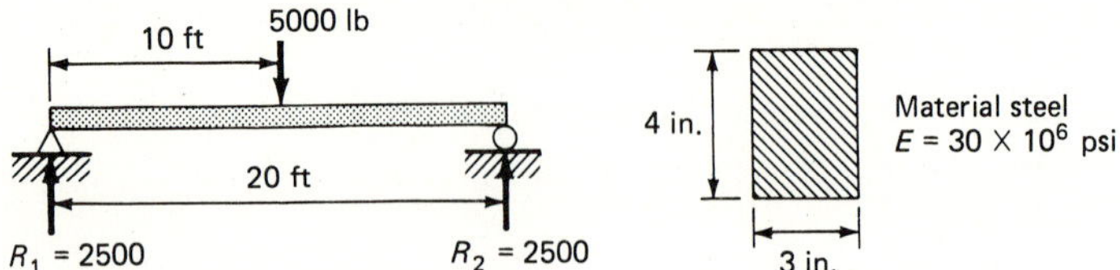

FIGURE 13.2   Illustrative Problem 13.1.

SOLUTION

The moment at the center of the beam $M = R_1 \times L/2 = 2500 \times 10 =$ 25 000 ft·lb $= 300\,000$ in.·lb. The moment of inertia $I = bh^3/12 = 3 \times 4^3/12 = 16$ in.^{4}. Using Eq. (13.6),

$$R = \frac{EI}{M} = \frac{30 \times 10^6 \times 16}{300\,000} = 1600 \text{ in.}$$

The radius of curvature is quite large, indicating that the curvature of the beam is relatively small, that is, the deformed beam is not curved a great deal.

SOLUTION

This problem can be solved in two ways. The first is simply to use Eq. (13.4) directly as follows:

$$S = \frac{CE}{R} = \frac{(0.001/2) \times 210 \times 10^9}{1/2} = 210 \text{ MPa}$$

Notice that the width of the saw blade (18 mm) did not enter into the solution.

The alternate approach is to use the flexure formula. The procedure is as follows:

$$M = \frac{EI}{R} = \frac{210 \times 10^9 \times 0.018 \times 0.001^3/12}{1/2} = 0.630 \text{ N·m}$$

                                    The Deflection of Beams

By the flexure formula

$$S = \frac{MC}{I} = \frac{0.630 \times 0.001/2}{0.018 \times 0.001^3/12} = 210 \text{ MPa}$$

Both methods yield equivalent results, but the second procedure shows that the direct load on the saw is $0.63/(\frac{1}{2}) = 1.26$ N $[M/R = (F \times R)/R = F]$. The direct stress in the blade is F/A, which equals $1.26/0.001 \times 0.018 = 70$ kPa, which obviously is not the critical design stress in the saw blade.

13.3 THE MOMENT AREA PROPOSITIONS

Consider a beam that has been bent by the loading applied to it. Such a case is shown in Fig. 13.3 with the curvature of the bent beam greatly exaggerated. Let us draw tangents to the elastic curve at points ① and ②. The angle that the tangent to the elastic curve makes with the horizontal (undeformed curve) is denoted by θ, the angle at ① by θ_1 and the angle at ② by θ_2. We can see from Fig. 13.3 that the difference in angle between the tangents drawn at points ① and ② is

$$\theta = \theta_2 - \theta_1 \tag{13.9}$$

and we may define the angle of curvature of a segment of a bent beam to be the angle formed by the two end tangents to the elastic curve.

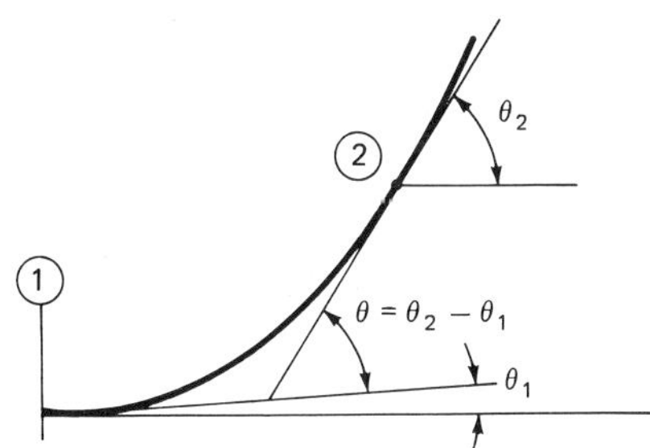

FIGURE 13.3 Beam curvature.

Let us now study the situation shown in Fig. 13.4. Due to some applied constant moment M_1, the segment of the neutral axis $(\Delta x)_1$ deflects to form the elastic curve. We shall assume that the length of the elastic curve AB is very nearly equal to the undeformed section of the neutral axis. This conclusion is based upon the developments of Section 13.2. Thus AB and $(\Delta x)_1$ are taken to be essentially equal. If an adjacent section of the beam $(\Delta x)_2$ is subjected to a constant moment M_2, the elastic curve will undergo further curvature indicated by point C. The change in curvature from A to C equals $(\Delta\theta)_B + (\Delta\theta)_C$, that is,

$$\Delta\theta = (\Delta\theta)_B + (\Delta\theta)_C \tag{13.10}$$

However, in Eq. (13.8) we had already derived an expression relating the change in angle to the applied moment, the modulus of elasticity of the material, and the properties of the cross section of the beam. Using this relation with Eq. (13.10) yields

$$\Delta\theta = \frac{M_1}{EI}(\Delta x)_1 + \frac{M_2}{EI}(\Delta x)_2 \qquad (13.11)$$

For the present E and I will be taken to be constants. If we continue this summation to some point D on the beam,

$$\theta\Big|_A^D = \theta_D - \theta_A = \sum_A^D \frac{M}{EI}(\Delta x) \qquad (13.12)$$

where Σ_A^D is used to denote the sum of $M/EI(\Delta x)$ terms from A to D. Equation (13.12) is the mathematical statement of the first moment area principle. The product term EI represents the *flexural rigidity* of the beam.

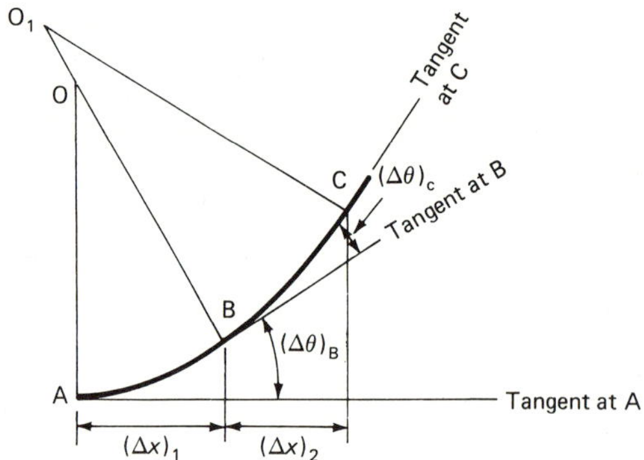

FIGURE 13.4   The elastic curve.

Principle 1:   *The difference in the slope angles (in radians) between tangents drawn to any two points on the elastic curve of a beam is equal to the area of the moment curve between the two points divided by EI, the flexural rigidity of the beam.*

In order to maintain the conventions already adopted for moments in beams, we will take areas due to positive moments to be positive and areas due to negative moments to be negative. Thus a net positive area of the M/EI summation indicates that the right-hand tangent (point C in Fig. 13.4) makes a positive (or counterclockwise) angle with respect to the left-hand tangent (point A in Fig. 13.4).

ILLUSTRATIVE PROBLEM 13.3

A cantilever beam is fixed at one end and is loaded by a concentrated force at the free end. Determine the angular deflection of the beam if E and I are constant.

SOLUTION

As shown in Fig. 13.5(b), the moment is negative and increases (in the

                                                    The Deflection of Beams

absolute sense) from zero at the free end to $-FL$ at the fixed end. The area between A and B of the moment curve is the area of the triangle whose base is L and whose altitude is $-FL$. This area is $-FL^2/2$, and the angle between tangents to the elastic curve at points A and B is

$$\theta = \frac{-FL^2}{2EI} \text{ rad}$$

with the negative value indicating a positive rotation of point B with respect to point A. Since the beam is built in at the fixed end, the tangent at point A is horizontal.

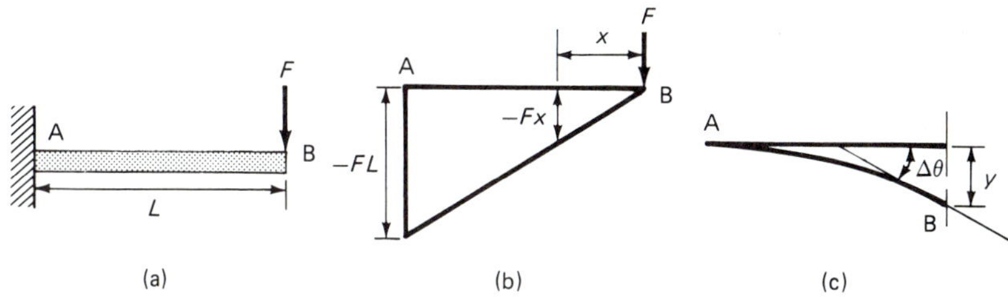

FIGURE 13.5   Illustrative Problem 13.3. (a) Load diagram. (b) Moment diagram. (c) Deflection diagram.

The next step is to develop a procedure that will enable us to determine the deflection of a beam at any point along the beam. By definition we shall take the deflection of a beam at a given section to be equal to the distance between the elastic curve and the neutral axis measured normal to the neutral axis. In the development of Principle 1 it was considered that the elastic curve of the beam was composed of a series of small arcs having varying curvature; we shall now proceed to extend this concept to determine beam deflections. For this purpose refer to Fig. 13.6, which is essentially the same as Fig. 13.4 with certain additions. Specifically $(\Delta y)_1$ and $(\Delta y)_2$ are shown as well as the x_1 and x_2 distances. For convenience we will take the tangent at A to be horizontal and to coincide with the undeformed neutral axis of the beam. Due to this assumption, the chords AB and BC will coincide with the tangents at B and C since the arcs are taken to be very small. If the line y–y is made normal to the neutral axis (and therefore vertical in this case) at D, the intercept of the tangents at B and C on y–y can be taken to be the tangent of the arc corresponding to the angle made by the tangents. Thus for the intercept due to the tangent at B

$$(\Delta y)_1 = (x_1)(\Delta\theta)_1 \tag{13.13}$$

and due to the tangent at C

$$(y)_2 = (x_2)(\Delta\theta)_2 \tag{13.14}$$

The Moment Area Propositions

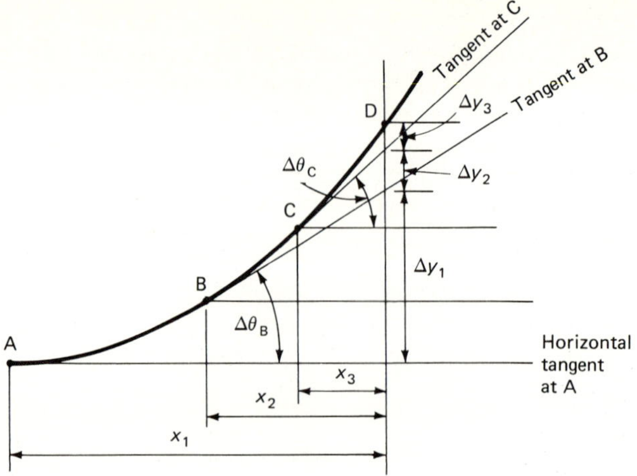

FIGURE 13.6   Beam deflections.

The total displacement of D from the tangent at A is denoted by y. Thus

$$y = (\Delta y_1) + (\Delta y_2) = x_1(\Delta\theta)_1 + x_2(\Delta\theta)_2 \tag{13.15}$$

Using Eq. (13.8) with Eq. (13.15) yields

$$y = x_1\frac{M_1\Delta x}{EI} + x_2\frac{M_2\Delta x}{EI} \tag{13.16}$$

Equation (13.16) allows us to state Principle 2 of the moment area method as follows:

Principle 2:   *The displacement of a point A on the elastic curve of a beam from the tangent at another point B on the curve measured normal to the unstrained neutral axis is equal to the moment of the area of the M/EI diagram included between the two points taken about the first point, A.*

ILLUSTRATIVE PROBLEM 13.4

Determine the deflection of the free end of the cantilever beam shown in Fig. 13.7.

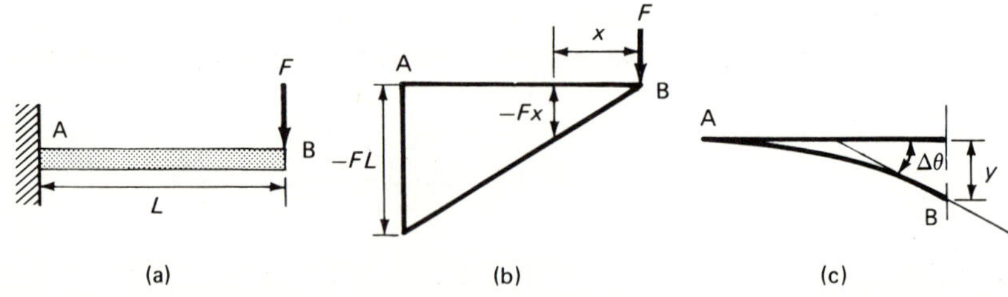

FIGURE 13.7   Illustrative Problem 13.4. (Repeated from Fig. 13.5.) (a) Load diagram. (b) Moment diagram. (c) Deflection diagram.

   The Deflection of Beams

SOLUTION

We have chosen the cantilever beam as the first illustration of both Principles 1 and 2 since the built-in end of the beam does not rotate, and therefore the tangent to the elastic curve at this end is horizontal. This condition simplifies the calculation considerably. Applying the second principle, we can state that the deflection at B (the free end) from the tangent to the elastic curve at A (the built-in end) is the moment of the area of the M/EI diagram between points A and B taken about point B. The area of the M/EI diagram (since E and I are constants) is $(-FL)(L/2)(1/EI) = -FL^2/2EI$. In order to obtain the moment of this area about point B, it is necessary to determine the distance of the centroid of this area from point B. This distance is the moment arm to use in determining the desired moment. For a triangle whose base is L, this distance is $\frac{2}{3}L$ (Fig. 13.8). Thus the moment is

$$\left(\frac{-FL^2}{2EI}\right)\left(\frac{2}{3}L\right) = \frac{-FL^3}{3EI}$$

The moment arm is taken as positive since it lies between A and B. The displacement $(-FL^3/3EI)$ of B from the tangent at A is therefore the deflection of B from its unloaded position, since the tangent at A is horizontal.

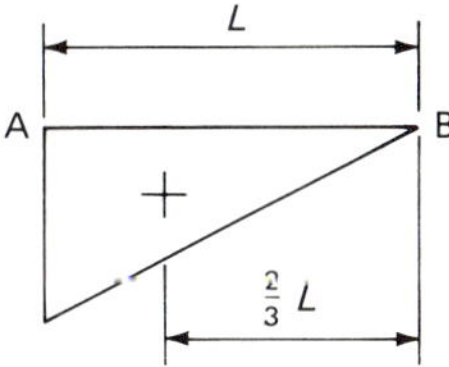

FIGURE 13.8   Centroid of a triangle.

ILLUSTRATIVE PROBLEM 13.5

A concentrated load F is placed at the midspan of a uniform cantilever beam, as shown in Fig. 13.9. Determine the deflection of the beam at the midpoint and at the free end.

SOLUTION

The deflection at the midspan is obtained exactly as it was in Illustrative Problem 13.4. Thus

$$y_{center} = \left(\frac{-FL}{2}\right)\left(\frac{L}{2}\right)\left(\frac{1}{2}\right)\left(\frac{L}{3}\right)\left(\frac{1}{EI}\right) = \frac{-FL^3}{24EI}$$

From the moment diagram we note that there is zero additional moment contributed at any section of the beam to the right of the load. From Principle 1 this is to be interpreted to mean that there is no additional angular change beyond the midpoint as one proceeds to the right of the midpoint. This is shown schematically

The Moment Area Propositions

in Fig. 13.9(c). We see that the deflection at the free end is $y_{center} + (L/2)\theta$, where we have substituted θ for tan θ. Using Principle 1, $\theta = (-FL/2)(L/2)(\frac{1}{2})(1/EI) = -FL^2/8EI$. Therefore $(L/2)\theta = (L/2)(-FL^2/8EI) = -FL^3/16EI$. The total free-end deflection is finally given as $y_{end} = y_{center} + (L/2)\theta = (-FL^3/24EI) + (-FL^3/16EI) = -5FL^3/48EI$. The student will note that this problem differs from the previous problem in that here the elastic curve consists of two parts rather than a single continuous element.

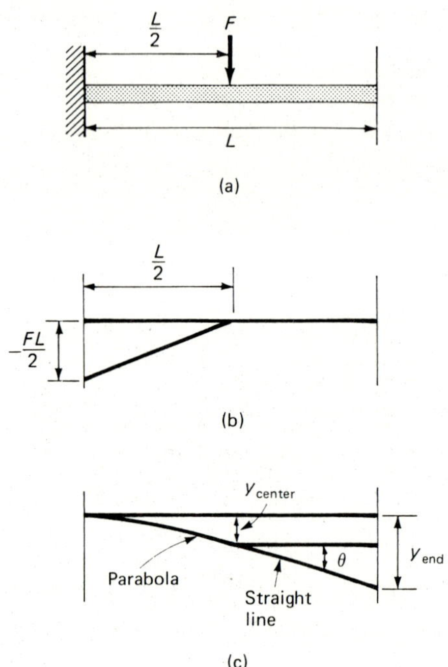

FIGURE 13.9   Illustrative Problem 13.5. (a) Load diagram. (b) Moment diagram. (c) Deflection diagram.

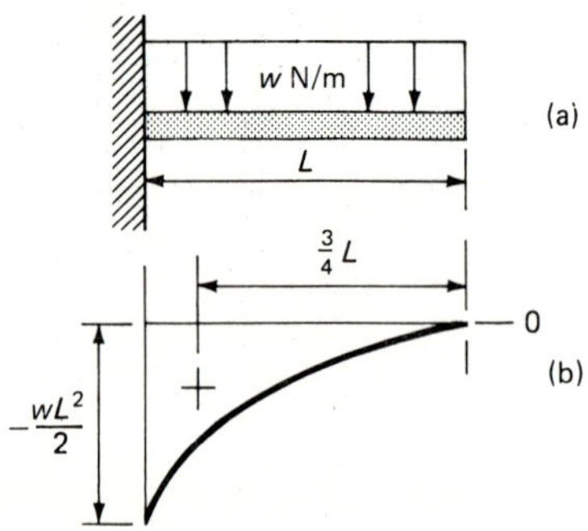

FIGURE 13.10   Illustrative Problem 13.6. (a) Load diagram. (b) Moment diagram.

                                    The Deflection of Beams

Section	Area	$\bar{x}$	$\bar{y}$
Square	b_2	$\dfrac{b}{2}$	$\dfrac{b}{2}$
Rectangle	bh	$\dfrac{b}{2}$	$\dfrac{h}{2}$
Triangle	$\frac{1}{2}\,bh$	$\dfrac{b}{3}$	$\dfrac{h}{3}$
Semicircle	$\dfrac{\pi R^2}{2}$	0	$\dfrac{4R}{3\pi}$ or $0.4244R$
Quarter circle	$\dfrac{\pi R^2}{4}$	$\dfrac{4R}{3\pi}$ or $0.4244R$	$\dfrac{4R}{3\pi}$ or $0.4244R$
Quadrant of ellipse	$\dfrac{\pi ab}{4}$	$\dfrac{4a}{3\pi}$ or $0.4244a$	$\dfrac{4b}{3\pi}$ or $0.4244b$
Spandrel	$\dfrac{bh}{n+1}$	$\dfrac{b}{n+2}$	$\left\{\dfrac{n+1}{4n+2}\right\}h$

The Moment Area Propositions

TABLE 13.2   Properties of Parabolas

Shape	Area	Centroid $\bar{x}$
(1) Parabola $y = ax^2$	$\dfrac{1}{3}bh$	$\dfrac{b}{4}$
(2) Cubic parabola $y = ax^3$	$\dfrac{1}{4}bh$	$\dfrac{b}{5}$
(3) Parabola $y^2 = \dfrac{h^2}{b}x$	$\dfrac{2}{3}bh$	$\dfrac{3}{8}b$
(4) mth order parabola (spandrel) $y = ax^m$	$\dfrac{bh}{m+1}$	$\dfrac{b}{m+2}$

ILLUSTRATIVE PROBLEM 13.6

A cantilever beam supports a uniformly distributed load of w N/m over its full length, as shown in Fig. 13.10. Determine the deflection of the free end if the beam has a constant cross section and a constant E.

SOLUTION

As shown in Fig. 13.10(b), the moment diagram due to the load is a parabola with a maximum negative value of $-wL^2/2$. The properties of various areas were

The Deflection of Beams

discussed in Chapter 8, and for convenience Table 8.1 is repeated at this time as Table 13.1. The spandrel noted in the table is also a parabola of the *n*th degree. Since the properties of the parabola are needed quite often in our present study, the properties of various parabolas are presented in Table 13.2. Using either Table 13.1 or Table 13.2 for the problem under discussion yields an area of $(1/3EI)(-wL^2/2)(L)$ for the moment diagram and a moment arm of $\frac{3}{4}L$. The deviation of the free end from the tangent at the fixed end (also the deflection of the free end in this case) is therefore

$$y_{\text{free end}} = \frac{1}{3EI}\left(\frac{-wL^2}{2}\right)L(\tfrac{3}{4}L) = \frac{-wL^4}{8EI} = \frac{-WL^3}{8EI}$$

where W, the total load, is equal to wL.

Later in this chapter we shall return to the cantilever beam as we develop methods of treating statically indeterminate beams. For the present we shall leave the cantilever beam and direct our discussion to the simply supported beam. We shall first develop a general procedure for handling this type of problem and then illustrate it with several examples. For this discussion refer to Fig. 13.11, which depicts a uniform beam, simply supported at its ends, and carrying any arbitrary vertical load.

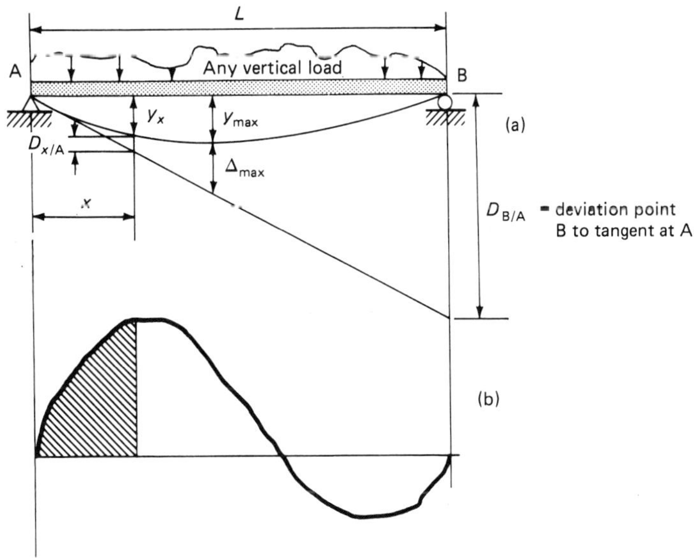

FIGURE 13.11 Simply supported beam. (a) Deflection curve. (b) Moment curve.

Referring to Fig. 13.11(a), we have from similar triangles at section x

$$\frac{y_x + D_{x/A}}{x} = \frac{D_{B/A}}{L} \tag{13.17}$$

The Moment Area Propositions   349 where $D_{B/A}$ is used to denote the deviation of point B to the tangent to the elastic curve at point A. Our concern at the moment is to evaluate the deflection y_x at x. Thus

$$y_x + D_{x/A} = \frac{x}{L} D_{B/A} \qquad (13.18)$$

The deviation $D_{B/A}$ is obtained as the moment of the M/EI diagram between points A and B about point B. Since $D_{x/A}$ is the deviation at x from a tangent to the elastic curve at A, it is evaluated as the moment of the M/EI curve between x and A [shaded in Fig. 13.11(b)] about x. Thus we can calculate all of the terms needed in Eq. (13.18) to determine y_x. We now note that the location of the section at which the deflection is maximum can be determined by two independent methods. Both methods use the fact that the tangent to the beam at the section where the deflection is maximum is horizontal. Thus in Fig. 13.12 we note that the angles θ_A and θ_x must be equal. This requirement can be evaluated by applying Principle 1 to yield

$$\theta_A = \theta_x \qquad \left(\text{area } \frac{M}{EI} \right)_{A-B} = \left(\text{area } \frac{M}{EI} \right)_{A-x_{max}} \qquad (13.19)$$

In words, the area of the M/EI diagram between the ends of the beam must equal the area between the left support and the section at which the maximum deflection occurs. This locates the section at which the maximum deflection occurs, and its value y_{max} is determined as previously outlined for any section x.

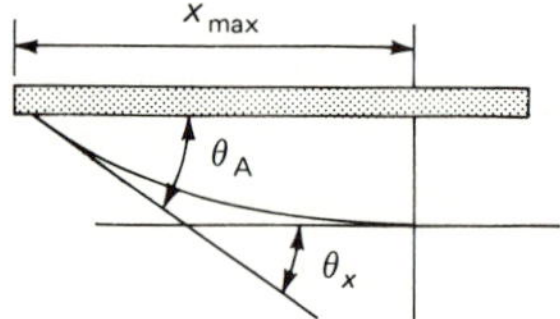

FIGURE 13.12   Simply supported beam—maximum deflection.

The second alternate (but equivalent) method of determining the location of the section at which the maximum deflection occurs is based upon the geometry in Fig. 13.13. The deviation of A from the tangent to the deflection curve at the section where the maximum deflection occurs must also equal the deviation of B from this same tangent since both deviations equal the maximum deflection. Both deviations are found from the second moment area principle, enabling us to determine the location of section x_{max}.

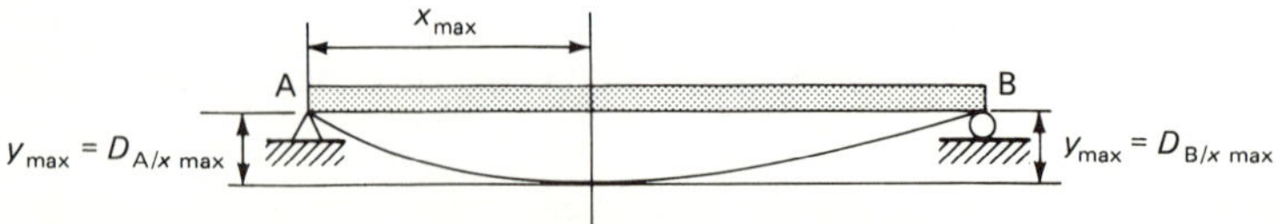

FIGURE 13.13   Maximum deflection of simple beams.

                                                The Deflection of Beams

A simply supported beam has a concentrated load F at its center, as shown in Fig. 13.14. If the length of the beam is L and the beam has a constant cross section, determine the maximum deflection of the beam.

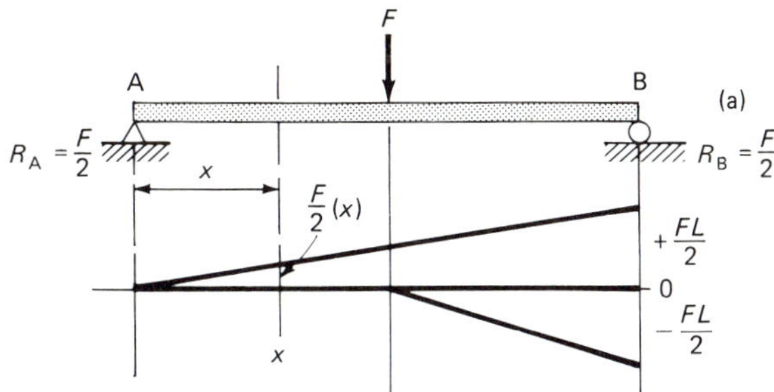

FIGURE 13.14   Illustrative Problem 13.7. (a) Load diagram. (b) Moment diagram.

SOLUTION

The beam is symmetrically loaded so that the maximum deflection occurs at its center. Taking advantage of this fact, we note that the deflection at the center can be obtained by our considering the center to be fixed and either end loaded by concentrated loads of $F/2$, as shown in Fig. 13.15. The end deflection of the cantilever is equal to the central deflection of this simple beam. The end load is $F/2$, and the length of beam is $L/2$ for the cantilever, yielding

$$(y_{end}) \text{ cantilever } = (y_{center}) \text{ simple } = \frac{(F/2)(L/2)^3}{3EI} = \frac{-FL^3}{48EI} = y_{max}$$

Had we not noted and used symmetry, we would first have had to calculate the location of the section of maximum deflection x, which is also the section at which the tangent to the deflection curve is horizontal. In order to simplify the work, the moment diagram of Fig. 13.14(b) is shown in parts. We have for θ_A

$$\theta_A = \frac{1}{EI}\left[\left(\frac{FL}{2}\right)\left(\frac{L}{2}\right) - \left(\frac{FL}{2}\right)\left(\frac{L}{4}\right)\right] = \frac{1}{EI}\left(\frac{FL^2}{8}\right)$$

If we assume that x is to the left of the center, as shown in Fig. 13.14,

$$\theta_x = \left[\left(\frac{F}{2}\right)x\right](x)\left(\frac{1}{EI}\right) = \left(\frac{1}{EI}\right)\left(\frac{Fx^2}{2}\right)$$

Therefore for $\theta_A = \theta_x$, we have,

$$\left(\frac{1}{EI}\right)\left(\frac{FL^2}{8}\right) = \left(\frac{1}{EI}\right)\left(\frac{Fx^2}{2}\right)$$

Solving, $x = L/2$, which we anticipated. We now utilize the fact that at this section a tangent is horizontal. Thus $y_{max} = -D_{A/x}$ (with the sense of the deviation re-

The Moment Area Propositions

versed giving rise to the negative sign before the right-hand term). The deviation $-D_{A/x} = -(1/EI)[(FL/4)(L/2)(\frac{1}{2})(\frac{2}{3})(L/2)] = -FL^3/48EI$. Thus, $y_{max} = -FL^3/48EI$, which agrees with our earlier result.

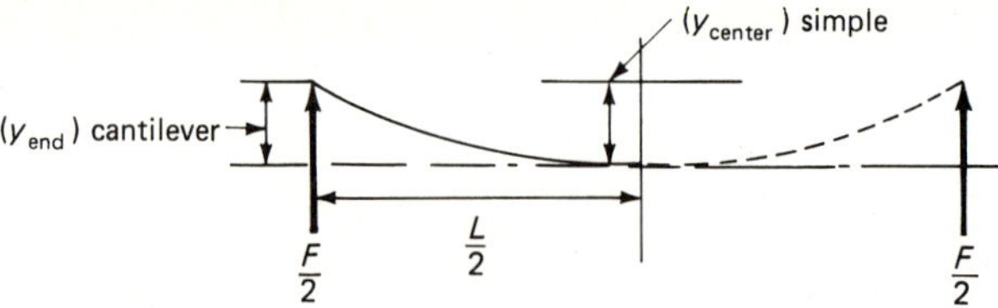

FIGURE 13.15   Illustrative Problem 13.7.

ILLUSTRATIVE PROBLEM 13.8

Determine the maximum deflection of a simply supported beam of constant cross section when loaded over its entire length by a uniform load of w lb/ft, as shown in Fig. 13.16.

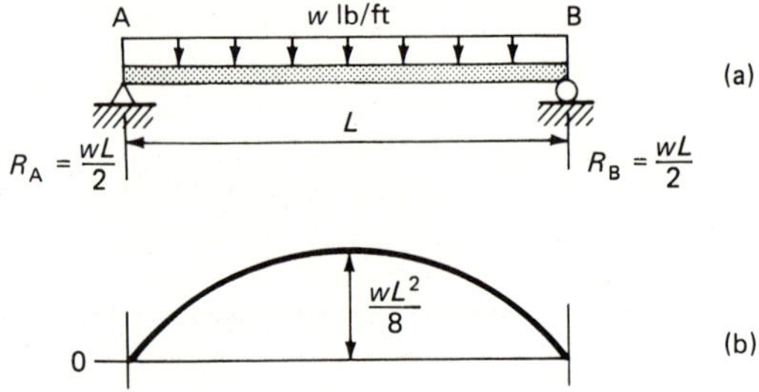

FIGURE 13.16   Illustrative Problem 13.8. (a) Load diagram. (b) Moment diagram.

SOLUTION

In the solution of this problem we shall utilize symmetry. Thus we conclude that the beam is symmetrical about its center, and the maximum deflection occurs at the center. From Chapter 11 we can conclude that the bending moment curve is a parabola with its vertex at the center and a maximum height of $wL^2/8$. From Table 13.2, case ③, the area of such a parabola is $\frac{2}{3}bh$, and the centroid of the area is located $\frac{3}{8}b$ from the right side ($\frac{5}{8}b$ from the left support). The moment of the M/EI curve between the center and left support about the left support yields $-y_{max}$:

$$y_{max} = -\frac{1}{EI}\left(\frac{2}{3}\frac{L}{2}\right)\frac{wL^2}{8}\left(\frac{5}{8}\frac{L}{2}\right) = \frac{-5wL^4}{384EI} = \frac{-5WL^3}{384EI}$$

where $W = wL$.

13.4   THE SUPERPOSITION METHOD AND ITS APPLICATION

At this point in our study let us note a very important fact. Both the slope and the deflection have been found to be directly proportional to the magnitude of the assumed

The Deflection of Beams

loads for every case that we have studied. In addition, the moment area method was predicated on the assumption that the deflections (both angular and linear) are small. Based upon the linearity between load and deflection and for the case of small beam deflections, it can be shown that the deflection and slope of a beam at a point where it is subjected to several loads can be obtained by determining the slope and deflection due to each load and then summing each item at the section in question. The particular importance of the foregoing is that it makes it possible to solve many deflection problems utilizing the results of a relatively few simple cases. Table 13.3 summarizes the cases studied in the previous sections of this chapter as well as several others that will prove useful. This table is based upon the extensive tabulations of beam diagrams and formulas given in the 8th edition of the AISC *Steel Construction Manual*. In addition to the shear and moment curves, general expressions are given for the deflection, the maximum deflection, and the section at which the maximum deflection occurs. It should be borne in mind that the superposition method is not an independent method for the determination of beam deflections, but rather a convenient, useful procedure which can be used in conjunction with other methods to give us solutions for complicated beam-load problems.

TABLE 13.3 Slope and Deflection Formulas for Statically Loaded Uniform Beams

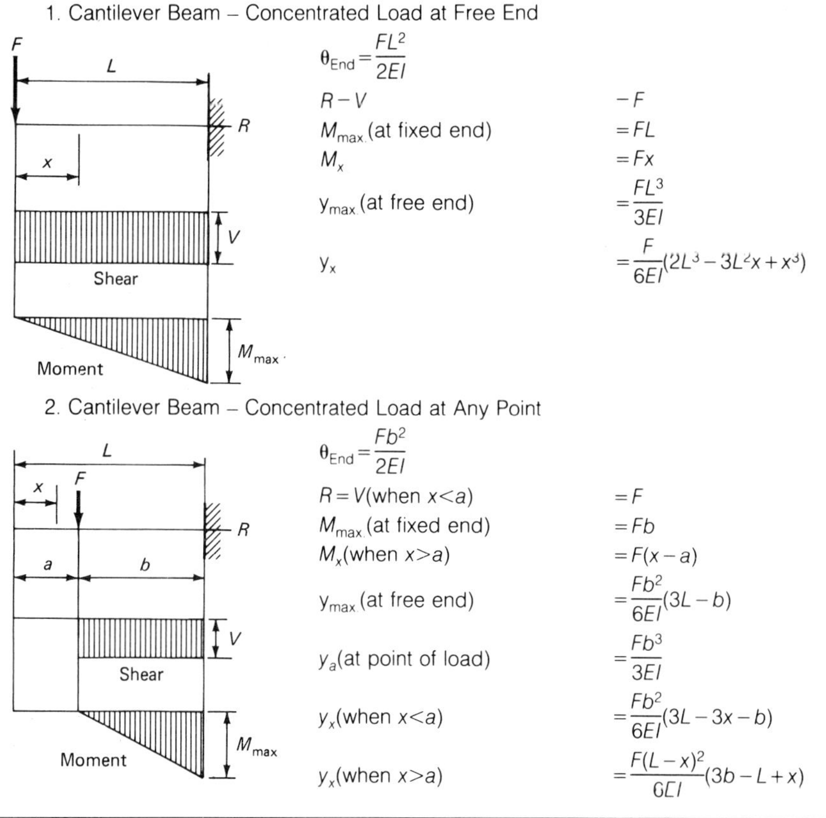

1. Cantilever Beam – Concentrated Load at Free End

$$\theta_{End} = \frac{FL^2}{2EI}$$

$R - V \qquad\qquad -F$

M_{max} (at fixed end) $\qquad = FL$

$M_x \qquad\qquad = Fx$

y_{max} (at free end) $\qquad = \dfrac{FL^3}{3EI}$

$y_x \qquad\qquad = \dfrac{F}{6EI}(2L^3 - 3L^2x + x^3)$

2. Cantilever Beam – Concentrated Load at Any Point

$$\theta_{End} = \frac{Fb^2}{2EI}$$

$R = V$ (when $x < a$) $\qquad = F$

M_{max} (at fixed end) $\qquad = Fb$

M_x (when $x > a$) $\qquad = F(x - a)$

y_{max} (at free end) $\qquad = \dfrac{Fb^2}{6EI}(3L - b)$

y_a (at point of load) $\qquad = \dfrac{Fb^3}{3EI}$

y_x (when $x < a$) $\qquad = \dfrac{Fb^2}{6EI}(3L - 3x - b)$

y_x (when $x > a$) $\qquad = \dfrac{F(L-x)^2}{6EI}(3b - L + x)$

The Superposition Method and Its Application

3. Cantilever Beam – Uniformly Distributed Load

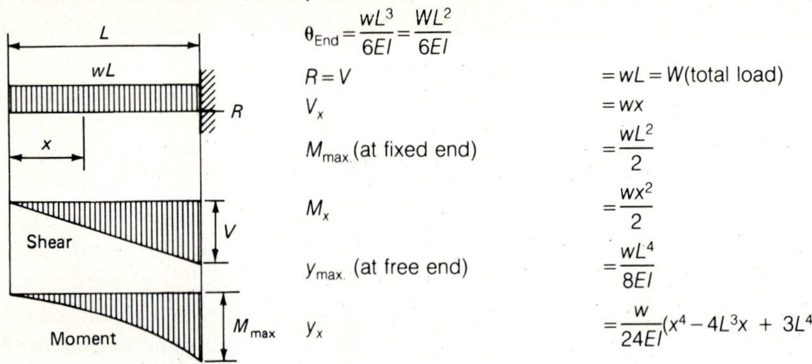

$$\theta_{End} = \frac{wL^3}{6EI} = \frac{WL^2}{6EI}$$

$R = V \qquad = wL = W\text{(total load)}$

$V_x \qquad = wx$

$M_{max.}\text{(at fixed end)} \qquad = \frac{wL^2}{2}$

$M_x \qquad = \frac{wx^2}{2}$

$y_{max.}\text{ (at free end)} \qquad = \frac{wL^4}{8EI}$

$y_x \qquad = \frac{w}{24EI}(x^4 - 4L^3x + 3L^4)$

4. Cantilever Beam – Load Increasing Uniformly to Fixed End

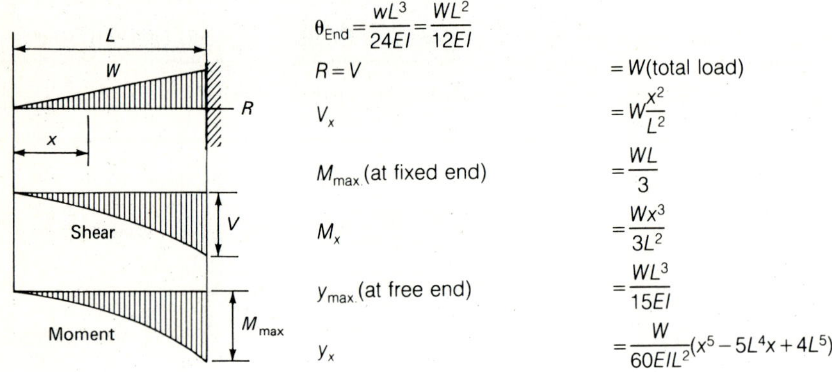

$$\theta_{End} = \frac{wL^3}{24EI} = \frac{WL^2}{12EI}$$

$R = V \qquad = W\text{(total load)}$

$V_x \qquad = W\frac{x^2}{L^2}$

$M_{max}\text{(at fixed end)} \qquad = \frac{WL}{3}$

$M_x \qquad = \frac{Wx^3}{3L^2}$

$y_{max.}\text{(at free end)} \qquad = \frac{WL^3}{15EI}$

$y_x \qquad = \frac{W}{60EIL^2}(x^5 - 5L^4x + 4L^5)$

5. Cantilever Beam – Moment at Free End

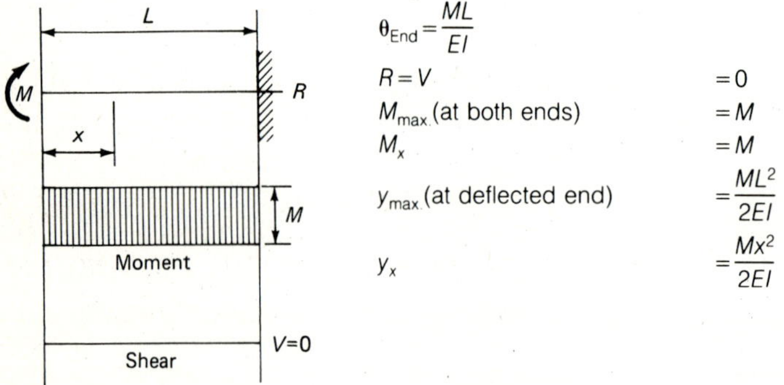

$$\theta_{End} = \frac{ML}{EI}$$

$R = V \qquad = 0$

$M_{max.}\text{(at both ends)} \qquad = M$

$M_x \qquad = M$

$y_{max.}\text{(at deflected end)} \qquad = \frac{ML^2}{2EI}$

$y_x \qquad = \frac{Mx^2}{2EI}$

6. Simple Beam – Concentrated Load at Center

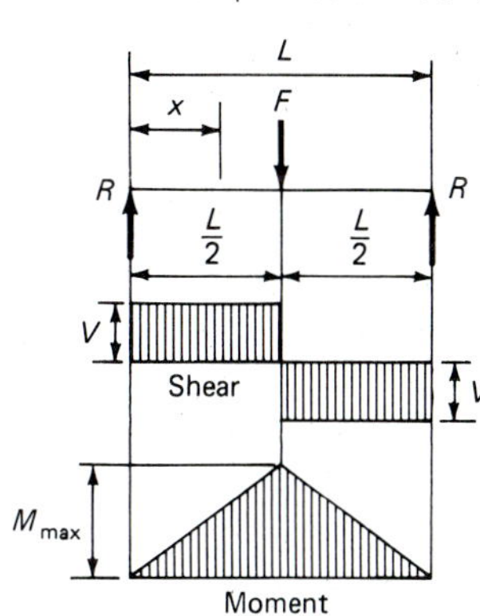

$$\theta = \frac{FL^2}{16EI}\text{(either end)}$$

$$R = V \qquad\qquad = \frac{F}{2}$$

$$M_{max}\text{(at point of load)} \qquad = \frac{FL}{4}$$

$$M_x\left(\text{when } x<\frac{L}{2}\right) \qquad = \frac{Fx}{2}$$

$$y_{max}\text{(at point of load)} \qquad = \frac{FL^3}{48EI}$$

$$y_x\left(\text{when } x<\frac{L}{2}\right) \qquad = \frac{Fx}{48EI}(3L^2 - 4x^2)$$

7. Simple Beam – Concentrated Load at Any Point

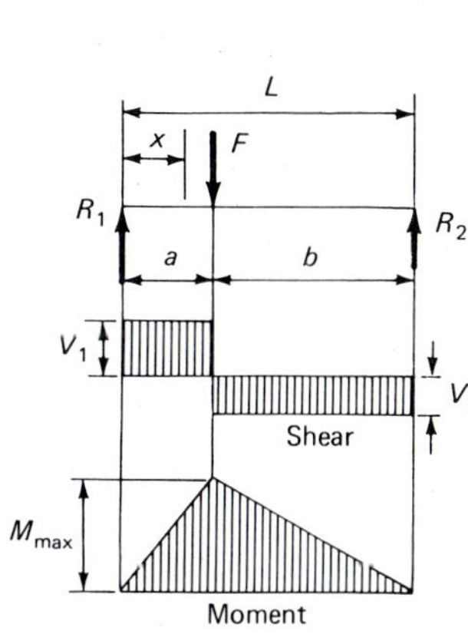

$$\theta_{Left} = \frac{Fb(L^2 - b^2)}{6EIL}; \quad \theta_{Right} = \frac{Fab\,(2L - b)}{6EIL}$$

$$R_1 = V_1\text{(max. when } a<b\text{)} \qquad = \frac{Fb}{L}$$

$$R_2 = V_2\text{(max. when } a >b\text{)} \qquad = \frac{Fa}{L}$$

$$M_{max}\text{(at point of load)} \qquad = \frac{Fab}{L}$$

$$M_x\text{(when } x<a\text{)} \qquad = \frac{Fbx}{L}$$

$$y_{max}\left(\text{at } x = \sqrt{\frac{a(a+2b)}{3}} \text{ when } a>b\right) = \frac{Fab(a+2b)\sqrt{3a(a+2b)}}{27EIL}$$

$$y_a\text{(at point of load)} \qquad = \frac{Fa^2b^2}{3EIL}$$

$$y_x\text{(when } x<a\text{)} \qquad = \frac{Fbx}{6EIL}(L^2 - b^2 - x^2)$$

$$y_x\text{(at center)}(a<b) \qquad = \frac{Fa}{48EI}(3L^2 - 4a^2)$$

8. Simple Beam – Two Equal Concentrated Loads Symmetrically Placed

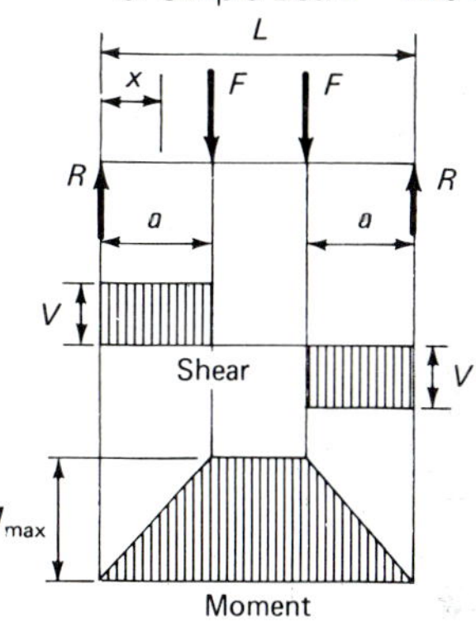

$$\theta = \frac{F}{2EI}(La - a^2)\text{(either end)}$$

$$R = V \qquad = F$$

$$M_{max}\text{(between loads)} \qquad = Fa$$

$$M_x\text{(when } x<a\text{)} \qquad = Fx$$

$$y_{max}\text{(at center)} \qquad = \frac{Fa}{24EI}(3L^2 - 4a^2)$$

$$y_x\text{(when } x<a\text{)} \qquad = \frac{Fx}{6EI}(3La - 3a^2 - x^2)$$

$$y_x\text{(when } x>a \text{ and } <(L-a)) \qquad = \frac{Fa}{6EI}(3Lx - 3x^2 - a^2)$$

The Superposition Method and Its Application

9. Simple Beam – Uniformly Distributed Load

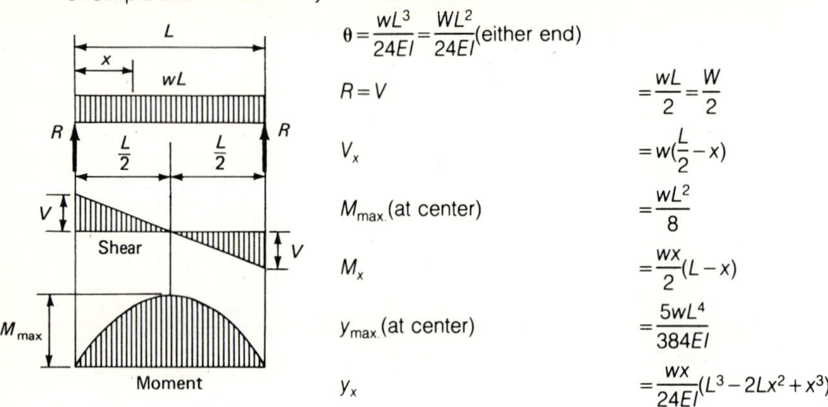

$$\theta = \frac{wL^3}{24EI} = \frac{WL^2}{24EI} \text{(either end)}$$

$$R = V \qquad = \frac{wL}{2} = \frac{W}{2}$$

$$V_x \qquad = w\left(\frac{L}{2} - x\right)$$

$$M_{max.} \text{(at center)} \qquad = \frac{wL^2}{8}$$

$$M_x \qquad = \frac{wx}{2}(L - x)$$

$$y_{max.} \text{(at center)} \qquad = \frac{5wL^4}{384EI}$$

$$y_x \qquad = \frac{wx}{24EI}(L^3 - 2Lx^2 + x^3)$$

10. Simple Beam – Load Increasing Uniformly to One End

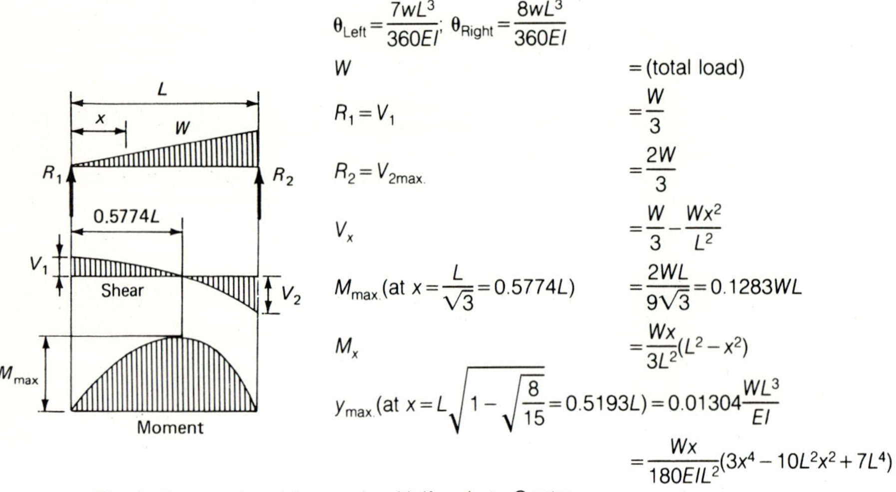

$$\theta_{Left} = \frac{7wL^3}{360EI}; \quad \theta_{Right} = \frac{8wL^3}{360EI}$$

$$W \qquad = \text{(total load)}$$

$$R_1 = V_1 \qquad = \frac{W}{3}$$

$$R_2 = V_{2max.} \qquad = \frac{2W}{3}$$

$$V_x \qquad = \frac{W}{3} - \frac{Wx^2}{L^2}$$

$$M_{max.}\left(\text{at } x = \frac{L}{\sqrt{3}} = 0.5774L\right) = \frac{2WL}{9\sqrt{3}} = 0.1283WL$$

$$M_x \qquad = \frac{Wx}{3L^2}(L^2 - x^2)$$

$$y_{max.}\left(\text{at } x = L\sqrt{1 - \sqrt{\frac{8}{15}}} = 0.5193L\right) = 0.01304\frac{WL^3}{EI}$$

$$= \frac{Wx}{180EIL^2}(3x^4 - 10L^2x^2 + 7L^4)$$

11. Simple Beam – Load Increasing Uniformly to Center

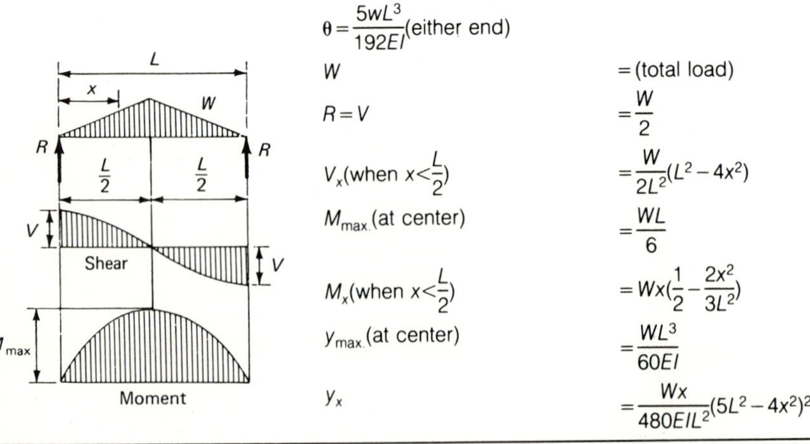

$$\theta = \frac{5wL^3}{192EI} \text{(either end)}$$

$$W \qquad = \text{(total load)}$$

$$R = V \qquad = \frac{W}{2}$$

$$V_x\left(\text{when } x < \frac{L}{2}\right) \qquad = \frac{W}{2L^2}(L^2 - 4x^2)$$

$$M_{max.} \text{(at center)} \qquad = \frac{WL}{6}$$

$$M_x\left(\text{when } x < \frac{L}{2}\right) \qquad = Wx\left(\frac{1}{2} - \frac{2x^2}{3L^2}\right)$$

$$y_{max.} \text{(at center)} \qquad = \frac{WL^3}{60EI}$$

$$y_x \qquad = \frac{Wx}{480EIL^2}(5L^2 - 4x^2)^2$$

The Deflection of Beams

12. Simple Beam – Moment at One End

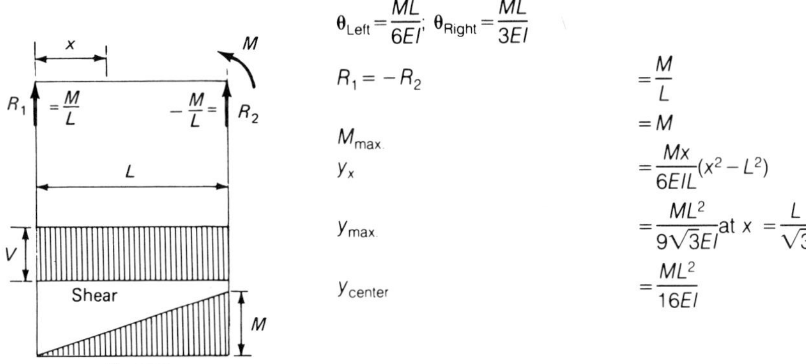

$$\theta_{Left} = \frac{ML}{6EI}; \quad \theta_{Right} = \frac{ML}{3EI}$$

$$R_1 = -R_2 \qquad = \frac{M}{L}$$

$$M_{max.} \qquad = M$$

$$y_x \qquad = \frac{Mx}{6EIL}(x^2 - L^2)$$

$$y_{max.} \qquad = \frac{ML^2}{9\sqrt{3}EI} \text{ at } x = \frac{L}{\sqrt{3}}$$

$$y_{center} \qquad = \frac{ML^2}{16EI}$$

13. Beam Fixed at One End, Supported at Other – Concentrated Load at Any Point

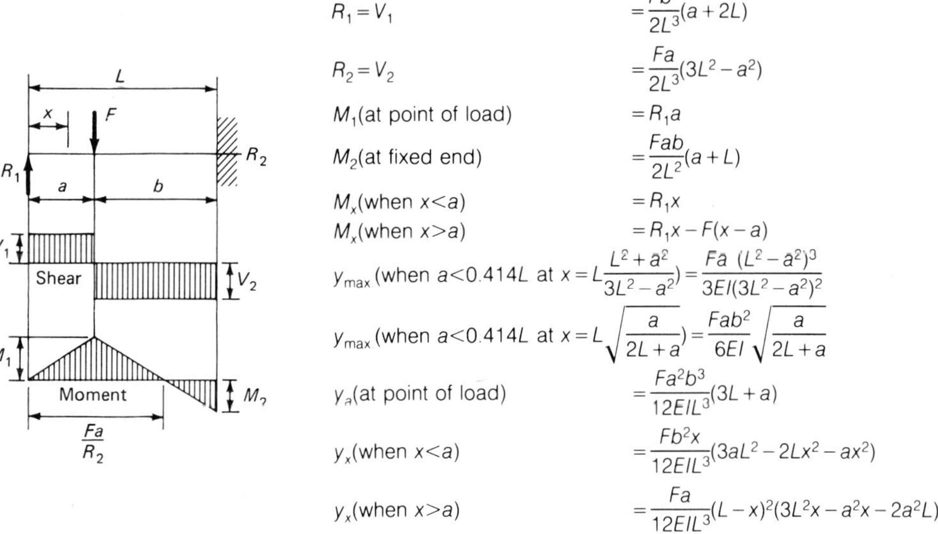

$$R_1 = V_1 \qquad = \frac{Fb^2}{2L^3}(a + 2L)$$

$$R_2 = V_2 \qquad = \frac{Fa}{2L^3}(3L^2 - a^2)$$

$$M_1\text{(at point of load)} \qquad = R_1 a$$

$$M_2\text{(at fixed end)} \qquad = \frac{Fab}{2L^2}(a + L)$$

$$M_x\text{(when } x<a) \qquad = R_1 x$$

$$M_x\text{(when } x>a) \qquad = R_1 x - F(x - a)$$

$$y_{max}\text{(when } a<0.414L \text{ at } x = L\frac{L^2 + a^2}{3L^2 - a^2}) = \frac{Fa\,(L^2 - a^2)^3}{3EI(3L^2 - a^2)^2}$$

$$y_{max}\text{(when } a<0.414L \text{ at } x = L\sqrt{\frac{a}{2L + a}}) = \frac{Fab^2}{6EI}\sqrt{\frac{a}{2L + a}}$$

$$y_a\text{(at point of load)} \qquad = \frac{Fa^2b^3}{12EIL^3}(3L + a)$$

$$y_x\text{(when } x<a) \qquad = \frac{Fb^2x}{12EIL^3}(3aL^2 - 2Lx^2 - ax^2)$$

$$y_x\text{(when } x>a) \qquad = \frac{Fa}{12EIL^3}(L - x)^2(3L^2x - a^2x - 2a^2L)$$

14. Beam Fixed at One End, Supported at Other – Uniformly Distributed Load

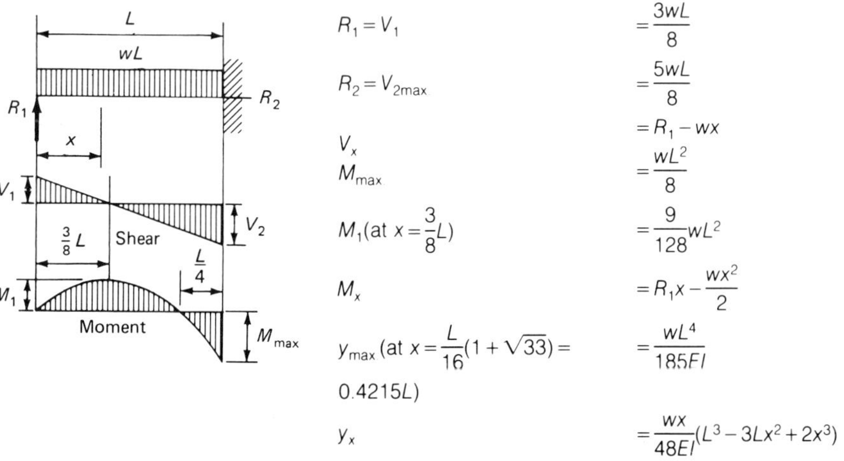

$$R_1 = V_1 \qquad = \frac{3wL}{8}$$

$$R_2 = V_{2max} \qquad = \frac{5wL}{8}$$

$$V_x \qquad = R_1 - wx$$

$$M_{max} \qquad = \frac{wL^2}{8}$$

$$M_1\text{(at } x = \tfrac{3}{8}L) \qquad = \frac{9}{128}wL^2$$

$$M_x \qquad = R_1 x - \frac{wx^2}{2}$$

$$y_{max}\text{(at } x = \frac{L}{16}(1 + \sqrt{33}) = 0.4215L) \qquad = \frac{wL^4}{185EI}$$

$$y_x \qquad = \frac{wx}{48EI}(L^3 - 3Lx^2 + 2x^3)$$

15. Beam Fixed at Both Ends – Concentrated Load at Any Point

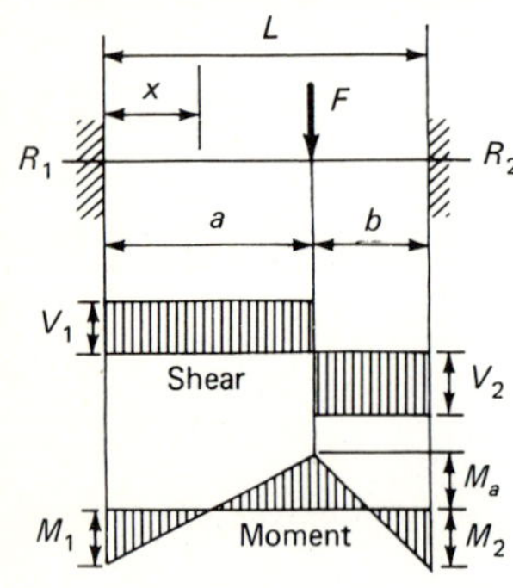

$$R_1 = V_1 \text{(max. when } a<b) = \frac{Fb^2}{L^3}(3a+b)$$

$$R_2 = V_2 \text{ (max. when } a>b) = \frac{Fa^2}{L^3}(a+3b)$$

$$M_1 \text{(max. when } a<b) = \frac{Fab^2}{L^2}$$

$$M_2 \text{(max. when } a>b) = \frac{Fa^2b}{L^2}$$

$$M_a \text{(at point of load)} = \frac{2Fa^2b^2}{L^3}$$

$$M_x \text{(when } x<a) = R_1 x - \frac{Fab^2}{L^2}$$

$$y_{max} \text{(when } a>b \text{ at } x = \frac{2aL}{3a+b}) = \frac{2Fa^3b^2}{3EI(3a+b)^2}$$

$$y_a \text{(at point of load)} = \frac{Fa^3b^3}{3EIL^3}$$

$$y_x \text{(when } x<a) = \frac{Fb^2x^2}{6EIL^3}(3aL - 3ax - bx)$$

16. Beam Fixed at Both Ends – Uniformly Distributed Load

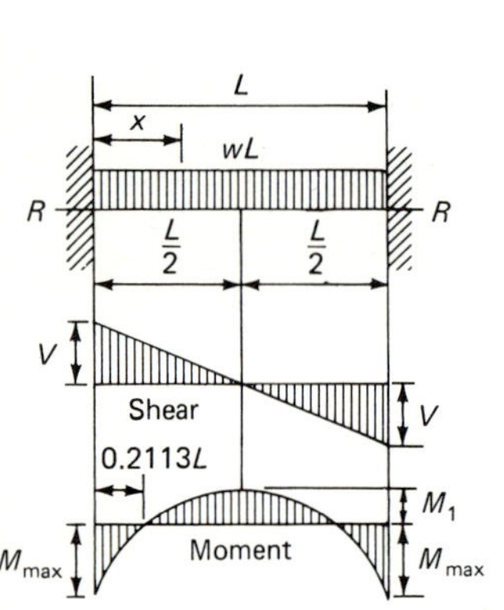

$$\text{Total load} = wL$$

$$R = V = \frac{wL}{2}$$

$$V_x = w\left(\frac{L}{2} - x\right)$$

$$M_{max} \text{(at ends)} = \frac{wL^2}{12}$$

$$M_1 \text{(at center)} = \frac{wL^2}{24}$$

$$M_x = \frac{w}{12}(6Lx - L^2 - 6x^2)$$

$$y_{max} \text{(at center)} = \frac{wL^4}{384EI}$$

$$y_x = \frac{wx^2}{24EI}(L - x)^2$$

ILLUSTRATIVE PROBLEM 13.9

A cantilever beam is loaded with both a concentrated load F at its free end and a uniformly distributed load of w N/m over its length L. Determine the deflection of the free end using the method of superposition. Also derive an equation for the deflection at any section x.

The Deflection of Beams

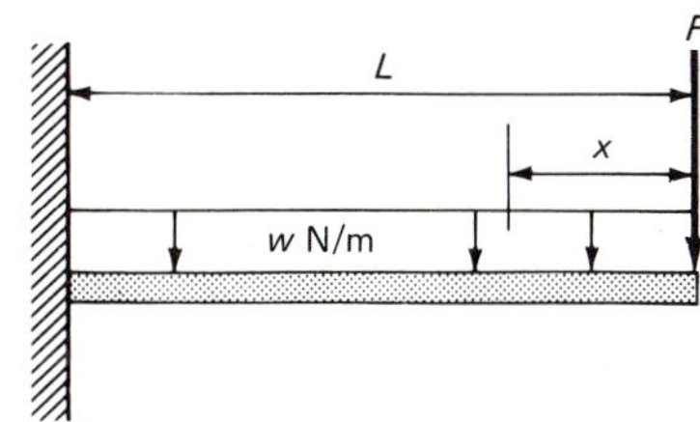

FIGURE 13.17   Illustrative Problem 13.9.

SOLUTION

Referring to Table 13.3, we note that this problem is a combined loading, cases 1 and 3 being used. Thus

$$y_{\text{free end}} = \frac{FL^3}{3EI} + \frac{wL^4}{8EI}$$

The deflection at x due to each load is summed to yield y_x. Thus

$$y_x = \frac{F}{6EI}(2L^3 - 3L^2x + x^3) + \frac{w}{24EI}(x^4 - 4L^3x + 3L^4)$$

and simplifying,

$$y_x = \frac{1}{24EI}(8FL^3 - 12FL^2x + 4Fx^3 + x^4 - 4L^3x + 3L^4)$$

ILLUSTRATIVE PROBLEM 13.10

As shown in Fig. 13.18, a simple beam is symmetrically loaded by two equal forces F (case 8 in Table 13.3). Using the results of case 7, derive the value of

$$y_{\text{max}} \text{ (at center)} = \frac{Fa}{24EI}(3L^2 - 4a^2)$$

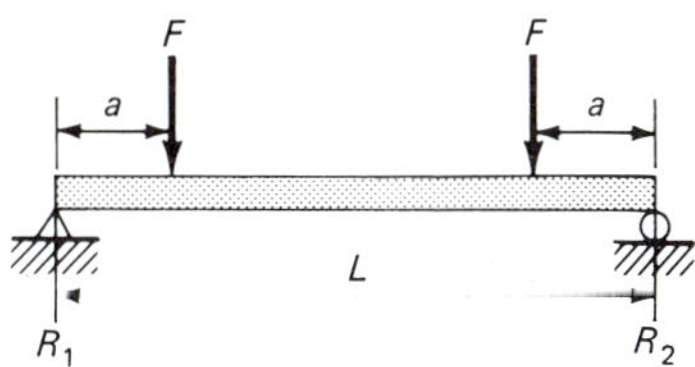

FIGURE 13.18   Illustrative Problem 13.10.

SOLUTION

Consider the beam to be loaded only with the left force F. The deflection at the center due to this load is (case 7)

$$y_{\text{center}} = \frac{Fa}{48EI}(3L^2 - 4a^2)$$

The Superposition Method and Its Application

By symmetry, y_{center} due to the right-hand load must equal the deflection of the center due to the left-hand load. Thus

$$y_{max} \text{ (at center) } = \frac{Fa}{24EI}(3L^2 - 4a^2)$$

Determine the deflection at any section between the supports for the beam shown in Fig. 13.19 when loaded at w N/m.

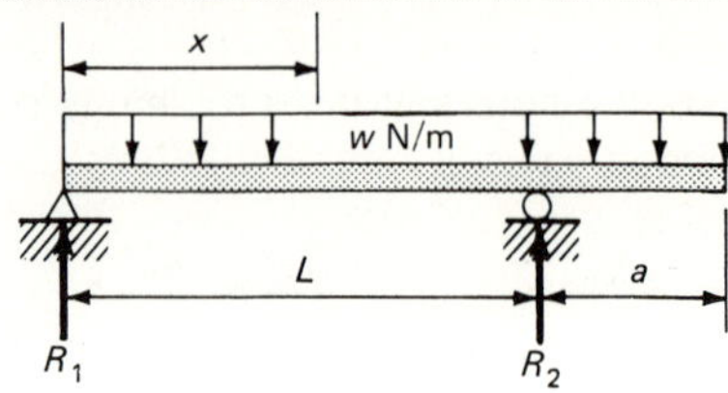

FIGURE 13.19   Illustrative Problem 13.11.

SOLUTION

Due to the overhang there is a moment of $wa^2/2$ on the right support, which serves to reduce our problem to one of a uniformly loaded simple beam with an end moment of $wa^2/2$, as shown in Fig. 13.20. This reduces the problem to the superposition of cases 9 and 12. For case 9,

$$y_x = \frac{wx}{24EI}(L^3 - 2Lx^2 + x^3)$$

and for case 12,

$$y_x = \frac{-Mx}{6EIL}(x^2 - L^2)$$

where the negative sign is used due to the direction of M, and $M = wa^2/2$. The total deflection is then

$$y_x = \frac{wx}{24EI}(L^3 - 2Lx^2 + x^3) - \frac{wa^2x}{12EIL}(x^2 - L^2)$$

or

$$y_x = \frac{wx}{24EI}\left[L^3 - 2Lx^2 + x^3 - \frac{2a^2}{L}(x^2 - L^2) \right]$$

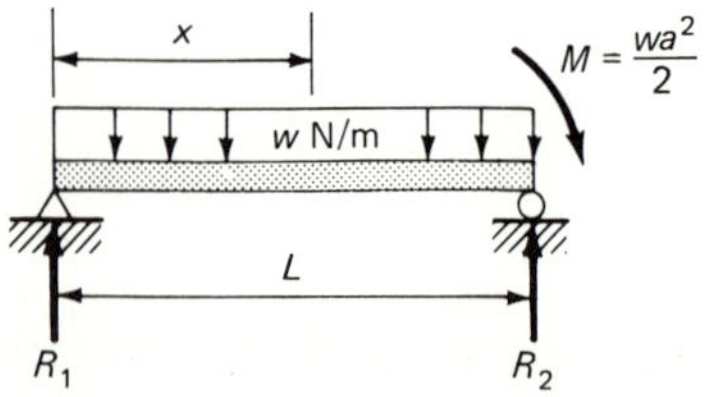

FIGURE 13.20   Illustrative Problem 13.11.

The Deflection of Beams

13.5 STATICALLY INDETERMINATE BEAMS

In Chapter 11 various types of beams were discussed and classified. At that time it was noted that beams having either more supports or more conditions at the supports than could be evaluated from the three conditions necessary for static equilibrium are called *statically indeterminate beams*. The three general types of beams falling under this heading are the *built-in beam,* the *propped beam,* and the *continuous beam.* The student should review the relevant sections of Chapter 11 which discussed these types of beams. In the following discussion we shall apply the superposition method to evaluate the additional information required to obtain a solution to statically indeterminate situations. We will use any condition available to us such as the deflection at a support being zero, the relation between the slopes at two positions, and so on, and a certain amount of ingenuity will be necessary. Most often the necessary condition is found in the statement of the problem.

The foregoing discussion is best illustrated by considering the propped beam. Figure 13.21 shows the propped beam as a cantilever beam with a support under the free end, as shown in Fig. 13.21(a). The equivalent beams, from a force and moment viewpoint, are shown in Fig. 13.21(b) and (c). It is interesting to note that the equivalent propped beam shown in Fig. 13.21(b) is (in principle) the same as was solved for in Illustrative Problem 13.11. Our problem is to evaluate the unknown moment at the built-in end or the magnitude of each of the reactions at the supports. Two conditions are available to us to perform this evaluation. Specifically, the deflection at the supported end is zero, and the slope at the built-in end is zero. The condition that the deflection at the propped end be zero is easier to apply to this case. The procedure consists of first removing the end support and determining the end deflection of the beam as if no support were present. The next step is to calculate the magnitude of the concentrated force applied to this end which is necessary to cause the unloaded beam to return to its undeflected initial position from the deflected position. This must be the value of the reaction at the support at the propped end. Once this is known, the requirements for static equilibrium, $\Sigma F = 0$, $\Sigma M = 0$, suffice to complete the solution of the problem.

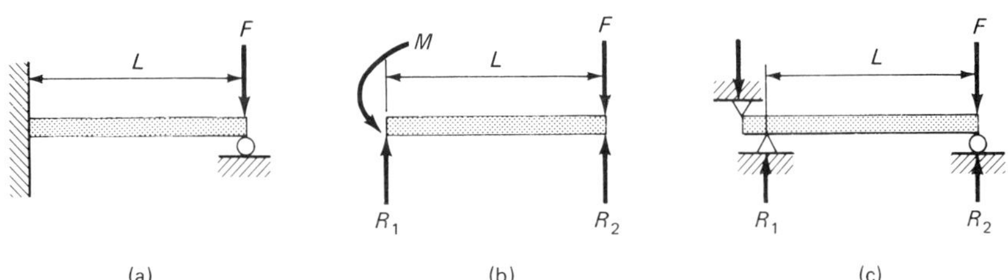

FIGURE 13.21   Propped beams. (Repeated from Fig. 11.5.)

ILLUSTRATIVE PROBLEM 13.12

Determine the reactions, end moment, and central deflection for the uniform simple beam shown in Fig. 13.22.

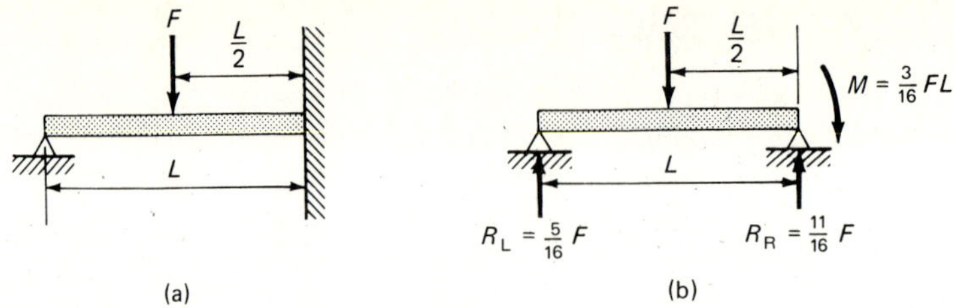

FIGURE 13.22   Illustrative Problem 13.12.

SOLUTION

As the first step let us remove the left end support. The deflection of the free end can be found from case 2 of Table 13.3 as

$$y_{\text{free end}} = \frac{Fb^2}{6EI}(3L - b) = \frac{F(L/2)^2}{6EI}\left(3L - \frac{L}{2}\right) = \frac{5FL^3}{48EI}$$

The force required at the end of the beam to return it to its initial undeflected position is found from case 1 to be

$$y = \frac{5FL^3}{48EI} = \frac{R_L L^3}{3EI} \qquad R_L = \frac{5}{16}F$$

The reaction at the left support is therefore $\frac{5}{16}F$, and at the right support it is $F - \frac{5}{16}F$ or $\frac{11}{16}F$. If we now invoke the condition that the moment at the left support be zero and take moments about the right support,

$$R_L(L) - F\left(\frac{L}{2}\right) - M = 0$$

$$M = \frac{-FL}{2} + R_L(L) = \frac{-FL}{2} + \frac{5FL}{16} = \frac{-3FL}{16}$$

The deflection at the center is found as in Illustrative Problem 13.11. The deflection at the center due to the central concentrated load is found from case 6 to be

$$y_{\text{center}} = \frac{FL^3}{48EI}$$

The propped beams which we have studied thus far can be considered a special case of beams that have both ends completely restrained or built in. As noted in Chapter 11, built-in beams can be thought of as double-ended cantilever beams, or simple beams with moments at both ends. These conditions are shown diagrammatically in Fig. 13.23.

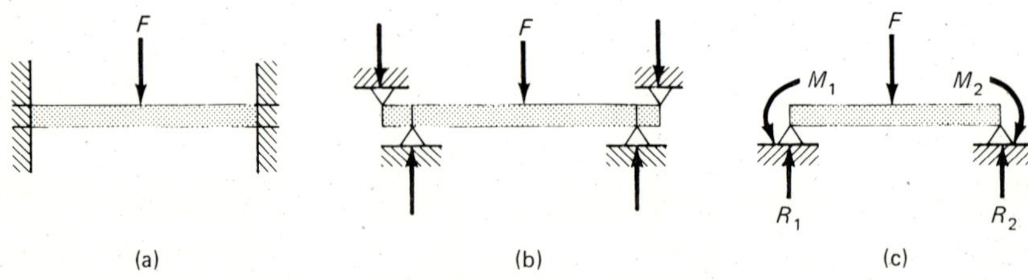

FIGURE 13.23   Built-in beams. (Repeated from Fig. 11.4.)

                    The Deflection of Beams

The method which we shall use to treat this category of beams is the same as that used for propped beams in the previous sections. However, where we required one extra condition to reach a solution for propped beams, we shall require two conditions to solve built-in beams. We can utilize both the shear and the moment at one end of the beam as unknowns, and by solving for them we obtain the required conditions. It is also possible to consider a symmetrical beam, as shown in Fig. 13.23, and by using the condition that $\theta_1 = \theta_2$ to solve for the end moments at each end of the beam. The following illustrative examples will serve to demonstrate the application of the superposition method to this type of structure.

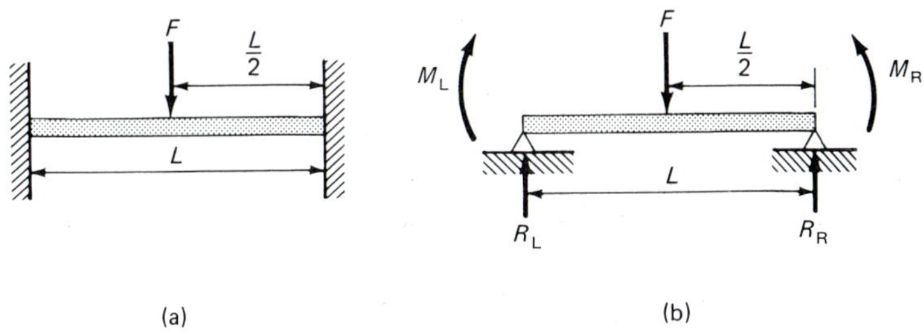

FIGURE 13.24   Illustrative Problem 13.13.

SOLUTION

As our first observation we note that the beam in question has complete symmetry and that $R_L = R_R = F/2$; $M_L = M_R$, and $\theta_L = \theta_R = 0$. From case 6 of Table 13.3, the angular deflection due to the central load only is given as

$$\theta_L = \theta_R = \frac{FL^2}{16EI}$$

From case 12 a moment applied on the right side of the beam will give rise to an angular deflection of $ML/3EI$ *on the right* and an angular deflection of half of this value *on the left,* or $ML/6EI$. Equal moments M at both ends of the beam cause an angular deflection of $ML/3EI + ML/6EI$ or $ML/2EI$ at each end. Thus the moment required at each end is found by the condition that $\theta_L = \theta_R = 0$ on the beam. The angular deflection due to the central load must be equal and opposite to that produced by the end moments

$$\frac{FL^2}{16EI} = \frac{ML}{2EI} \quad \text{and} \quad M = \frac{FL}{8}$$

The deflection at the center due to the concentrated load less the deflection at the center due to the end moments gives us the desired result. For the concentrated load $y_{center} = FL^3/48EI$ (from case 6), and the central deflection due to each moment is found from case 12 to be $ML^2/16EI$. When M is substituted for, it is $(FL/8)(L^2/16EI)$ or $FL^3/128EI$. Performing the required subtraction yields

$$y_{\text{center}} = \frac{FL^3}{48EI} - 2\left(\frac{FL^3}{128EI}\right) = \frac{FL^3}{192EI}$$

The student can verify this result using case 15 of Table 13.3.

13.6 CONTINUOUS BEAMS—THE THREE-MOMENT EQUATION

A *continuous beam* is a beam having three or more supports. This type of beam has more supports than are necessary for static equilibrium and is the last type of indeterminate beam that we shall study. Figure 13.25 shows an example of this type of beam. When the beam has three supports, we can use the procedure of the previous section of this chapter. When more than three supports exist, it is necessary to determine the bending moments at the supports using a general relation among the bending moments at any two sections of the beam, known as the three-moment equation. We shall first consider a beam having three supports and apply the methods already developed in this chapter. Then the three-moment equation will be used to solve the same problem.

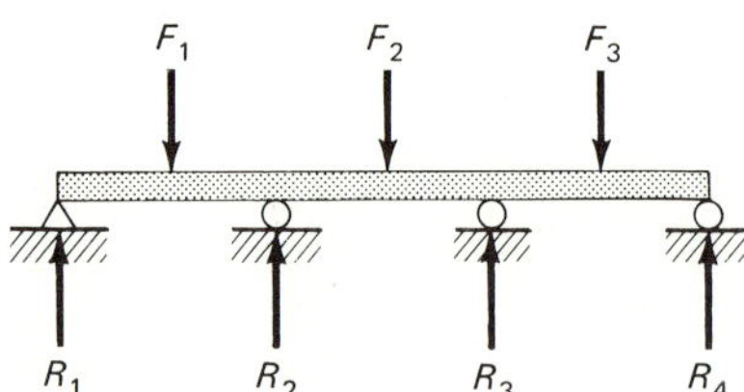

FIGURE 13.25   Continuous beam. (Repeated from Fig. 11.6.)

The illustration we shall use is Fig. 13.26. It consists of a two-span beam having equal spans and equal centrally placed, concentrated loads on each span. Using the method of superposition, we can consider the beam to be loaded as shown in Fig. 13.26(a) with the central support removed. For this beam the end reactions are equal to F and the central deflection is found from case 8 of Table 13.3 to be $(Fa/24EI)(3L^2 - 4a^2)$. The condition that the middle support imposes is that the deflection at the middle support be zero. Thus we have a simply supported beam with a central force acting upward to cause the net deflection at this force to be zero. From case 6 of Table 13.3,

$$y_{\text{center}} = \frac{F'L^3}{48EI}$$

where the F' is the required force at support R_2. Thus $F'L^3/48EI = 11FL^3/384EI$ and $F' = {}^{11}\!/_8 F$. Notice that half of this load, acting as shown in Fig. 13.26(c), must act

                                                    The Deflection of Beams

at each support. The net force at each support is the sum of the terms at each support, yielding a support reaction on each of the outer supports of $\frac{5}{16}F$. The complete shear diagram for the beam is shown in Fig. 13.26(d). With the beam's shear diagram known and the moment at the center span known, the complete moment diagram for the beam can be drawn (and checked) using the principles developed in Chapter 11. We can now calculate the deflection of the beam using the superposition method applied to each span. The loading on each span consists of the applied loads and the end moments due to the supports.

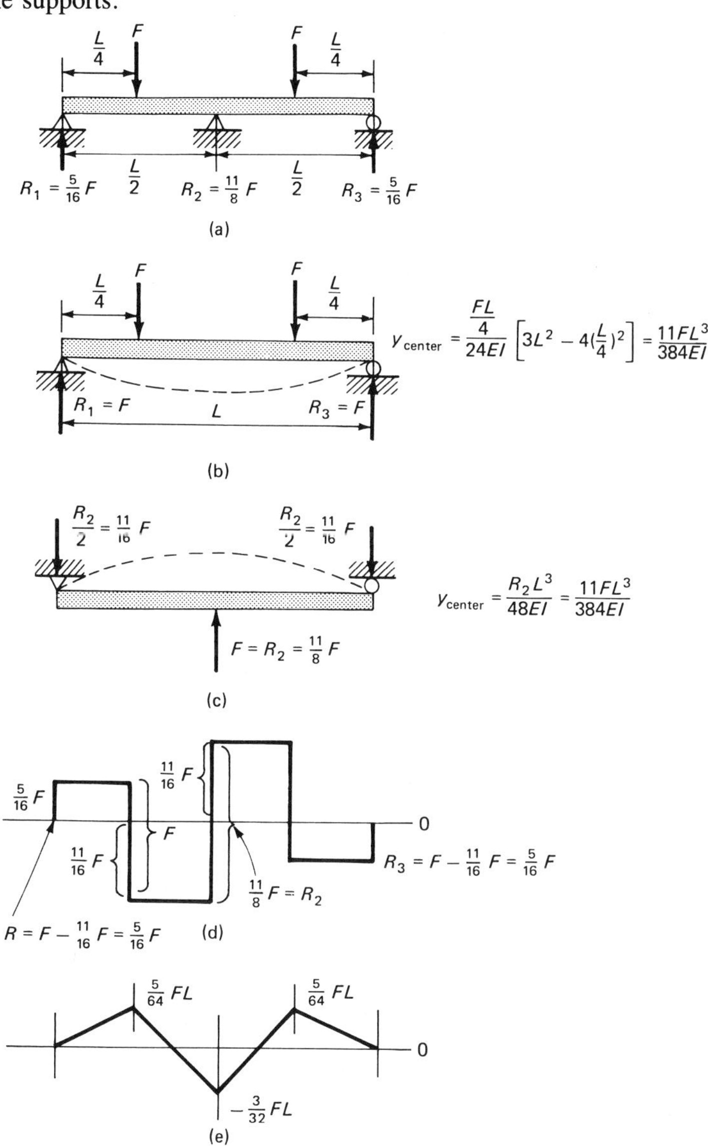

FIGURE 13.26 Two-span, three-support continuous beam. (a) Beam loading. (b) Beam deflection—central support removed. (c) Beam deflection—central support only. (d) Shear diagram. (e) Moment diagram.

A general relation known as the *three-moment equation* can be derived from the moment area method for the bending moments at any two sections (three supports) on a beam. Referring to Fig. 13.27 we can write the three-moment equation as follows, if we assume all of the supports to be at the same level:

$$M_1L_1 + 2M_2(L_1 + L_2) + M_3L_2 + (\text{load term})_{\text{left span}}$$

$$+ (\text{load term})_{\text{right span}} = 0 \quad (13.20)$$

Equation (13.20) is based upon the assumptions that all supports are on the same level, the beam is initially straight, E and I are constant throughout the beam length, and the supports remain in a straight line after loading has been applied. If the moment at any point is negative, it must be used as negative quantity in Eq. (13.20), and a negative result indicates a negative moment at a given point. The load terms to be used for various types of loading can be derived from the moment area method. When this is done, the results are those shown in Table 13.4. This table can be used for other loadings by noting that the load terms for any combination of loads in Table 13.4 are additive.

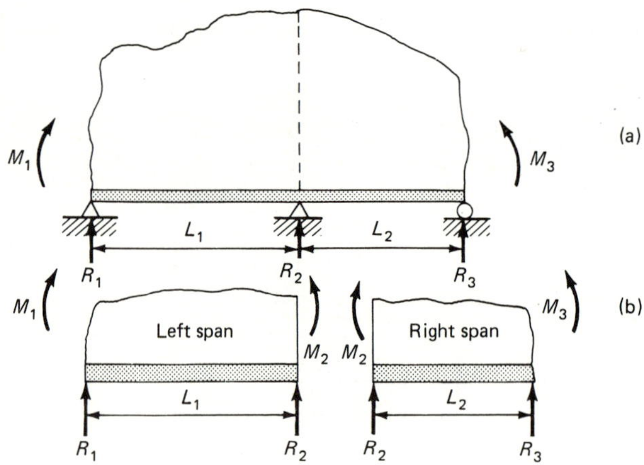

FIGURE 13.27   Generalized loads on a continuous beam. (a) Beam loading. (b) Free-body diagram for each span.

Let us now solve the previous problem by applying the three-moment equation as follows:

$$M_1L_1 + 2M_2(L_1 + L_2) + M_3L_2 + (\text{load term})_{\text{left span}} + (\text{load term})_{\text{right span}} = 0$$

$$M_1 = 0$$

$$L_1 = \frac{L}{2}$$

$$L_2 = \frac{L}{2}$$

$$M_3 = 0$$

$$(\text{load term})_{\text{left span}} = (\text{load term})_{\text{right span}}$$

The Deflection of Beams

TABLE 13.4 Load Terms for Use with the Three-Moment Equation*

w = uniform load (lb/ft); wL = total load on span = w

Left span (load term)	no.	Type of loading	Right span (load term)
$\dfrac{Fa}{L}(L^2 - a^2)$	1		$\dfrac{Fb}{L}(L^2 - b^2)$
$\dfrac{wL^3}{4} = \dfrac{WL^2}{4}$	2		$\dfrac{wL^3}{4} = \dfrac{WL^2}{4}$
$\dfrac{8}{60}wL^3 = \dfrac{8}{30}WL^2$	3		$\dfrac{7}{60}wL^3 = \dfrac{7}{30}WL^2$
$\dfrac{7}{60}wL^3 = \dfrac{7}{30}WL^2$	4		$\dfrac{8}{60}wL^3 = \dfrac{8}{30}WL^2$
$\dfrac{5}{32}wL^3 = \dfrac{5}{16}WL^2$	5		$\dfrac{5}{32}wL^3 = \dfrac{5}{16}WL^2$
$\dfrac{w(b^2 - a^2)}{4L}[2L^2 - (b^2 + a^2)]$	6		$\dfrac{w(d^2 - c^2)}{4L}[2L^2 - (d^2 + c^2)]$

* Reproduced with permission from A. Jensen and H. H. Chenoweth, *Applied Strength of Materials*, New York: McGraw-Hill, 1967, p. 251.

Therefore

$$2M_2(L) + 2\left\{\frac{F(L/4)}{L/2}\left[\left(\frac{L}{2}\right)^2 - \left(\frac{L}{4}\right)^2\right]\right\} = 0$$

and

$$M_2 = -\frac{3FL}{32}$$

Continuous Beams—The Three-Moment Equation

The same result was obtained by the earlier approach. Knowing M_2, we can calculate the deflection on each span using the superposition method.

Where there are more than two spans, it is necessary to apply the three-moment equation successively to each pair of spans, that is, spans 1 and 2, spans 2 and 3, spans 3 and 4, and so on, and then solve the resulting simultaneous equations for the unknown moments at the supports.

13.7 CLOSURE

It is not possible for any text to adequately cover all of the combinations of methods that can be applied to the solution of beam deflection problems. In this chapter we have considered the moment area method and the superposition method to obtain solutions to typical beam deflection situations. There is a great deal of arithmetic computation involved in these problems, and too often these computations mask the principle being used. The student should always try to qualitatively set up each problem, drawing the necessary sketches of beam shear, bending moment, and deflection, as well as clearly noting all quantities given in each problem on these diagrams. Attention must also be given to the proper use of dimensions since multiple powers of length occur in all of these problems.

The student may have noted that there are more illustrative problems in this chapter than in any other previous chapter. The reason for this is simply that there are more principles to be illustrated to more situations than before. Understand each problem and the situation illustrated before proceeding to the next one, and in this manner you will develop an understanding of the material as well as confidence in applying it to different situations.

REFERENCES

American Institute of Steel Construction, MANUAL OF STEEL CONSTRUCTION, 8th ed., 1980.

Amirikian, A., ANALYSIS OF RIGID FRAMES. U.S. Government Printing Office.

Arges, K. P., and A. E. Palmer, MECHANICS OF MATERIALS. New York: McGraw-Hill, 1963.

Bassin, M. E., S. M. Brodsky, and H. Wolkoff, STATICS AND STRENGTH OF MATERIALS, 3rd ed. New York: McGraw-Hill, 1979.

Breneman, J. W., STRENGTH OF MATERIALS, 3rd ed. New York: McGraw-Hill, 1965.

Conway, H. D., MECHANICS OF MATERIALS. Englewood Cliffs, NJ: Prentice-Hall.

Jensen, A., and H. H. Chenoweth, APPLIED STRENGTH OF MATERIALS, 3rd ed. New York: McGraw-Hill, 1971.

Levinson, I. J., MECHANICS OF MATERIALS, 2nd ed. Englewood Cliffs, NJ: Prentice-Hall, 1970.

Lisarelli, F. R., ESSENTIAL STRENGTH OF MATERIALS. New York: McGraw-Hill.

Singer, F. L., STRENGTH OF MATERIALS, 2nd ed. New York: Harper and Row, 1962.

Timoshenko, S., and D. H. Young, ELEMENTS OF STRENGTH OF MATERIALS, 5th ed. New York: D. Van Nostrand, 1968.

　The Deflection of Beams

PROBLEMS

Use $E = 210 \times 10^9$ Pa or 30×10^6 psi for steel in all problems.

13.1   A beam having a span of 15 ft is loaded with a uniform load of 200 lb/ft over its entire length. If the beam is an S10 × 35 section, determine its radius of curvature at the center.

13.2   A cantilever beam has a span of 10 ft, and a concentrated load of 1000 lb is placed at its free end. What is the radius of curvature at the built-in end if the beam is a W12 × 14 beam?

13.3   A band saw has a blade 1 in. wide and 0.05 in. thick. If it wraps around a pulley that has a radius of 24 in., determine the bending stresses in the blade.

13.4   A flagpole may be considered to be a cantilever beam fastened at the ground and subjected to a uniform wind load. At 120 mi/h the loading is estimated to be 75 lb/ft on a 3-in. diameter rod. If the flagpole is 20 ft long, determine the deflection of the free end.

13.5   A simple beam has a span of 6 m and a uniform load of 6000 N/m over its entire length. If the beam is a W8 × 31 section, what is its radius of curvature?

13.6   A cantilever beam has a span of 10 m, and a concentrated load of 1000 N is placed at its free end. What is the radius of curvature at the built-in end if the beam is a W12 × 14 beam?

13.7   A cantilever beam has a span of 12 m, and a concentrated load of 2 kN is placed at its free end. What is the radius of curvature of the beam at its built-in end if the beam has a moment of inertia of a W8 × 31 beam?

13.8   A cantilever beam has a span of 6 m and has a uniform load of 6500 N/m over its entire length. What is the radius of curvature of the beam at its built-in end if the beam has a moment of inertia of 100 in.^{4}?

13.9   An S10 × 35 beam is used as a cantilever having a span of 18 ft and carrying a load of 1500 lb at its free end. What are the slope and deflection at the free end?

13.10   If instead of the 1500-lb load, the beam in Problem 13.9 had a 1500-ft·lb moment at its free end, determine the deflection and slope its free end would have.

13.11   What concentrated load F is necessary, acting upward on the free end of a cantilever beam, to keep the deflection of the free end zero if there is a uniform downward load of w lb/ft acting on the beam?

13.12   Derive the expressions for the maximum deflection and slope at the free end of the cantilever beam of case 4 of Table 13.3 using the moment area method.

13.13   A cantilever beam carries a uniform load of 100 lb/ft as shown. Using the moment area method, determine the slope and deflection of the free end.

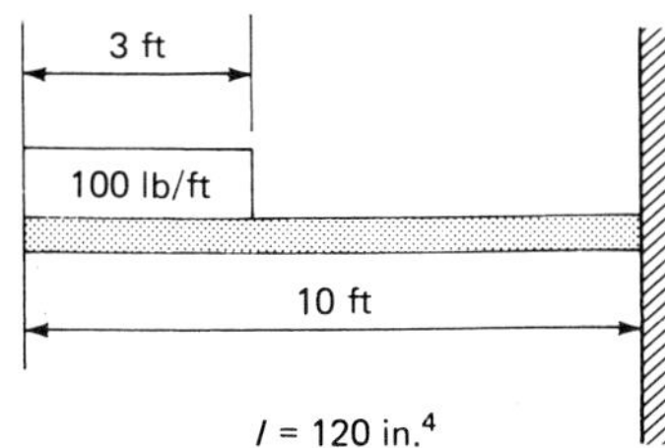

FIGURE P 13.13

13.14   Using the moment area method, determine the slope at each end of the beam shown and the maximum deflection.

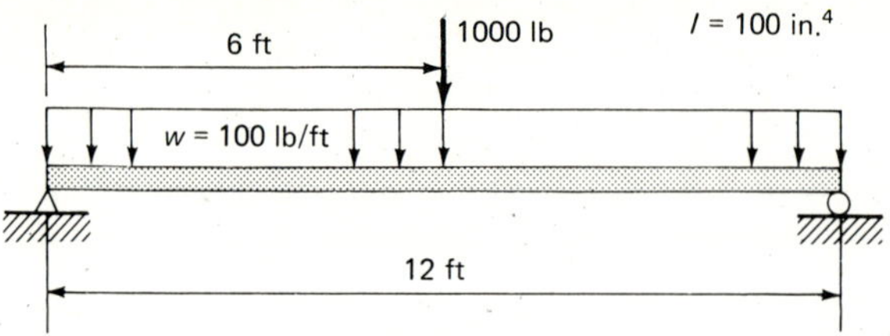

FIGURE P 13.14

13.15   Solve Problem 13.14 using Table 13.3.

13.16   A shaft having a 2-in. diameter is supported by bearings 12 ft apart. Consider the bearings to be equivalent to simple supports and calculate the maximum deflection of the shaft due to its own weight. Steel weighs 500 lb/ft^3.

13.17   A simply supported beam is 16 ft long between supports and has a concentrated load of 1000 lb placed 4 ft to the right of the left support. If $I = 119$ in.^{4}, determine the deflection at the load and the slope at each end of the beam.

13.18   Using the moment area method, determine the deflection at the center of the beam shown if $I = 49$ in.^{4}.

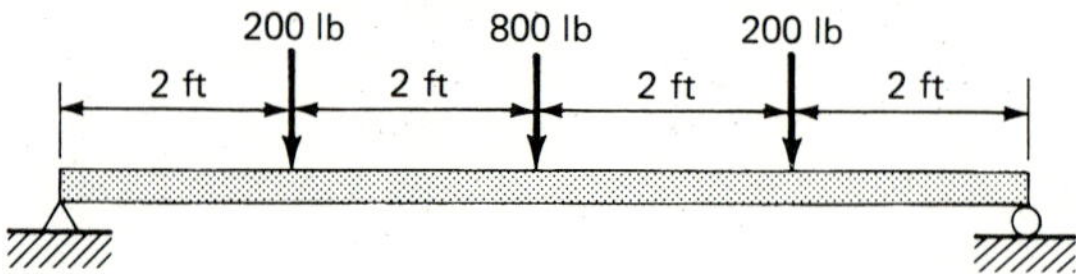

FIGURE P. 13.18

13.19   Solve Problem 13.18 using Table 13.3.

13.20   A cantilever beam carries a load of w lb/ft. If a support is placed under the free end to limit the deflection to one half of that which would occur if no support were present, determine the moment at the built-in end. The length of the beam is L.

13.21   A cantilever beam carries a uniform load of 100 N/m as shown. Using the moment area method, determine the slope and deflection of the free end.

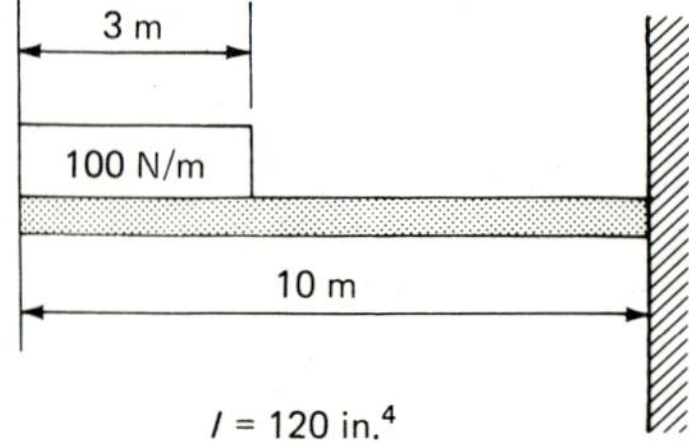

FIGURE P 13.21

                                          The Deflection of Beams

13.22 Calculate the central deflection of the simple beam shown.

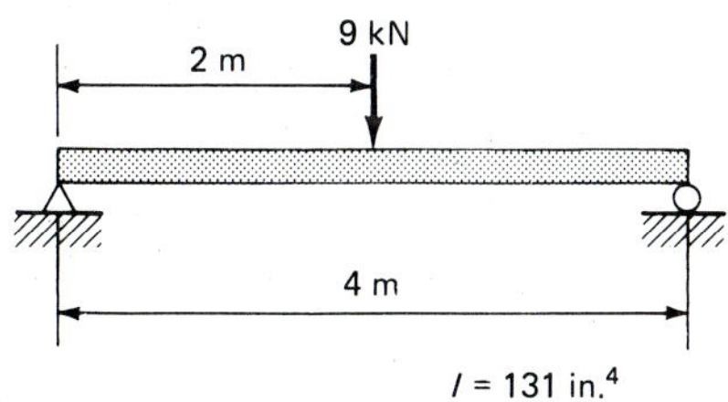

FIGURE P 13.22

13.23 Calculate the angular deflection at either end of the beam shown in Problem 13.22.

13.24 Select a W beam to support a uniform load of 150 lb/ft over the 20-ft span of a simple beam. Assume that the maximum deflection of the beam is limited to 1/360 of the span and the allowable bending stress is 20 000 psi.

13.25 Select a W beam to support a uniform load of 2250 N/m over the 6-m span of a simple beam. Assume that the maximum deflection of the beam is limited to 1/360 of the span and the allowable bending stress is 140 MPa.

13.26 Calculate the central deflection for the simple beam shown.

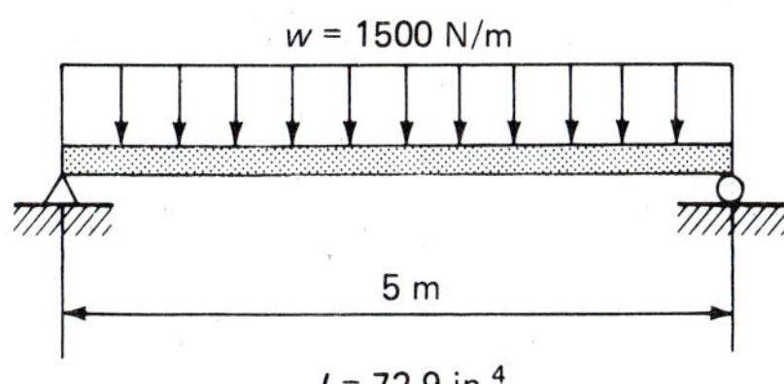

FIGURE P 13.26

13.27 Using the moment area method, determine the slope at each end of the beam shown and the maximum deflection.

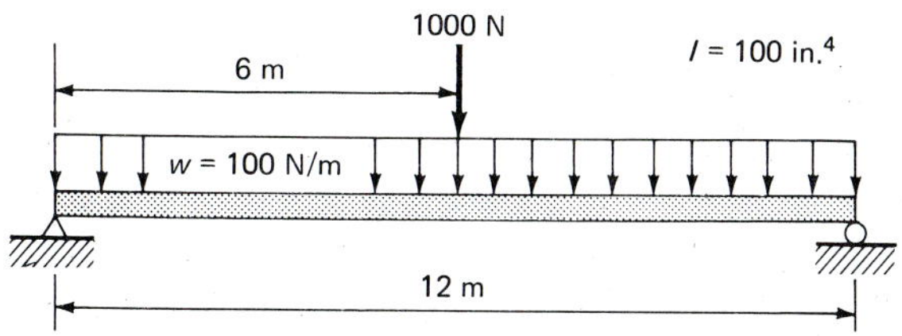

FIGURE P 13.27

Problems

13.28 Solve Problem 13.27 using Table 13.3.

13.29 Using Table 13.3, determine the central deflection of the beam shown.

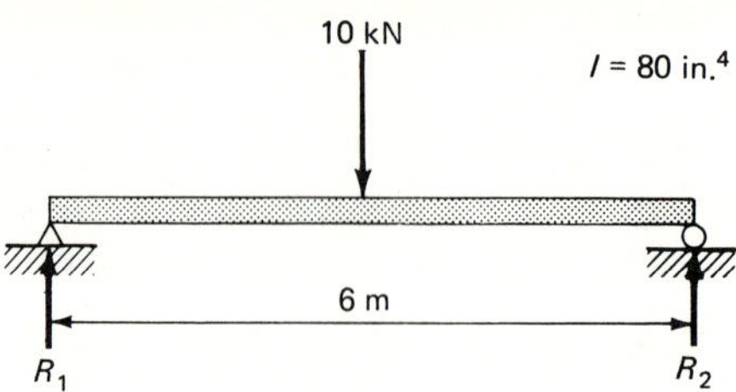

FIGURE P 13.29

13.30 Using Table 13.3, determine the slope at each end of the beam shown in Problem 13.29.

13.31 Determine the deflection of the center of the beam shown.

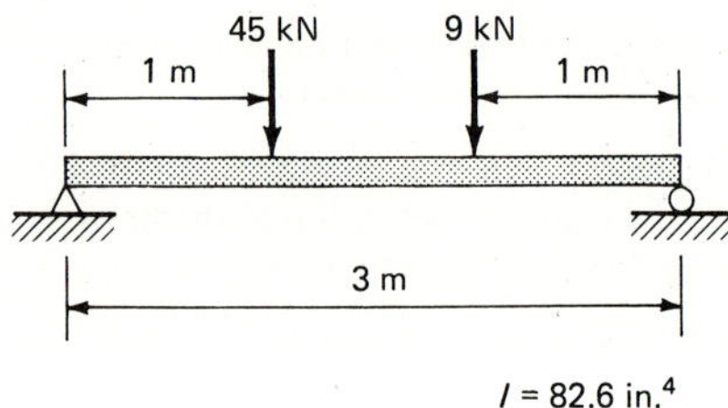

FIGURE P 13.31

13.32 Determine the angular deflection at the left support for the beam shown in Problem 13.31.

13.33 A shaft having a 50-mm diameter is supported by bearings 3.5 m apart. Consider the bearings to be equivalent to simple supports and calculate the maximum deflection of the shaft due to its own weight. Steel weighs 78 540 N/m^3.

13.34 A simply supported beam is 16 m long between supports and has a concentrated load of 1000 N placed 4 m to the right of the left support. If $I = 119$ in.^{4}, determine the deflection at the load and the slope at each end of the beam.

13.35 Using Table 13.3, solve for the deflection at the center of the beam shown.

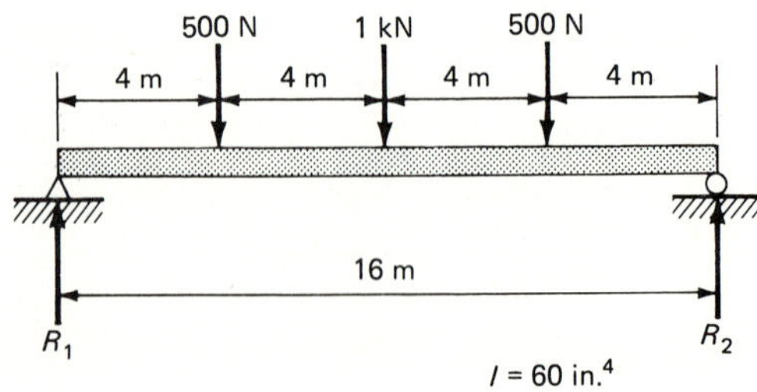

FIGURE P 13.35

                                                        The Deflection of Beams

13.36 Determine the slope at each end of the beam of Problem 13.35 using Table 13.3.

13.37 Using the moment area method, determine the deflection at the center of the beam shown in Fig. P 13.37 if $I = 49$ in.^{4}.

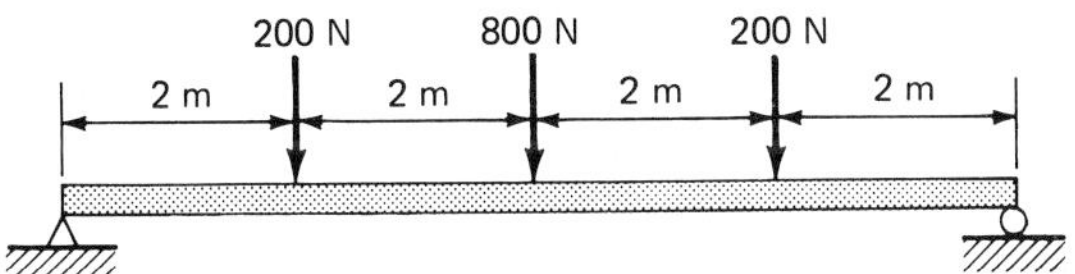

FIGURE P 13.37

13.38 Solve Problem 13.37 using Table 13.3.

13.39 A cantilever beam carries a load of w N/m. If a support is placed under the free end to limit the deflection to one half of that which would occur if no support were present, determine the moment at the built-in end. The length of the beam is L.

13.40 A simply supported beam has an overhang as shown. Determine the deflection at the 4000-N load.

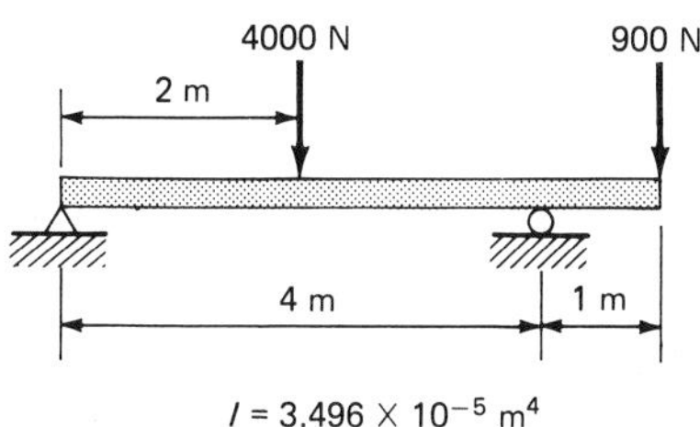

FIGURE P 13.40

13.41 Determine the deflection at the midpoint of the span for the beam shown.

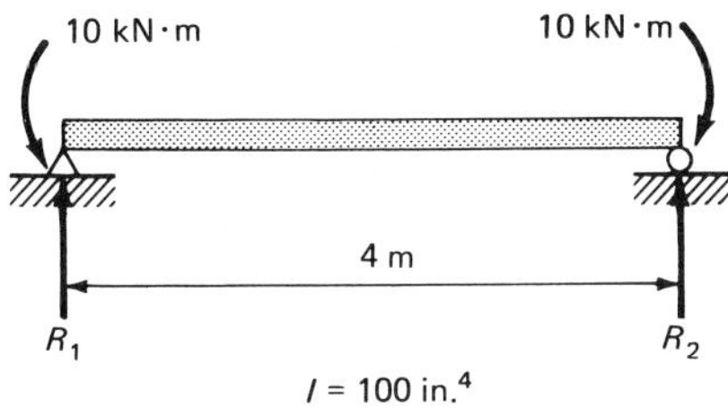

FIGURE P 13.41

13.42 Determine the deflection at the midpoint of the central span for the beam shown.

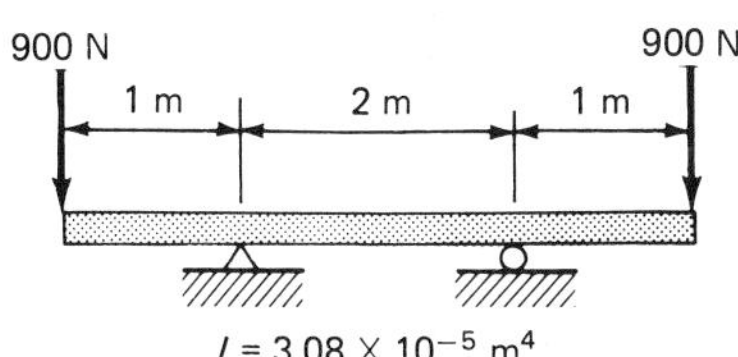

FIGURE P 13.42

Problems

13.43 Determine the deflection at the midpoint of the center span of the beam shown if $I = 3.0 \times 10^{-5}$ m⁴.

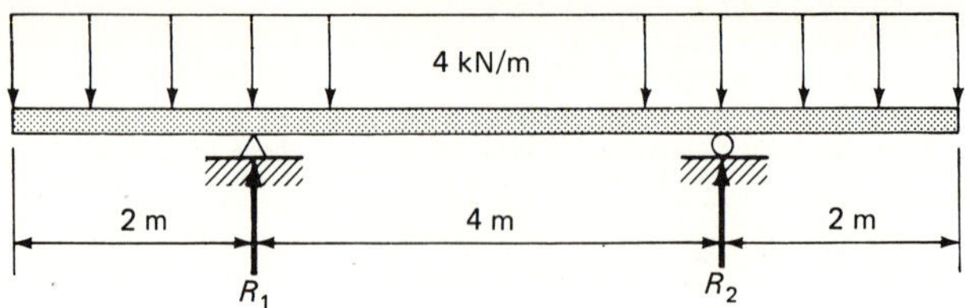

FIGURE P 13.43

13.44 Determine the deflection at the center of the span of the beam shown if $I = 3.183 \times 10^{-5}$ m⁴.

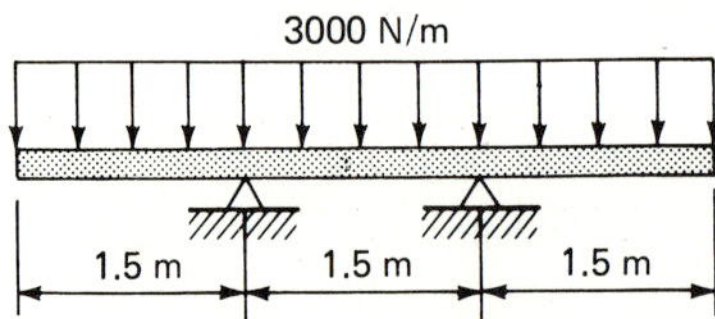

FIGURE P 13.44

13.45 Determine the maximum deflection for the beam shown if $I = 55.3$ in.⁴.

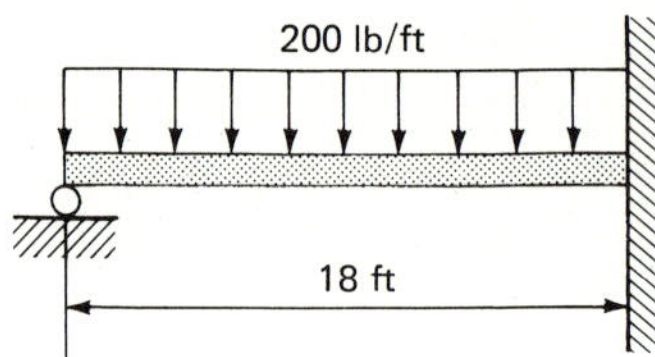

FIGURE P 13.45

13.46 What is the deflection at the center of the beam shown if $I = 114$ in.⁴.

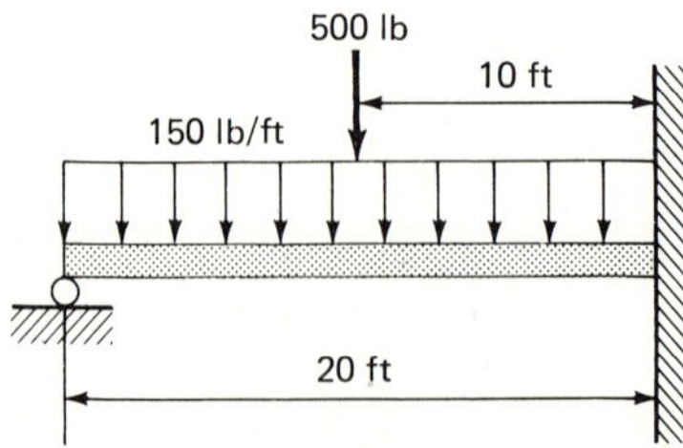

FIGURE P 13.46

                                    The Deflection of Beams

13.47   Determine the deflection at section *x* for the beam shown if $I = 147.5$ in.^{4}.

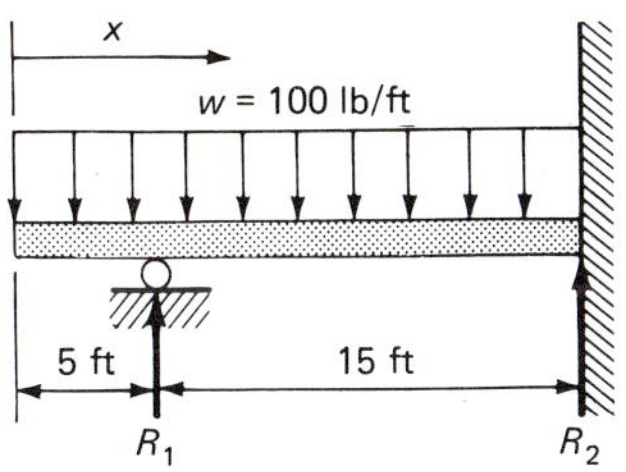

FIGURE P 13.47

13.48   What moment of inertia is required for the beam loaded as shown. Assume that the deflection is limited to 1/360 of the span.

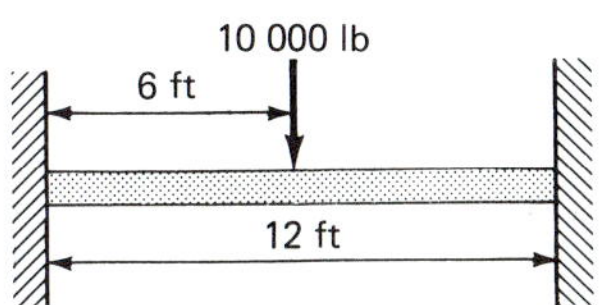

FIGURE P 13.48

13.49   If the moment of inertia determined in Problem 13.48 is for a rectangle whose height is twice its width, determine the maximum stress in the beam.

13.50   Determine the reaction at each support for the beam shown. Use the superposition method.

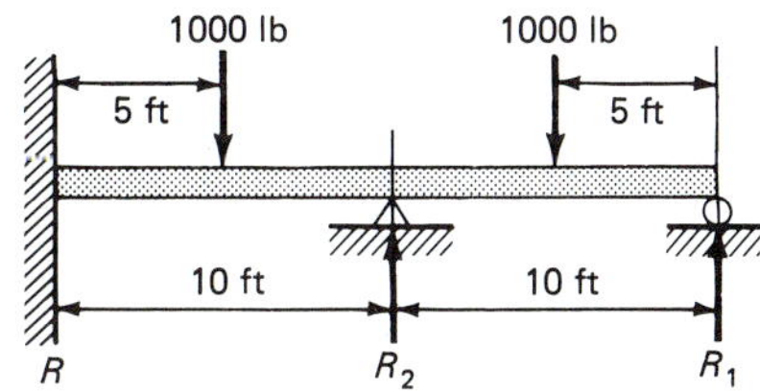

FIGURE P 13.50

13.51   Solve Problem 13.50 using the three-moment equation.

13.52   Determine the moment at the center support of the beam shown, using the superposition method. Also determine R_1, R_2, and R_3.

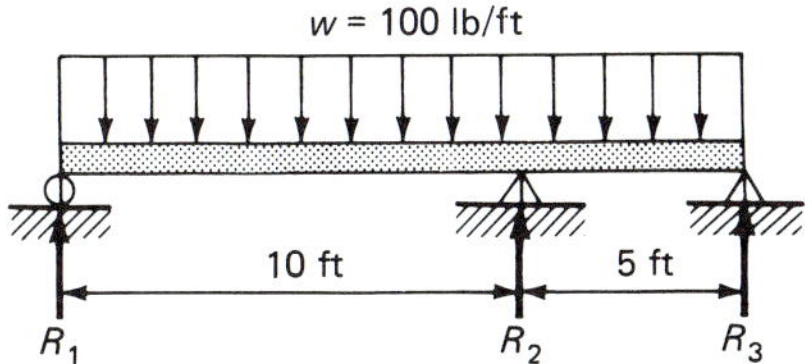

FIGURE P 13.52

Problems   375

13.53 Use the three-moment equation to determine the moment at the center support of the beam in Problem 13.52.

13.54 Determine the moment at the center support for the beam shown, using the superposition method.

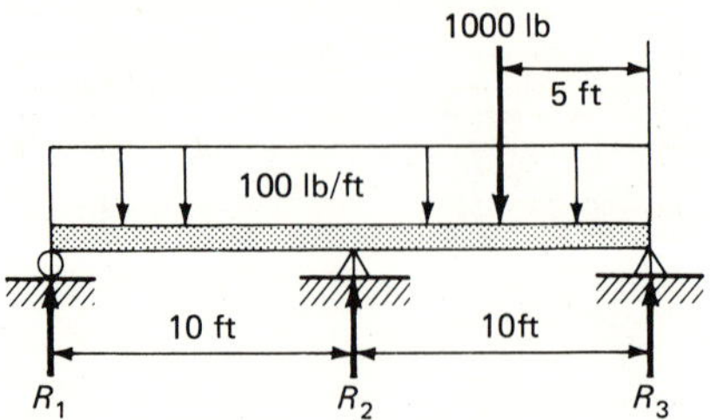

FIGURE P 13.54

13.55 Determine the moment at the center support of the beam shown, using the superposition method. Also determine R_1, R_2, and R_3.

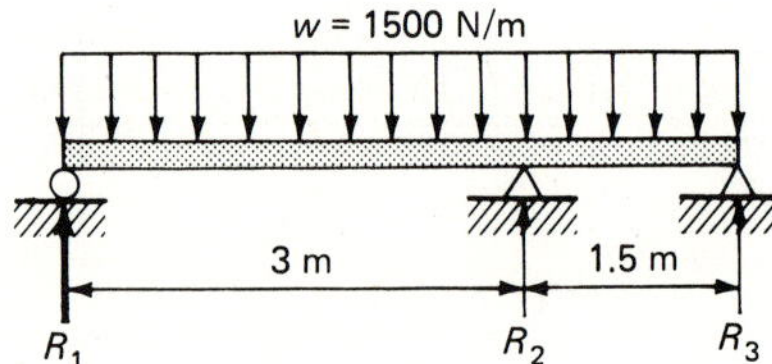

FIGURE P 13.55

13.56 Use the three-moment equation to determine the moment at the center support of the beam in Problem 13.55.

13.57 Use the three-moment equation to determine the moment at each of the supports of the beam shown.

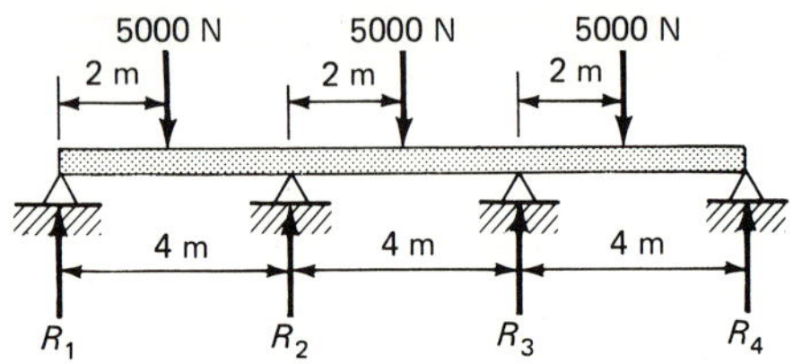

FIGURE P 13.57

13.58 A multispan beam carries a uniform load of 3000 N/m. Determine the moment at each support.

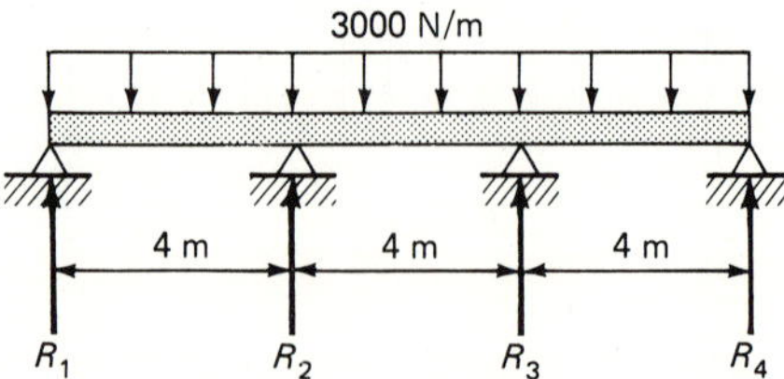

FIGURE P 13.58

                                                        The Deflection of Beams

13.59   A multispan beam carries the loads shown. Determine the moment at each support. Compare your result to that of Problem 13.58.

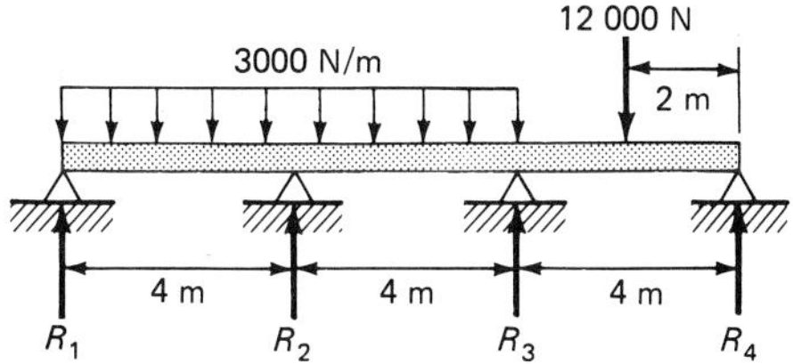

FIGURE P 13.59

13.60   Determine the moment at each support for the beam shown.

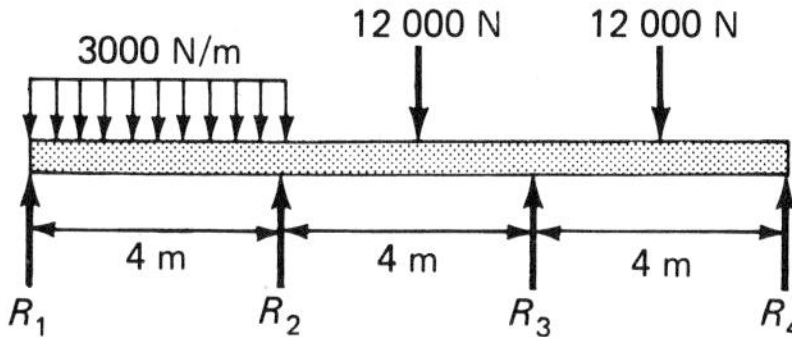

FIGURE P 13.60

13.61   Two beams are placed at right angles to each other and are loaded at their central intersection by a load F. If both beams have the same moment of inertia and they are made of the same material, determine the amount of the load F that each beam carries.

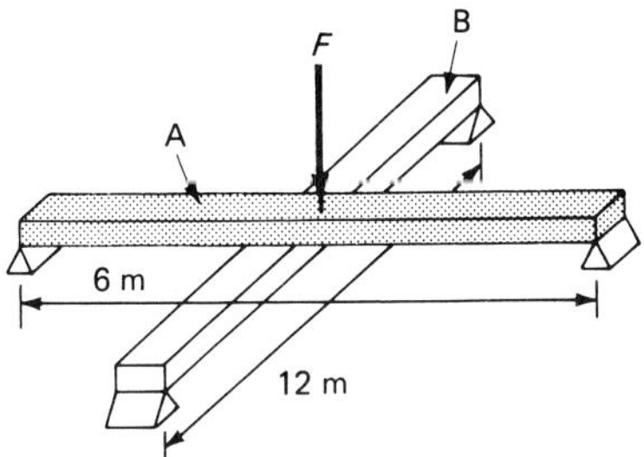

FIGURE P 13.61

14

Columns

14.1 INTRODUCTION

When a structural member was subjected to a compressive axial load, we assumed that it could be treated as an elastic member that would fail at a loading dependent upon its elastic properties, that is, its yield point. However, when the member is long relative to its lateral dimensions, it is called a *column,* and columns fail primarily due to lateral bending leading to lateral instability. This type of failure, called *buckling,* does not depend upon the strength of the material, but is found to be a function only of the dimensions of the member and its modulus of elasticity. Thus two identical columns will fail (buckle) at the same load even though one may be made of a high-strength alloy steel and the other of carbon steel. Unfortunately the point at which a compression member becomes a column is not clearly defined. There is a region where the member acts as both a column and a compression member before it can be said to be a true column. In the subsequent sections of this chapter we will classify columns as *short columns, intermediate columns,* and *long columns.* It will also be necessary to discuss the character of the loading of a column (centric or eccentric), and the type and action of the end conditions in the column. All of these variables are of such nature as to be only partly amenable to mathematical analysis. As a consequence we shall resort to empirical relations and rules to a degree only exceeded by our usage of such relations when we dealt with riveted joints in Chapter 10.

14.2 COLUMNS—GENERAL

Before studying columns, let us first discuss the failure of a column of cross-sectional area A when subjected to an axial load F. Assume that the column in question is centrally loaded, is of a given material, and has end conditions that are kept constant. The only variables for our initial consideration will be the load and unsupported length of the column L. If the column is short, it can be loaded until the stress F/A reaches the yield stress of the material. If we make the column longer, failure will still be characterized by the yield stress of the material. Note that a property of the material (its yield stress) is the limiting condition for failure.

If we continue our experiment, we find some unsupported column length where the column becomes unstable (fails) and where the stress F/A at failure is greater than the proportional limit, but does not exceed the yield stress of the material. Continued increase in the length of the column gives the same type of action, but an ever-decreasing failure value of F/A is needed to cause it to occur.

Let us consider the action of the column as the load F is applied at some still larger value of unsupported column length. While F/A is still within the elastic range of the material, the column will fail due to buckling or lateral bending. This condition is one of elastic instability and is not a function of the properties of the material. The continued increase in length L results in smaller and smaller values of the applied load F causing buckling to occur.

If the experiment just described is carried out for differing cross sections (all other conditions such as material and restraints being kept constant), it will be found that a single failure curve can be plotted on which the failure stress F/A is plotted as a function of L/r, where L is the unsupported length and r is the least radius of gyration of the column.* Because of the universal appearance of the term L/r in column formulas, it is called the *slenderness ratio*.

Figure 14.1 shows a typical plot of F/A against L/r for the conditions we have described. For L/r values up to approximately 60, we can consider the column to be a *compression member;* for L/r values between approximately 60 and 120, we shall call the column an *intermediate column; L/r* values above 120 characterize the column as a *slender column*. The student should again note that L is the effective length of the column, and r is the least radius of gyration of the column about the bending axis.

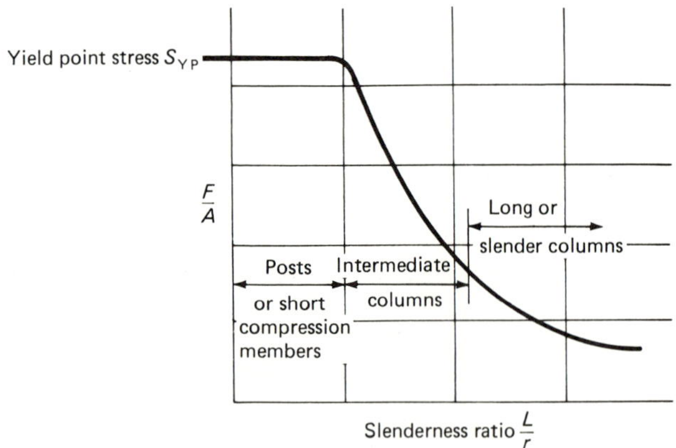

FIGURE 14.1 Typical plot of F/A versus L/r.

*The use of r for the radius of gyration is common in column design. Note that this is k as defined in Chapter 8.

Columns

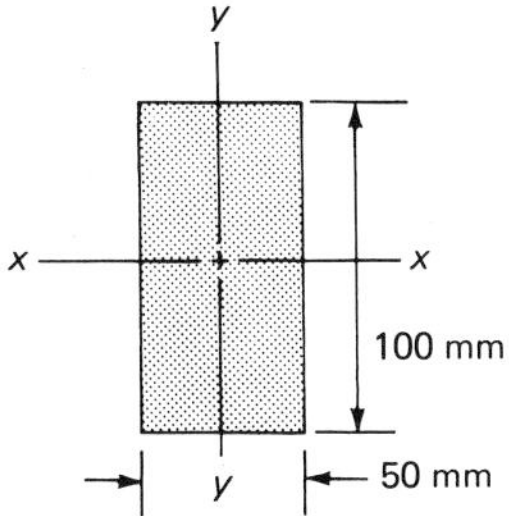

FIGURE 14.2   Illustrative Problem 14.1.

SOLUTION

The moment of inertia of a rectangle is $bh^3/12$. Applying this to both the x–x and the y–y axes, and noting that the area is $0.050 \times 0.100 = 0.005$ m²,

$$I_{xx} = \frac{0.050 \times 0.100^3}{12} = 4.17 \times 10^{-6} \text{ m}^4$$

$$I_{yy} = \frac{0.100 \times 0.050^3}{12} = 1.04 \times 10^{-6} \text{ m}^4$$

$$r_{xx} = \sqrt{\frac{I_{xx}}{A}} = \sqrt{\frac{4.17 \times 10^{-6}}{0.5 \times 10^{-3}}} = 2.89 \times 10^{-2} \text{ m}$$

$$r_{yy} = \sqrt{\frac{I_{yy}}{A}} = \sqrt{\frac{1.04 \times 10^{-6}}{5 \times 10^{-3}}} = 1.44 \times 10^{-2} \text{ m}$$

r_{yy} is least and limits the design. Therefore $L/r = 1.5/1.44 \times 10^{-2} = 104.2$; the column is intermediate.

SOLUTION

From Appendix B this section has a least value of r equal to 1.51 in. for the y–y axis. The length L is therefore

$$L = 1.51 \times 0.0254 \times 120 = 4.60 \text{ m}$$

14.3   LONG COLUMNS

Buckling is defined as the state of loading that will cause the column to undergo excessively large deflections while stressed within the elastic limits of the material.

Long Columns

The load that will just cause buckling to occur is known as the *critical load*. We can illustrate this condition by reference to Fig. 14.3. The column shown in Fig. 14.3(a) is loaded by a central force F, which is below the critical load. If the load F is initially zero, the column will be straight, and the application of the load F causes the column to deflect as shown (exaggerated). Assuming the load to be below the critical load, we find that the withdrawal of F will return the column to its initial straight condition. A similar action will take place by the horizontal application of the force F'. By keeping F' sufficiently small, we ensure that its removal would cause the column to return to its original shape. If the load F were applied vertically and exceeded the critical load, the column would undergo a large deformation, as shown in Fig. 14.3(c), and would not return to its initial straight condition upon removal of the load. If the load F were near the critical load, the application of a lateral force [similar to F' in Fig. 14.3(b)] would cause an excessive deflection (buckling or elastic instability), which would in turn cause the column to take on a permanent set.

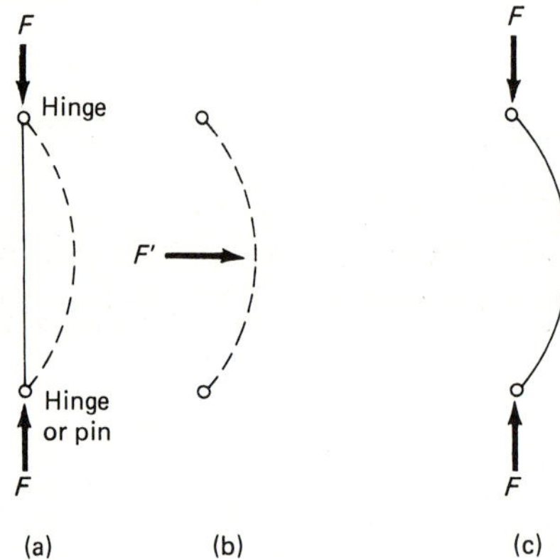

FIGURE 14.3   Long column behavior under load.

The action of long columns within the elastic limits of their materials was first investigated in the mid-1700s by a Swiss mathematician named Leonhard Euler. The following are a consequence of his analysis.

1. The *critical stress,* defined as the ratio of the critical load to the cross-sectional area of the column F_{cr}/A is independent of the strength of the material of the column. As noted earlier, a column of alloy steel having a given cross-sectional area will have the same critical stress as a column of carbon steel having the same cross-sectional area.
2. The critical stress is directly proportional to E, the modulus of elasticity of the column material.
3. The critical stress varies inversely as the square of the slenderness ratio of the column, that is, $1/(L/r)^2$.
4. The critical stress is a function of the method by which the ends of the column are restrained.

                                                          Columns

We can express these results mathematically as

$$S_{cr} = \frac{F_{cr}}{A} = \frac{n\pi^2 E}{(L/r)^2} \tag{14.1}$$

where S_{cr} is the critical stress (not a maximum stress), n is a constant for a given type of end restraint, E is the modulus of elasticity, and L/r is the slenderness ratio of the column, with L being the effective length and r the least radius of gyration of the column.

Equation (14.1) is known as *Euler's equation,* and it is the rationale behind most column formulas. Explicit in its derivation is the assumption that the critical stress shall not exceed the yield point of the material (more properly the stress at the proportional limit, but the yield point is usually used to determine the critical stress). This enables us to evaluate the maximum value of L/r for any material.

SOLUTION

$$S_{cr} = \frac{n\pi^2 E}{(L/r)^2}$$

or

$$L/r = \sqrt{\frac{n\pi^2 E}{S_{cr}}} = \sqrt{\frac{1 \times \pi^2 \times 203 \times 10^9}{175 \times 10^6}} = 107$$

In Illustrative Problem 14.3 we used a value of $n = 1$ as part of the statement of the problem. Since n (and the effective column length as well) is dependent upon the end conditions of the column, let us briefly consider the end fixity of columns. The column shown in Fig. 14.3 had ends that were pinned or hinged, which permitted them to rotate, but they were not permitted to translate. In effect this condition (used to derive Euler's equation) determines the value of L to be used in Eq. (14.1). The effective length L is the length between points of zero moment on the deflection curve of the column. Figure 14.4 shows four conditions of end restraint for a column:

1. Both ends hinged, $L_e = L$ $n = 1$
2. Both ends fixed (built in), $L_e = L/2$ $n = 4$
3. Hinged at one end, fixed at the other, $L_e = 0.7L$ $n = 2$
4. One end free, one end fixed, $L_e - 2L$ $n - \frac{1}{4}$

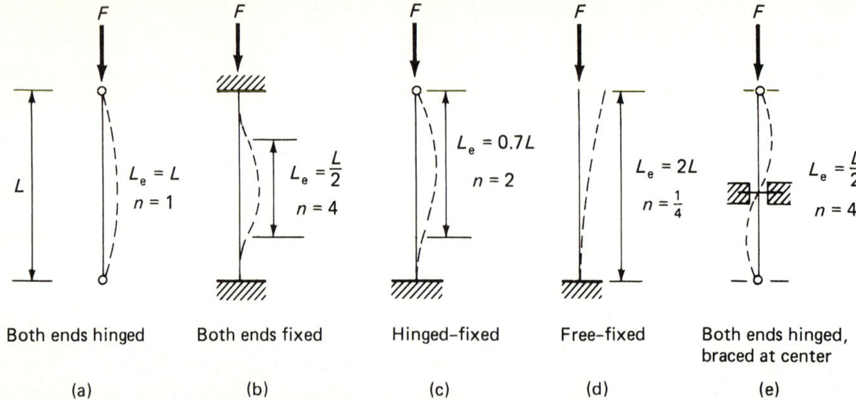

FIGURE 14.4   End restraint of columns.

For end conditions that afford partial restraint, the effective length can be satisfactorily taken to be between $0.5L$ and L, which gives values of n between 1 and 4. Also, where a column is braced [as shown in Fig. 14.4(e)], the effective length is halved if the bracing is at the center; for bracing at the third points, $L_e = L/3$ and $n = 9$. The student should note that since $n = 1$ for both ends hinged, n can be thought of as a factor of comparison that indicates the relative strength of a column of a given end fixity compared to a column having both ends hinged.

ILLUSTRATIVE PROBLEM 14.4

An S5 × 10 beam is to be used as a 6-m long column. If $E = 210$ GPa, and one end is free while the other is fixed, determine the value of S_{cr}. Compare this stress to the 210-MPa yield stress of the material.

SOLUTION

For this case $n = \frac{1}{4}$ and $S_{cr} = n\pi^2 E/(L/r)^2$. For an S5 × 10 beam $r_{xx} = 2.05$ and $r_{yy} = 0.643$. Thus the least value of L/r is

$$\frac{L}{r} = \frac{6}{2.05 \times 0.0254} = 115.2$$

and hence from Eq. (14.1),

$$S_{cr} = \frac{\frac{1}{4}\pi^2 \times 210 \times 10^9}{115.2^2} = 39.04 \times 10^6 \text{ Pa}$$

Since S at the yield point is 210×10^6 Pa, it is obvious that failure will occur long before the yield point of the material is reached.

In the design of any structural member there are many unknowns, which have been noted elsewhere in this text. For this reason it is common to modify Euler's equation by introducing a *factor of safety f*, which effectively decreases the value of S_{cr} in Eq. (14.1) and yields

$$S_{cr} = \frac{n\pi^2 E}{f(L/r)^2} \tag{14.2}$$

The load factor of safety thus defined is the ratio of the critical stress to the applied stress (load). For steady loads f is commonly taken to be 2.5; its values may vary from 2 to 3.

A word of caution at this point. For practical reasons the AISC code prohibits the use of compression members beyond an L/r value of 200. In the case of main members that are subjected to shock and/or vibration, L/r is further limited to a value of 120 or less. As will be discussed later on in this chapter, main members whose slenderness ratios lie between 120 and 200 have their allowable stress decreased by a factor that is a function of L/r. Also, a column can be braced at unequal lengths along either axis, requiring that both axes be investigated.

14.4 INTERMEDIATE STEEL COLUMNS—HISTORICAL

Prior to the early 1960s a large body of empirical data had been accumulated, and as a consequence, design formulas were proposed for the intermediate column range ($L/r < 120$). The problem that arises when this regime of column design is studied is that the strength of the material is a governing parameter. Because of this, one finds that there exist differing design formulas, each based upon empirical data, which depend upon the various legal jurisdictions (cities, states, federal). After 1961 the AISC code was changed to a form having a variable factor of safety, which will be more fully discussed in the next section of this chapter. Since the earlier formulas are still widely used and are part of the building codes of many municipalities, we shall first study them and their applications.

The formulas proposed and used can be broken into the following categories based upon their mathematical form:

1. The *straight-line formula*
2. The *parabolic formula*
3. The *Rankine–Gordon formula* (so named for the English engineers who first proposed it)

The straight-line formula, first proposed in about 1890, gives the critical stress on a column in the following form:

$$\left(\frac{F}{A}\right)_{\text{cr}} = S_{\text{cr}} = a - b\left(\frac{L}{r}\right) \tag{14.3}$$

where a and b are constants of the material. For most practical purposes the yield point is the value taken for S_{cr}. In design it is more customary to use an equation of the form of Eq. (14.3), based upon the allowable stress. One of the most common of these formulas is the one known as the *Chicago formula,* since it was part of the Chicago Building Code for many years. In terms of the allowable stress S_{all},

$$\left(\frac{F}{A}\right)_{\text{all}} = S_{\text{all}} = 16\,000 - 70\left(\frac{L}{r}\right) \text{ psi}$$

$$= 110 - 0.483\left(\frac{L}{r}\right) \text{ MPa} \tag{14.4}$$

and the formula can be used for $30 < L/r < 120$ for main members. For values of $L/r < 30$, $S_{all} = 14\,000$ psi (96.5 MPa). This equation has been plotted on Fig. 14.5, where it will be noted that it intersects the Euler curve at $L/r = 126$, when the Euler curve is based upon a factor of safety of 2.5. Below L/r values of 126 the allowable stress increases to an L/r of 30, where it is $14\,000$ psi (96.5 MPa), and remains constant for all values of $L/r < 30$. Since Eq. (14.4) contains both the area and the radius of gyration, its use (and the use of most column equations) usually entails a trial-and-error procedure. Note that the conversion factor (multiplier) to convert from psi to pascal is 6895.

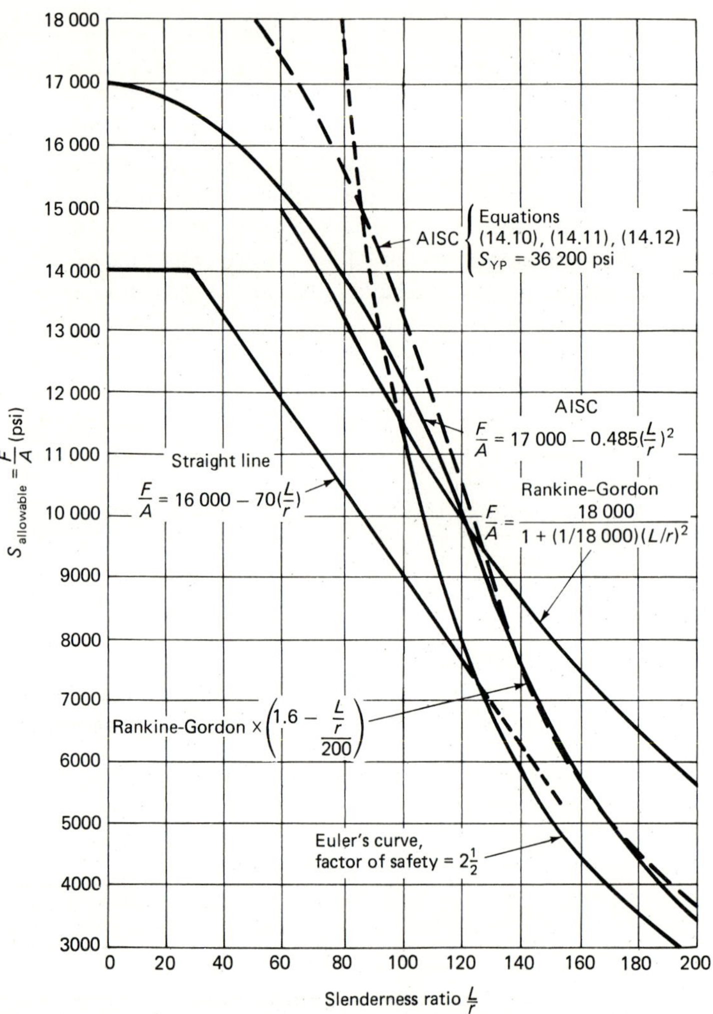

FIGURE 14.5 Comparison of unit loads of various column formulas.

Columns

The parabolic formula expresses the critical stress of an intermediate column in the form

$$\left(\frac{F}{A}\right)_{cr} = S_{cr} = a - b\left(\frac{L}{r}\right)^2 \tag{14.5}$$

where a and b are constants depending upon the physical properties of the material. A working equation of the form of Eq. (14.5) specified in the earlier (pre-1960) editions of the AISC *Manual of Steel Construction* gives the allowable (working) stress on columns that have $L/r < 120$ as

$$\left(\frac{F}{A}\right)_{all} = S_{all} = 17\ 000 - 0.485\left(\frac{L}{r}\right)^2 \text{ psi} \tag{14.6}$$

Equation (14.6) is also shown on Fig. 14.5. This curve gives a value of $S_{all} = 17\ 000$ at $L/r = 0$, and 10 000 at $L/r = 120$.

The third equation, the Rankine–Gordon equation, has the form

$$\left(\frac{F}{A}\right)_{cr} = S_{cr} = \frac{a}{1 + b(L/r)^2} \tag{14.7}$$

where a and b are (once more) constants depending upon the properties of the material. The AISC code specified that axial-loaded columns which are used for bracing, or as other secondary members, should have allowable stresses that do not exceed those given by Eq. (14.8), which is of the Rankine–Gordon form:

$$\left(\frac{F}{A}\right)_{all} = S_{all} = \frac{18\ 000}{1 + (1/18\ 000)\ (L/r)^2} \tag{14.8}$$

Equation (14.8) can be used for values of $120 < L/r < 200$ for bracing or secondary members, that is, those members that are used to lend stiffness and/or stability to a structure but do not carry the permanent load of the structure.

The AISC code permitted the further use of a Rankine–Gordon type equation [Eq. (14.7)] for main compression members for values of $120 < L/r < 200$, provided that they would not ordinarily be subject to shock or vibratory loads if a derating factor

Intermediate Steel Columns—Historical

were used in conjunction with Eq. (14.8) for this condition. The derating factor is

$$\left(1.6 - \frac{L/r}{200}\right) \tag{14.9}$$

and multiplies the right-hand side of Eq. (14.8) by a factor effectively less than unity. Equations (14.8) and (14.9) are also shown in Fig. 14.5. The AISC code prohibited the use of columns in all cases with L/r values greater than 200.

14.5 INTERMEDIATE STEEL COLUMNS—AISC DESIGN PROCEDURE

In 1961 the AISC adopted a set of empirical design formulas for intermediate steel columns after extensive studies of all available column data. The results of these studies were incorporated into the 1963 edition of the AISC *Manual of Steel Construction* and the AISC design specifications. In addition, the 6th edition of the manual (1963) provided for the use of structural steel having a specified minimum yield point up to, but not exceeding, 50 000 psi (344.8 MPa). The 7th edition of this manual (1970) has approved for use high-strength steels with yield strengths of up to 100 000 psi (689.5 MPa). However, the 8th edition of the AISC *Manual of Steel Construction* (1980) gives tables of allowable concentric loads for effective lengths KL for only 36 000 and 50 000 psi. Formulas are given for yield strengths up to 100 000 psi.

Before considering the present design procedure, a further discussion of the end restraints of columns is in order since the AISC manual now contains provisions for various end restraints. The basic column considered is one that has pinned or hinged ends and is braced against side (lateral) displacement. This corresponds to Fig. 14.4(a), where $n = 1$ and $L_e = L$. Where the ends of the column are restrained, Fig. 14.4(b) and (c) indicates effective lengths less than the actual length of the column. For these cases the decrease in effective length should be equivalent to permitting a larger value of allowable loading on the column. Where there is no end restraint but only lateral restraint [such as at the free end of the cantilevered column shown in Fig. 14.4(d)], the effective length is greater than the actual length of the column, and this effect should be equivalent to decreasing the allowable loading on the column. The AISC design procedure allows for this effect by defining a slenderness ratio KL/r, where K is the ratio of effective column length to actual unbraced length. Thus for the cases indicated in Fig. 14.4(a), (b), and (c), $K \leq 1$, while for the column shown in Fig. 14.4(d), $K = 1$. Figure 14.6 shows both the theoretical K values for six idealized conditions and the recommended design values for use when ideal conditions are approached in actual design. Since the end fixity of a column is not readily apparent, we shall retain K in the design procedure, but take its value to be unity (i.e., $K = 1$) in all problems.

The AISC design procedure for main and secondary members whose slenderness ratios KL/r are less than 120 recommends the following formula:

$$\left(\frac{F}{A}\right)_{\text{all}} = S_{\text{all}} = \frac{1 - (KL/r)^2/2C_c^2}{f}S_{\text{yp}} \tag{14.10}$$

Columns

where the factor of safety f is given by

$$f = \frac{5}{3} + \frac{3(KL/r)}{8C_c} - \frac{(KL/r)^3}{8C_c^3} \tag{14.11}$$

and

$$C_c = \sqrt{\frac{2\pi^2 E}{S_{yp}}} \tag{14.12}$$

The term S_{yp} is the yield point stress of the material, while S_{all} is the allowable axial stress. It is interesting to note that the factor of safety f has a minimum value of $\frac{5}{3}$ for a slenderness ratio of zero and increases as the slenderness ratio increases. Thus the maximum allowable stress from Eq. (14.10) is $\frac{3}{5}S_{yp}$ and it decreases as the slenderness ratio increases. Equations (14.10), (14.11), and (14.12) can be used for all main and secondary members whose slenderness ratios are 120 or less ($KL/r \leq 120$) and for main members whose slenderness ratios are equal to or less than the value of C_c, that is, $KL/r \leq C_c$.

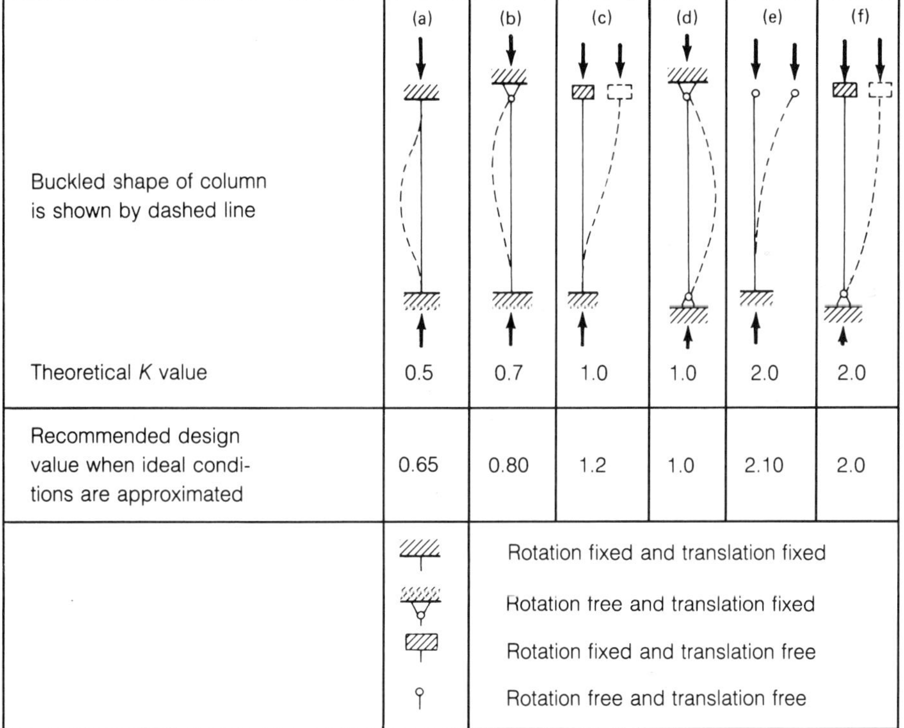

FIGURE 14.6 Values of K for column design. (Reproduced with permission from AISC, *Manual of Steel Construction,* 8th ed. New York, NY: American Institute of Steel Construction, 1980.)

If the slenderness ratio KL/r exceeds C_c, the recommended equation for the allowable stress is

$$\left(\frac{F}{A}\right)_{all} = S_{all} = \frac{12\pi^2 E}{23(KL/r)^2} \tag{14.13}$$

For secondary members or braces subjected to compression loads with slenderness ratios in excess of C_c, $KL/r > C_c$, Eqs. (14.10) and (14.13) are divided by the derating factor given in Eq. (14.9), which yields Eq. (14.14). This procedure, in effect, gives a larger permissible value of S_{all} for those members. Thus

$$\left(\frac{F}{A}\right)_{all} = S_{all} = [\text{Eq. (14.10) or (14.13)}]\left[\frac{1}{1.6 - (L/r)/200}\right] \tag{14.14}$$

It should be noted that K does not appear in the derating term, which is equivalent to assuming $K = 1$ in this term. The more liberal working stress allowed for these members has been justified by the relative unimportance of such members and also by the greater effectiveness of end restraint likely to be present at their ends. Due to this, the full unbraced length of the member should always be used, and Eq. (14.14) should be restricted to those columns whose ends are more or less fixed against rotation and translation at end points. It is interesting to note that Eq. (14.13), which is used for columns that can fail by elastic buckling, is Euler's equation with a constant factor of safety of 23/12.

The AISC has revised the tables of allowable concentric loads in the 8th edition (1980). Allowable concentric loads in the tables that follow are given for the effective lengths KL in feet, indicated at the left of each table. They are applicable to axially loaded members with respect to their minor axis in accordance with the AISC specification. Two strengths are covered, $F_y = 36\ 000$ psi and $F_y = 50\ 000$ psi. The heavy horizontal lines appearing within the tables indicate $Kl/r = 200$. No values are listed beyond $Kl/r = 200$.

Table 14.1 gives vaues of the allowable axial load on some W12 members having yield points of 36 000 and 50 000 psi (36 and 50 ksi). It should be noted again that the maximum value of KL/r for any member is 200. Similar tables for other shapes are given in Appendix C. Table 14.2 gives values of C_c as a function of the yield stress of the material. For this table E has been taken to be 29×10^6 psi.

Figure 14.7 shows the allowable stress for columns that have $KL/r \leq 120$ as a function of the yield point of the steel. Reference is again made to Fig. 14.5, where Eqs. (14.10), (14.11), and (14.12) have been plotted for a steel having a yield point of 36 000 psi. It is quite apparent that the allowable stress (for slenderness ratios ≤ 120) based upon the latest AISC formulas exceeds those previously permitted by earlier codes and formulas. Also, for short struts the present AISC procedure allows a limiting stress of 21 600 psi against a maximum value of 15 000 psi by earlier AISC manuals and the value of 14 000 psi allowed by the Chicago formula. For $L/r > 126$, the allowable stress by the latest procedure yields a value of allowable stress very close to that obtained by earlier AISC procedures.

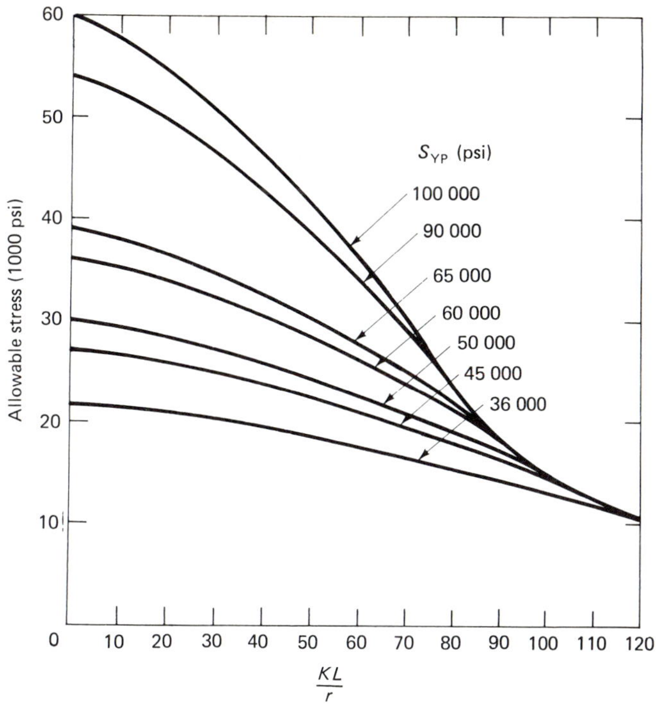

FIGURE 14.7   Allowable stress for columns.

ILLUSTRATIVE PROBLEM 14.5

Determine the maximum load that a W12 × 65 column can carry if it is 35 ft long. Assume it to be: (a) a main member; (b) a secondary member. Use S_{yp} = 36 000 psi (248.2 MPa) and $E = 29 \times 10^6$ psi (200 GPa).

SOLUTION

The greatest L/r for this column is $35 \times 12/3.02 = 139$, with $K = 1$. For this steel $C_c = \sqrt{\dfrac{2\pi^2 E}{S_{yp}}} = \sqrt{(2\pi^2 \times 29 \times 10^6)/36\,000} = 126.1$. Since the slenderness ratio exceeds C_c we shall have to use Eq. (14.13) for a main member or Eq. (14.14) for a secondary member. For this steel we can also use Appendix C or Fig. 14.5. Therefore,

(a)   For a main member at an effective KL of 35, the table of Appendix C is interpolated to give an allowable load of 147 500 lb (656.1 kN).

(b)   For a secondary member we use the value found in (a), multiplied by the factor given by Eq. (14.14). Thus

$$147\,500 \left(\frac{1}{1.6 - 139/200} \right) = 162\,980 \text{ lb } (725.0 \text{ kN})$$

Intermediate Steel Columns—AISC Design Procedure

TABLE 14.1   Allowable Load on Concentrically Loaded W12 Columns Having Yield Points of 36 000 psi and 50 000 psi*

$F_y = 36$ ksi
$F_y = 50$ ksi

COLUMNS — W shapes

Allowable axial loads in kips

Desig-nation	W12									
Wt./ft.	336		305		279		252		230	
F_y	36	50	36	50	36	50	36	50	36	50
0	2134	2964	1935	2688	1769	2457	1601	2223	1462	2031
6	2031	2788	1840	2526	1681	2306	1519	2085	1387	1903
7	2009	2751	1820	2491	1662	2274	1502	2055	1371	1876
8	1986	2711	1799	2454	1642	2240	1484	2023	1355	1847
9	1962	2669	1777	2415	1622	2204	1465	1990	1337	1816
10	1937	2625	1753	2375	1600	2166	1445	1955	1319	1784
11	1911	2579	1729	2332	1578	2126	1425	1919	1300	1750
12	1884	2531	1704	2288	1554	2085	1403	1881	1280	1715
13	1856	2482	1678	2242	1530	2042	1381	1842	1259	1678
14	1827	2430	1651	2194	1505	1998	1358	1801	1238	1641
15	1797	2377	1623	2145	1479	1952	1334	1759	1216	1601
16	1766	2322	1594	2094	1452	1905	1309	1715	1193	1561
17	1733	2265	1565	2041	1425	1856	1284	1670	1169	1519
18	1701	2206	1534	1987	1396	1805	1258	1623	1145	1476
19	1667	2146	1503	1931	1367	1753	1231	1575	1120	1431
20	1632	2084	1471	1873	1337	1699	1203	1526	1095	1386
22	1560	1955	1404	1753	1275	1588	1146	1423	1041	1290
24	1484	1819	1333	1627	1209	1470	1085	1314	985	1190
26	1404	1675	1260	1494	1141	1346	1022	1200	927	1084
28	1321	1525	1183	1354	1069	1216	956	1079	866	972
30	1235	1366	1102	1206	994	1078	887	952	801	855
32	1144	1205	1018	1061	916	948	815	837	734	751
34	1050	1067	930	940	834	839	739	742	664	665
36	951	952	839	839	749	749	661	661	594	594
38	854	854	753	753	672	672	594	594	533	533
40	771	771	679	679	606	606	536	536	481	481

Row label: Effective length in ft. KL with respect to least radius of gyration r_y

*Reproduced with permission from AISC, *Manual of Steel Construction*, 8th ed. New York, NY: American Institute of Steel Construction, 1980.

S_{yp} (psi)	C_c	S_{yp} (psi)	C_c
33 000	131.7	46 000	111.6
35 000	129.7	50 000	107.0
36 000	126.1	55 000	102.0
39 000	121.2	60 000	97.7
42 000	116.7	65 000	93.8
45 000	112.8	90 000	79.8
		100 000	75.7

14.6 ECCENTRICALLY LOADED COLUMNS

In the preceding sections of this chapter we have considered only the centric loading of columns, that is, the resultant force acting through the centroid of the column section. From a practical standpoint it is nearly impossible to avoid small eccentricities in the application of central loads on a column. These effects have been considered, and the empirical equations discussed in Sections 14.4 and 14.5 incorporate, in their safe load values, factors of safety against such unavoidable load eccentricities. There are cases, however, where a column is loaded eccentrically due to its usage. Figure 14.8 shows such a situation where the column is eccentrically loaded about the x–x axis by a nonaxial load on a bracket attached to the side of the column. It is possible to extend the theoretical considerations that lead to the Euler equation to take into account the eccentricity of the load. The resulting equation is known as the *secant formula,* due to the appearance of the trigonometric secant function in the equation's final form. This equation is cumbersome to use, and direct solutions are not possible. Frequently design curves are generated to facilitate the numerical calculations.

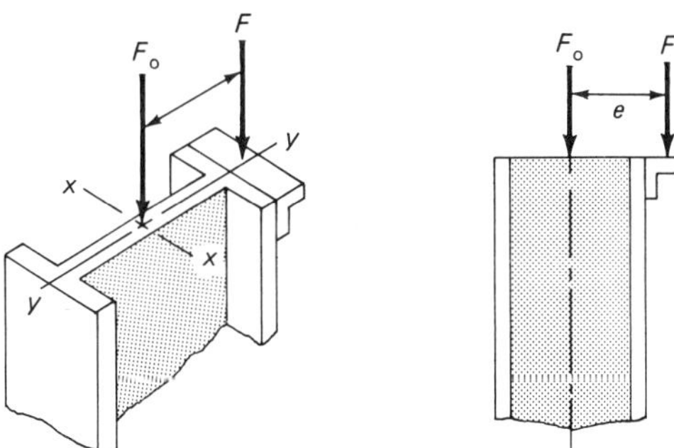

FIGURE 14.8 Eccentrically loaded columns.

An alternate design method can be obtained by considering the action of the central load F_O and the eccentric load F separately. The load on the column is taken to be a direct load due to the sum of $F + F_O$ and a moment Fe. The stresses due to the axial loads and the moment are assumed to be additive, and their arithmetic sum does not exceed the allowable stress. Thus

$$\frac{F + F_O}{A} + \frac{(Fe)C}{I} = S_{all} \qquad (14.15)$$

However, the allowable stress for a column is a function of its slenderness ratio, while the allowable stress in bending is not a function of length. In order to account for this condition, both sides of Eq. (14.15) are divided by S_{all}, yielding

$$\frac{(F + F_O)/A}{S_{all}} + \frac{(Fe)C/I}{S_{all}} = 1 \qquad (14.16)$$

If we now denote

$$\frac{F + F_O}{A} = \text{actual axial stress,} = (S_{act})_A$$

$$S_{all} = \text{allowable stress permitted if only axial stress existed} = (S_{all})_A$$

$$\frac{(Fe)C}{I} = \text{actual bending unit stress,} = (S_{act})_B$$

$$S_{all} = \text{allowable stress permitted if only bending existed} = (S_{all})_B = 0.6S_{yp}$$

we can rewrite Eq. (14.16) in the following form:

$$\frac{(S_{act})_A}{(S_{all})_A} + \frac{(S_{act})_B}{(S_{all})_B} = 1 \qquad (14.17)$$

Equation (14.17) is the form given by the AISC. For mild steel the allowable permitted stress in bending can be taken as 20 000 psi, and the allowable permitted stress due to the axial load is determined by one of the suitable column formulas for the slenderness ratio of the column. When calculating the bending stress, the moment of inertia to be used is the one that is about the axis on which the eccentric load causes bending, but the slenderness ratio to be used should be based upon the least radius of gyration, regardless of the axis about which bending may occur. The AISC manual should be consulted for specific details of the recommended design procedure.

The design of columns, other than steel columns, is usually done by empirical equations for the specific material, established either by the local building code authority or, in its absence, by a federal or nationally known engineering society or association. Thus timber columns are usually designed to the requirements of the Forest Products Laboratory or the National Lumber Manufacuturers' Association. Aluminum columns are often designed to requirements of the American Society of Civil Engineers or *Alcoa Structural Handbook* formulas. Due to many such organizations, as well as the large number of materials used as structural columns, it would not be possible to do more than just list formulas without the necessary precautions or detailed discussions of each formula that should be given. For any column design, the student (as well as the practicing engineer) should consult the governing authority.

Columns

 CLOSURE

Long slender columns can be designed using Euler's equation as a basis; intermediate columns cannot be designed satisfactorily by any single rational approach. Once again, in the realm of actual design, we find that we are faced with the reality that design equations are mandated by the legal authority that has jurisdiction. However, the student should note that the members of code committees that are charged with the responsibility of formulating the design rules are invariably engineers with the resources of an entire industry at their disposal. Their responsibility is really quite simple—the public must be protected at all costs. For this reason, code rules are invariably conservative, and only after adequate tests or operational experience will these code committees alter their rules. It is most unusual to hear of a structural failure in a structure designed to code. Unfortunately this same conservative approach does tend to limit original designs not contemplated when a specific code was formulated. For this reason (among many others) codes are constantly being revised and updated.

At best, this chapter is designed to give the student a brief introduction to those approaches most widely used in the design of columns. It is not expected that the expertise gained from this chapter will make anyone a column designer; on the contrary, it is hoped that the work of this chapter will put students on the alert to go to the proper authority (AISC code, etc.) if and when they must execute a column design.

REFERENCES

American Institute of Steel Construction, MANUAL OF STEEL CONSTRUCTION, 8th ed. 1980.

Arges, K. P., and A. E. Palmer, MECHANICS OF MATERIALS. New York: McGraw-Hill, 1963.

Bassin, M. E., S. M. Brodsky, and H. Wolkoff, STATICS AND STRENGTH OF MATERIALS, 3rd ed. New York: McGraw-Hill, 1979.

Doughty, V. L., and A. Vallance, DESIGN OF MACHINE MEMBERS, 4th ed. New York: McGraw-Hill, 1964.

Higdon, A., E. E. Ohlsen, and W. B. Stiles, MECHANICS OF MATERIALS, 3rd ed. New York: John Wiley, 1976.

Jensen, A., and H. H. Chenoweth, APPLIED STRENGTH OF MATERIALS, 3rd ed. New York: McGraw-Hill, 1971.

Levinson, J., MECHANICS OF MATERIALS. 2nd ed. Englewood Cliffs, NJ: Prentice-Hall, 1970.

Olsen, G. A., ELEMENTS OF MECHANICS OF MATERIALS, 3rd ed. Englewood Cliffs, NJ: Prentice-Hall, 1974.

Singer, F. L., STRENGTH OF MATERIALS, 2nd ed. New York: Harper and Row, 1962.

Timoshenko, S., and D. H. Young, ELEMENTS OF STRENGTH OF MATERIALS, 5th ed. New York: D. Van Nostrand, 1968.

PROBLEMS

14.1 Determine the maximum dimension of a square column if it is 10 ft long, and the slenderness ratio L/r is not to exceed 120.

14.2 If a column is a rectangle having one side twice the other, determine the dimensions of the column if it is 8 ft long and L/r is not to exceed 120. Check both the x–x and the y–y axes.

14.3 If a solid circular cylinder is to be used as a column, what maximum length should it have if its diameter is 4 in. and L/r is to be limited to 120?

14.4 Solve Problem 14.2 if L/r is permitted to be 200, as for a secondary member.

14.5 Four 2-in. square bars are to be used to support a compressive load as shown. Using a limiting L/r value for the assembly of 120, determine the maximum length for this combination. Is this for the x–x or the y–y axis?

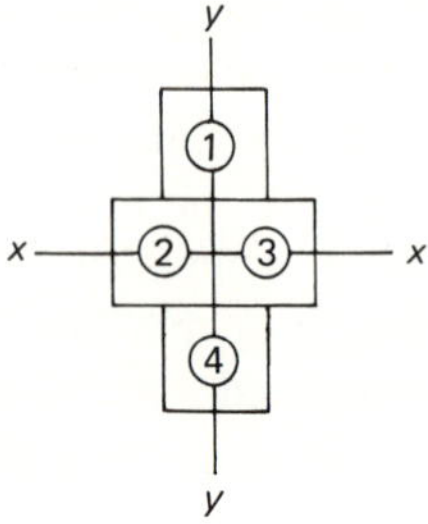

FIGURE P 14.5

14.6 Can a 12-in. Schedule 80 pipe be used as a main member if it is 25 ft long?

14.7 A solid circular cylinder is to be used as a structural member, with L/r not to exceed 120. If it is 7 ft long, determine the required diameter of the cylinder.

14.8 Determine the required diameter of a solid circular cylinder that is used as a main member. (Use Euler's equation.) Both ends of the cylinder are fixed. It is made of a steel alloy having a yield stress of 45 000 psi and $E = 30 \times 10^6$ psi. A factor of safety of 2½ is taken to be appropriate for the application. The column is 10 ft long.

14.9 Solve Problem 14.8 if the yield stress of the steel is 100 000 psi.

14.10 Determine the required diameter of the column in Problem 14.8 if both ends are hinged.

14.11 If a 4-in. Schedule 40 steel pipe is used as a column and its L/r is to be 150, determine the safe load it will carry. Use Euler's equation and a factor of safety of 2½ and assume the ends to be hinged. $E = 30 \times 10^6$ psi.

14.12 If the column in Problem 14.11 has both ends built in, determine its safe load.

14.13 Solve Problem 14.11 if one end is hinged and the other end is built in.

14.14 A cantilevered beam is subject to a direct axial load. What maximum value of load can this member support? Use a factor of safety of 3; $E = 10 \times 10^6$. Assume that Euler's equation applies.

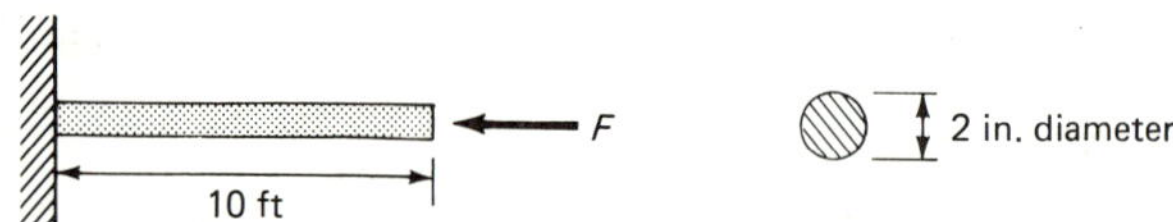

FIGURE P 14.14

14.15 A W8 × 31 beam is to be used as a column. What is the allowable load that it can carry if it is hinged at both ends of its 25-ft length? Use a factor of safety of 2½ and $E = 30 \times 10^6$. Assume that Euler's equation is applicable.

14.16 If the column in Problem 14.15 has one end fixed and one end hinged, what allowable load can it carry?

                                                                        Columns

14.17 Solve Problem 14.16 if both ends of the column are fixed.

14.18 A W10 $\times$ 33 column, 50 ft long and hinged at both ends, is braced at its midpoint in its weakest direction and is unbraced in its strongest direction. Using a factor of safety of 2 in Euler's equation, determine the maximum load that it will carry. $E = 30 \times 10^6$ psi.

14.19 Using the Chicago formula, determine the maximum load that a 4-in. Schedule 40 pipe can sustain as a main member.

14.20 Use the Chicago formula to determine the maximum load that a solid 4-in. diameter steel bar can sustain as a main member.

Use the latest AISC procedures and the tables in Appendix C to solve Problems 14.21 through 14.26. The methods of Section 14.5 and Figure 14.7 may also be used to solve these problems. Use $K = 1$.

14.21 A W8 $\times$ 31 structural member is 25 ft long and is to be used as a column. Using a material having a yield point stress of 50 000 psi, determine the maximum load that it can support. It is a main member.

14.22 A W12 $\times$ 58 structural member is 20 ft long and is to be used as a column. If the yield point stress of the material is 36 000 psi, determine the maximum load that it will support.

14.23 Determine the maximum load that a W10 $\times$ 88 structural member made of 50 000 psi yield point material can carry if it is 15 ft long.

14.24 A column is to support a centric load of 75 000 lb. If it is 11 ft long, select a suitable M section for this application. Assume the material to have a yield point stress of 36 000 psi.

14.25 Select a column for Problem 14.24 if the yield point of the material is 50 000 psi.

14.26 Select an S beam suitable to carry a 15 000 lb centric load if the material has a yield point of 50 000 psi and the column is 9 ft long.

Use the method of Section 14.5, Fig. 14.7 and/or the tables in Appendixes B and C for the following problems. The material can be taken to have a yield stress of 50 000 psi and an allowable bending stress of 30 000 psi. Note that the allowable tensile stress equals $0.6 \times S_{yp}$. Use $K = 1$.

14.27 A W14 $\times$ 109 column is to carry a concentric load of 200 000 lb and an eccentric load of 100 000 lb. If the column is 10 ft long, what is the maximum eccentricity from the x–x axis for which this column can be designed?

14.28 Can a W12 $\times$ 22 column support an 80 000-lb load eccentrically applied 8 in. from the x–x axis of the column? The column is to be 13 ft long.

14.29 In addition to a 50 000-lb eccentrically applied load at 6 in. from the x–x axis of a W6 $\times$ 25 beam, there is a centric load. If the column is 10 ft long, what maximum centric load can it support?

14.30 A W12 $\times$ 79 structural section is to be used to carry a centric load of 500 000 lb and an eccentric load of 100 000 lb applied 4 in. from the x–x axis. Determine whether this section is suitable for use as a column 9 ft long.

15

Combined Stresses

15.1 INTRODUCTION

In Chapters 9, 10, 12, and 14 we briefly looked at several loading conditions that give rise to a combination of shear, tension, and/or compression. It is rare indeed to find actual structural or machine parts which are loaded by a single, well-defined load that gives rise to a simple, single state of stress. More frequently, the member in question is loaded in such a manner as to induce a combination of stresses, such as shear, tension, and so on. Under certain types of loading it is permissible simply to add algebraically the resulting stresses to obtain the maximum stress at a point, but this is generally not possible. Our task in this chapter will be to investigate combinations of stresses that exist at a given section and the resultant effects induced.

The formulas, which we have derived elsewhere in this book, for the most part, enable us to determine the stresses on certain planes passing through a point. In this chapter we shall further evaluate the maximum stress conditions as well as determine the planes on which they occur. While it is possible, in principle, to mathematically determine the state of stress at a given section in a member, it is much more difficult to incorporate these results into meaningful theories or criteria of failure for such members under combined loads. Even simple cases of loading cause internal stress distributions sufficiently complex so as to prevent us from using the data from simple tension tests to predict the onset of failure (inelastic action).

15.2 BENDING AND AXIAL LOADS COMBINED

In Section 14.6 we considered the eccentrically loaded column and developed a rational approach to the design of such structural members by considering the columns to be loaded by a central load F_O and an eccentric load F, as shown in Fig. 15.1. Basically we considered each load separately and added the effects with proper consideration of the sense of each effect, that is, compression or tension. Due to axial loading, the member is in compression and the resulting compression stress is

$$S_{\text{axial}} = \frac{F_O + F}{A} \tag{15.1}$$

The eccentric load F gives rise to a moment about the x–x axis causing flexural stresses to exist, which have maximum values given by the flexure formula as

$$S_{\text{flexure}} = \frac{MC}{I} = \frac{(Fe)C}{I} \tag{15.2}$$

At plane B the element is in compression and at plane A it is in tension. We now assume that it is possible to algebraically sum the axial and flexural stresses to obtain

$$S = \pm\frac{F_O + F}{A} \pm \frac{MC}{I} = \pm\frac{F_O + F}{A} \pm \frac{(Fe)C}{I} \tag{15.3}$$

where the $\pm$ notation indicates that compression is to be considered (algebraically) as a negative effect, and tension is to be considered as a positive effect. Equation (15.3) is satisfactory for those cases where the member can be considered to be rigid, such as posts or short compression members. Where the member is long and slender and its deflection under load is significant, the method of superposition of stresses used above is not applicable.

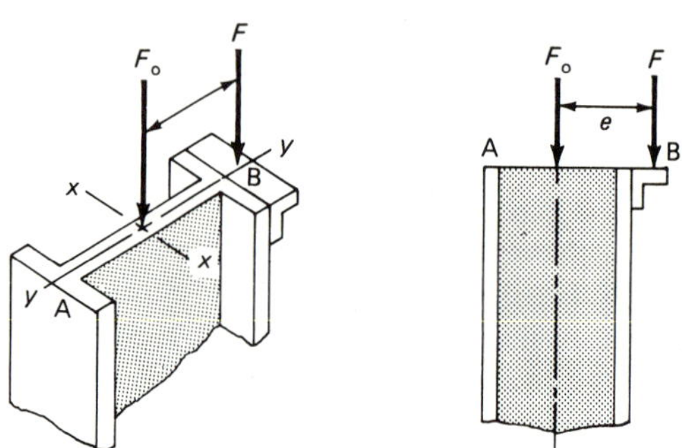

FIGURE 15.1   Eccentrically loaded columns. (Repeated from Fig. 14.8.)

ILLUSTRATIVE PROBLEM 15.1

A clamp is made as shown in Fig. 15.2. If the jaws are tightened so that the clamp is subjected to a load of 500 lb, determine the stresses induced on plane z–z.

                                                        Combined Stresses

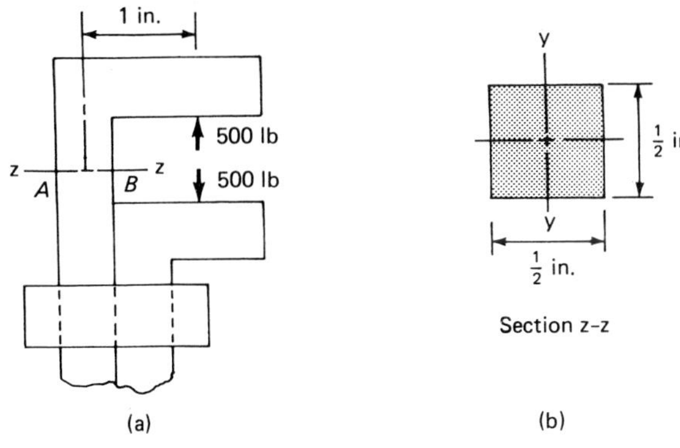

FIGURE 15.2   Illustrative Problem 15.1.

SOLUTION

The clamp is made of ½ in. × ½ in. stock, its area A is ¼ in.², and its moment of inertia about x–x or y–y is $bh^3/12 = ½(½)^3/12 = 1/192$. Evaluating the direct load and the flexural load, we find

$$S = \frac{F}{A} = \frac{500}{¼} = 2000 \text{ psi (tension)}$$

$$S = \frac{MC}{I} = \frac{500 \times 1 \times ¼}{¹⁄_{192}} = 24\,000 \text{ psi}$$

At A, bending causes compression. Therefore

$$S = +2000 - 24\,000 = -22\,000 \text{ psi (compression)}$$

At B, bending causes tension. Therefore

$$S = +2000 + 24\,000 = +26\,000 \text{ psi (tension)}$$

ILLUSTRATIVE PROBLEM 15.2

A hanger is designed as shown in Fig. 15.3, using an S5 × 10 beam. Determine the stresses at sections A and B if the hung load weighs 20 000 N.

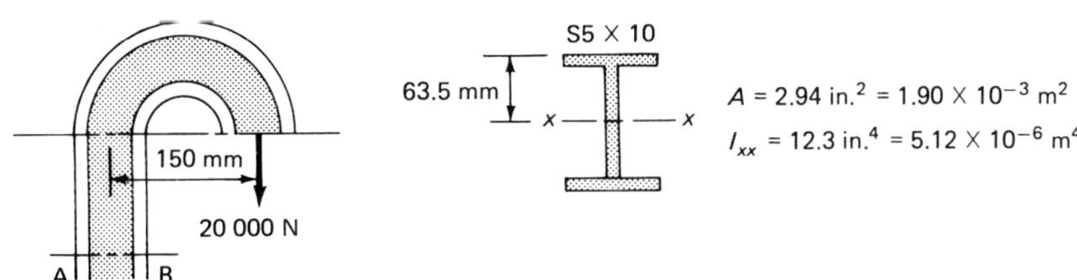

FIGURE 15.3   Illustrative Problem 15.2.

Bending and Axial Loads Combined

The direct stress on the section due to the 2-ton load is

$$S = \frac{F}{A} = \frac{20\ 000}{1.90 \times 10^{-3}} = 10.53 \text{ MPa (compression)}$$

The bending stress is $S = MC/I = 0.150 \times 20\ 000 \times 0.0635/5.12 \times 10^{-6} = 37.2$ MPa. At A the section is in tension due to bending. Therefore

$$S = -10.53 + 37.21 = 26.68 \text{ MPa (tension)}$$

At B the section is in compression. Therefore

$$S = -10.53 - 37.21 = -47.74 \text{ MPa (compression)}$$

15.3 PURE SHEAR

Let us now reexamine the conclusions we reached in Chapter 12 concerning the existence of a shear stress on a plane through a given point in a body. We showed earlier that if a shear stress occurs on a plane, a shear stress of equal intensity exists on a plane at right angles to the first plane and passing through the same point as the first plane. When shearing stresses occur only on two perpendicular planes at a point in the body, that point is said to be "subject to pure shear." Referring to Fig. 15.4, we can write

$$S_s = S_s' \tag{15.4}$$

If any plane is now passed through the point, there will be both normal and shear stresses acting on this oblique plane. In order to determine the stress conditions on any oblique plane, we shall use the following technique. An imaginary plane will be passed through the body at the point in question. We shall then isolate the resulting small free body, and on it all *stresses* will be shown. Since a free-body diagram shows all of the forces acting upon the body, it becomes necessary to multiply the stresses by the areas over which they act and then to apply the conditions for static equilibrium. Regardless of the distribution of the stresses at a point in a body, there exist three mutually perpendicular planes through the point on which the stresses are normal to the planes. These planes, therefore, have no shearing stresses on them, and the normal stresses on these planes are called *principal stresses*. The importance of this concept of principal stress lies in the fact that the maximum normal stress at a point is always a principal stress. We shall return to this point later in this chapter.

Figure 15.5 shows a free-body diagram of an element cut by an oblique plane. We shall endeavor to determine the angle φ for the plane of principal stress as well

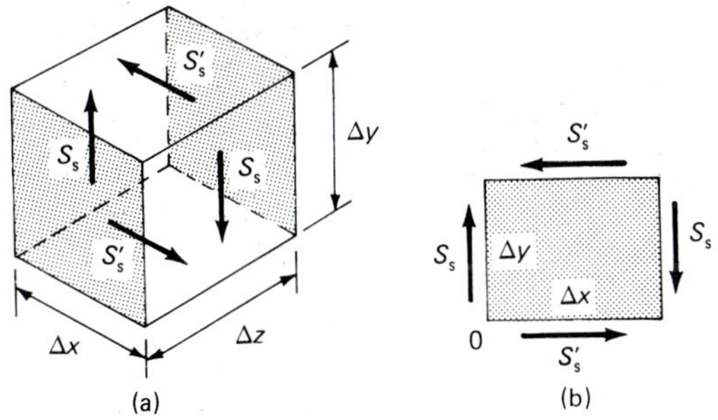

FIGURE 15.4   Stresses on an element. (Repeated from Fig. 12.11.)

as the value of the principal stress S_N. Since S_N is a principal stress, there will be zero shear on this plane. The free-body diagram [Fig. 15.5(b)] therefore shows only S_N on this oblique plane, and the force $S_N A$, due to S_N, has been resolved into its component parts as indicated. Taking a force summation in the horizontal direction yields

$$S_s A \sin \varphi = S_N A \cos \varphi \tag{15.5}$$

and a force summation in the vertical direction gives us

$$S_s A \cos \varphi = S_N A \sin \varphi \tag{15.6}$$

From Eq. (15.5),

$$S_N = S_s \frac{\sin \varphi}{\cos \varphi} \tag{15.7}$$

and from Eq. (15.6),

$$S_N = S_s \frac{\cos \varphi}{\sin \varphi} \tag{15.8}$$

From the simultaneous conditions shown by Eqs. (15.7) and (15.8) we conclude that $\sin \varphi = \cos \varphi$, requiring φ to be 45° and that $S_N = S_s$. Note that S_N is a compressive stress. Had we cut the element as shown in Fig. 15.5(c), we would also have concluded φ to be 45° and $S_N = S_s$, where S_N is a tensile stress.

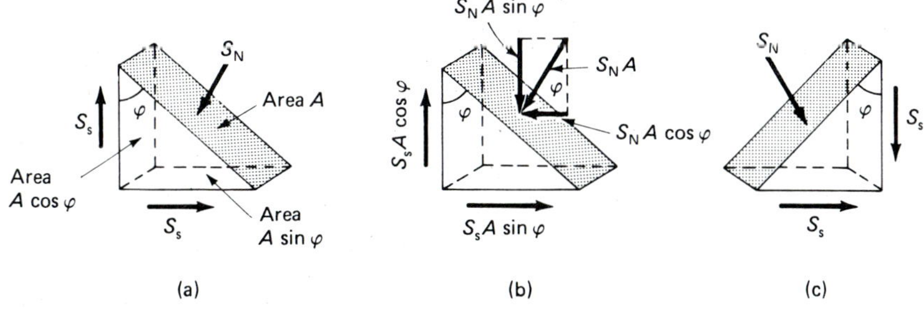

FIGURE 15.5   Pure shear.

Pure Shear

We can summarize the foregoing by stating that if a body is subjected to a pure shear at a point, there exist planes having only normal stresses (principal stresses) numerically equal to the applied shearing stress. These planes make angles of 45° to both the horizontal and the vertical and are at right angles to each other. One important conclusion reached is that a material which is weak in tension will fail in tension on a plane that is inclined at 45° to the plane of shear when the body is subject to pure shear. Figure 15.6 shows the conditions and types of failure one obtains when a brittle material is subject to pure shear.

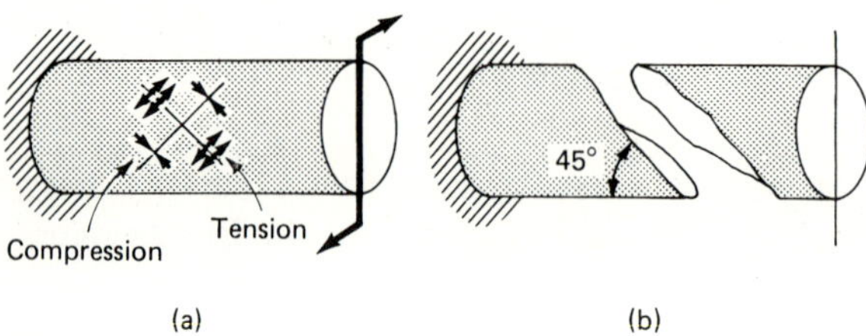

FIGURE 15.6   Brittle failure in pure shear.

The inverse of the situation discussed above occurs when an element of area is subjected to a vertical compressive stress and an equal horizontal tensile stress. This situation is illustrated in Fig. 15.7(a). Due to these applied stresses, there must exist planes on which pure shear stresses act. These planes are rotated at 45° to the planes of tension and compression, as shown in Fig. 15.7(b), and the stresses on these planes are known as *equivalent shearing stresses*.

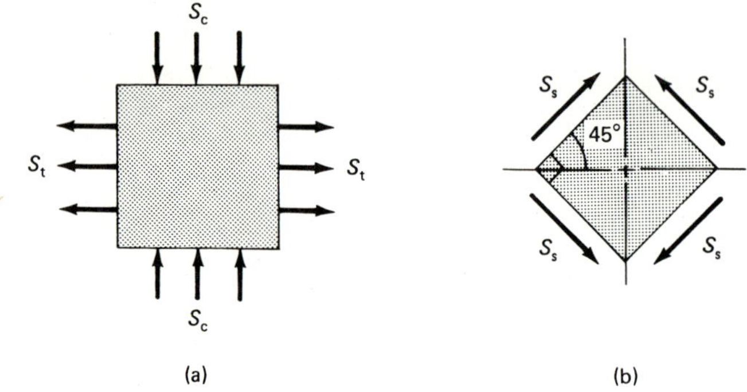

FIGURE 15.7   Equivalent shearing stresses. (a) $S_t = S_c$. (b) Induced shear $S_s = S_t = S_c$.

Combined Stresses

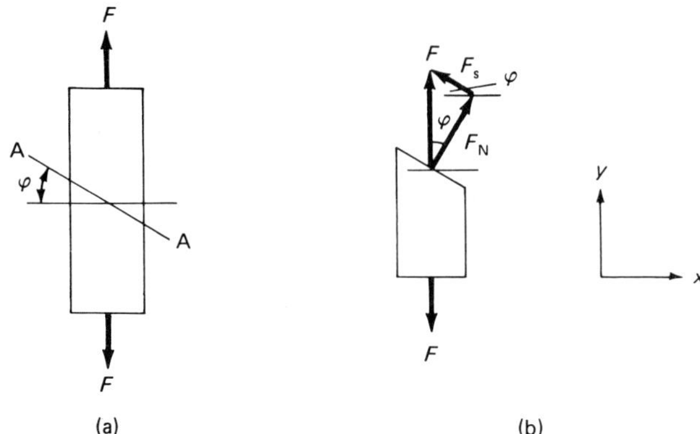

FIGURE 15.8   Illustrative Problem 15.3.

SOLUTION

In order to solve this problem, we shall consider that a hypothetical cutting plane is passed through the bar coinciding with plane A–A. In order for the bar to be in equilibrium, let us postulate that a force F_N exists normal to the plane in question, and a force F_s exists parallel to the plane. In order for the cut bar to be in equilibrium, it is necessary that $\Sigma F_y = 0$ and $\Sigma F_x = 0$. Thus

$$F_s \cos \varphi = F_N \sin \varphi$$

and

$$F = F_s \sin \varphi + F_N \cos \varphi$$

Solving for F_N from the first equation and substituting it into the second equation yields

$$F = F_s \sin \varphi + \left(F_s \frac{\cos \varphi}{\sin \varphi}\right) \cos \varphi = \frac{F_s (\sin^2 \varphi + \cos^2 \varphi)}{\sin \varphi} = \frac{F_s}{\sin \varphi}$$

yielding

$$F_s = F \sin \varphi$$

Solving for F_N,

$$F_N = F \cos \varphi$$

The stress that each force causes equals the force divided by the area over which the force acts. If the area normal to F is denoted by A, the area of plane A–A is $A/\cos \varphi$. The shear stress S_s is given by

$$S_s = \frac{F_s}{A_s} = \frac{F \sin \varphi}{A/\cos \varphi} = \frac{F}{A} \sin \varphi \cos \varphi$$

Pure Shear

Using the trigonometric relation $\sin 2\varphi = 2 \sin \varphi \cos \varphi$,

$$S_s = \frac{F \sin 2\varphi}{2A}$$

Similarly, $S_N = F_N/A_N = F \cos \varphi/(A/\cos \varphi) = (F/A) \cos^2 \varphi$. Using the trigonometric identity

$$\cos^2 \varphi = \frac{1 + \cos 2\varphi}{2}$$

we have

$$S_N = \frac{F}{A}\left(\frac{1 + \cos 2\varphi}{2}\right)$$

We can check our results in this problem by noting that S_N is maximum when $\varphi = 0$ and equals F/A (the principal stress). Also, the maximum value of S_s occurs when $\varphi = 45°$ ($2\varphi = 90°$) and equals $F/2A$ or $S_1/2$. These results agree with our previous general conclusions. Note that for this problem φ is the angle the plane makes with the horizontal and is the angle that the normal to the plane makes with the vertical.

15.4 GENERAL CASE OF PLANE STRESS

If we consider an element subjected to both shear and normal stresses on its faces, as shown in Fig. 15.9, we have the general conditions of plane stress. For this case we shall adopt the conventions that S_s is the shear stress on any inclined plane, and that S_N is the normal stress on this plane. S_{xy} will be used to denote the shear stress acting on the x plane in a direction parallel to the y plane. Thus S_{yx} denotes the shear stress on the y plane parallel to the x plane. Since $S_{xy} = S_{yx}$, we will use S_{xy} for both stresses. Figure 15.9(b) is a free-body diagram of a wedge cut out of the element shown in Fig. 15.9(a). The area of the inclined plane has been denoted by A, making $A_x = A \sin \varphi$ and $A_y = A \cos \varphi$. Note that on the free-body diagram forces, not stresses, are shown. Resolving forces into components in the x and y directions and applying the condition for equilibrium yield, in the x direction,

$$S_x A \cos \varphi - S_{xy} A \sin \varphi - AS_N \cos \varphi - AS_s \sin \varphi = 0 \qquad (15.9)$$

and in the y direction,

$$S_{xy} A \cos \varphi - S_y A \sin \varphi + AS_N \sin \varphi - AS_s \cos \varphi = 0 \qquad (15.10)$$

The relations involving the double angle are $\cos^2 \varphi = (1 + \cos 2\varphi)/2$ and $\sin^2 \varphi = (1 - \cos 2\varphi)/2$. Therefore we can obtain the relations for the normal and shear

Combined Stresses

stresses on the cutting plane of Fig. 15.9 as

$$S_N = \frac{S_x + S_y}{2} + \frac{S_x - S_y}{2} (\cos 2\varphi) - S_{xy} \sin 2\varphi \qquad (15.11)$$

and

$$S_s = \frac{S_x - S_y}{2} (\sin 2\varphi) + S_{xy} \cos 2\varphi \qquad (15.12)$$

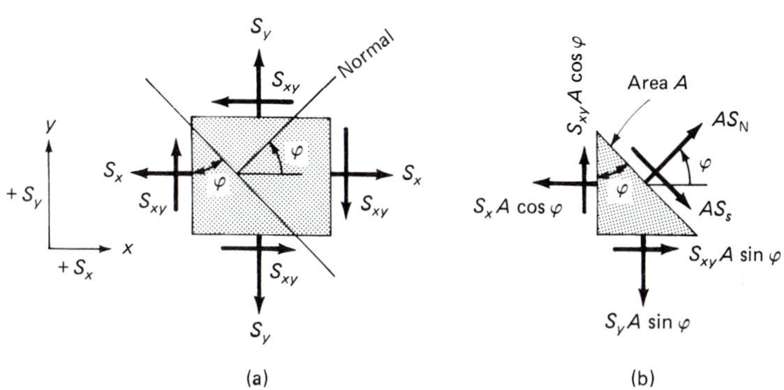

FIGURE 15.9   General case of plane stress. (a) Stress diagram. (b) Free-body diagram.

The planes of maximum and minimum normal stresses can be obtained by the methods of calculus, giving

$$\tan 2\varphi_N = -\frac{2S_{xy}}{S_x - S_y} \qquad (15.13)$$

By similar techniques we can obtain the planes of maximum shear as

$$\tan 2\varphi_s = \frac{S_x - S_y}{2S_{xy}} \qquad (15.14)$$

From these equations we conclude the following.

1. Planes of maximum and minimum normal stresses occur 90° apart.
2. Planes of maximum shear occur 90° apart.
3. Maximum and minimum normal stresses occur on planes of zero shear and are therefore principal stresses.
4. Planes of maximum shearing stress form angles of 45° with the planes of principal stress.

By substitution of the angles defined by Eqs. (15.13) and (15.14) into Eqs. (15.11)

General Case of Plane Stress

and (15.12), respectively, we can obtain the maximum stresses as

$$S_{N \text{ max or min}} = \frac{S_x + S_y}{2} \pm \sqrt{\left(\frac{S_x - S_y}{2}\right)^2 + S_{xy}^2} \qquad (15.15)$$

$$S_{s \text{ max}} = \pm \sqrt{\left(\frac{S_x - S_y}{2}\right)^2 + S_{xy}^2} \qquad (15.16)$$

At this time we shall introduce a graphical construction, known as *Mohr's circle,* which greatly simplifies the calculation of combined stresses once the construction has been fully understood. Basically, the construction is a graphical portrayal of Eqs. (15.13), (15.14), (15.15), and (15.16). In order to apply Mohr's circle we shall adopt one further convention at this time, namely, that shearing stress is positive when it causes a clockwise movement about the center of an element. We can now proceed to construct Mohr's circle for the case of generalized plane stresses by the following steps.

1. Locate the points on the plot of S_N versus S_s of points having the coordinates S_x, S_{xy} and S_y, S_{yx}, as shown in Fig. 15.10.

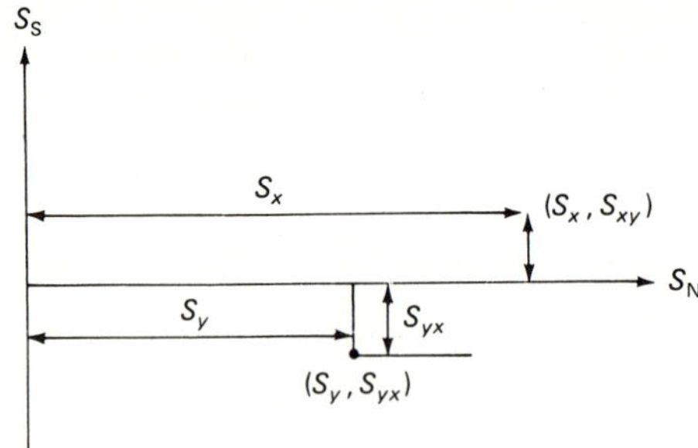

FIGURE 15.10   Step 1 in the construction of generalized Mohr's circle.

2. The diameter of Mohr's circle is the line joining S_x, S_{xy} and S_y, S_{yx}.
3. The principal maximum and minimum stresses occur on planes of zero shear. Thus, points A and B on the S_N axis of Fig. 15.11 correspond to the maximum and minimum principal stresses.

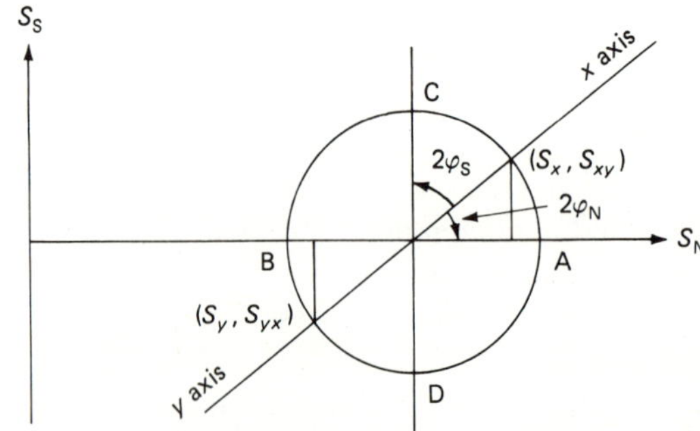

FIGURE 15.11   Steps 2 and 3 in the construction of generalized Mohr's circle.

                                    Combined Stresses

4. $\tan 2\varphi_N = -\dfrac{S_{xy}}{(S_x - S_y)/2} = -\dfrac{2S_{xy}}{S_x - S_y}$

5. The maximum shear stress occurs at points C and D.

6. $\tan 2\varphi_s = \dfrac{S_x - S_y}{2S_{xy}}$

ILLUSTRATIVE PROBLEM 15.4

A plane element is subjected to the stresses shown in Fig. 15.12. Determine the maximum and minimum normal stresses, the maximum shear stress, and locate the planes of maximum shear stress and the principal planes.

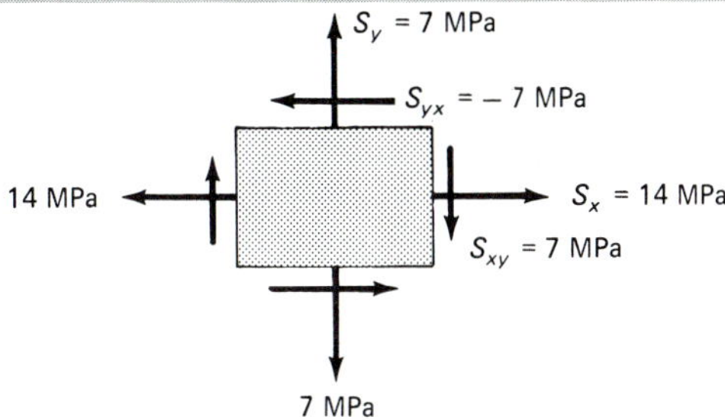

FIGURE 15.12   Illustrative Problem 15.4.

SOLUTION

We construct Mohr's circle using the rules developed previously (Fig. 15.13).

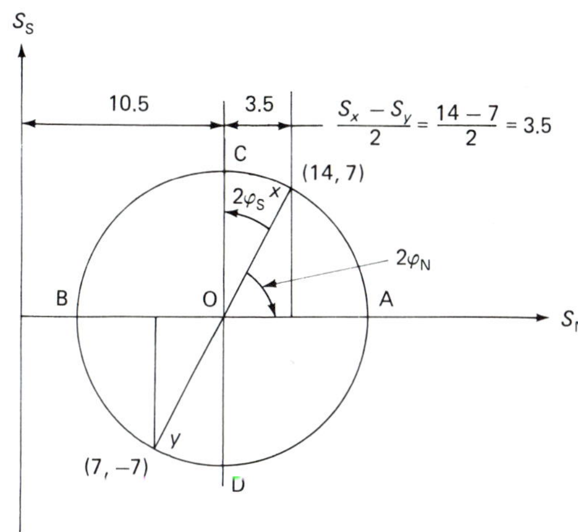

FIGURE 15.13   Mohr's circle for Illustrative Problem 15.4.

radius of circle = $\sqrt{7^2 + 3.5^2}$ = 7.826 MPa

point A = $S_{N\ max}$ = 10.5 + 7.826 = 18.326 MPa

point B = $S_{N\ min}$ = 10.5 − 7.826 = 2.674 MPa

point C = $S_{s\ max}$ = +7.826 MPa on a plane having S_N = 10.5 MPa point D $= S_{s\ min} = -7.826$ MPa on a plane having $S_N = 10.5$ MPa

$$\tan 2\varphi_N = 7/3.5 = 2;$$

$$2\varphi_N = 63.5° \qquad \varphi_N = 31.75° \text{ and } (31.75 + 90) = 121.75°$$

The angle φ_N is measured clockwise from O–x, the positive x axis to the S_N axis.

The plane of maximum shear stress is 45° from the plane of maximum normal stress, that is, $\varphi_N + \varphi_s = 45°$. Therefore $\varphi_s = 45 - 31.75 = 13.25°$ counterclockwise from the x axis.

The complete solution is shown in Fig. 15.14.

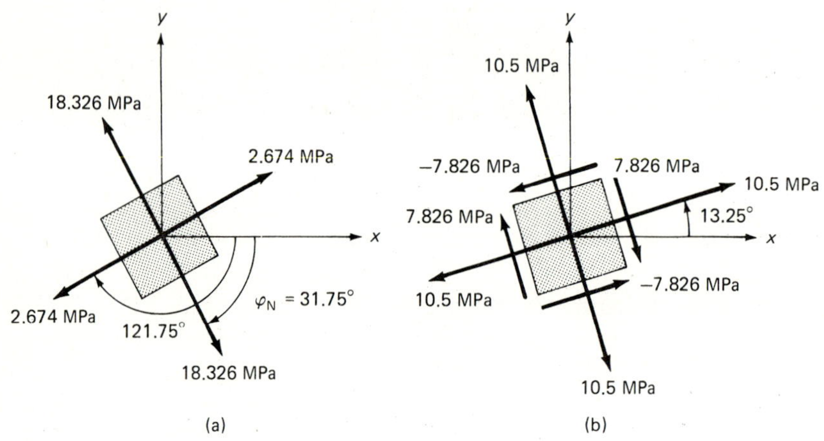

FIGURE 15.14 Solution to Illustrative Problem 15.4. (a) Principal stresses. (b) Maximum shear stresses.

15.5 COMBINED STRESSES IN CIRCULAR SHAFTS

In our discussion of pure shear we noted that a body subjected to pure shear has principal stresses in tension and compression equal numerically to the applied shear stress acting on planes inclined at 45° to the shaft axis. If in addition to being subjected to the pure shear, the body is also acted upon by a bending moment, there will be a longitudinal stress induced in the body. Consider that the body is a circular shaft subject to a torque of T in.lb and a bending moment of M in.lb. For this case the principal stresses will occur at the top and bottom of the shaft. Due to the applied torque T, $S_s = TC/J$; and due to the bending moment M, $S_b = MC/I$. The bending action causes either tension or compression at the top or bottom of the shaft, respectively. The principal stress due to this action, which is shown in Fig. 15.15, can be obtained from Eq. (15.15) as

$$S_{N\ max\ or\ min} = \frac{S_b}{2} \pm \frac{1}{2}\sqrt{S_b^2 + 4S_s^2} \tag{15.17}$$

Let us now recall that the polar moment of inertia J equals the sum of the rectangular moments of inertia $I_x + I_y$. For a symmetrical shape $J = 2I$, where I is the rectangular moment of inertia about either axis of symmetry. We can therefore write

$$S_b = \frac{MC}{I} \qquad S_s = \frac{TC}{J} \qquad J = 2I$$

and

$$\frac{S_b}{S_s} = \frac{MC/I}{TC/2I} = \frac{2M}{T} \qquad \text{or} \qquad S_s = S_b\left(\frac{T}{2M}\right) \qquad (15.18)$$

If the results of Eq. (15.18) are combined with Eq. (15.17),

$$S_N = \frac{S_b}{2} \pm \frac{1}{2}\sqrt{S_b^2 + 4S_b^2\left(\frac{T}{2M}\right)^2}$$

$$= \frac{S_b}{2} \pm \frac{1}{2}\sqrt{\frac{(M^2 + T^2)S_b^2}{M^2}}$$

and

$$S_N = \frac{S_b}{2}\left(1 \pm \sqrt{\frac{M^2 + T^2}{M^2}}\right) \qquad (15.19)$$

At this point we will define an *equivalent bending moment* M_e as that moment which, if it acted alone on the shaft, would produce the same maximum principal stress in the shaft as that produced when the shaft is subjected to bending and twisting. Using this definition,

$$S_N = \frac{M_e C}{I} \qquad (15.20)$$

Substitution into Eq. (15.19) yields

$$\frac{M_e C}{I} = \frac{MC}{2I}\left(1 + \sqrt{\frac{M^2 + T^2}{M^2}}\right) \qquad (15.21)$$

from which M_e is obtained as

$$M_e - \frac{1}{2}(M + \sqrt{M^2 + T^2}) \qquad (15.22)$$

The maximum shearing stress on the element can be directly obtained by use of Eq. (15.16),

$$S_{s\;max} = \pm\frac{1}{2}\sqrt{S_b^2 + 4S_s^2} = \pm\frac{1}{2}S_b\sqrt{\frac{M^2 + T^2}{M^2}} \qquad (15.23)$$

Defining T_e as the equivalent torque, which would produce the same maximum shearing stress as is given by Eq. (15.23), yields

$$\frac{T_e C}{J} = \frac{S_b T_e}{2M} = \frac{S_b}{2M} \sqrt{M^2 + T^2} \tag{15.24}$$

where Eq. (15.18) gave us the relation between S_b and S_s. Simplifying Eq. (15.24),

$$T_e = \sqrt{M^2 + T^2} \tag{15.25}$$

Combining Eq. (15.25) with Eq. (15.22) yields

$$M_e = \frac{1}{2}(M + T_e) \tag{15.26}$$

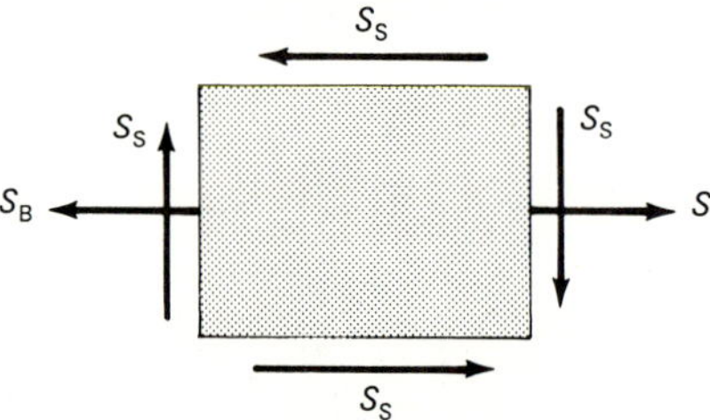

FIGURE 15.15 Shaft subject to bending and twisting.

ILLUSTRATIVE PROBLEM 15.5

A shaft having a 2-in. diameter is subjected to a torque of 9000 in.·lb and a bending moment of 12 000 in.·lb. Determine the maximum shear and bending stresses in the shaft.

SOLUTION

The equivalent torque and equivalent bending moment are determined from Eqs. (15.25) and (15.26), respectively, as

$$T_e = \sqrt{M^2 + T^2} = \sqrt{12\ 000^2 + 9000^2} = 15\ 000 \text{ in.·lb}$$

$$M_e = \tfrac{1}{2}(M + T_e) = \tfrac{1}{2}(12\ 000 + 15\ 000) = 13\ 500 \text{ in.·lb}$$

The maximum shear stress is given by

$$S_s = \frac{T_e C}{J} = \frac{15\ 000 \times 1}{\pi 2^4/32} = 9550 \text{ psi}$$

The maximum principal stress is

$$S = \frac{M_e C}{I} = \frac{13\ 500 \times 1}{\pi 2^4/64} = 17\ 200 \text{ psi}$$

                                    Combined Stresses

SOLUTION

In order to solve this problem we shall have to size the shaft for bending and
shear, and use the condition that gives us the greater shaft size. The equivalent
torque and bending moment are

$$T_e = \sqrt{M^2 + T^2} = \sqrt{15\,000^2 + 10\,000^2} = 18\,000 \text{ in.·lb}$$

$$M_e = \tfrac{1}{2}(M + T_e) = \tfrac{1}{2}(15\,000 + 18\,000) = 16\,500 \text{ in.·lb}$$

For shear,

$$S_s = 12\,000 = \frac{T_e C}{J} = \frac{18\,000 \times d/2}{\pi d^4/32} = \frac{91\,700}{d^3}$$

$$d^3 = \frac{91\,700}{12\,000} = 7.65$$

$$d = 1.97 \text{ in.}$$

For bending,

$$S_b = 20\,000 = \frac{M_e C}{I} = \frac{16\,500 \times d/2}{\pi d^4/64} = \frac{168\,000}{d^3}$$

$$d^3 = \frac{168\,000}{20\,000} = 8.4$$

$$d = 2.03 \text{ in.}$$

The shaft size to use is approximately $2\tfrac{1}{32}$ in.

There are many cases in which shafts are subjected to twisting and axial loads.
Propeller shafts are just one example of this type of loading. The state of stress in the
element of such a shaft will appear to be the same as that shown in Fig. 15.15, and
we can use our foregoing analysis for the calculation of the principal stresses and the
maximum shear stresses. Thus

$$S_N = \frac{F}{2A} \pm \frac{1}{2}\sqrt{\left(\frac{F}{A}\right)^2 + 4\left(\frac{TC}{J}\right)^2} \tag{15.27}$$

$$S_{s\,\text{max}} = \pm\frac{1}{2}\sqrt{\left(\frac{F}{A}\right)^2 + 4\left(\frac{TC}{J}\right)^2} \tag{15.28}$$

where F/A is the direct axial stress. In the case of axial compression it may be necessary to consider the shaft as a long column. However, we shall not consider it as such in our present discussion. Mohr's circle is instructive in visualizing the state of stress for the situation we have when a shaft is subject to an axial load as well as to twisting. The same diagram can also be used for bending and twisting. Figure 15.16 shows the construction of Mohr's circle for this case. Point A locates $S_{N\ max}$ as

$$S_{N\ max} = \frac{S_x}{2} + \frac{1}{2}\sqrt{S_x^2 + 4S_s^2} \tag{15.15a}$$

Points C and D locate the maximum shear stresses as

$$S_{s\ max} = R = \pm\frac{1}{2}\sqrt{S_x^2 + 4S_s^2} \tag{15.16a}$$

Equations (15.15a) and (15.16a) are obviously in agreement with Eqs. (15.15) and (15.16).

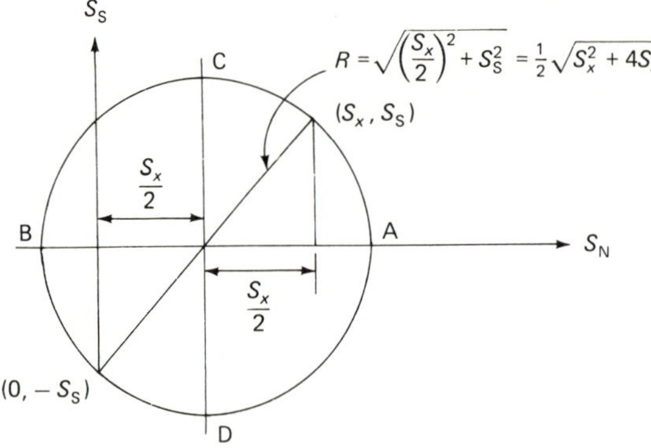

FIGURE 15.16 Mohr's circle for combined axial loading and twisting.

ILLUSTRATIVE PROBLEM 15.7

A solid propeller shaft must transmit 800 hp at 300 rev/min at the same time it is subjected to a 10 000-lb axial load. If the maximum shear stress is considered to be the critical stress and is limited to 14 000 psi, compute the necessary shaft size.

SOLUTION

The direct axial load is

$$\frac{F}{A} = \frac{10\ 000}{\pi d^2/4} = \frac{12\ 740}{d^2}$$

The shear due to twisting is found from the torque equation. The torque to be transmitted is

$$T = \frac{63\ 000\ hp}{N} = \frac{63\ 000 \times 800}{300} = 168\ 000\ \text{in.·lb}$$

                    Combined Stresses

The stress S_s is

$$S_s = \frac{TC}{J} = \frac{168\,000 \times d/2}{\pi d^4/32} = \frac{168\,000(16/\pi)}{d^3} = \frac{856\,000}{d^3}$$

$$S_{s\,max} = 14\,000 = \frac{1}{2}\sqrt{\left(\frac{12\,740}{d^2}\right)^2 + \left(\frac{856\,000}{d^3}\right)^2 4}$$

This equation cannot be solved directly, but must be solved by a trial-and-error procedure. When this has been done, it will be found that the required diameter is approximately 4 in.

When several forces acting on a shaft cause bending in different planes, it is necessary to determine the component bending moments in the horizontal and vertical directions. The resultant moment at any section is given by $\sqrt{(M_{vertical})^2 + (M_{horizontal})^2}$. We proceed with combining this resultant moment with the shear stress, as we did earlier.

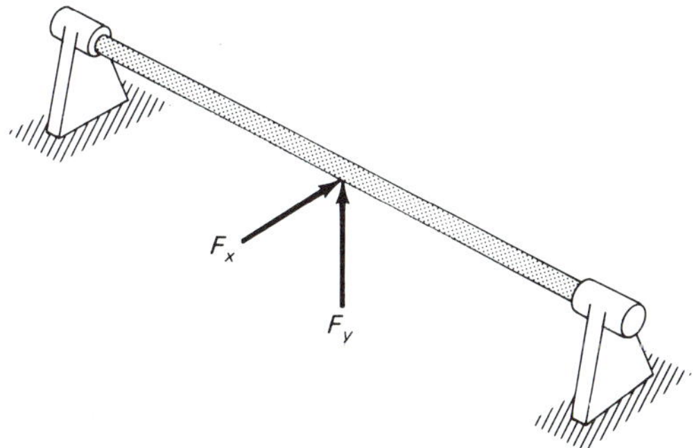

FIGURE 15.17   Combined horizontal and vertical bending.

ILLUSTRATIVE PROBLEM 15.8

At a given section of a shaft there is a horizontal bending moment of 1400 N·m and a vertical bending moment of 800 N·m. If the shaft is 50 mm in diameter and must transmit a torque of 2500 N·m, determine the principal stress and shear stress in the shaft.

SOLUTION

The resultant bending moment is $M = \sqrt{1400^2 + 800^2} = 1612.5$ N·m.

$$T_e = \sqrt{M^2 + T^2} = \sqrt{1612.5^2 + 2500^2} = 2974.9 \text{ N·m}$$

$$S_{s\,max} = \frac{T_e C}{J} = \frac{2974.9 \times 0.050/2}{\pi \times 0.050^4/32} = 121.2 \text{ MPa}$$

$$M_e = \frac{1}{2}(M + T_e) = \frac{1}{2}(1612.5 + 2974.9) = 2293.7 \text{ N·m}$$

$$S_{N\,max} = \frac{M_e C}{I} = \frac{2293.7 \times 0.050/2}{\pi \times 0.050^4/64} = 186.9 \text{ MPa}$$

Combined Stresses in Circular Shafts

15.6 CLOSURE

This chapter has been divided into two basic parts, the mathematical derivation of principal and shear stresses, and the use of Mohr's circle to accomplish the same end. It is always tempting to the student to solve problems by substituting numbers into equations. Unfortunately this procedure does not lend itself to the learning process and very often is self-defeating. The construction of Mohr's circle is more time-consuming, but if it is used frequently, students will find that they will have greater and greater facility in the use of the construction and will have a better grasp of the principles studied in this chapter.

One further caution must be stated at this time. The use of the free-body diagram requires that all of the *forces* acting on a body be properly shown. It is an all too common error to show the stresses acting on a body and to apply the equations of equilibrium to the stresses rather than the forces. By systematically showing the stresses and the areas upon which these stresses act, we can avoid this error. The drawing and sketching of loads and areas should be done with care, and angles should be carefully labeled to avoid errors in problem solution.

REFERENCES

Arges, K. P., and A. E. Palmer, MECHANICS OF MATERIALS. New York: McGraw-Hill, 1963.

Bassin, M. E., S. M. Brodsky, and H. Wolkoff, STATICS AND STRENGTH OF MATERIALS, 3rd ed. New York: McGraw-Hill, 1979.

Conway, H. D., MECHANICS OF MATERIALS. Englewood Cliffs, NJ: Prentice-Hall.

Jensen, A., and H. H. Chenoweth, APPLIED STRENGTH OF MATERIALS, 3rd ed. New York: McGraw-Hill, 1971.

Levinson, I. J., MECHANICS OF MATERIALS, 2nd ed. Englewood Cliffs, NJ: Prentice-Hall, 1970.

Olsen, G. A., STRENGTH OF MATERIALS. Englewood Cliffs, NJ: Prentice-Hall.

Seely, F. B., and J. O. Smith, ADVANCED MECHANICS OF MATERIALS, 2nd ed. New York: John Wiley, 1952.

Singer, F. L., STRENGTH OF MATERIALS, 2nd ed. New York: Harper and Row, 1962.

Timoshenko, S., and D. H. Young, ELEMENTS OF STRENGTH OF MATERIALS, 5th ed. New York: D. Van Nostrand, 1968.

PROBLEMS

15.1 A square bar is subjected to an axial load and a transverse load as shown. Determine the maximum and minimum stress that occur at the built-in end.

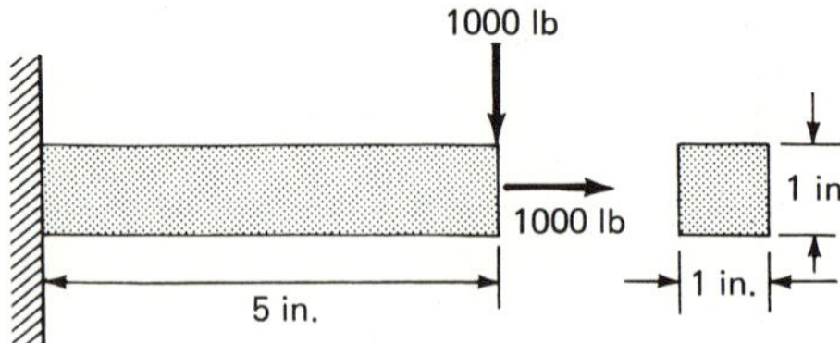

FIGURE P 15.1

15.2 For the bar shown, determine the maximum and minimum stresses at the built-in end.

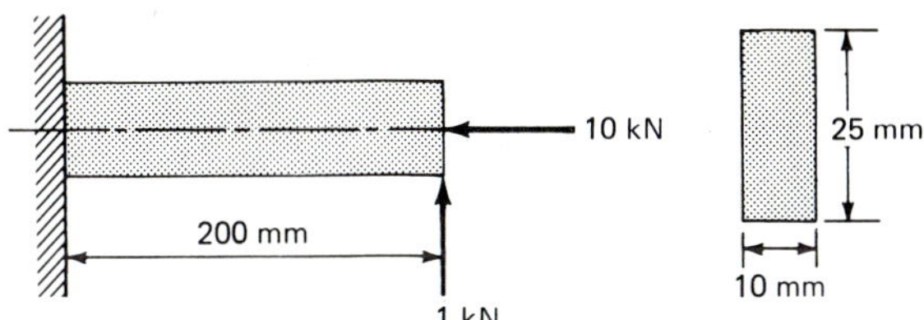

FIGURE P 15.2

15.3 A short 125-mm diameter solid steel bar is used to carry a 90-kN centric compression load and a 45-kN eccentric load located 150 mm from its center. Determine the maximum stress in the bar.

15.4 A 4-in. Schedule 40 pipe is used as a hanger. What is the maximum stress at the built-in end?

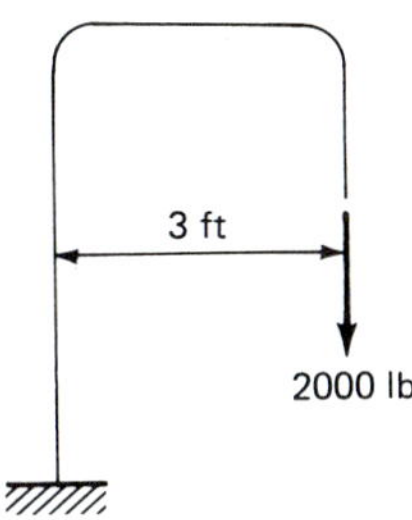

FIGURE P 15.4

15.5 A short strut is subjected to a load through its centroid as shown. Determine the maximum and minimum stresses at its base.

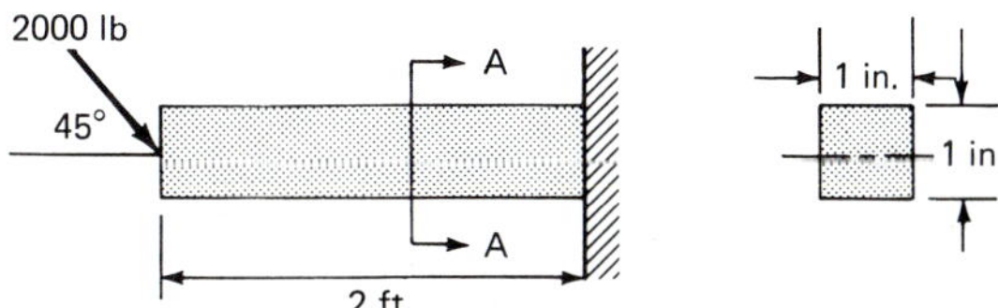

FIGURE P 15.5

15.6 A W14 × 109 short strut is to carry a concentric load of 200 000 lb and an eccentric load of 100 000 lb having an eccentricity of 8 in., giving a net moment about the x–x axis as shown in Fig. 15.1. Determine the maximum and minimum combined stress in the member.

15.7 An S5 × 10 short strut is loaded as shown. Determine the maximum stress at the base of the strut.

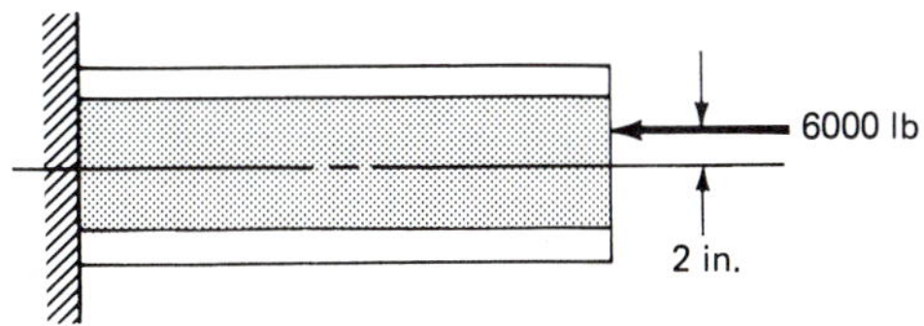

FIGURE P 15.7

Problems

15.8   If an S5 × 10 strut is loaded as shown, determine the maximum stress at the base of
the strut.

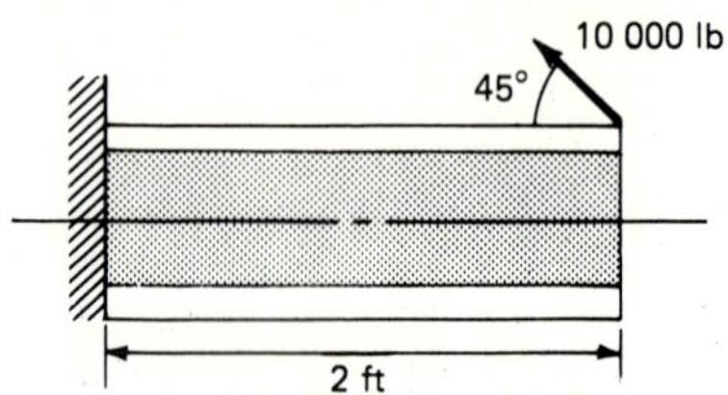

FIGURE P 15.8

15.9   If the C clamp shown is loaded with a force of $F = 1000$ lb, determine the maximum
stress across section A–A.

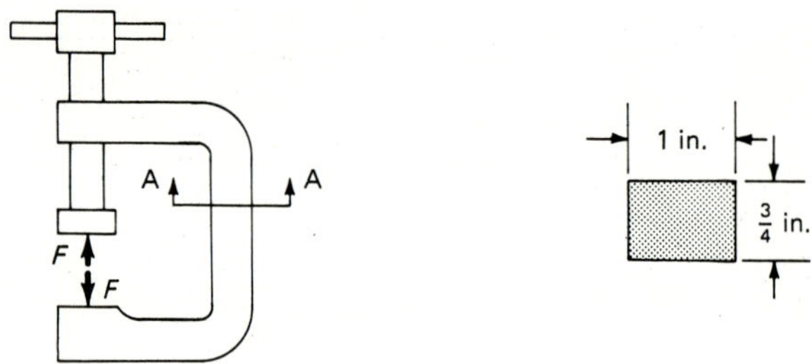

FIGURE P 15.9

15.10   If section A–A of Problem 15.9 is a T section, determine the maximum stress across
section A–A.

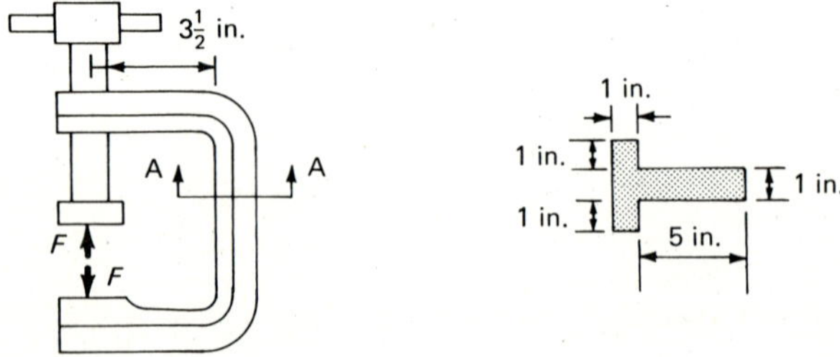

FIGURE P 15.10

418   Combined Stresses

15.11   A centric load of 5000 lb acts as shown. Determine the normal and shear stresses on planes making 0°, 30°, 45°, 60°, and 90° to the horizontal.

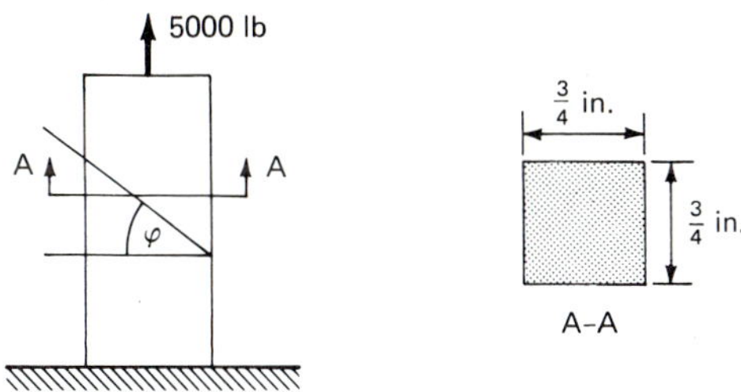

FIGURE P 15.11

15.12   Solve Problem 15.11 using Mohr's circle.

15.13   A centric load of 22 kN acts as shown. Determine the normal and shear stresses on planes making 0°, 30°, 45°, 60°, and 90° to the horizontal.

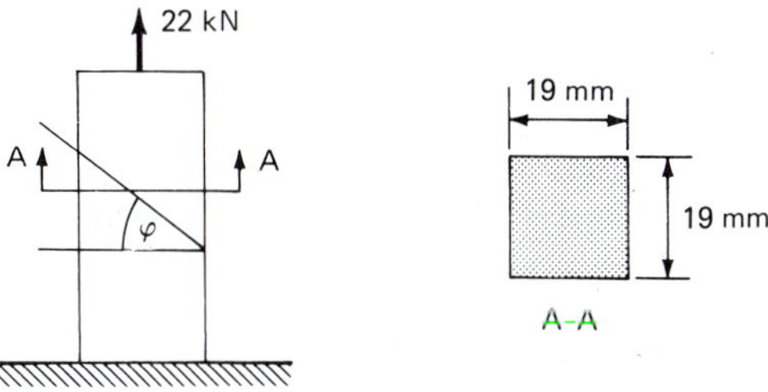

FIGURE P 15.13

15.14   Solve Problem 15.13 using Mohr's circle.

15.15   A short square bar with a side of 25 mm has a maximum allowable compressive stress of 140 MPa and a maximum allowable shear stress of 70 MPa. What axial load can this strut support?

15.16   A short strut having a rectangular cross section of 25 mm × 50 mm has a maximum allowable compressive stress of 200 MPa and a maximum allowable shear stress of 100 MPa. What axial load can this strut support?

15.17   A short square bar with a side of 1 in. has a maximum allowable compressive stress of 20 000 psi and a maximum allowable shear stress of 10 000 psi. What axial load can this strut support?

15.18   If an element is loaded so that $S_x = 10\,000$ psi and $S_y = 5000$ psi (both tension and both principal stresses), determine the normal and shear stresses on a plane whose normal makes an angle of $+30°$ with the x plane (angle φ in Fig. 15.9). Use Eqs. (15.11) and (15.12) to obtain the solution. $S_{xy} = 0$.

15.19   Solve Problem 15.18 if S_y is compressive.

15.20   Solve Problem 15.18 using Mohr's circle.

15.21   Solve Problem 15.19 using Mohr's circle.

15.22 An element is subjected to the stresses shown. Determine the stress components S_N and S_s along the diagonal 1–1. Use Eqs. (15.11) and (15.12).

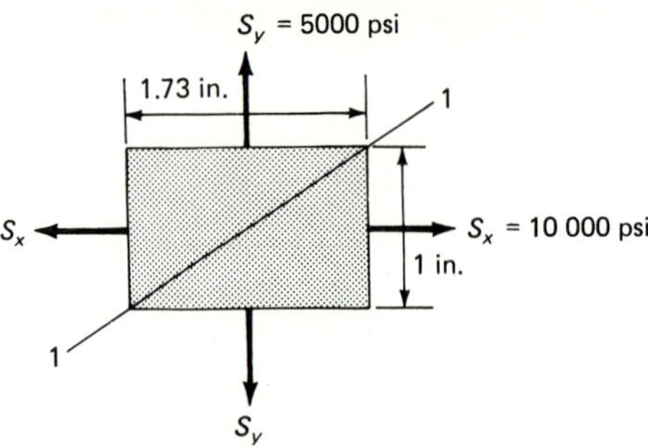

FIGURE P 15.22

15.23 Solve Problem 15.22 if S_y is compressive.

15.24 Solve Problem 15.22 using Mohr's circle.

15.25 If an element is loaded so that $S_x = 100$ MPa and $S_y = 50$ MPa (both tension and both principal stresses), determine the normal and shear stresses on a plane whose normal makes an angle of $+30°$ with the x plane (angle φ in Fig. 15.9). Use Eqs. (15.11) and (15.12) to obtain the solution.

15.26 Solve Problem 15.25 if S_y is compressive.

15.27 Solve Problem 15.25 using Mohr's circle.

15.28 Solve Problem 15.26 using Mohr's circle.

15.29 A centric load of 45 kN acts on a 12.5-mm $\times$ 12.5-mm bar as shown. Determine the normal and shearing stresses acting on a plane making an angle of 35° with the x plane (angle φ in Fig. 15.9). Use Eqs. (15.11) and (15.12).

FIGURE P 15.29

15.30 Solve Problem 15.29 using an angle of 25°.

15.31 Solve Problem 15.29 using Mohr's circle.

15.32 An element is loaded so that $S_x = 75$ MPa and $S_y = 50$ MPa. Determine the normal and shearing stresses on a plane making an angle of 15° with the x plane (angle φ in Fig. 15.9). Use Eqs. (15.11) and (15.12).

15.33 Solve Problem 15.32 using Mohr's circle.

15.34 Determine the maximum principal stress and the maximum shear stress on the element of Problem 15.32 using Eqs. (15.11), (15.12), and (15.13).

15.35 Solve Problem 15.34 using Mohr's circle.

                                                    Combined Stresses

15.36 For the element shown determine the principal stresses, the maximum shear stress, and the orientation of the planes of maximum normal stress. Use Eqs. (15.11), (15.12), and (15.13) to obtain the solution.

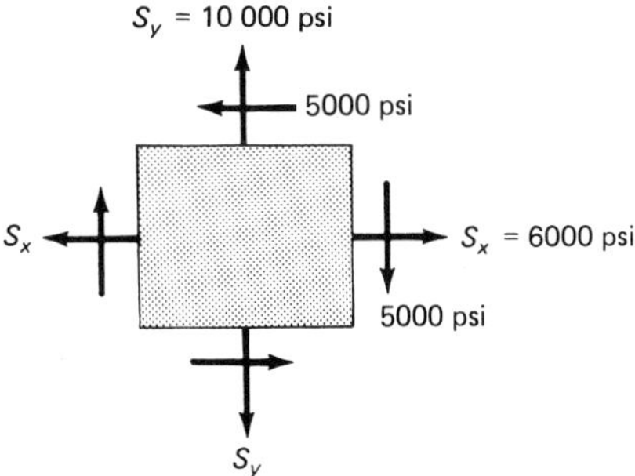

FIGURE P 15.36

15.37 Solve Problem 15.36 using Mohr's circle.

15.38 For the element shown determine the principal stresses, the maximum shear stress, and the orientation of the plane of maximum normal stress. Use Eqs. (15.11), (15.12), and (15.13) to obtain the solution.

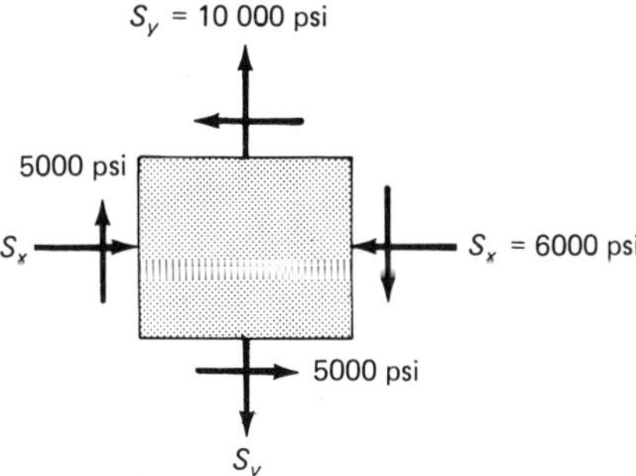

FIGURE P 15.38

15.39 Solve Problem 15.38 using Mohr's circle.

15.40 For the element shown determine the principal stresses, the maximum shear stress, and the orientation of the plane of maximum normal stress. Use Eqs. (15.11), (15.12), and (15.13).

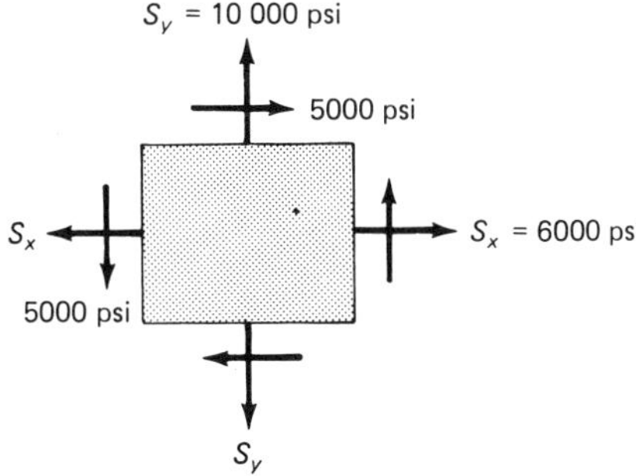

FIGURE P 15.40

Problems

15.41   Solve Problem 15.40 using Mohr's circle.

15.42   A 1-in. diameter shaft is subjected to the simultaneous action of a 400-in./lb torque and an 800-in.·lb bending moment. Determine the maximum shear and bending stresses in the shaft.

15.43   A 3-in. diameter shaft must transmit 500 hp at 500 rev/min, while it is concurrently subjected to an axial load of 100 000 lb. Determine the maximum shear stress in the shaft.

15.44   The shaft shown in Fig. 15.17 is 10 ft long between supports. If $F_x = F_y = 1000$ lb and if the beam is considered to be simply supported, determine the principal stress and shear stress in the shaft. The shaft must simultaneously transmit a torque of 15 000 in.·lb and is 1½ in. in diameter. Assume that the loads act at the center of the shaft.

15.45   A shaft is loaded as shown. Determine the maximum shear and principal stresses in the shaft. The bearings provide simple support. Neglect the weight of the pulleys and shaft.

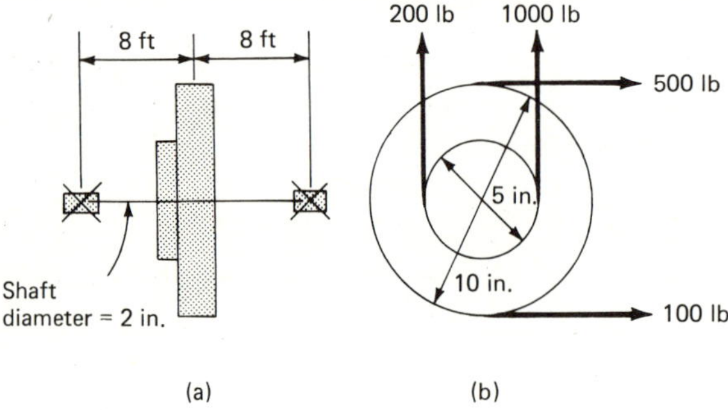

FIGURE P 15.45

15.46   Two pulleys, each 10 in. in diameter, are placed at the third points of a shaft, one pulley being the driver and the other the driven. If the belts are taken off so that the belt tensions are in the horizontal direction, determine the normal and shear stresses in the shaft. The shaft is to transmit 100 hp at 500 rev/min and is 2 in. in diameter and 12 ft long. Since one pulley is the driver and the other the driven, each will give an equal, but opposite net tensile force acting in the horizontal plane due to the unbalanced belt tensions.

                                                    Combined Stresses

Answers to Even-Numbered Problems

CHAPTER 1

1.2 12 in.
1.4 53.13°; 36,87°.
1.6 30.02 ft.
1.8 2.24 mi.
1.10 8.66 in.
1.12 43.75 mm; 101.95 mm.
1.14 6.70 cm; 47.19°, 77.81°.
1.16 40.89°.
1.18 77.16°; 59.35°; 43.49°.

CHAPTER 2

2.2 350 mi/h (west).
2.4 $R = 65$ N (right).
2.6 $R = 108.4$ N.
2.8 $R = 5.83$ mi/h;
$\theta = 59.04°$.
2.10 $R = 173.2$ lb.
2.12 $R = 108.4$ N.
2.14 $R = 5.83$ mi/h;
$\theta = 59.04°$.
2.16 $R = 173.2$ lb.
2.18 $R = 241.4$ N.
2.20 $R = 29.86$ N; $\theta = 3.65°$.
2.22 $R = 86.3$ N.
2.24 $R = 120$ N.
2.26 $R = 7.81$ km.
2.28 70.7 lb.
2.30 $R = 110.2$ N; $\theta = 55.69°$.
2.32 -600 in.·lb clockwise.
2.34 6.33 ft from left.
2.36 $\Sigma F_y = 0$; $M_A = 0$.
2.38 $M_A = -850$ ft·lb (left
side); $R =$ couple =
-850 ft·lb.
2.40 $R =$ moment $= 1800$ ft·lb
counterclockwise.

2.42 $\cos \theta_x = 0.408$; $\cos \theta_y = 0.408$; $\cos \theta_z = 0.817$.
2.44 $\cos \theta_x = -0.873$; $\cos \theta_y = -0.436$; $\cos \theta_z = 0.218$.

CHAPTER 3

3.2 Figure.
3.4 $M = 10\,000$ N·m;
$R = 1000$ N.
3.6 Figure.
3.8 $R_1 = 1458$ N;
$R_2 = 1792$ N.
3.10 $T = 500$ lb.
3.12 $T_1 = 896.5$ N;
$T_2 = 731.9$ N.
3.14 $R_A = 375$ lb;
$R_B = 1500$ lb;
$R_C = 625$ lb;
$R_D = 375$ lb.
3.16 $A_x = 25.0$ lb;
$A_y = 75.0$ lb;
$B = 35.4$ lb.
3.18 $A_x = 0$; $B_x = 0$; $A_y = 233.3$ N; $B_y = 266.7$ N.
3.20 $T = 592.9$ lb.
3.22 $A_x = -1588$ lb;
$A_y = 1000$ lb;
$B_x = 1588$ lb; $B_y = 0$.
3.24 $T = 1767$ lb;
$A_x = 1249$ lb;
$A_y = -251$ lb.
3.26 $A_x = 1111$ lb;
$A_y = 1000$ lb;
$D_x = 555.5$ lb; $C_x = 555.5$ lb; $C = 677.4$ lb; $D = 677.4$ lb.

3.28 $T_1 = 1187.6$ lb; $T_2 = 50.0$ lb; $A_x = 350$ lb; $A_y = -500$ lb; $A_z = 1000$ lb.

CHAPTER 4

4.2 $R_A = 750$ lb;
$R_C = 750$ lb;
AD = CD = 1167 lb (T);
AB = BC = 894 lb (C);
DB = 1500 lb (T).
4.4 $R_A = 250$ lb;
$R_D = 750$ lb;
BC = 500 lb (T),
AE = 250 lb (C);
CE = 353.6 lb (C);
BE = 353.6 lb (C);
ED = 750 lb (T);
CD = 1061 lb (C);
AB = 353.6 lb (T).
4.6 GC = 0;
BC = 4000 N (T);
BG = 2828 N (C).
4.8 FE = 1000 lb (C);
FC = 707 lb (T);
BC = 500 lb (T).
4.10 $T = 500$ lb; $A_x = 250$ lb;
$A_y = 1866$ lb.
4.12 FC = 707 lb (T);
BC = 2122 lb (C);
EF = 1000 lb (T).
4.14 ID = 565.8 lb (C);
IH = 2800 lb (C);
CD = 3200 lb (T).
4.16 $B_x = 288.7$ lb;
$B_y = 500$ lb.
4.18 $B_y = 1600$ lb;
$B_x = 1850$ lb.

$A_y = 600$ lb;
$A_x = 1850$ lb;
$D_x = 1850$ lb;
$D_y = 2100$ lb

4.20 No change at B.

CHAPTER 5

5.2 $\mu = 0.466$.
5.4 $\mu = 0.452$.
5.6 $F = 150$ lb.
5.8 $F = 1204.4$ lb.
5.10 To slide = 3800 lb; to roll = 16.7 lb.
5.12 $T_2 = 82.17$ lb.
5.14 Torque = 2200 in.·lb.
5.16 $T_2 = W = 60.8$ lb.
5.18 Torque to lower = 62.3 in.·lb.
5.20 29.1 in.·lb.

CHAPTER 6

6.2 448.2 MPa.
6.4 11 318 psi.
6.6 12 732 psi.
6.8 49 000 lb.
6.10 11 318 psi; 13 570 psi; 4650 psi.
6.12 30 630 lb.
6.14 16 980 psi; 694 psi; 174 psi.
6.16 0.066 84 in.
6.18 1.3 MPa.
6.20 $\epsilon = 0.000\ 85$; $S = 25\ 464$ psi; $E = 30.17 \times 10^6$ psi.
6.22 $S_s = 9.4$ MPa; $S_c = 3.13$ MPa.
6.24 1.7188×10^{-3} m.
6.26 0.000 208 rad.
6.28 0.0019 in.^{2}
6.30 $\mu = 0.263$.
6.32 Aluminum 0.17; iron 0.31; copper 0.44.
6.34 5.55×10^{-4} rad.
6.36 0.195 in.
6.38 $\alpha = 9 \times 10^{-6}$ in./in.
6.40 32.2 MPa.
6.42 97.5 kN.

6.44 $F_{\text{steel}} = F_{\text{copper}} = 3628.5$ lb; $S_{\text{steel}} = 4620$ psi; $S_{\text{copper}} = 8213$ psi.
6.46 0.22 in.
6.48 15 mm.
6.50 92.6 psi.
6.52 $S_{\text{thin (inside)}} = 4000$ psi; $S_{\text{mean}} = 4500$ psi; $S_{\text{thick}} = 5600$ psi.

CHAPTER 7

7.2 $E = 509.3$ MPa.
7.4 $\mu = 0.32$.
7.6 $\mu = 0.40$.
7.8 40% elongation.
7.10 Proportional limit = 407.5 MPa; failure = 1.386 GPa.
7.12 $S = 135\ 870$ psi; $\left(\dfrac{\Delta A}{A}\right) = 64\%$.
7.14 $\dfrac{\Delta A}{A} = 64.4\%$; $\dfrac{\Delta L}{L} = 37.5\%$.
7.16 $\dfrac{\Delta L}{L} = 39.5\%$; $\dfrac{\Delta A}{A} = 61.8\%$.
7.18 $F = 8053$ N. $\epsilon = 0.0003125$ mm/mm; 65.63 MPa
7.20 $E = 169.7$ GPa; $\mu = 0.267$.
7.22 $E = 50$ GPa.
7.24 Proportional limit = 33 783 psi; yield stress = 45 045 psi; breaking stress = 90 090 psi; true breaking stress = 216 200 psi; $\dfrac{\Delta A}{A} = 58.3\%$.
7.26 $\dfrac{\Delta A}{A} = 64.2\%$; ultimate stress = 60 000 psi; true ultimate stress = 167 590 psi; breaking stress

= 46 000 psi; true breaking stress = 128 490 psi.
7.28 Student exercise.

CHAPTER 8

8.2 $\bar{y} = 88.69$ mm (from base).
8.4 $\bar{x} = 0$; $\bar{y} = 68.75$ mm (from base).
8.6 $\bar{x} = 2.43$ in.; $\bar{y} = 1.64$ in.
8.8 $\bar{x} = 22.66$ mm.
8.10 $\bar{x} = 0$; $\bar{y} = 5.97$ in.
8.12 $\bar{x} = 2.53$ in.; $\bar{y} = 5.13$ in.
8.14 $\bar{x} = 0$; $\bar{y} = 10.88$ in.
8.16 $R = 118.35$ mm.
8.18 $\bar{y} = 159.79$ mm.
8.20 $\bar{y} = 3.25$ in.; $I_{\bar{y}} = 18.16$ in.^{4}.
8.22 $k_o = 4.48$ in.; $J_o = 602.5$ in.^{4}.
8.24 $k_o = 4.43$ in.; $J_o = 410.3$ in.^{4}.
8.26 $I_{xx} = 71.2$ in.^{4}.
8.28 $W = 1.85$ in.
8.30 $\bar{I}_y = 74.2$ in.^{4}.
8.32 $I = 34.488 \times 10^6$ mm^4.
8.34 $I = 5111.0$ in.^{4}; $k = 6.36$ in.

CHAPTER 9

9.2 16.1 hp.
9.4 $T = 4200$ in.·lb A $\rightarrow$ B; $T = 6300$ in.·lb B $\rightarrow$ C; $T = 10\ 500$ in.·lb C $\rightarrow$ D.
9.6 $d = 57.93$ mm.
9.8 $d = 2.09$ in.
9.10 $G = 7.99 \times 10^6$ psi.
9.12 $d = 77.81$ mm.
9.14 $\theta = 0.43$ rad.
9.16 2.57 deg.
9.18 Use $d = 62.7$ mm.
9.20 280 hp.
9.22 $T = 8819$ N·m.
9.24 $D = 16.72$ in.

CHAPTER 10

10.2 1.16 in. (bearing).

10.4 $S_s = 213.8$ MPa.

10.6 131.29 kN (tension).

10.8 $\eta = 25.3\%$ (top plate); 18 985 lb.

10.10 79 500 lb (double shear).

10.12 57.91 kN.

10.14 8333 lb.

10.16 Same maximum stress.

10.18 Discuss.

10.20 B and D will have maximum stress.

CHAPTER 11

11.2 $M = Fa\left(1 - \dfrac{x}{a+b}\right);$
$-R_2 = -F[a/(a+b)]$

11.4 Shear $= 1000$ N; moment $= 4400$ N·m (neg).

11.6 Shear $= 5000$ N; moment $= 26\,000$ N·m (neg).

11.8 Shear $= 3500$ lb; moment $= 16\,250$ ft·lb.

11.10 Diagram.

11.12 $R_1 = 800$ N; $R_2 = 700$ N; $M_{max} = 2800$ N·m.

11.14 $M_{max} = 13\,000$ N·m; $R_1 = R_2 = 4000$ N.

11.16 $M_{max} = 1035$ ft·lb; $R_1 = 845$ lb; $R_2 = 455$ lb.

11.18 $M_{max} = \pm 250$ ft·lb; $R_1 = -50$ lb; $R_2 = 50$ lb.

11.20 $M_{max} = 2800$ ft·lb; $R_1 = 1600$ lb; $R_2 = 1000$ lb.

11.22 $M_{max} = 6250$ ft·lb; $R_1 = 1907$ lb; $R_2 = 3293$ lb.

11.24 Diagram.

11.26 Diagram.

11.28 $M_{max} = 16\,810$ ft·lb; diagram.

11.30 $M_{max} = 137\,400$ ft·lb; $S_{max} = 12\,840$ lb.

CHAPTER 12

12.2 763 300 in.·lb.

12.4 $S = 216.0$ MPa.

12.6 Load $= 280$ lb/ft.

12.8 $S = 143.98$ MPa.

12.10 $S = 211.2$ MPa.

12.12 S8 $\times$ 18.4 or larger.

12.14 $S = 229\,840$ ft·lb; 187% increase.

12.16 $S = 9830$ psi (steel); $S = 276$ psi (wood).

12.18 $d = 20.5$ in.; $A_s = 2.44$ in.^{2}.

12.20 $d = 24.8$ in.; $A_s = 1.88$ in.^{2}.

12.22 Concrete governs, $M = 907\,460$ in.·lb.

12.24 Bending governs, $F = 4.667$ kN.

12.26 Bending governs, $F = 11\,690$ lb.

12.28 $e = 3.77$ in.

12.30 $V = 39\,850$ lb.

CHAPTER 13

13.2 $R = 22\,000$ in.

13.4 $y = 21.7$ in.

13.6 $R = 769.2$ m.

13.8 $R = 74.7$ m.

13.10 $\theta = 8.82 \times 10^{-4}$ rad; $y = 0.095$ in.

13.12 $y_{max} = \dfrac{WL^3}{15EI}.$

13.14 $y_{max} = 0.036$ in.; $\theta = 7.8 \times 10^{-4}$ rad.

13.16 $y_{max} = 0.216$ in.

13.18 $y_{max} = 0.0135$ in.

13.20 $M = \dfrac{5}{16}wL^2.$

13.22 $y_{max} = 1.05 \times 10^{-3}$ m.

13.24 $I = 26.87$, $I/C = 4.5$, use W6 $\times$ 16.

13.26 $y = 1.916 \times 10^{-3}$ m.

13.28 $\theta = 1.853 \times 10^{-3}$ rad; $y_{max} = 7.21 \times 10^{-3}$ m.

13.30 $\theta = 0.0045$ rad.

13.32 $\theta_L = 7.62 \times 10^{-4}$ rad.

13.34 $\theta_L = 0.151$ rad; $\theta_R = 5.02 \times 10^{-2}$ rad; $y = 0.0646$ m

13.36 $\theta = 5.34 \times 10^{-3}$ rad (both sides).

13.38 $y_{center} = 2.68 \times 10^{-3}$ m.

13.40 $y = 6.04 \times 10^{-4}$ m.

13.42 $y = 6.96 \times 10^{-5}$ m (up).

13.44 $I = 1.124 \times 10^{-4}$ m (up).

13.46 deflection $= 0.0815$ in.

13.48 $I = 12.96$ in.^{4}.

13.50 $R_1 = 339$ lb, $R = 447$ lb; $R_2 = 834$ lb (up).

13.52 $R_1 = 406$ lb; $R_2 = 63$ lb; $M_{center} = -935$ ft·lb.

13.54 $R_2 = 1938$ lb; $R_{center} = 281$ lb; $R_3 = 781$ lb; $M_{center} = -2190$ ft·lb.

13.56 $M_{center} = -1265.6$ N·m.

13.58 $M_{R_1} = 0$; $M_{R_2} = -4800$ N·m; $M_{R_3} = -6400$ N·m; $M_{R_4} = 0$.

13.60 $M_{R_1} = 0$; $M_{R_2} = -5600$; $M_{R_3} = -7600$; $M_{R_4} = 0$.

CHAPTER 14

14.2 $b = 2.77$ in.; $2b = 5.54$ in.

14.4 $b = 4.62$ in.; $2b = 9.23$ in.

14.6 Yes.

14.8 $d = 4.68$ in.

14.10 $d = 9.36$ in.

14.12 66 808 lb.

14.14 1346 lb.

14.16 98 074 lb.

14.18 $F_{CR} = 70\,101$ lb.

14.20 $F_{all} = 23\,876$ lb.

14.22 $F = 230\,000$ lb.

14.24 M6 $\times$ 20.

14.26 S5 $\times$ 10.

14.28 Satisfactory.

14.30 Not satisfactory.

CHAPTER 15

15.2 $S_{max} = 232$ MPa; $S_{min} = 152$ MPa.

15.4 $S_{max} = 23\,031$ psi.

15.6 13 994 psi (C); 4756 psi (T).

15.8 $S_{max} = 40\,510$ psi.

15.10 $S_{max} = 425$ psi (T).

15.12

Angle°	S_s psi	S_N psi
0	0	8880
30	3840	6660
45	4440	4440
60	3840	2220
90	0	0

15.14

Angle°	S_s MPa	S_N MPa
0	0	60.94
30	26.39	45.71
45	30.47	30.47
60	26.39	15.24
90	0	0

15.16 $F = 250$ kN.

15.18 $S_N = 8750$ psi; $S_s = 2170$ psi.

15.20 $S_N = 8750$ psi; $S_s = 2170$ psi.

15.22 $S_N = 6250$ psi; $S_s = -2165$ psi.

15.24 $S_N = 6250$ psi; $S_s = -2165$ psi.

15.26 $S_N = 62.5$ MPa; $S_s = 65.0$ MPa.

15.28 $S_N = 62.5$ MPa; $S_s = 65.0$ MPa.

15.30 $S_N = 236.6$ MPa; $S_s = 110.3$ MPa.

15.32 $S_N = 73.3$ MPa; $S_s = 6.25$ MPa.

15.34 $S_{N\,max} = 50$ MPa; $S_{s\,max} = 12.5$ MPa.

15.36 $2\varphi_N = 68.2°$; $2\varphi_S = -21.8°$; $S_{N\,max} = 13\,385$ psi; $S_{s\,max} = 5385$ psi.

15.38 $S_N = -7430$ psi (minimum); $S_N = 11\,430$ psi (maximum); $S_s = \pm 9430$ psi; $2\varphi_N = 31.9°$; $2\varphi_s = -58°$.

15.40 $S_N = 13\,390$ psi (maximum); $S_N = 2615$ psi (minimum); $S_s = \pm 5390$ psi; $2\varphi_N = -68.4°$; $2\varphi_s = 21.8°$.

15.42 $S_s = 4560$ psi; $S_b = 8640$ psi.

15.44 $S_{s\,max} = 67\,950$ psi; $S_{N\,max} = 131\,950$ psi.

15.46 $S_{s\,max} = 39\,400$ psi; $S_{N\,max} = 77\,900$ psi.

Average Mechanical Properties of Selected Materials

TABLE A-1   Average Mechanical Properties of Selected Materials

Material	U.S. Customary Units			SI Units			Coefficient of Thermal Expansion 10^{-6} m/m·°C	μ
	Yield Stress (psi)	Modulus of Elasticity (psi)		Yield Stress (MPa)	Modulus of Elasticity (GPa)			
		Tension, E	Shear, G		Tension, E	Shear, G		
Steel (SAE 1020)	36 000	30 000 000	12 000 000	250	210	85	11.7	0.27
Steel (SAE 1090)	65 000	30 000 000	12 000 000	450	210	85	11.7	0.27
Cast iron	80 000	15 000 000	6 000 000	560	105	40	12.1	0.20
Aluminum, 6061 alloy	35 000	10 000 000	4 000 000	240	70	30	23.6	0.33
Brass	15 000	15 000 000	6 000 000	105	105	40	18.9	0.35
Bronze	20 000	15 000 000	6 500 000	140	105	45		
Copper, hard-drawn	35 000	17 000 000	6 000 000	240	115	40	32.4	
Douglas fir timber (air dry, parallel to grain)		1 700 000			12		5.4	
Lead	2 000	2 000 000	700 000	14	14	4.9	29.5	0.43
Stainless steel	80 000	28 000 000	9 500 000	560	193	66.5	17.3	0.30
Titanium alloy	120 000	15 900 000	6 200 000	840	110	43.4	8.3	0.34

Properties of Structural Shapes*

*Reprinted with permission from AISC, *Manual of Steel Construction*, 8th ed. New York: American Institute of Steel Construction, 1980.

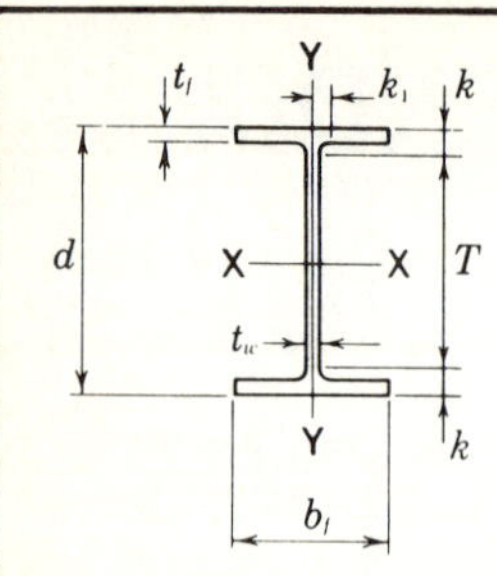

W SHAPES
Dimensions

Properties

Designation	Area A	Depth d		Web Thickness t_w		$\dfrac{t_w}{2}$	Flange Width b_f		Thickness t_f		Elastic Properties Axis X-X I	Z	k	Axis Y-Y I	Z	k
	In.²	In.		In.		In.	In.		In.		In.⁴	In.³	In.	In.⁴	In.³	In.
W 36x300	88.3	36.74	36³⁄₄	0.945	¹⁵⁄₁₆	¹⁄₂	16.655	16⁵⁄₈	1.680	1¹¹⁄₁₆	20300	1110	15.2	1300	156	3.83
x280	82.4	36.52	36¹⁄₂	0.885	⁷⁄₈	⁷⁄₁₆	16.595	16⁵⁄₈	1.570	1⁹⁄₁₆	18900	1030	15.1	1200	144	3.81
x260	76.5	36.26	36¹⁄₄	0.840	¹³⁄₁₆	⁷⁄₁₆	16.550	16¹⁄₂	1.440	1⁷⁄₁₆	17300	953	15.0	1090	132	3.78
x245	72.1	36.08	36¹⁄₈	0.800	¹³⁄₁₆	⁷⁄₁₆	16.510	16¹⁄₂	1.350	1³⁄₈	16100	895	15.0	1010	123	3.75
x230	67.6	35.90	35⁷⁄₈	0.760	³⁄₄	³⁄₈	16.470	16¹⁄₂	1.260	1¹⁄₄	15000	837	14.9	940	114	3.73
W 36x210	61.8	36.69	36³⁄₄	0.830	¹³⁄₁₆	⁷⁄₁₆	12.180	12¹⁄₈	1.360	1³⁄₈	13200	719	14.6	411	67.5	2.58
x194	57.0	36.49	36¹⁄₂	0.765	³⁄₄	³⁄₈	12.115	12¹⁄₈	1.260	1¹⁄₄	12100	664	14.6	375	61.9	2.56
x182	53.6	36.33	36³⁄₈	0.725	³⁄₄	³⁄₈	12.075	12¹⁄₈	1.180	1³⁄₁₆	11300	623	14.5	347	57.6	2.55
x170	50.0	36.17	36¹⁄₈	0.680	¹¹⁄₁₆	³⁄₈	12.030	12	1.100	1¹⁄₈	10500	580	14.5	320	53.2	2.53
x160	47.0	36.01	36	0.650	⁵⁄₈	⁵⁄₁₆	12.000	12	1.020	1	9750	542	14.4	295	49.1	2.50
x150	44.2	35.85	35⁷⁄₈	0.625	⁵⁄₈	⁵⁄₁₆	11.975	12	0.940	¹⁵⁄₁₆	9040	504	14.3	270	45.1	2.47
x135	39.7	35.55	35¹⁄₂	0.600	⁵⁄₈	⁵⁄₁₆	11.950	12	0.790	¹³⁄₁₆	7800	439	14.0	225	37.7	2.38
W 33x241	70.9	34.18	34¹⁄₈	0.830	¹³⁄₁₆	⁷⁄₁₆	15.860	15⁷⁄₈	1.400	1³⁄₈	14200	829	14.1	932	118	3.63
x221	65.0	33.93	33⁷⁄₈	0.775	³⁄₄	³⁄₈	15.805	15³⁄₄	1.275	1¹⁄₄	12800	757	14.1	840	106	3.59
x201	59.1	33.68	33⁵⁄₈	0.715	¹¹⁄₁₆	³⁄₈	15.745	15³⁄₄	1.150	1¹⁄₈	11500	684	14.0	749	95.2	3.56
W 33x152	44.7	33.49	33¹⁄₂	0.635	⁵⁄₈	⁵⁄₁₆	11.565	11⁵⁄₈	1.055	1¹⁄₁₆	8160	487	13.5	273	47.2	2.47
x141	41.6	33.30	33¹⁄₄	0.605	⁵⁄₈	⁵⁄₁₆	11.535	11¹⁄₂	0.960	¹⁵⁄₁₆	7450	448	13.4	246	42.7	2.43
x130	38.3	33.09	33¹⁄₈	0.580	⁹⁄₁₆	⁵⁄₁₆	11.510	11¹⁄₂	0.855	⁷⁄₈	6710	406	13.2	218	37.9	2.39
x118	34.7	32.86	32⁷⁄₈	0.550	⁹⁄₁₆	⁵⁄₁₆	11.480	11¹⁄₂	0.740	³⁄₄	5900	359	13.0	187	32.6	2.32
W 30x211	62.0	30.94	31	0.775	³⁄₄	³⁄₈	15.105	15¹⁄₈	1.315	1⁵⁄₁₆	10300	663	12.9	757	100	3.49
x191	56.1	30.68	30⁵⁄₈	0.710	¹¹⁄₁₆	³⁄₈	15.040	15	1.185	1³⁄₁₆	9170	598	12.8	673	89.5	3.46
x173	50.8	30.44	30¹⁄₂	0.655	⁵⁄₈	⁵⁄₁₆	14.985	15	1.065	1¹⁄₁₆	8200	539	12.7	598	79.8	3.43
W 30x132	38.9	30.31	30¹⁄₄	0.615	⁵⁄₈	⁵⁄₁₆	10.545	10¹⁄₂	1.000	1	5770	380	12.2	196	37.2	2.25
x124	36.5	30.17	30¹⁄₈	0.585	⁹⁄₁₆	⁵⁄₁₆	10.515	10¹⁄₂	0.930	¹⁵⁄₁₆	5360	355	12.1	181	34.4	2.23
x116	34.2	30.01	30	0.565	⁹⁄₁₆	⁵⁄₁₆	10.495	10¹⁄₂	0.850	⁷⁄₈	4930	329	12.0	164	31.3	2.19
x108	31.7	29.83	29⁷⁄₈	0.545	⁹⁄₁₆	⁵⁄₁₆	10.475	10¹⁄₂	0.760	³⁄₄	4470	299	11.9	146	27.9	2.15
x 99	29.1	29.65	29⁵⁄₈	0.520	¹⁄₂	¹⁄₄	10.450	10¹⁄₂	0.670	¹¹⁄₁₆	3990	269	11.7	128	24.5	2.10

Properties of Structural Shapes

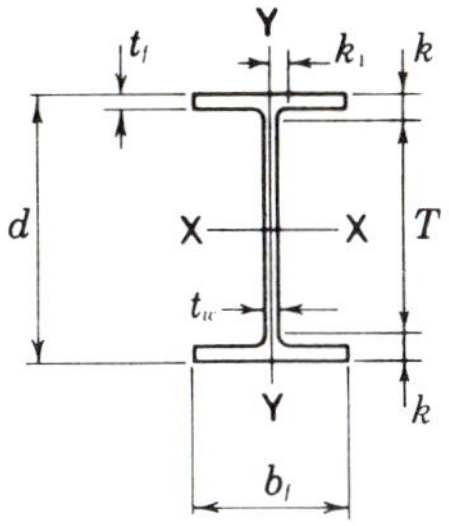

W SHAPES

Dimensions

Properties

Designation	Area A	Depth d		Web: Thickness t_w		Web: $\frac{t_w}{2}$	Flange: Width b_f		Flange: Thickness t_f		Elastic Properties: Axis X-X: I	Axis X-X: Z	Axis X-X: k	Axis Y-Y: I	Axis Y-Y: Z	Axis Y-Y: k
	In.^{2}	In.		In.		In.	In.		In.		In.^{4}	In.^{3}	In.	In.^{4}	In.^{3}	In.
W 27x178	52.3	27.81	27 3/4	0.725	3/4	3/8	14.085	14 1/8	1.190	1 3/16	6990	502	11.6	555	78.8	3.26
x161	47.4	27.59	27 5/8	0.660	11/16	3/8	14.020	14	1.080	1 1/16	6280	455	11.5	497	70.9	3.24
x146	42.9	27.38	27 3/8	0.605	5/8	5/16	13.965	14	0.975	1	5630	411	11.4	443	63.5	3.21
W 27x114	33.5	27.29	27 1/4	0.570	9/16	5/16	10.070	10 1/8	0.930	15/16	4090	299	11.0	159	31.5	2.18
x102	30.0	27.09	27 1/8	0.515	1/2	1/4	10.015	10	0.830	13/16	3620	267	11.0	139	27.8	2.15
x 94	27.7	26.92	26 7/8	0.490	1/2	1/4	9.990	10	0.745	3/4	3270	243	10.9	124	24.8	2.12
x 84	24.8	26.71	26 3/4	0.460	7/16	1/4	9.960	10	0.640	5/8	2850	213	10.7	106	21.2	2.07
W 24x162	47.7	25.00	25	0.705	11/16	3/8	12.955	13	1.220	1 1/4	5170	414	10.4	443	68.4	3.05
x146	43.0	24.74	24 3/4	0.650	5/8	5/16	12.900	12 7/8	1.090	1 1/16	4580	371	10.3	391	60.5	3.01
x131	38.5	24.48	24 1/2	0.605	5/8	5/16	12.855	12 7/8	0.960	15/16	4020	329	10.2	340	53.0	2.97
x117	34.4	24.26	24 1/4	0.550	9/16	5/16	12.800	12 3/4	0.850	7/8	3540	291	10.1	297	46.5	2.94
x104	30.6	24.06	24	0.500	1/2	1/4	12.750	12 3/4	0.750	3/4	3100	258	10.1	259	40.7	2.91
W 24x 94	27.7	24.31	24 1/4	0.515	1/2	1/4	9.065	9 1/8	0.875	7/8	2700	222	9.87	109	24.0	1.98
x 84	24.7	24.10	24 1/8	0.470	1/2	1/4	9.020	9	0.770	3/4	2370	196	9.79	94.4	20.9	1.95
x 76	22.4	23.92	23 7/8	0.440	7/16	1/4	8.990	9	0.680	11/16	2100	176	9.69	82.5	18.4	1.92
x 68	20.1	23.73	23 3/4	0.415	7/16	1/4	8.965	9	0.585	9/16	1830	154	9.55	70.4	15.7	1.87
W 24x 62	18.2	23.74	23 3/4	0.430	7/16	1/4	7.040	7	0.590	9/16	1550	131	9.23	34.5	9.80	1.38
x 55	16.2	23.57	23 5/8	0.395	3/8	3/16	7.005	7	0.505	1/2	1350	114	9.11	29.1	8.30	1.34
W 21x147	43.2	22.06	22	0.720	3/4	3/8	12.510	12 1/2	1.150	1 1/8	3630	329	9.17	376	60.1	2.95
x132	38.8	21.83	21 7/8	0.650	5/8	5/16	12.440	12 1/2	1.035	1 1/16	3220	295	9.12	333	53.5	2.93
x122	35.9	21.68	21 5/8	0.600	5/8	5/16	12.390	12 3/8	0.960	15/16	2960	273	9.09	305	49.2	2.92
x111	32.7	21.51	21 1/2	0.550	9/16	5/16	12.340	12 3/8	0.875	7/8	2670	249	9.05	274	44.5	2.90
x101	29.8	21.36	21 3/8	0.500	1/2	1/4	12.290	12 1/4	0.800	13/16	2420	227	9.02	248	40.3	2.89
W 21x 93	27.3	21.62	21 5/8	0.580	9/16	5/16	8.420	8 3/8	0.930	15/16	2070	192	8.70	92.9	22.1	1.84
x 83	24.3	21.43	21 3/8	0.515	1/2	1/4	8.355	8 3/8	0.835	13/16	1830	171	8.67	81.4	19.5	1.83
x 73	21.5	21.24	21 1/4	0.455	7/16	1/4	8.295	8 1/4	0.740	3/4	1600	151	8.64	70.6	17.0	1.81
x 68	20.0	21.13	21 1/8	0.430	7/16	1/4	8.270	8 1/4	0.685	11/16	1480	140	8.60	64.7	15.7	1.80
x 62	18.3	20.99	21	0.400	3/8	3/16	8.240	8 1/4	0.615	5/8	1330	127	8.54	57.5	13.9	1.77
W 21x 57	16.7	21.06	21	0.405	3/8	3/16	6.555	6 1/2	0.650	5/8	1170	111	8.36	30.6	9.35	1.35
x 50	14.7	20.83	20 7/8	0.380	3/8	3/16	6.530	6 1/2	0.535	9/16	984	94.5	8.18	24.9	7.64	1.30
x 44	13.0	20.66	20 5/8	0.350	3/8	3/16	6.500	6 1/2	0.450	7/16	843	81.6	8.06	20.7	6.36	1.26

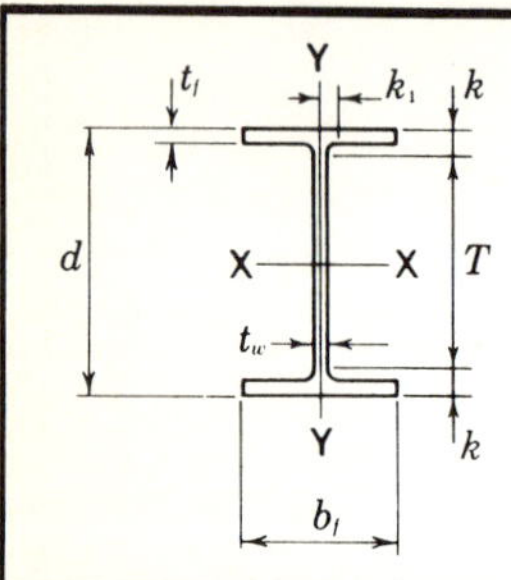

W SHAPES

Dimensions

Properties

Designation	Area A	Depth d		Web Thickness t_w		$\dfrac{t_w}{2}$	Flange Width b_f		Flange Thickness t_f		Axis X-X I	Z	k	Axis Y-Y I	Z	k
	In.²	In.		In.		In.	In.		In.		In.⁴	In.³	In.	In.⁴	In.³	In.
W 18x119	35.1	18.97	19	0.655	⅝	⁵⁄₁₆	11.265	11¼	1.060	1¹⁄₁₆	2190	231	7.90	253	44.9	2.69
x106	31.1	18.73	18¾	0.590	⁹⁄₁₆	⁵⁄₁₆	11.200	11¼	0.940	¹⁵⁄₁₆	1910	204	7.84	220	39.4	2.66
x 97	28.5	18.59	18⅝	0.535	⁹⁄₁₆	⁵⁄₁₆	11.145	11⅛	0.870	⅞	1750	188	7.82	201	36.1	2.65
x 86	25.3	18.39	18⅜	0.480	½	¼	11.090	11⅛	0.770	¾	1530	166	7.77	175	31.6	2.63
x 76	22.3	18.21	18¼	0.425	⁷⁄₁₆	¼	11.035	11	0.680	¹¹⁄₁₆	1330	146	7.73	152	27.6	2.61
W 18x 71	20.8	18.47	18½	0.495	½	¼	7.635	7⅝	0.810	¹³⁄₁₆	1170	127	7.50	60.3	15.8	1.70
x 65	19.1	18.35	18⅜	0.450	⁷⁄₁₆	¼	7.590	7⅝	0.750	¾	1070	117	7.49	54.8	14.4	1.69
x 60	17.6	18.24	18¼	0.415	⁷⁄₁₆	¼	7.555	7½	0.695	¹¹⁄₁₆	984	108	7.47	50.1	13.3	1.69
x 55	16.2	18.11	18⅛	0.390	⅜	³⁄₁₆	7.530	7½	0.630	⅝	890	98.3	7.41	44.9	11.9	1.67
x 50	14.7	17.99	18	0.355	⅜	³⁄₁₆	7.495	7½	0.570	⁹⁄₁₆	800	88.9	7.38	40.1	10.7	1.65
W 18x 46	13.5	18.06	18	0.360	⅜	³⁄₁₆	6.060	6	0.605	⅝	712	78.8	7.25	22.5	7.43	1.29
x 40	11.8	17.90	17⅞	0.315	⁵⁄₁₆	³⁄₁₆	6.015	6	0.525	½	612	68.4	7.21	19.1	6.35	1.27
x 35	10.3	17.70	17¾	0.300	⁵⁄₁₆	³⁄₁₆	6.000	6	0.425	⁷⁄₁₆	510	57.6	7.04	15.3	5.12	1.22
W 16x100	29.4	16.97	17	0.585	⁹⁄₁₆	⁵⁄₁₆	10.425	10⅜	0.985	1	1490	175	7.10	186	35.7	2.51
x 89	26.2	16.75	16¾	0.525	½	¼	10.365	10⅜	0.875	⅞	1300	155	7.05	163	31.4	2.49
x 77	22.6	16.52	16½	0.455	⁷⁄₁₆	¼	10.295	10¼	0.760	¾	1110	134	7.00	138	26.9	2.47
x 67	19.7	16.33	16⅜	0.395	⅜	³⁄₁₆	10.235	10¼	0.665	¹¹⁄₁₆	954	117	6.96	119	23.2	2.46
W 16x 57	16.8	16.43	16⅜	0.430	⁷⁄₁₆	¼	7.120	7⅛	0.715	¹¹⁄₁₆	758	92.2	6.72	43.1	12.1	1.60
x 50	14.7	16.26	16¼	0.380	⅜	³⁄₁₆	7.070	7⅛	0.630	⅝	659	81.0	6.68	37.2	10.5	1.59
x 45	13.3	16.13	16⅛	0.345	⅜	³⁄₁₆	7.035	7	0.565	⁹⁄₁₆	586	72.7	6.65	32.8	9.34	1.57
x 40	11.8	16.01	16	0.305	⁵⁄₁₆	³⁄₁₆	6.995	7	0.505	½	518	64.7	6.63	28.9	8.25	1.57
x 36	10.6	15.86	15⅞	0.295	⁵⁄₁₆	³⁄₁₆	6.985	7	0.430	⁷⁄₁₆	448	56.5	6.51	24.5	7.00	1.52
W 16x 31	9.12	15.88	15⅞	0.275	¼	⅛	5.525	5½	0.440	⁷⁄₁₆	375	47.2	6.41	12.4	4.49	1.17
x 26	7.68	15.69	15¾	0.250	¼	⅛	5.500	5½	0.345	⅜	301	38.4	6.26	9.59	3.49	1.12

Properties of Structural Shapes

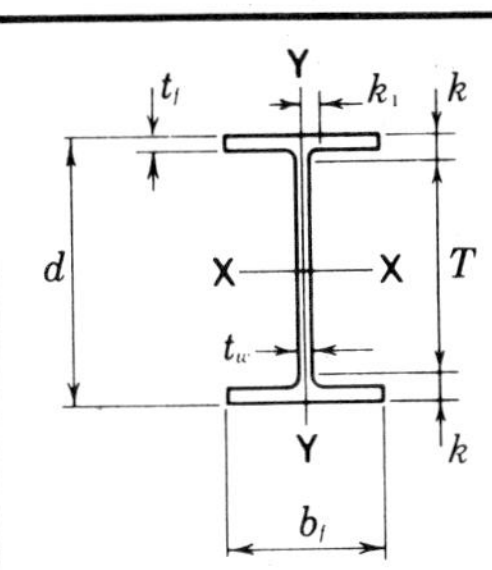

W SHAPES
Dimensions Properties

Designation	Area A	Depth d		Web Thickness t_w		$\dfrac{t_w}{2}$	Flange Width b_f		Thickness t_f		Elastic Properties Axis X-X			Axis Y-Y		
	$I_n.^2$	$I_n.$		$I_n.$		$I_n.$	$I_n.$		$I_n.$		$I_n.^4$	$I_n.^3$	$I_n.$	$I_n.^4$	$I_n.^3$	$I_n.$
											I	Z	k	I	Z	k
W 14x730	215.0	22.42	$22\frac{3}{8}$	3.070	$3\frac{1}{16}$	$1\frac{9}{16}$	17.890	$17\frac{7}{8}$	4.910	$4\frac{15}{16}$	14300	1280	8.17	4720	527	4.69
x665	196.0	21.64	$21\frac{5}{8}$	2.830	$2\frac{13}{16}$	$1\frac{7}{16}$	17.650	$17\frac{5}{8}$	4.520	$4\frac{1}{2}$	12400	1150	7.98	4170	472	4.62
x605	178.0	20.92	$20\frac{7}{8}$	2.595	$2\frac{5}{8}$	$1\frac{5}{16}$	17.415	$17\frac{3}{8}$	4.160	$4\frac{3}{16}$	10800	1040	7.80	3680	423	4.55
x550	162.0	20.24	$20\frac{1}{4}$	2.380	$2\frac{3}{8}$	$1\frac{3}{16}$	17.200	$17\frac{1}{4}$	3.820	$3\frac{13}{16}$	9430	931	7.63	3250	378	4.49
x500	147.0	19.60	$19\frac{5}{8}$	2.190	$2\frac{3}{16}$	$1\frac{1}{8}$	17.010	17	3.500	$3\frac{1}{2}$	8210	838	7.48	2880	339	4.43
x455	134.0	19.02	19	2.015	2	1	16.835	$16\frac{7}{8}$	3.210	$3\frac{3}{16}$	7190	756	7.33	2560	304	4.38
W 14x426	125.0	18.67	$18\frac{5}{8}$	1.875	$1\frac{7}{8}$	$\frac{15}{16}$	16.695	$16\frac{3}{4}$	3.035	$3\frac{1}{16}$	6600	707	7.26	2360	283	4.34
x398	117.0	18.29	$18\frac{1}{4}$	1.770	$1\frac{3}{4}$	$\frac{7}{8}$	16.590	$16\frac{5}{8}$	2.845	$2\frac{7}{8}$	6000	656	7.16	2170	262	4.31
x370	109.0	17.92	$17\frac{7}{8}$	1.655	$1\frac{5}{8}$	$\frac{13}{16}$	16.475	$16\frac{1}{2}$	2.660	$2\frac{11}{16}$	5440	607	7.07	1990	241	4.27
x342	101.0	17.54	$17\frac{1}{2}$	1.540	$1\frac{9}{16}$	$\frac{13}{16}$	16.360	$16\frac{3}{8}$	2.470	$2\frac{1}{2}$	4900	559	6.98	1810	221	4.24
x311	91.4	17.12	$17\frac{1}{8}$	1.410	$1\frac{7}{16}$	$\frac{3}{4}$	16.230	$16\frac{1}{4}$	2.260	$2\frac{1}{4}$	4330	506	6.88	1610	199	4.20
x283	83.3	16.74	$16\frac{3}{4}$	1.290	$1\frac{5}{16}$	$\frac{11}{16}$	16.110	$16\frac{1}{8}$	2.070	$2\frac{1}{16}$	3840	459	6.75	1440	179	4.17
x257	75.6	16.38	$16\frac{3}{8}$	1.175	$1\frac{3}{16}$	$\frac{5}{8}$	15.995	16	1.890	$1\frac{7}{8}$	3400	415	6.71	1290	161	4.13
x233	68.5	16.04	16	1.070	$1\frac{1}{16}$	$\frac{9}{16}$	15.890	$15\frac{7}{8}$	1.720	$1\frac{3}{4}$	3010	375	6.63	1150	145	4.10
x211	62.0	15.72	$15\frac{3}{4}$	0.980	1	$\frac{1}{2}$	15.800	$15\frac{3}{4}$	1.560	$1\frac{9}{16}$	2660	338	6.55	1030	130	4.07
x193	56.8	15.48	$15\frac{1}{2}$	0.890	$\frac{7}{8}$	$\frac{7}{16}$	15.710	$15\frac{3}{4}$	1.440	$1\frac{7}{16}$	2400	310	6.50	931	119	4.05
x176	51.8	15.22	$15\frac{1}{4}$	0.830	$\frac{13}{16}$	$\frac{7}{16}$	15.650	$15\frac{5}{8}$	1.310	$1\frac{5}{16}$	2140	281	6.43	838	107	4.02
x159	46.7	14.98	15	0.745	$\frac{3}{4}$	$\frac{3}{8}$	15.565	$15\frac{5}{8}$	1.190	$1\frac{3}{16}$	1900	254	6.38	748	96.2	4.00
x145	42.7	14.78	$14\frac{3}{4}$	0.680	$\frac{11}{16}$	$\frac{3}{8}$	15.500	$15\frac{1}{2}$	1.090	$1\frac{1}{16}$	1710	232	6.33	677	87.3	3.98

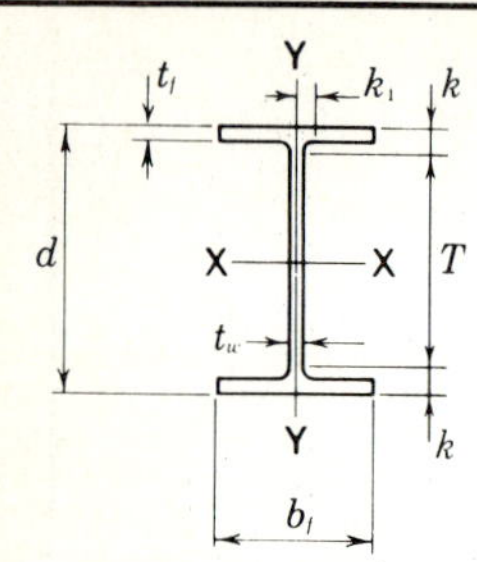

W SHAPES
Dimensions
Properties

Designation	Area A	Depth d		Web Thickness t_w		$\frac{t_w}{2}$	Flange Width b_f		Flange Thickness t_f		Axis X-X I	Axis X-X Z	Axis X-X k	Axis Y-Y I	Axis Y-Y Z	Axis Y-Y k
	In.²	In.		In.		In.	In.		In.		In.⁴	In.³	In.	In.⁴	In.³	In.
W 14x132	38.8	14.66	14⅝	0.645	⅝	5/16	14.725	14¾	1.030	1	1530	209	6.28	548	74.5	3.76
x120	35.3	14.48	14½	0.590	9/16	5/16	14.670	14⅝	0.940	15/16	1380	190	6.24	495	67.5	3.74
x109	32.0	14.32	14⅜	0.525	½	¼	14.605	14⅝	0.860	⅞	1240	173	6.22	447	61.2	3.73
x 99	29.1	14.16	14⅛	0.485	½	¼	14.565	14⅝	0.780	¾	1110	157	6.17	402	55.2	3.71
x 90	26.5	14.02	14	0.440	7/16	¼	14.520	14½	0.710	11/16	999	143	6.14	362	49.9	3.70
W 14x 82	24.1	14.31	14¼	0.510	½	¼	10.130	10⅛	0.855	⅞	882	123	6.05	148	29.3	2.48
x 74	21.8	14.17	14⅛	0.450	7/16	¼	10.070	10⅛	0.785	13/16	796	112	6.04	134	26.6	2.48
x 68	20.0	14.04	14	0.415	7/16	¼	10.035	10	0.720	¾	723	103	6.01	121	24.2	2.46
x 61	17.9	13.89	13⅞	0.375	⅜	3/16	9.995	10	0.645	⅝	640	92.2	5.98	107	21.5	2.45
W 14x 53	15.6	13.92	13⅞	0.370	⅜	3/16	8.060	8	0.660	11/16	541	77.8	5.89	57.7	14.3	1.92
x 48	14.1	13.79	13¾	0.340	5/16	3/16	8.030	8	0.595	⅝	485	70.3	5.85	51.4	12.8	1.91
x 43	12.6	13.66	13⅝	0.305	5/16	3/16	7.995	8	0.530	½	428	62.7	5.82	45.2	11.3	1.89
W 14x 38	11.2	14.10	14⅛	0.310	5/16	3/16	6.770	6¾	0.515	½	385	54.6	5.87	26.7	7.88	1.55
x 34	10.0	13.98	14	0.285	5/16	3/16	6.745	6¾	0.455	7/16	340	48.6	5.83	23.3	6.91	1.53
x 30	8.85	13.84	13⅞	0.270	¼	⅛	6.730	6¾	0.385	⅜	291	42.0	5.73	19.6	5.82	1.49
W 14x 26	7.69	13.91	13⅞	0.255	¼	⅛	5.025	5	0.420	7/16	245	35.3	5.65	8.91	3.54	1.08
x 22	6.49	13.74	13¾	0.230	¼	⅛	5.000	5	0.335	5/16	199	29.0	5.54	7.00	2.80	1.04

Properties of Structural Shapes

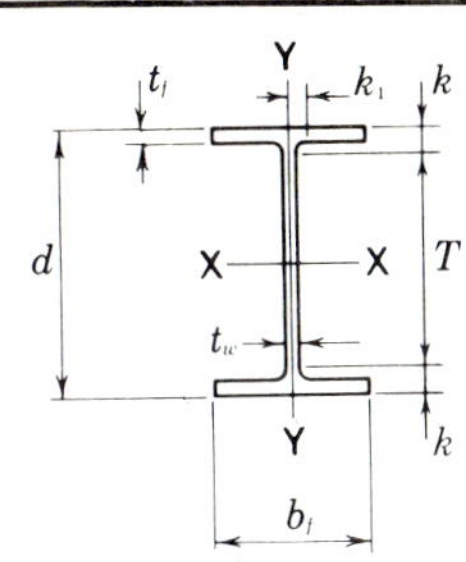

W SHAPES

Dimensions

Properties

Designation	Area A	Depth d		Web Thickness t_w		$\dfrac{t_w}{2}$	Flange Width b_f		Thickness t_f		Axis X-X I	Z	k	Axis Y-Y I	Z	k
	In.²	In.		In.		In.	In.		In.		In.⁴	In.³	In.	In.⁴	In.³	In.
W 12x336	98.8	16.82	$16\frac{7}{8}$	1.775	$1\frac{3}{4}$	$\frac{7}{8}$	13.385	$13\frac{3}{8}$	2.955	$2\frac{15}{16}$	4060	483	6.41	1190	177	3.47
x305	89.6	16.32	$16\frac{3}{8}$	1.625	$1\frac{5}{8}$	$\frac{13}{16}$	13.235	$13\frac{1}{4}$	2.705	$2\frac{11}{16}$	3550	435	6.29	1050	159	3.42
x279	81.9	15.85	$15\frac{7}{8}$	1.530	$1\frac{1}{2}$	$\frac{3}{4}$	13.140	$13\frac{1}{8}$	2.470	$2\frac{1}{2}$	3110	393	6.16	937	143	3.38
x252	74.1	15.41	$15\frac{3}{8}$	1.395	$1\frac{3}{8}$	$\frac{11}{16}$	13.005	13	2.250	$2\frac{1}{4}$	2720	353	6.06	828	127	3.34
x230	67.7	15.05	15	1.285	$1\frac{5}{16}$	$\frac{11}{16}$	12.895	$12\frac{7}{8}$	2.070	$2\frac{1}{16}$	2420	321	5.97	742	115	3.31
x210	61.8	14.71	$14\frac{3}{4}$	1.180	$1\frac{3}{16}$	$\frac{5}{8}$	12.790	$12\frac{3}{4}$	1.900	$1\frac{7}{8}$	2140	292	5.89	664	104	3.28
x190	55.8	14.38	$14\frac{3}{8}$	1.060	$1\frac{1}{16}$	$\frac{9}{16}$	12.670	$12\frac{5}{8}$	1.735	$1\frac{3}{4}$	1890	263	5.82	589	93.0	3.25
x170	50.0	14.03	14	0.960	$\frac{15}{16}$	$\frac{1}{2}$	12.570	$12\frac{5}{8}$	1.560	$1\frac{9}{16}$	1650	235	5.74	517	82.3	3.22
x152	44.7	13.71	$13\frac{3}{4}$	0.870	$\frac{7}{8}$	$\frac{7}{16}$	12.480	$12\frac{1}{2}$	1.400	$1\frac{3}{8}$	1430	209	5.66	454	72.8	3.19
x136	39.9	13.41	$13\frac{3}{8}$	0.790	$\frac{13}{16}$	$\frac{7}{16}$	12.400	$12\frac{3}{8}$	1.250	$1\frac{1}{4}$	1240	186	5.58	398	64.2	3.16
x120	35.3	13.12	$13\frac{1}{8}$	0.710	$\frac{11}{16}$	$\frac{3}{8}$	12.320	$12\frac{3}{8}$	1.105	$1\frac{1}{8}$	1070	163	5.51	345	56.0	3.13
x106	31.2	12.89	$12\frac{7}{8}$	0.610	$\frac{5}{8}$	$\frac{5}{16}$	12.220	$12\frac{1}{4}$	0.990	1	933	145	5.47	301	49.3	3.11
x 96	28.2	12.71	$12\frac{3}{4}$	0.550	$\frac{9}{16}$	$\frac{5}{16}$	12.160	$12\frac{1}{8}$	0.900	$\frac{7}{8}$	833	131	5.44	270	44.4	3.09
x 87	25.6	12.53	$12\frac{1}{2}$	0.515	$\frac{1}{2}$	$\frac{1}{4}$	12.125	$12\frac{1}{8}$	0.810	$\frac{13}{16}$	740	118	5.38	241	39.7	3.07
x 79	23.2	12.38	$12\frac{3}{8}$	0.470	$\frac{1}{2}$	$\frac{1}{4}$	12.080	$12\frac{1}{8}$	0.735	$\frac{3}{4}$	662	107	5.34	216	35.8	3.05
x 72	21.1	12.25	$12\frac{1}{4}$	0.430	$\frac{7}{16}$	$\frac{1}{4}$	12.040	12	0.670	$\frac{11}{16}$	597	97.4	5.31	195	32.4	3.04
x 65	19.1	12.12	$12\frac{1}{8}$	0.390	$\frac{3}{8}$	$\frac{3}{16}$	12.000	12	0.605	$\frac{5}{8}$	533	87.9	5.28	174	29.1	3.02
W 12x 58	17.0	12.19	$12\frac{1}{4}$	0.360	$\frac{3}{8}$	$\frac{3}{16}$	10.010	10	0.640	$\frac{5}{8}$	475	78.0	5.28	107	21.4	2.51
x 53	15.6	12.06	12	0.345	$\frac{3}{8}$	$\frac{3}{16}$	9.995	10	0.575	$\frac{9}{16}$	425	70.6	5.23	95.8	19.2	2.48
W 12x 50	14.7	12.19	$12\frac{1}{4}$	0.370	$\frac{3}{8}$	$\frac{3}{16}$	8.080	$8\frac{1}{8}$	0.640	$\frac{5}{8}$	394	64.7	5.18	56.3	13.9	1.96
x 45	13.2	12.06	12	0.335	$\frac{5}{16}$	$\frac{3}{16}$	8.045	8	0.575	$\frac{9}{16}$	350	58.1	5.15	50.0	12.4	1.94
x 40	11.8	11.94	12	0.295	$\frac{5}{16}$	$\frac{3}{16}$	8.005	8	0.515	$\frac{1}{2}$	310	51.9	5.13	44.1	11.0	1.93
W 12x 35	10.3	12.50	$12\frac{1}{2}$	0.300	$\frac{5}{16}$	$\frac{3}{16}$	6.560	$6\frac{1}{2}$	0.520	$\frac{1}{2}$	285	45.6	5.25	24.5	7.47	1.54
x 30	8.79	12.34	$12\frac{3}{8}$	0.260	$\frac{1}{4}$	$\frac{1}{8}$	6.520	$6\frac{1}{2}$	0.440	$\frac{7}{16}$	238	38.6	5.21	20.3	6.24	1.52
x 26	7.65	12.22	$12\frac{1}{4}$	0.230	$\frac{1}{4}$	$\frac{1}{8}$	6.490	$6\frac{1}{2}$	0.380	$\frac{3}{8}$	204	33.4	5.17	17.3	5.34	1.51
W 12x 22	6.48	12.31	$12\frac{1}{4}$	0.260	$\frac{1}{4}$	$\frac{1}{8}$	4.030	4	0.425	$\frac{7}{16}$	156	25.4	4.91	4.66	2.31	0.847
x 19	5.57	12.16	$12\frac{1}{8}$	0.235	$\frac{1}{4}$	$\frac{1}{8}$	4.005	4	0.350	$\frac{3}{8}$	130	21.3	4.82	3.76	1.88	0.822
x 16	4.71	11.99	12	0.220	$\frac{1}{4}$	$\frac{1}{8}$	3.990	4	0.265	$\frac{1}{4}$	103	17.1	4.67	2.82	1.41	0.773
x 14	4.16	11.91	$11\frac{7}{8}$	0.200	$\frac{3}{16}$	$\frac{1}{8}$	3.970	4	0.225	$\frac{1}{4}$	88.6	14.9	4.62	2.36	1.19	0.753

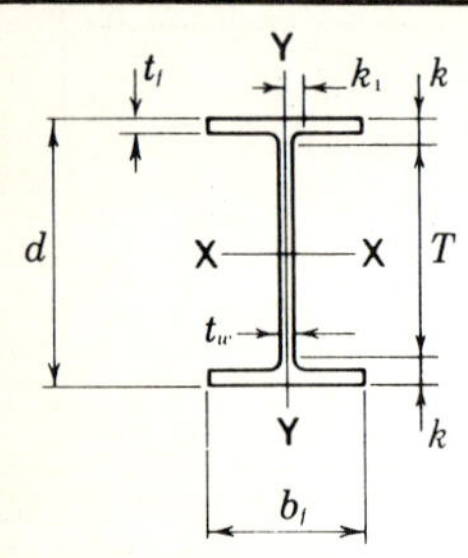

W SHAPES
Dimensions   ## Properties

Designation	Area A	Depth d		Web Thickness t_w		$\dfrac{t_w}{2}$	Flange Width b_f		Flange Thickness t_f		Axis X-X I	Z	k	Axis Y-Y I	Z	k
	In.²	In.		In.		In.	In.		In.		In.⁴	In.³	In.	In.⁴	In.³	In.
W 10x112	32.9	11.36	11³⁄₈	0.755	³⁄₄	³⁄₈	10.415	10³⁄₈	1.250	1¹⁄₄	716	126	4.66	236	45.3	2.68
x100	29.4	11.10	11¹⁄₈	0.680	¹¹⁄₁₆	³⁄₈	10.340	10³⁄₈	1.120	1¹⁄₈	623	112	4.60	207	40.0	2.65
x 88	25.9	10.84	10⁷⁄₈	0.605	⁵⁄₈	⁵⁄₁₆	10.265	10¹⁄₄	0.990	1	534	98.5	4.54	179	34.8	2.63
x 77	22.6	10.60	10⁵⁄₈	0.530	¹⁄₂	¹⁄₄	10.190	10¹⁄₄	0.870	⁷⁄₈	455	85.9	4.49	154	30.1	2.60
x 68	20.0	10.40	10³⁄₈	0.470	¹⁄₂	¹⁄₄	10.130	10¹⁄₈	0.770	³⁄₄	394	75.7	4.44	134	26.4	2.59
x 60	17.6	10.22	10¹⁄₄	0.420	⁷⁄₁₆	¹⁄₄	10.080	10¹⁄₈	0.680	¹¹⁄₁₆	341	66.7	4.39	116	23.0	2.57
x 54	15.8	10.09	10¹⁄₈	0.370	³⁄₈	³⁄₁₆	10.030	10	0.615	⁵⁄₈	303	60.0	4.37	103	20.6	2.56
x 49	14.4	9.98	10	0.340	⁵⁄₁₆	³⁄₁₆	10.000	10	0.560	⁹⁄₁₆	272	54.6	4.35	93.4	18.7	2.54
W 10x 45	13.3	10.10	10¹⁄₈	0.350	³⁄₈	³⁄₁₆	8.020	8	0.620	⁵⁄₈	248	49.1	4.32	53.4	13.3	2.01
x 39	11.5	9.92	9⁷⁄₈	0.315	⁵⁄₁₆	³⁄₁₆	7.985	8	0.530	¹⁄₂	209	42.1	4.27	45.0	11.3	1.98
x 33	9.71	9.73	9³⁄₄	0.290	⁵⁄₁₆	³⁄₁₆	7.960	8	0.435	⁷⁄₁₆	170	35.0	4.19	36.6	9.20	1.94
W 10x 30	8.84	10.47	10¹⁄₂	0.300	⁵⁄₁₆	³⁄₁₆	5.810	5³⁄₄	0.510	¹⁄₂	170	32.4	4.38	16.7	5.75	1.37
x 26	7.61	10.33	10³⁄₈	0.260	¹⁄₄	¹⁄₈	5.770	5³⁄₄	0.440	⁷⁄₁₆	144	27.9	4.35	14.1	4.89	1.36
x 22	6.49	10.17	10¹⁄₈	0.240	¹⁄₄	¹⁄₈	5.750	5³⁄₄	0.360	³⁄₈	118	23.2	4.27	11.4	3.97	1.33
W 10x 19	5.62	10.24	10¹⁄₄	0.250	¹⁄₄	¹⁄₈	4.020	4	0.395	³⁄₈	96.3	18.8	4.14	4.29	2.14	0.874
x 17	4.99	10.11	10¹⁄₈	0.240	¹⁄₄	¹⁄₈	4.010	4	0.330	⁵⁄₁₆	81.9	16.2	4.05	3.56	1.78	0.844
x 15	4.41	9.99	10	0.230	¹⁄₄	¹⁄₈	4.000	4	0.270	¹⁄₄	68.9	13.8	3.95	2.89	1.45	0.810
x 12	3.54	9.87	9⁷⁄₈	0.190	³⁄₁₆	¹⁄₈	3.960	4	0.210	³⁄₁₆	53.8	10.9	3.90	2.18	1.10	0.785

                                    Properties of Structural Shapes

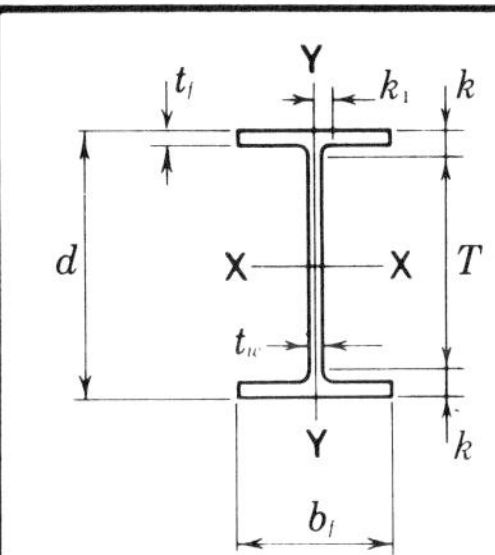

W SHAPES

Dimensions

Properties

Designation	Area A	Depth d		Web Thickness t_w		$\dfrac{t_w}{2}$	Flange Width b_f		Thickness t_f		Axis X-X I	Z	k	Axis Y-Y I	Z	k
	In.²	In.		In.		In.	In.		In.		In.⁴	In.³	In.	In.⁴	In.³	In.
W 8x67	19.7	9.00	9	0.570	9/16	5/16	8.280	8¼	0.935	15/16	272	60.4	3.72	88.6	21.4	2.12
x58	17.1	8.75	8¾	0.510	½	¼	8.220	8¼	0.810	13/16	228	52.0	3.65	75.1	18.3	2.10
x48	14.1	8.50	8½	0.400	⅜	3/16	8.110	8⅛	0.685	11/16	184	43.3	3.61	60.9	15.0	2.08
x40	11.7	8.25	8¼	0.360	⅜	3/16	8.070	8⅛	0.560	9/16	146	35.5	3.53	49.1	12.2	2.04
x35	10.3	8.12	8⅛	0.310	5/16	3/16	8.020	8	0.495	½	127	31.2	3.51	42.6	10.6	2.03
x31	9.13	8.00	8	0.285	5/16	3/16	7.995	8	0.435	7/16	110	27.5	3.47	37.1	9.27	2.02
W 8x28	8.25	8.06	8	0.285	5/16	3/16	6.535	6½	0.465	7/16	98.0	24.3	3.45	21.7	6.63	1.62
x24	7.08	7.93	7⅞	0.245	¼	⅛	6.495	6½	0.400	⅜	82.8	20.9	3.42	18.3	5.63	1.61
W 8x21	6.16	8.28	8¼	0.250	¼	⅛	5.270	5¼	0.400	⅜	75.3	18.2	3.49	9.77	3.71	1.26
x18	5.26	8.14	8⅛	0.230	¼	⅛	5.250	5¼	0.330	5/16	61.9	15.2	3.43	7.97	3.04	1.23
W 8x15	4.44	8.11	8⅛	0.245	¼	⅛	4.015	4	0.315	5/16	48.0	11.8	3.29	3.41	1.70	0.876
x13	3.84	7.99	8	0.230	¼	⅛	4.000	4	0.255	¼	39.6	9.91	3.21	2.73	1.37	0.843
x10	2.96	7.89	7⅞	0.170	3/16	⅛	3.940	4	0.205	3/16	30.8	7.81	3.22	2.09	1.06	0.841
W 6x25	7.34	6.38	6⅜	0.320	5/16	3/16	6.080	6⅛	0.455	7/16	53.4	16.7	2.70	17.1	5.61	1.52
x20	5.87	6.20	6¼	0.260	¼	⅛	6.020	6	0.365	⅜	41.4	13.4	2.66	13.3	4.41	1.50
x15	4.43	5.99	6	0.230	¼	⅛	5.990	6	0.260	¼	29.1	9.72	2.56	9.32	3.11	1.46
W 6x16	4.74	6.28	6¼	0.260	¼	⅛	4.030	4	0.405	⅜	32.1	10.2	2.60	4.43	2.20	0.966
x12	3.55	6.03	6	0.230	¼	⅛	4.000	4	0.280	¼	22.1	7.31	2.49	2.99	1.50	0.918
x 9	2.68	5.90	5⅞	0.170	3/16	⅛	3.940	4	0.215	3/16	16.4	5.56	2.47	2.19	1.11	0.905
W 5x19	5.54	5.15	5⅛	0.270	¼	⅛	5.030	5	0.430	7/16	26.2	10.2	2.17	9.13	3.63	1.28
x16	4.68	5.01	5	0.240	¼	⅛	5.000	5	0.360	⅜	21.3	8.51	2.13	7.51	3.00	1.27
W 4x13	3.83	4.16	4⅛	0.280	¼	⅛	4.060	4	0.345	⅜	11.3	5.46	1.72	3.86	1.90	1.00

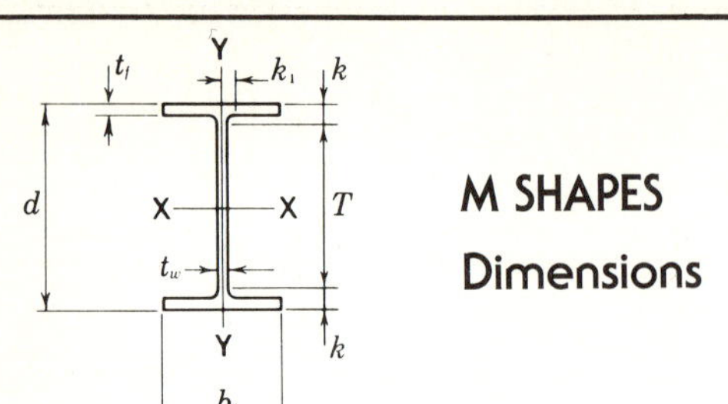

M SHAPES
Dimensions   Properties

Designation	Area A	Depth d	Web Thickness t_w		$\frac{t_w}{2}$	Flange Width b_f		Thickness t_f		Axis X-X I	Z	k	Axis Y-Y I	Z	k
	In.²	In.	In.		In.	In.		In.		In.⁴	In.³	In.	In.⁴	In.³	In.
M 14x18	5.10	14.00	14	0.215	$\frac{3}{16}$	$\frac{1}{8}$	4.000	4	0.270	$\frac{1}{4}$	148	21.1	5.38	2.64	1.32	0.719
M 12x11.8	3.47	12.00	12	0.177	$\frac{3}{16}$	$\frac{1}{8}$	3.065	$3\frac{1}{8}$	0.225	$\frac{1}{4}$	71.9	12.0	4.55	0.980	0.639	0.532
M 10x9	2.65	10.00	10	0.157	$\frac{3}{16}$	$\frac{1}{8}$	2.690	$2\frac{3}{4}$	0.206	$\frac{3}{16}$	38.8	7.76	3.83	0.609	0.453	0.480
M 8x6.5	1.92	8.00	8	0.135	$\frac{1}{8}$	$\frac{1}{16}$	2.281	$2\frac{1}{4}$	0.189	$\frac{3}{16}$	18.5	4.62	3.10	0.343	0.301	0.423
M 6x20	5.89	6.00	6	0.250	$\frac{1}{4}$	$\frac{1}{8}$	5.938	6	0.379	$\frac{3}{8}$	39.0	13.0	2.57	11.6	3.90	1.40
M 6x4.4	1.29	6.00	6	0.114	$\frac{1}{8}$	$\frac{1}{16}$	1.844	$1\frac{7}{8}$	0.171	$\frac{3}{16}$	7.20	2.40	2.36	0.165	0.179	0.358
M 5x18.9	5.55	5.00	5	0.316	$\frac{5}{16}$	$\frac{3}{16}$	5.003	5	0.416	$\frac{7}{16}$	24.1	9.63	2.08	7.86	3.14	1.19
M 4x13	3.81	4.00	4	0.254	$\frac{1}{4}$	$\frac{1}{8}$	3.940	4	0.371	$\frac{3}{8}$	10.5	5.24	1.66	3.36	1.71	0.939

          Properties of Structural Shapes

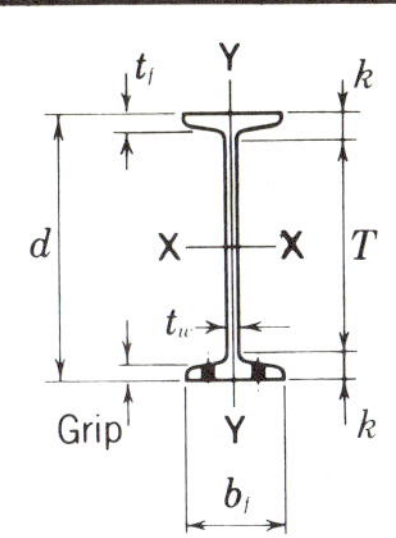

S SHAPES

Dimensions   Properties

Designation	Area A	Depth d		Web Thickness t_w		$\dfrac{t_w}{2}$	Flange Width b_f		Flange Thickness t_f		Axis X-X I	Z	k	Axis Y-Y I	Z	k
	In.²	In.		In.		In.	In.		In.		In.⁴	In.³	In.	In.⁴	In.³	In.
S 24x121	35.6	24.50	24½	0.800	13/16	7/16	8.050	8	1.090	1 1/16	3160	258	9.43	83.3	20.7	1.53
x106	31.2	24.50	24½	0.620	5/8	5/16	7.870	7 7/8	1.090	1 1/16	2940	240	9.71	77.1	19.6	1.57
S 24x100	29.3	24.00	24	0.745	3/4	3/8	7.245	7 1/4	0.870	7/8	2390	199	9.02	47.7	13.2	1.27
x90	26.5	24.00	24	0.625	5/8	5/16	7.125	7 1/8	0.870	7/8	2250	187	9.21	44.9	12.6	1.30
x80	23.5	24.00	24	0.500	1/2	1/4	7.000	7	0.870	7/8	2100	175	9.47	42.2	12.1	1.34
S 20x96	28.2	20.30	20¼	0.800	13/16	7/16	7.200	7 1/4	0.920	15/16	1670	165	7.71	50.2	13.9	1.33
x86	25.3	20.30	20¼	0.660	11/16	3/8	7.060	7	0.920	15/16	1580	155	7.89	46.8	13.3	1.36
S 20x75	22.0	20.00	20	0.635	5/8	5/16	6.385	6 3/8	0.795	13/16	1280	128	7.62	29.8	9.32	1.16
x66	19.4	20.00	20	0.505	1/2	1/4	6.255	6 1/4	0.795	13/16	1190	119	7.83	27.7	8.85	1.19
S 18x70	20.6	18.00	18	0.711	11/16	3/8	6.251	6 1/4	0.691	11/16	926	103	6.71	24.1	7.72	1.08
x54.7	16.1	18.00	18	0.461	7/16	1/4	6.001	6	0.691	11/16	804	89.4	7.07	20.8	6.94	1.14
S 15x50	14.7	15.00	15	0.550	9/16	5/16	5.640	5 5/8	0.622	5/8	486	64.8	5.75	15.7	5.57	1.03
x42.9	12.6	15.00	15	0.411	7/16	1/4	5.501	5 1/2	0.622	5/8	447	59.6	5.95	14.4	5.23	1.07
S 12x50	14.7	12.00	12	0.687	11/16	3/8	5.477	5 1/2	0.659	11/16	305	50.8	4.55	15.7	5.74	1.03
x40.8	12.0	12.00	12	0.462	7/16	1/4	5.252	5 1/4	0.659	11/16	272	45.4	4.77	13.6	5.16	1.06
S 12x35	10.3	12.00	12	0.428	7/16	1/4	5.078	5 1/8	0.544	9/16	229	38.2	4.72	9.87	3.89	0.980
x31.8	9.35	12.00	12	0.350	3/8	3/16	5.000	5	0.544	9/16	218	36.4	4.83	9.36	3.74	1.00
S 10x35	10.3	10.00	10	0.594	5/8	5/16	4.944	5	0.491	1/2	147	29.4	3.78	8.36	3.38	0.901
x25.4	7.46	10.00	10	0.311	5/16	3/16	4.661	4 5/8	0.491	1/2	124	24.7	4.07	6.79	2.91	0.954
S 8x23	6.77	8.00	8	0.441	7/16	1/4	4.171	4 1/8	0.426	7/16	64.9	16.2	3.10	4.31	2.07	0.798
x18.4	5.41	8.00	8	0.271	1/4	1/8	4.001	4	0.426	7/16	57.6	14.4	3.26	3.73	1.86	0.831
S 7x20	5.88	7.00	7	0.450	7/16	1/4	3.860	3 7/8	0.392	3/8	42.4	12.1	2.69	3.17	1.64	0.734
x15.3	4.50	7.00	7	0.252	1/4	1/8	3.662	3 5/8	0.392	3/8	36.7	10.5	2.86	2.64	1.44	0.766
S 6x17.25	5.07	6.00	6	0.465	7/16	1/4	3.565	3 5/8	0.359	3/8	26.3	8.77	2.28	2.31	1.30	0.675
x12.5	3.67	6.00	6	0.232	1/4	1/8	3.332	3 3/8	0.359	3/8	22.1	7.37	2.45	1.82	1.09	0.705
S 5x14.75	4.34	5.00	5	0.494	1/2	1/4	3.284	3 1/4	0.326	5/16	15.2	6.09	1.87	1.67	1.01	0.620
x10	2.94	5.00	5	0.214	3/16	1/8	3.004	3	0.326	5/16	12.3	4.92	2.05	1.22	0.809	0.643
S 4x9.5	2.79	4.00	4	0.326	5/16	3/16	2.796	2 3/4	0.293	5/16	6.79	3.39	1.56	0.903	0.646	0.569
x7.7	2.26	4.00	4	0.193	3/16	1/8	2.663	2 5/8	0.293	5/16	6.08	3.04	1.64	0.764	0.574	0.581
S 3x7.5	2.21	3.00	3	0.349	3/8	3/16	2.509	2 1/2	0.260	1/4	2.93	1.95	1.15	0.586	0.468	0.516
x5.7	1.67	3.00	3	0.170	3/16	1/8	2.330	2 3/8	0.260	1/4	2.52	1.68	1.23	0.455	0.390	0.522

Properties of Structural Shapes

CHANNELS
AMERICAN STANDARD

Dimensions   Properties

Designation	Area A	Depth d	Web Thickness t_w	Web $\frac{t_w}{2}$	Flange Width b_f	Flange Average thickness t_f	Nominal Weight per Ft.	$\bar{x}$	Shear Center Location e_o	$\frac{d}{A_f}$	Axis X-X I	Axis X-X Z	Axis X-X k	Axis Y-Y I	Axis Y-Y Z	Axis Y-Y k
	In.^{2}	In.	In.	In.	In.	In.		In.	In.		In.^{4}	In.^{3}	In.	In.^{4}	In.^{3}	In.
C 15x50	14.7	15.00	0.716	$^{11}/_{16}$	$^3/_8$	3.716 $3^3/_4$	0.650 $^5/_8$	50	0.798	0.583	6.21	404	53.8	5.24	11.0	3.78	0.867
x40	11.8	15.00	0.520	$^1/_2$	$^1/_4$	3.520 $3^1/_2$	0.650 $^5/_8$	40	0.777	0.767	6.56	349	46.5	5.44	9.23	3.37	0.886
x33.9	9.96	15.00	0.400	$^3/_8$	$^3/_{16}$	3.400 $3^3/_8$	0.650 $^5/_8$	33.9	0.787	0.896	6.79	315	42.0	5.62	8.13	3.11	0.904
C 12x30	8.82	12.00	0.510	$^1/_2$	$^1/_4$	3.170 $3^1/_8$	0.501 $^1/_2$	30	0.674	0.618	7.55	162	27.0	4.29	5.14	2.06	0.763
x25	7.35	12.00	0.387	$^3/_8$	$^3/_{16}$	3.047 3	0.501 $^1/_2$	25	0.674	0.746	7.85	144	24.1	4.43	4.47	1.88	0.780
x20.7	6.09	12.00	0.282	$^5/_{16}$	$^1/_8$	2.942 3	0.501 $^1/_2$	20.7	0.698	0.870	8.13	129	21.5	4.61	3.88	1.73	0.799
C 10x30	8.82	10.00	0.673	$^{11}/_{16}$	$^5/_{16}$	3.033 3	0.436 $^7/_{16}$	30	0.649	0.369	7.55	103	20.7	3.42	3.94	1.65	0.669
x25	7.35	10.00	0.526	$^1/_2$	$^1/_4$	2.886 $2^7/_8$	0.436 $^7/_{16}$	25	0.617	0.494	7.94	91.2	18.2	3.52	3.36	1.48	0.676
x20	5.88	10.00	0.379	$^3/_8$	$^3/_{16}$	2.739 $2^3/_4$	0.436 $^7/_{16}$	20	0.606	0.637	8.36	78.9	15.8	3.66	2.81	1.32	0.692
x15.3	4.49	10.00	0.240	$^1/_4$	$^1/_8$	2.600 $2^5/_8$	0.436 $^7/_{16}$	15.3	0.634	0.796	8.81	67.4	13.5	3.87	2.28	1.16	0.713
C 9x20	5.88	9.00	0.448	$^7/_{16}$	$^1/_4$	2.648 $2^5/_8$	0.413 $^7/_{16}$	20	0.583	0.515	8.22	60.9	13.5	3.22	2.42	1.17	0.642
x15	4.41	9.00	0.285	$^5/_{16}$	$^1/_8$	2.485 $2^1/_2$	0.413 $^7/_{16}$	15	0.586	0.682	8.76	51.0	11.3	3.40	1.93	1.01	0.661
x13.4	3.94	9.00	0.233	$^1/_4$	$^1/_8$	2.433 $2^3/_8$	0.413 $^7/_{16}$	13.4	0.601	0.743	8.95	47.9	10.6	3.48	1.76	0.962	0.669
C 8x18.75	5.51	8.00	0.487	$^1/_2$	$^1/_4$	2.527 $2^1/_2$	0.390 $^3/_8$	18.75	0.565	0.431	8.12	44.0	11.0	2.82	1.98	1.01	0.599
x13.75	4.04	8.00	0.303	$^5/_{16}$	$^1/_8$	2.343 $2^3/_8$	0.390 $^3/_8$	13.75	0.553	0.604	8.75	36.1	9.03	2.99	1.53	0.854	0.615
x11.5	3.38	8.00	0.220	$^1/_4$	$^1/_8$	2.260 $2^1/_4$	0.390 $^3/_8$	11.5	0.571	0.697	9.08	32.6	8.14	3.11	1.32	0.781	0.625
C 7x14.75	4.33	7.00	0.419	$^7/_{16}$	$^3/_{16}$	2.299 $2^1/_4$	0.366 $^3/_8$	14.75	0.532	0.441	8.31	27.2	7.78	2.51	1.38	0.779	0.564
x12.25	3.60	7.00	0.314	$^5/_{16}$	$^3/_{16}$	2.194 $2^1/_4$	0.366 $^3/_8$	12.25	0.525	0.538	8.71	24.2	6.93	2.60	1.17	0.703	0.571
x 9.8	2.87	7.00	0.210	$^3/_{16}$	$^1/_8$	2.090 $2^1/_8$	0.366 $^3/_8$	9.8	0.540	0.647	9.14	21.3	6.08	2.72	0.968	0.625	0.581
C 6x13	3.83	6.00	0.437	$^7/_{16}$	$^3/_{16}$	2.157 $2^1/_8$	0.343 $^5/_{16}$	13	0.514	0.380	8.10	17.4	5.80	2.13	1.05	0.642	0.525
x10.5	3.09	6.00	0.314	$^5/_{16}$	$^3/_{16}$	2.034 2	0.343 $^5/_{16}$	10.5	0.499	0.486	8.59	15.2	5.06	2.22	0.866	0.564	0.529
x 8.2	2.40	6.00	0.200	$^3/_{16}$	$^1/_8$	1.920 $1^7/_8$	0.343 $^5/_{16}$	8.2	0.511	0.599	9.10	13.1	4.38	2.34	0.693	0.492	0.537
C 5x 9	2.64	5.00	0.325	$^5/_{16}$	$^3/_{16}$	1.885 $1^7/_8$	0.320 $^5/_{16}$	9	0.478	0.427	8.29	8.90	3.56	1.83	0.632	0.450	0.489
x 6.7	1.97	5.00	0.190	$^3/_{16}$	$^1/_8$	1.750 $1^3/_4$	0.320 $^5/_{16}$	6.7	0.484	0.552	8.93	7.49	3.00	1.95	0.479	0.378	0.493
C 4x 7.25	2.13	4.00	0.321	$^5/_{16}$	$^3/_{16}$	1.721 $1^3/_4$	0.296 $^5/_{16}$	7.25	0.459	0.386	7.84	4.59	2.29	1.47	0.433	0.343	0.450
x 5.4	1.59	4.00	0.184	$^3/_{16}$	$^1/_{16}$	1.584 $1^5/_8$	0.296 $^5/_{16}$	5.4	0.457	0.502	8.52	3.85	1.93	1.56	0.319	0.283	0.449
C 3x 6	1.76	3.00	0.356	$^3/_8$	$^3/_{16}$	1.596 $1^5/_8$	0.273 $^1/_4$	6	0.455	0.322	6.87	2.07	1.38	1.08	0.305	0.268	0.416
x 5	1.47	3.00	0.258	$^1/_4$	$^1/_8$	1.498 $1^1/_2$	0.273 $^1/_4$	5	0.438	0.392	7.32	1.85	1.24	1.12	0.247	0.233	0.410
x 4.1	1.21	3.00	0.170	$^3/_{16}$	$^1/_{16}$	1.410 $1^3/_8$	0.273 $^1/_4$	4.1	0.436	0.461	7.78	1.66	1.10	1.17	0.197	0.202	0.404

CHANNELS MISCELLANEOUS

Dimensions

Properties

Designation	Area A (In.²)	Depth d (In.)	Web Thickness t_w (In.)		$\frac{t_w}{2}$ (In.)	Flange Width b_f (In.)		Average thickness t_f (In.)		Nominal Weight per Ft.	$\bar{x}$ (In.)	Shear Center Location e_o (In.)	$\frac{d}{A_f}$	Axis X-X I (In.⁴)	Axis X-X Z (In.³)	Axis X-X k (In.)	Axis Y-Y I (In.⁴)	Axis Y-Y Z (In.³)	Axis Y-Y k (In.)
MC 18x58	17.1	18.00	0.700	11/16	3/8	4.200	4¼	0.625	5/8	58	0.862	0.695	6.86	676	75.1	6.29	17.8	5.32	1.02
x51.9	15.3	18.00	0.600	5/8	5/16	4.100	4⅛	0.625	5/8	51.9	0.858	0.797	7.02	627	69.7	6.41	16.4	5.07	1.04
x45.8	13.5	18.00	0.500	1/2	1/4	4.000	4	0.625	5/8	45.8	0.866	0.909	7.20	578	64.3	6.56	15.1	4.82	1.06
x42.7	12.6	18.00	0.450	7/16	1/4	3.950	4	0.625	5/8	42.7	0.877	0.969	7.29	554	61.6	6.64	14.4	4.69	1.07
MC 13x50	14.7	13.00	0.787	13/16	3/8	4.412	4⅜	0.610	5/8	50	0.974	0.815	4.83	314	48.4	4.62	16.5	4.79	1.06
x40	11.8	13.00	0.560	9/16	1/4	4.185	4⅛	0.610	5/8	40	0.963	1.03	5.09	273	42.0	4.82	13.7	4.26	1.08
x35	10.3	13.00	0.447	7/16	1/4	4.072	4⅛	0.610	5/8	35	0.980	1.16	5.23	252	38.8	4.95	12.3	3.99	1.10
x31.8	9.35	13.00	0.375	3/8	3/16	4.000	4	0.610	5/8	31.8	1.00	1.24	5.33	239	36.8	5.06	11.4	3.81	1.11
MC 12x50	14.7	12.00	0.835	13/16	7/16	4.135	4⅛	0.700	11/16	50	1.05	0.741	4.15	269	44.9	4.28	17.4	5.65	1.09
x45	13.2	12.00	0.712	11/16	3/8	4.012	4	0.700	11/16	45	1.04	0.844	4.27	252	42.0	4.36	15.8	5.33	1.09
x40	11.8	12.00	0.590	9/16	5/16	3.890	3⅞	0.700	11/16	40	1.04	0.952	4.41	234	39.0	4.46	14.3	5.00	1.10
x35	10.3	12.00	0.467	7/16	1/4	3.767	3¾	0.700	11/16	35	1.05	1.07	4.55	216	36.1	4.59	12.7	4.67	1.11
MC 12x37	10.9	12.00	0.600	5/8	5/16	3.600	3⅝	0.600	5/8	37	0.866	0.747	5.56	205	34.2	4.34	9.81	3.59	0.950
x32.9	9.67	12.00	0.500	1/2	1/4	3.500	3½	0.600	5/8	32.9	0.867	0.843	5.71	191	31.8	4.44	8.91	3.39	0.960
x30.9	9.07	12.00	0.450	7/16	1/4	3.450	3½	0.600	5/8	30.9	0.873	0.893	5.80	183	30.6	4.50	8.46	3.28	0.966
MC 12x10.6	3.10	12.00	0.190	3/16	1/8	1.500	1½	0.309	5/16	10.6	0.269	0.284	25.9	55.4	9.23	4.22	0.382	0.310	0.351
MC 10x41.1	12.1	10.00	0.796	13/16	3/8	4.321	4⅜	0.575	9/16	41.1	1.09	0.864	4.02	158	31.5	3.61	15.8	4.88	1.14
x33.6	9.87	10.00	0.575	9/16	5/16	4.100	4⅛	0.575	9/16	33.6	1.08	1.06	4.24	139	27.8	3.75	13.2	4.38	1.16
x28.5	8.37	10.00	0.425	7/16	3/16	3.950	4	0.575	9/16	28.5	1.12	1.21	4.40	127	25.3	3.89	11.4	4.02	1.17
MC 10x28.3	8.32	10.00	0.477	1/2	1/4	3.502	3½	0.575	9/16	28.3	0.933	0.928	4.97	118	23.6	3.77	8.21	3.20	0.993
x25.3	7.43	10.00	0.425	7/16	3/16	3.550	3½	0.500	1/2	25.3	0.918	0.977	5.63	107	21.4	3.79	7.61	2.89	1.01
x24.9	7.32	10.00	0.377	3/8	3/16	3.402	3⅜	0.575	9/16	24.9	0.954	1.03	5.11	110	22.0	3.87	7.32	2.99	1.00
x21.9	6.43	10.00	0.325	5/16	3/16	3.450	3½	0.500	1/2	21.9	0.954	1.09	5.80	98.5	19.7	3.91	6.74	2.70	1.02
MC 10x 8.4	2.46	10.00	0.170	3/16	1/16	1.500	1½	0.280	1/4	8.4	0.284	0.332	23.8	32.0	6.40	3.61	0.328	0.270	0.365
MC 10x 6.5	1.91	10.00	0.152	1/8	1/16	1.127	1⅛	0.202	3/16	6.5	0.180	0.167	43.8	22.1	4.42	3.40	0.112	0.118	0.242

CHANNELS
MISCELLANEOUS

Dimensions

Properties

Designation	Area A	Depth d	Web Thickness t_w	$\frac{t_w}{2}$	Flange Width b_f		Average thickness t_f		Nominal Weight per Ft.	$\bar{x}$	Shear Center Location e_o	$\frac{d}{A_f}$	Axis X-X I	Z	k	Axis Y-Y I	Z	k
	In.^{2}	In.	In.	In.	In.		In.			In.	In.		In.^{4}	In.^{3}	In.	In.^{4}	In.^{3}	In.
MC 9x25.4	7.47	9.00	0.450	$^7/_{16}$	$^1/_4$	3.500	$3^1/_2$	0.550	$^9/_{16}$	25.4	0.970	0.986	4.68	88.0	19.6	3.43	7.65	3.02	1.01
x23.9	7.02	9.00	0.400	$^3/_8$	$^3/_{16}$	3.450	$3^1/_2$	0.550	$^9/_{16}$	23.9	0.981	1.04	4.74	85.0	18.9	3.48	7.22	2.93	1.01
MC 8x22.8	6.70	8.00	0.427	$^7/_{16}$	$^3/_{16}$	3.502	$3^1/_2$	0.525	$^1/_2$	22.8	1.01	1.04	4.35	63.8	16.0	3.09	7.07	2.84	1.03
x21.4	6.28	8.00	0.375	$^3/_8$	$^3/_{16}$	3.450	$3^1/_2$	0.525	$^1/_2$	21.4	1.02	1.09	4.42	61.6	15.4	3.13	6.64	2.74	1.03
MC 8x20	5.88	8.00	0.400	$^3/_8$	$^3/_{16}$	3.025	3	0.500	$^1/_2$	20	0.840	0.843	5.29	54.5	13.6	3.05	4.47	2.05	0.872
x18.7	5.50	8.00	0.353	$^3/_8$	$^3/_{16}$	2.978	3	0.500	$^1/_2$	18.7	0.849	0.889	5.37	52.5	13.1	3.09	4.20	1.97	0.874
MC 8x 8.5	2.50	8.00	0.179	$^3/_{16}$	$^1/_{16}$	1.874	$1^7/_8$	0.311	$^5/_{16}$	8.5	0.428	0.542	13.7	23.3	5.83	3.05	0.628	0.434	0.501
MC 7x22.7	6.67	7.00	0.503	$^1/_2$	$^1/_4$	3.603	$3^5/_8$	0.500	$^1/_2$	22.7	1.04	1.01	3.89	47.5	13.6	2.67	7.29	2.85	1.05
x19.1	5.61	7.00	0.352	$^3/_8$	$^3/_{16}$	3.452	$3^1/_2$	0.500	$^1/_2$	19.1	1.08	1.15	4.06	43.2	12.3	2.77	6.11	2.57	1.04
MC 7x17.6	5.17	7.00	0.375	$^3/_8$	$^3/_{16}$	3.000	3	0.475	$^1/_2$	17.6	0.873	0.890	4.91	37.6	10.8	2.70	4.01	1.89	0.881
MC 6x18	5.29	6.00	0.379	$^3/_8$	$^3/_{16}$	3.504	$3^1/_2$	0.475	$^1/_2$	18	1.12	1.17	3.60	29.7	9.91	2.37	5.93	2.48	1.06
x15.3	4.50	6.00	0.340	$^5/_{16}$	$^3/_{16}$	3.500	$3^1/_2$	0.385	$^3/_8$	15.3	1.05	1.16	4.45	25.4	8.47	2.38	4.97	2.03	1.05
MC 6x16.3	4.79	6.00	0.375	$^3/_8$	$^3/_{16}$	3.000	3	0.475	$^1/_2$	16.3	0.927	0.930	4.21	26.0	8.68	2.33	3.82	1.84	0.892
x15.1	4.44	6.00	0.316	$^5/_{16}$	$^3/_{16}$	2.941	3	0.475	$^1/_2$	15.1	0.940	0.982	4.29	25.0	8.32	2.37	3.51	1.75	0.889
MC 6x12	3.53	6.00	0.310	$^5/_{16}$	$^1/_8$	2.497	$2^1/_2$	0.375	$^3/_8$	12	0.704	0.725	6.41	18.7	6.24	2.30	1.87	1.04	0.728

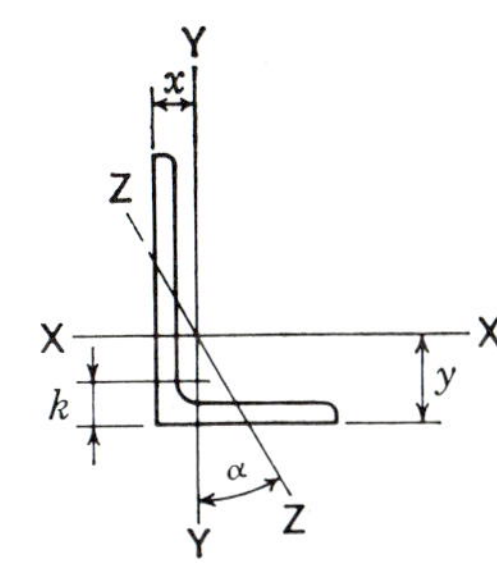

ANGLES

Equal Legs and Unequal Legs
Properties for Designing

Size and Thickness	k	Weight per Foot	Area	AXIS X-X.				AXIS Y-Y				AXIS Z-Z	
				I	Z	k	y	I	Z	k	x	k	Tan
In.	In.	Lb.	In.^{2}	In.^{4}	In.^{3}	In.	In.	In.^{4}	In.^{3}	In.	In.	In.	α
L 9 x4 x $\frac{5}{8}$	$1\frac{1}{8}$	26.3	7.73	64.9	11.5	2.90	3.36	8.32	2.65	1.04	0.858	.847	0.216
$\frac{9}{16}$	$1\frac{1}{16}$	23.8	7.00	59.1	10.4	2.91	3.33	7.63	2.41	1.04	0.834	.850	0.218
$\frac{1}{2}$	1	21.3	6.25	53.2	9.34	2.92	3.31	6.92	2.17	1.05	0.810	.854	0.220
L 8 x8 x$1\frac{1}{8}$	$1\frac{3}{4}$	56.9	16.7	98.0	17.5	2.42	2.41	98.0	17.5	2.42	2.41	1.56	1.000
1	$1\frac{5}{8}$	51.0	15.0	89.0	15.8	2.44	2.37	89.0	15.8	2.44	2.37	1.56	1.000
$\frac{7}{8}$	$1\frac{1}{2}$	45.0	13.2	79.6	14.0	2.45	2.32	79.6	14.0	2.45	2.32	1.57	1.000
$\frac{3}{4}$	$1\frac{3}{8}$	38.9	11.4	69.7	12.2	2.47	2.28	69.7	12.2	2.47	2.28	1.58	1.000
$\frac{5}{8}$	$1\frac{1}{4}$	32.7	9.61	59.4	10.3	2.49	2.23	59.4	10.3	2.49	2.23	1.58	1.000
$\frac{9}{16}$	$1\frac{3}{16}$	29.6	8.68	54.1	9.34	2.50	2.21	54.1	9.34	2.50	2.21	1.59	1.000
$\frac{1}{2}$	$1\frac{1}{8}$	26.4	7.75	48.6	8.36	2.50	2.19	48.6	8.36	2.50	2.19	1.59	1.000
L 8 x6 x1	$1\frac{1}{2}$	44.2	13.0	80.8	15.1	2.49	2.65	38.8	8.92	1.73	1.65	1.28	0.543
$\frac{7}{8}$	$1\frac{3}{8}$	39.1	11.5	72.3	13.4	2.51	2.61	34.9	7.94	1.74	1.61	1.28	0.547
$\frac{3}{4}$	$1\frac{1}{4}$	33.8	9.94	63.4	11.7	2.53	2.56	30.7	6.92	1.76	1.56	1.29	0.551
$\frac{5}{8}$	$1\frac{1}{8}$	28.5	8.36	54.1	9.87	2.54	2.52	26.3	5.88	1.77	1.52	1.29	0.554
$\frac{9}{16}$	$1\frac{1}{16}$	25.7	7.56	49.3	8.95	2.55	2.50	24.0	5.34	1.78	1.50	1.30	0.556
$\frac{1}{2}$	1	23.0	6.75	44.3	8.02	2.56	2.47	21.7	4.79	1.79	1.47	1.30	0.558
$\frac{7}{16}$	$\frac{15}{16}$	20.2	5.93	39.2	7.07	2.57	2.45	19.3	4.23	1.80	1.45	1.31	0.560
L 8 x4 x1	$1\frac{1}{2}$	37.4	11.0	69.6	14.1	2.52	3.05	11.6	3.94	1.03	1.05	0.846	0.247
$\frac{3}{4}$	$1\frac{1}{4}$	28.7	8.44	54.9	10.9	2.55	2.95	9.36	3.07	1.05	0.953	0.852	0.258
$\frac{9}{16}$	$1\frac{1}{16}$	21.9	6.43	42.8	8.35	2.58	2.88	7.43	2.38	1.07	0.882	0.861	0.265
$\frac{1}{2}$	1	19.6	5.75	38.5	7.49	2.59	2.86	6.74	2.15	1.08	0.859	0.865	0.267
L 7 x4 x $\frac{3}{4}$	$1\frac{1}{4}$	26.2	7.69	37.8	8.42	2.22	2.51	9.05	3.03	1.09	1.01	0.860	0.324
$\frac{5}{8}$	$1\frac{1}{8}$	22.1	6.48	32.4	7.14	2.24	2.46	7.84	2.58	1.10	0.963	0.865	0.329
$\frac{1}{2}$	1	17.9	5.25	26.7	5.81	2.25	2.42	6.53	2.12	1.11	0.917	0.872	0.335
$\frac{3}{8}$	$\frac{7}{8}$	13.6	3.98	20.6	4.44	2.27	2.37	5.10	1.63	1.13	0.870	0.880	0.340

Angles in shaded rows may not be readily available.   Availability is subject to rolling accumulation and geographical location, and should be checked with material suppliers.

ANGLES
Equal Legs and Unequal Legs
Properties for Designing

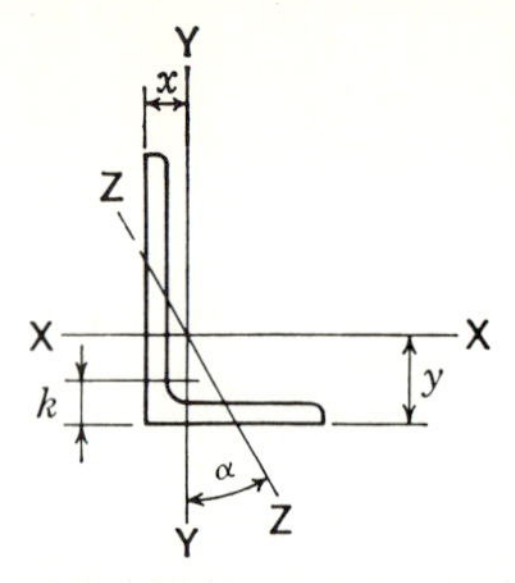

Size and Thickness	k	Weight per Foot	Area	AXIS X-X				AXIS Y-Y				AXIS Z-Z	
				I	Z	k	y	I	Z	k	x	k	Tan
In.	In.	Lb.	In.^{2}	In.^{4}	In.^{3}	In.	In.	In.^{4}	In.^{3}	In.	In.	In.	α
L 6 x6 x1	$1\frac{1}{2}$	37.4	11.0	35.5	8.57	1.80	1.86	35.5	8.57	1.80	1.86	1.17	1.000
$\frac{7}{8}$	$1\frac{3}{8}$	33.1	9.73	31.9	7.63	1.81	1.82	31.9	7.63	1.81	1.82	1.17	1.000
$\frac{3}{4}$	$1\frac{1}{4}$	28.7	8.44	28.2	6.66	1.83	1.78	28.2	6.66	1.83	1.78	1.17	1.000
$\frac{5}{8}$	$1\frac{1}{8}$	24.2	7.11	24.2	5.66	1.84	1.73	24.2	5.66	1.84	1.73	1.18	1.000
$\frac{9}{16}$	$1\frac{1}{16}$	21.9	6.43	22.1	5.14	1.85	1.71	22.1	5.14	1.85	1.71	1.18	1.000
$\frac{1}{2}$	1	19.6	5.75	19.9	4.61	1.86	1.68	19.9	4.61	1.86	1.68	1.18	1.000
$\frac{7}{16}$	$\frac{15}{16}$	17.2	5.06	17.7	4.08	1.87	1.66	17.7	4.08	1.87	1.66	1.19	1.000
$\frac{3}{8}$	$\frac{7}{8}$	14.9	4.36	15.4	3.53	1.88	1.64	15.4	3.53	1.88	1.64	1.19	1.000
$\frac{5}{16}$	$\frac{13}{16}$	12.4	3.65	13.0	2.97	1.89	1.62	13.0	2.97	1.89	1.62	1.20	1.000
L 6 x4 x $\frac{7}{8}$	$1\frac{3}{8}$	27.2	7.98	27.7	7.15	1.86	2.12	9.75	3.39	1.11	1.12	0.857	0.421
$\frac{3}{4}$	$1\frac{1}{4}$	23.6	6.94	24.5	6.25	1.88	2.08	8.68	2.97	1.12	1.08	0.860	0.428
$\frac{5}{8}$	$1\frac{1}{8}$	20.0	5.86	21.1	5.31	1.90	2.03	7.52	2.54	1.13	1.03	0.864	0.435
$\frac{9}{16}$	$1\frac{1}{16}$	18.1	5.31	19.3	4.83	1.90	2.01	6.91	2.31	1.14	1.01	0.866	0.438
$\frac{1}{2}$	1	16.2	4.75	17.4	4.33	1.91	1.99	6.27	2.08	1.15	0.987	0.870	0.440
$\frac{7}{16}$	$\frac{15}{16}$	14.3	4.18	15.5	3.83	1.92	1.96	5.60	1.85	1.16	0.964	0.873	0.443
$\frac{3}{8}$	$\frac{7}{8}$	12.3	3.61	13.5	3.32	1.93	1.94	4.90	1.60	1.17	0.941	0.877	0.446
$\frac{5}{16}$	$\frac{13}{16}$	10.3	3.03	11.4	2.79	1.94	1.92	4.18	1.35	1.17	0.918	0.882	0.448
L 6 x3$\frac{1}{2}$x $\frac{1}{2}$	1	15.3	4.50	16.6	4.24	1.92	2.08	4.25	1.59	0.972	0.833	0.759	0.344
$\frac{3}{8}$	$\frac{7}{8}$	11.7	3.42	12.9	3.24	1.94	2.04	3.34	1.23	0.988	0.787	0.767	0.350
$\frac{5}{16}$	$\frac{13}{16}$	9.8	2.87	10.9	2.73	1.95	2.01	2.85	1.04	0.996	0.763	0.772	0.352
L 5 x5 x $\frac{7}{8}$	$1\frac{3}{8}$	27.2	7.98	17.8	5.17	1.49	1.57	17.8	5.17	1.49	1.57	0.973	1.000
$\frac{3}{4}$	$1\frac{1}{4}$	23.6	6.94	15.7	4.53	1.51	1.52	15.7	4.53	1.51	1.52	0.975	1.000
$\frac{5}{8}$	$1\frac{1}{8}$	20.0	5.86	13.6	3.86	1.52	1.48	13.6	3.86	1.52	1.48	0.978	1.000
$\frac{1}{2}$	1	16.2	4.75	11.3	3.16	1.54	1.43	11.3	3.16	1.54	1.43	0.983	1.000
$\frac{7}{16}$	$\frac{15}{16}$	14.3	4.18	10.0	2.79	1.55	1.41	10.0	2.79	1.55	1.41	0.986	1.000
$\frac{3}{8}$	$\frac{7}{8}$	12.3	3.61	8.74	2.42	1.56	1.39	8.74	2.42	1.56	1.39	0.990	1.000
$\frac{5}{16}$	$\frac{13}{16}$	10.3	3.03	7.42	2.04	1.57	1.37	7.42	2.04	1.57	1.37	0.994	1.000

Angles in shaded rows may not be readily available. Availability is subject to rolling accumulation and geographical location, and should be checked with material suppliers.

Properties of Structural Shapes

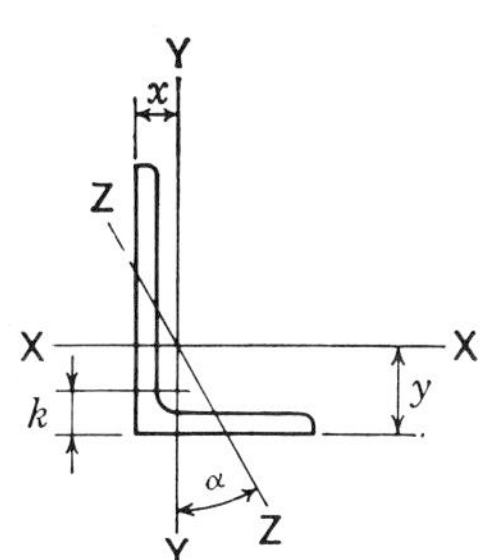

ANGLES
Equal Legs and Unequal Legs
Properties for Designing

Size and Thickness	k	Weight per Foot	Area	AXIS X-X				AXIS Y-Y				AXIS Z-Z	
				I	Z	k	y	I	Z	k	x	k	Tan
In.	In.	Lb.	In.^{2}	In.^{4}	In.^{3}	In.	In.	In.^{4}	In.^{3}	In.	In.	In.	α
L 5 x3½x ¾	1¼	19.8	5.81	13.9	4.28	1.55	1.75	5.55	2.22	0.977	0.996	0.748	0.464
⅝	1⅛	16.8	4.92	12.0	3.65	1.56	1.70	4.83	1.90	0.991	0.951	0.751	0.472
½	1	13.6	4.00	9.99	2.99	1.58	1.66	4.05	1.56	1.01	0.906	0.755	0.479
⁷⁄₁₆	¹⁵⁄₁₆	12.0	3.53	8.90	2.64	1.59	1.63	3.63	1.39	1.01	0.883	0.758	0.482
⅜	⅞	10.4	3.05	7.78	2.29	1.60	1.61	3.18	1.21	1.02	0.861	0.762	0.486
⁵⁄₁₆	¹³⁄₁₆	8.7	2.56	6.60	1.94	1.61	1.59	2.72	1.02	1.03	0.838	0.766	0.489
¼	¾	7.0	2.06	5.39	1.57	1.62	1.56	2.23	0.830	1.04	0.814	0.770	0.492
L 5 x3 x ⅝	1	15.7	4.61	11.4	3.55	1.57	1.80	3.06	1.39	0.815	0.796	0.644	0.349
½	1	12.8	3.75	9.45	2.91	1.59	1.75	2.58	1.15	0.829	0.750	0.648	0.357
⁷⁄₁₆	¹⁵⁄₁₆	11.3	3.31	8.43	2.58	1.60	1.73	2.32	1.02	0.837	0.727	0.651	0.361
⅜	⅞	9.8	2.86	7.37	2.24	1.61	1.70	2.04	0.888	0.845	0.704	0.654	0.364
⁵⁄₁₆	¹³⁄₁₆	8.2	2.40	6.26	1.89	1.61	1.68	1.75	0.753	0.853	0.681	0.658	0.368
¼	¾	6.6	1.94	5.11	1.53	1.62	1.66	1.44	0.614	0.861	0.657	0.663	0.371
L 4 x4 x ¾	1⅛	18.5	5.44	7.67	2.81	1.19	1.27	7.67	2.81	1.19	1.27	0.778	1.000
⅝	1	15.7	4.61	6.66	2.40	1.20	1.23	6.66	2.40	1.20	1.23	0.779	1.000
½	⅞	12.8	3.75	5.56	1.97	1.22	1.18	5.56	1.97	1.22	1.18	0.782	1.000
⁷⁄₁₆	¹³⁄₁₆	11.3	3.31	4.97	1.75	1.23	1.16	4.97	1.75	1.23	1.16	0.785	1.000
⅜	¾	9.8	2.86	4.36	1.52	1.23	1.14	4.36	1.52	1.23	1.14	0.788	1.000
⁵⁄₁₆	¹¹⁄₁₆	8.2	2.40	3.71	1.29	1.24	1.12	3.71	1.29	1.24	1.12	0.791	1.000
¼	⅝	6.6	1.94	3.04	1.05	1.25	1.09	3.04	1.05	1.25	1.09	0.795	1.000
L 4 x3½x ⅝	1¹⁄₁₆	14.7	4.30	6.37	2.35	1.22	1.29	4.52	1.84	1.03	1.04	0.719	0.745
½	¹⁵⁄₁₆	11.9	3.50	5.32	1.94	1.23	1.25	3.79	1.52	1.04	1.00	0.722	0.750
⁷⁄₁₆	⅞	10.6	3.09	4.76	1.72	1.24	1.23	3.40	1.35	1.05	0.978	0.724	0.753
⅜	¹³⁄₁₆	9.1	2.67	4.18	1.49	1.25	1.21	2.95	1.17	1.06	0.955	0.727	0.755
⁵⁄₁₆	¾	7.7	2.25	3.56	1.26	1.26	1.18	2.55	0.994	1.07	0.932	0.730	0.757
¼	¹¹⁄₁₆	6.2	1.81	2.91	1.03	1.27	1.16	2.09	0.808	1.07	0.909	0.734	0.759

Angles in shaded rows may not be readily available. Availability is subject to rolling accumulation and geographical location, and should be checked with material suppliers.

Properties of Structural Shapes

ANGLES
Equal Legs and Unequal Legs
Properties for Designing

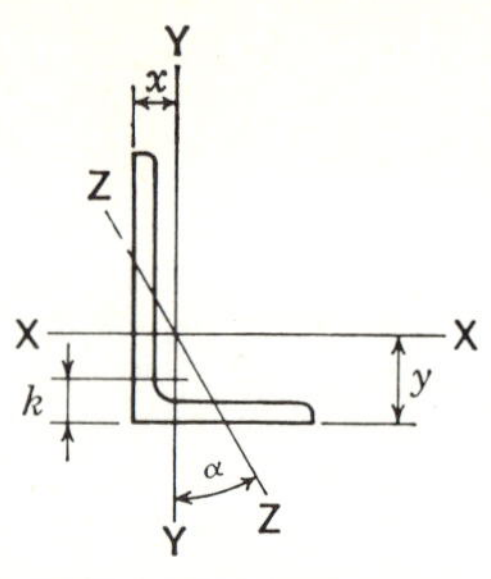

Size and Thickness	k	Weight per Foot	Area	AXIS X-X				AXIS Y-Y				AXIS Z-Z	
				I	Z	k	y	I	Z	k	x	k	Tan
In.	In.	Lb.	In.^{2}	In.^{4}	In.^{3}	In.	In.	In.^{4}	In.^{3}	In.	In.	In.	α
L 4 x3 x 5/8	1 1/16	13.6	3.98	6.03	2.30	1.23	1.37	2.87	1.35	0.849	0.871	0.637	0.534
1/2	15/16	11.1	3.25	5.05	1.89	1.25	1.33	2.42	1.12	0.864	0.827	0.639	0.543
7/16	7/8	9.8	2.87	4.52	1.68	1.25	1.30	2.18	0.992	0.871	0.804	0.641	0.547
3/8	13/16	8.5	2.48	3.96	1.46	1.26	1.28	1.92	0.866	0.879	0.782	0.644	0.551
5/16	3/4	7.2	2.09	3.38	1.23	1.27	1.26	1.65	0.734	0.887	0.759	0.647	0.554
1/4	11/16	5.8	1.69	2.77	1.00	1.28	1.24	1.36	0.599	0.896	0.736	0.651	0.558
L 3 1/2 x3 1/2 x 1/2	7/8	11.1	3.25	3.64	1.49	1.06	1.06	3.64	1.49	1.06	1.06	0.683	1.000
7/16	13/16	9.8	2.87	3.26	1.32	1.07	1.04	3.26	1.32	1.07	1.04	0.684	1.000
3/8	3/4	8.5	2.48	2.87	1.15	1.07	1.01	2.87	1.15	1.07	1.01	0.687	1.000
5/16	11/16	7.2	2.09	2.45	0.976	1.08	0.990	2.45	0.976	1.08	0.990	0.690	1.000
1/4	5/8	5.8	1.69	2.01	0.794	1.09	0.968	2.01	0.794	1.09	0.968	0.694	1.000
L 3 1/2 x3 x 1/2	15/16	10.2	3.00	3.45	1.45	1.07	1.13	2.33	1.10	0.881	0.875	0.621	0.714
7/16	7/8	9.1	2.65	3.10	1.29	1.08	1.10	2.09	0.975	0.889	0.853	0.622	0.718
3/8	13/16	7.9	2.30	2.72	1.13	1.09	1.08	1.85	0.851	0.897	0.830	0.625	0.721
5/16	3/4	6.6	1.93	2.33	0.954	1.10	1.06	1.58	0.722	0.905	0.808	0.627	0.724
1/4	11/16	5.4	1.56	1.91	0.776	1.11	1.04	1.30	0.589	0.914	0.785	0.631	0.727
L 3 1/2 x2 1/2 x 1/2	15/16	9.4	2.75	3.24	1.41	1.09	1.20	1.36	0.760	0.704	0.705	0.534	0.486
7/16	7/8	8.3	2.43	2.91	1.26	1.09	1.18	1.23	0.677	0.711	0.682	0.535	0.491
3/8	13/16	7.2	2.11	2.56	1.09	1.10	1.16	1.09	0.592	0.719	0.660	0.537	0.496
5/16	3/4	6.1	1.78	2.19	0.927	1.11	1.14	0.939	0.504	0.727	0.637	0.540	0.501
1/4	11/16	4.9	1.44	1.80	0.755	1.12	1.11	0.777	0.412	0.735	0.614	0.544	0.506
L 3 x3 x 1/2	13/16	9.4	2.75	2.22	1.07	0.898	0.932	2.22	1.07	0.898	0.932	0.584	1.000
7/16	3/4	8.3	2.43	1.99	0.954	0.905	0.910	1.99	0.954	0.905	0.910	0.585	1.000
3/8	11/16	7.2	2.11	1.76	0.833	0.913	0.888	1.76	0.833	0.913	0.888	0.587	1.000
5/16	5/8	6.1	1.78	1.51	0.707	0.922	0.865	1.51	0.707	0.922	0.865	0.589	1.000
1/4	9/16	4.9	1.44	1.24	0.577	0.930	0.842	1.24	0.577	0.930	0.842	0.592	1.000
3/16	1/2	3.71	1.09	0.962	0.441	0.939	0.820	0.962	0.441	0.939	0.820	0.596	1.000

Angles in shaded rows may not be readily available. Availability is subject to rolling accumulation and geographical location, and should be checked with material suppliers.

Properties of Structural Shapes

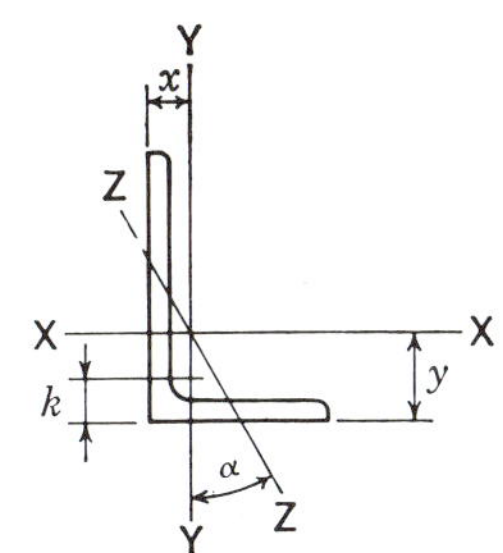

ANGLES
Equal Legs and Unequal Legs
Properties for Designing

Size and Thickness	k	Weight per Foot	Area	AXIS X-X				AXIS Y-Y				AXIS Z-Z	
				I	Z	k	y	Z	S	k	x	k	Tan
In.	In.	Lb.	In.^{2}	In.^{4}	In.^{3}	In.	In.	In.^{4}	In.^{3}	In.	In.	In.	α
L 3 x2½x ½	⅞	8.5	2.50	2.08	1.04	0.913	1.00	1.30	0.744	0.722	0.750	0.520	0.667
⁷⁄₁₆	¹³⁄₁₆	7.6	2.21	1.88	0.928	0.920	0.978	1.18	0.664	0.729	0.728	0.521	0.672
⅜	¾	6.6	1.92	1.66	0.810	0.928	0.956	1.04	0.581	0.736	0.706	0.522	0.676
⁵⁄₁₆	¹¹⁄₁₆	5.6	1.62	1.42	0.688	0.937	0.933	0.898	0.494	0.744	0.683	0.525	0.680
¼	⅝	4.5	1.31	1.17	0.561	0.945	0.911	0.743	0.404	0.753	0.661	0.528	0.684
³⁄₁₆	⁹⁄₁₆	3.39	0.996	0.907	0.430	0.954	0.888	0.577	0.310	0.761	0.638	0.533	0.688
L 3 x2 x ½	¹³⁄₁₆	7.7	2.25	1.92	1.00	0.924	1.08	0.672	0.474	0.546	0.583	0.428	0.414
⁷⁄₁₆	¾	6.8	2.00	1.73	0.894	0.932	1.06	0.609	0.424	0.553	0.561	0.429	0.421
⅜	¹¹⁄₁₆	5.9	1.73	1.53	0.781	0.940	1.04	0.543	0.371	0.559	0.539	0.430	0.428
⁵⁄₁₆	⅝	5.0	1.46	1.32	0.664	0.948	1.02	0.470	0.317	0.567	0.516	0.432	0.435
¼	⁹⁄₁₆	4.1	1.19	1.09	0.542	0.957	0.993	0.392	0.260	0.574	0.493	0.435	0.440
³⁄₁₆	½	3.07	0.902	0.842	0.415	0.966	0.970	0.307	0.200	0.583	0.470	0.439	0.446
L 2½x2½x ½	¹³⁄₁₆	7.7	2.25	1.23	0.724	0.739	0.806	1.23	0.724	0.739	0.806	0.487	1.000
⅜	¹¹⁄₁₆	5.9	1.73	0.984	0.566	0.753	0.762	0.984	0.566	0.753	0.762	0.487	1.000
⁵⁄₁₆	⅝	5.0	1.46	0.849	0.482	0.761	0.740	0.849	0.482	0.761	0.740	0.489	1.000
¼	⁹⁄₁₆	4.1	1.19	0.703	0.394	0.769	0.717	0.703	0.394	0.769	0.717	0.491	1.000
³⁄₁₆	½	3.07	0.902	0.547	0.303	0.778	0.694	0.547	0.303	0.778	0.694	0.495	1.000
L 2½x2 x ⅜	¹¹⁄₁₆	5.3	1.55	0.912	0.547	0.768	0.831	0.514	0.363	0.577	0.581	0.420	0.614
⁵⁄₁₆	⅝	4.5	1.31	0.788	0.466	0.776	0.809	0.446	0.310	0.584	0.559	0.422	0.620
¼	⁹⁄₁₆	3.62	1.06	0.654	0.381	0.784	0.787	0.372	0.254	0.592	0.537	0.424	0.626
³⁄₁₆	½	2.75	0.809	0.509	0.293	0.793	0.764	0.291	0.196	0.600	0.514	0.427	0.631
L 2 x2 x ⅜	¹¹⁄₁₆	4.7	1.36	0.479	0.351	0.594	0.636	0.479	0.351	0.594	0.636	0.389	1.000
⁵⁄₁₆	⅝	3.92	1.15	0.416	0.300	0.601	0.614	0.416	0.300	0.601	0.614	0.390	1.000
¼	⁹⁄₁₆	3.19	0.938	0.348	0.247	0.609	0.592	0.348	0.247	0.609	0.592	0.391	1.000
³⁄₁₆	½	2.44	0.715	0.272	0.190	0.617	0.569	0.272	0.190	0.617	0.569	0.394	1.000
⅛	⁷⁄₁₆	1.65	0.484	0.190	0.131	0.626	0.546	0.190	0.131	0.626	0.546	0.398	1.000

Angles in shaded rows may not be readily available. Availability is subject to rolling accumulation and geographical location, and should be checked with material suppliers.

APPENDIX C
Column Tables*

*Reprinted with permission from AISC, *Manual of Steel Construction*, 8th ed. New York: American Institute of Steel Construction, 1980.

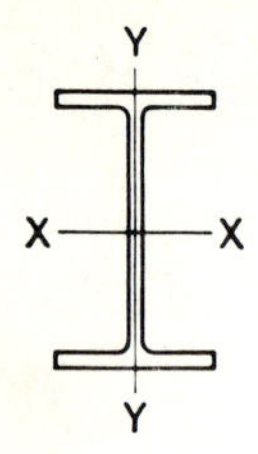

COLUMNS
W shapes

F_y = 36 ksi
F_y = 50 ksi

Allowable axial loads in kips

Desig-nation		W14									
Wt./ft.		730		665		605		550		500	
F_y		36	50	36	50	36	50	36	50	36	50
	0	4644	6450	4234	5880	3845	5340	3499	4860	3175	4410
	11	4315	5887	3928	5356	3562	4855	3237	4411	2933	3995
	12	4277	5819	3892	5293	3529	4797	3206	4357	2905	3945
	13	4237	5749	3855	5228	3494	4736	3175	4301	2875	3893
	14	4196	5677	3817	5161	3459	4674	3142	4243	2845	3840
	15	4153	5602	3777	5092	3422	4609	3108	4183	2813	3784
	16	4110	5525	3737	5020	3384	4543	3073	4121	2781	3727
	17	4065	5446	3695	4946	3345	4474	3037	4057	2748	3668
	18	4019	5365	3652	4870	3306	4404	3000	3992	2714	3608
	19	3971	5282	3608	4793	3265	4331	2962	3925	2678	3546
	20	3923	5196	3563	4713	3223	4257	2923	3856	2642	3482
	22	3823	5018	3469	4547	3136	4103	2842	3713	2568	3350
	24	3718	4832	3372	4374	3045	3942	2758	3564	2490	3211
	26	3609	4638	3270	4193	2951	3774	2670	3407	2409	3066
	28	3496	4436	3164	4004	2853	3598	2579	3244	2324	2915
	30	3378	4225	3055	3807	2751	3415	2484	3074	2236	2758
	32	3256	4006	2941	3603	2645	3225	2386	2897	2145	2594
	34	3130	3779	2823	3391	2535	3028	2284	2714	2051	2423
	36	3000	3543	2702	3170	2422	2822	2179	2522	1954	2246
	38	2865	3298	2576	2941	2305	2609	2070	2324	1853	2061
	40	2726	3044	2446	2703	2184	2387	1958	2117	1748	1870
	42	2582	2780	2312	2459	2059	2166	1841	1920	1640	1696
	44	2434	2533	2173	2241	1930	1974	1721	1749	1529	1545
	46	2281	2318	2030	2050	1797	1806	1597	1601	1413	1414
	48	2123	2129	1882	1883	1659	1659	1470	1470	1298	1298
	50	1962	1962	1735	1735	1529	1529	1355	1355	1197	1197

Effective length in ft. KL with respect to least radius of gyration r_y

Column Tables

COLUMNS
W shapes

Allowable axial loads in kips

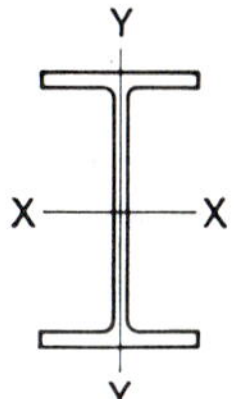

Desig-nation		W14									
Wt./ft.		455		426		398		370		342	
F_y		36	50	36	50	36	50	36	50	36	50
	0	2894	4020	2700	3750	2527	3510	2354	3270	2182	3030
	11	2671	3636	2489	3388	2328	3168	2167	2947	2006	2728
	12	2644	3590	2464	3344	2304	3126	2144	2908	1985	2692
	13	2617	3542	2438	3298	2280	3083	2121	2868	1963	2654
	14	2589	3492	2411	3251	2255	3039	2097	2826	1941	2614
	15	2560	3441	2384	3203	2229	2993	2073	2782	1918	2574
	16	2530	3388	2356	3153	2202	2946	2047	2738	1894	2532
	17	2499	3333	2326	3101	2174	2897	2021	2692	1870	2489
	18	2467	3277	2296	3048	2146	2847	1995	2644	1845	2444
	19	2435	3220	2266	2994	2117	2795	1967	2596	1819	2399
	20	2401	3161	2234	2938	2087	2743	1939	2546	1793	2352
	22	2332	3038	2169	2822	2025	2633	1881	2442	1738	2255
	24	2260	2909	2100	2700	1961	2518	1820	2333	1681	2153
	26	2184	2775	2029	2573	1893	2397	1756	2220	1621	2047
	28	2106	2635	1955	2440	1823	2272	1690	2101	1559	1935
	30	2025	2488	1878	2302	1750	2141	1621	1977	1495	1819
	32	1940	2336	1798	2158	1674	2004	1549	1848	1428	1698
	34	1853	2178	1715	2008	1596	1862	1475	1713	1358	1572
	36	1762	2013	1630	1852	1515	1714	1398	1573	1286	1441
	38	1668	1841	1541	1689	1431	1560	1319	1427	1212	1304
	40	1571	1666	1449	1526	1344	1409	1237	1288	1135	1177
	42	1471	1511	1354	1384	1254	1278	1151	1168	1053	1067
	44	1367	1377	1255	1261	1161	1164	1063	1065	972	973
	46	1260	1260	1154	1154	1065	1065	974	974	890	890
	48	1157	1157	1060	1060	978	978	895	895	817	817
	50	1066	1066	977	977	902	902	824	824	753	753

Effective length in ft. KL with respect to least radius of gyration r_y

COLUMNS
W shapes

Allowable axial loads in kips

Desig-nation		W14									
Wt./ft.		311		283		257		233		211	
F_y		36	50	36	50	36	50	36	50	36	50
	0	1974	2742	1799	2499	1633	2268	1480	2055	1339	1860
	6	1898	2613	1729	2381	1569	2159	1421	1956	1286	1769
	7	1883	2587	1715	2356	1556	2137	1409	1935	1275	1750
	8	1867	2558	1700	2330	1542	2113	1396	1913	1263	1730
	9	1850	2529	1685	2303	1528	2088	1383	1890	1251	1709
	10	1832	2498	1668	2274	1513	2062	1370	1866	1239	1687
	11	1813	2465	1651	2245	1497	2034	1355	1841	1226	1665
	12	1794	2432	1634	2214	1481	2006	1340	1815	1212	1641
	13	1774	2397	1615	2182	1464	1976	1325	1788	1198	1616
	14	1754	2361	1597	2148	1447	1946	1309	1760	1183	1590
	15	1733	2324	1577	2114	1429	1914	1293	1731	1168	1564
	16	1711	2285	1557	2079	1410	1882	1276	1701	1153	1537
	17	1689	2246	1536	2042	1391	1848	1258	1671	1137	1509
	18	1666	2205	1515	2005	1372	1814	1241	1639	1121	1480
	19	1642	2163	1494	1966	1352	1778	1222	1607	1104	1450
	20	1618	2120	1471	1927	1331	1742	1204	1573	1087	1419
	22	1568	2031	1425	1845	1289	1666	1165	1504	1051	1356
	24	1515	1938	1377	1758	1244	1587	1124	1431	1014	1289
	26	1460	1840	1326	1668	1198	1504	1081	1355	975	1220
	28	1403	1738	1274	1574	1149	1417	1037	1275	934	1147
	30	1344	1631	1219	1476	1099	1326	991	1192	892	1071
	32	1283	1520	1162	1373	1047	1232	943	1106	848	991
	34	1219	1404	1104	1266	993	1133	893	1015	803	908
	36	1153	1283	1043	1155	936	1030	842	921	755	822
	38	1084	1158	980	1040	878	926	788	827	707	738
	40	1013	1045	914	939	818	836	733	746	656	666

Effective length in ft. KL with respect to least radius of gyration r_y

<table>
<tr><td colspan="2" rowspan="2">F_y = 36 ksi
F_y = 50 ksi</td><td colspan="8"><h1>COLUMNS
W shapes</h1></td></tr>
<tr><td colspan="8">Allowable axial loads in kips</td></tr>
</table>

Desig-nation		W14							
Wt./ft.		193		176		159		145	
F_y		36	50	36	50	36	50	36	50
	0	1227	1704	1119	1554	1009	1401	922	1281
	6	1178	1620	1074	1477	968	1331	884	1217
	7	1167	1603	1064	1461	959	1317	877	1203
	8	1157	1584	1054	1444	950	1301	869	1189
	9	1146	1565	1044	1426	941	1285	860	1174
	10	1134	1545	1034	1407	931	1268	851	1159
	11	1122	1524	1022	1388	921	1250	842	1142
	12	1110	1502	1011	1368	911	1232	832	1125
	13	1097	1479	999	1347	900	1213	822	1108
	14	1083	1455	987	1325	889	1193	812	1090
	15	1069	1431	974	1302	877	1173	801	1071
	16	1055	1406	961	1279	865	1152	790	1051
	17	1040	1380	947	1255	853	1130	779	1031
	18	1025	1353	933	1231	840	1107	767	1011
	19	1010	1326	919	1205	827	1085	755	990
	20	994	1298	904	1179	814	1061	743	968
	22	961	1239	874	1125	786	1012	718	923
	24	927	1178	842	1069	758	960	691	875
	26	891	1113	809	1009	727	906	663	825
	28	853	1046	775	947	696	850	634	773
	30	814	976	739	882	663	791	604	719
	32	774	902	701	815	629	729	573	662
	34	732	826	662	744	594	665	540	603
	36	688	745	622	670	558	598	507	541
	38	643	669	580	601	520	537	472	486
	40	596	604	537	543	480	484	435	438

Effective length in ft. KL with respect to least radius of gyration r_y

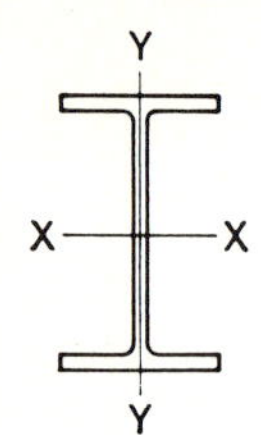

COLUMNS
W shapes

Allowable axial loads in kips

F_y = 36 ksi

F_y = 50 ksi

Desig-nation		W14									
Wt./ft.		132		120		109		99		90	
F_y		36	50	36	50	36	50	36	50[†]	36	50[†]
0		838	1164	762	1059	691	960	629	873	572	795
6		801	1101	729	1002	661	908	600	825	547	751
7		794	1088	722	990	654	897	595	815	541	742
8		786	1074	714	977	647	885	589	805	536	732
9		777	1060	707	963	640	873	582	793	530	722
10		768	1044	699	949	633	860	575	782	524	711
11		759	1028	690	935	626	847	568	769	517	700
12		750	1011	682	919	618	833	561	757	511	689
13		740	994	673	903	609	818	554	743	504	676
14		730	976	663	887	601	803	546	730	497	664
15		719	958	654	870	592	788	538	715	489	651
16		708	938	644	852	583	772	529	701	482	637
17		697	919	633	834	574	755	521	685	474	624
18		686	898	623	815	564	738	512	670	466	609
19		674	877	612	796	554	721	503	654	458	595
20		662	856	601	776	544	703	494	637	449	580
22		637	811	578	735	523	665	475	603	432	548
24		610	764	554	692	501	626	454	567	413	515
26		583	714	528	647	478	585	433	529	394	481
28		554	663	502	599	454	541	411	489	374	444
30		524	608	475	549	429	496	388	448	353	406
32		493	551	446	497	403	449	365	404	331	366
34		461	492	416	443	376	399	340	359	308	325
36		427	439	385	395	348	356	314	320	285	290
38		392	394	353	355	319	320	287	288	260	261

Effective length in ft. KL with respect to least radius of gyration r_y

Column Tables

COLUMNS
W shapes

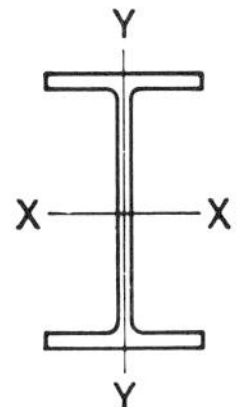

Allowable axial loads in kips

Desig-nation		W14												
Wt./ft.		82		74		68		61		53		48		43
F_y		36	50	36	50	36	50	36	50‡	36	50‡	36	50‡	36‡	50‡
	0	521	723	471	654	432	600	387	537	337	468	305	423	272	377
	6	482	657	436	595	400	545	358	487	302	408	273	369	244	329
	7	474	643	429	581	393	533	351	476	295	395	266	356	237	317
	8	465	627	421	567	385	519	345	464	286	380	258	343	230	305
	9	456	610	412	552	377	505	338	452	277	364	250	329	223	292
	10	446	593	403	536	369	491	330	439	268	348	242	313	215	279
	11	435	575	394	520	360	475	322	425	258	330	233	298	207	264
	12	425	555	384	502	351	459	314	410	248	312	224	281	199	249
	14	402	515	363	465	332	425	297	379	226	273	204	245	181	216
	16	377	471	341	426	311	388	278	346	202	229	182	206	161	181
	18	351	423	317	383	289	348	258	310	177	184	159	165	140	144
	20	323	372	292	337	266	305	237	272	149	149	133	133	117	117
	22	293	318	265	287	241	259	214	230	123	123	110	110	96	96
	24	261	267	236	241	214	218	190	193	104	104	93	93	81	81
	26	227	227	206	206	186	186	165	165	88	88	79	79	69	69
	28	196	196	177	177	160	160	142	142	76	76	68	68	60	60
	30	171	171	154	154	139	139	124	124	66	66	59	59	52	52
	31	160	160	145	145	131	131	116	116	62	62	56	56	49	49
	32	150	150	136	136	123	123	109	109	58	58				
	34	133	133	120	120	109	109	96	96						
	36	119	119	107	107	97	97	86	86						
	38	106	106	96	96	87	87	77	77						

Effective length in ft. KL with respect to least radius of gyration r_y

COLUMNS
W shapes

Allowable axial loads in kips

$F_y = 36$ ksi
$F_y = 50$ ksi

Desig-nation		W12									
Wt./ft.		336		305		279		252		230	
F_y		36	50	36	50	36	50	36	50	36	50
Effective length in ft. KL with respect to least radius of gyration r_y	0	2134	2964	1935	2688	1769	2457	1601	2223	1462	2031
	6	2031	2788	1840	2526	1681	2306	1519	2085	1387	1903
	7	2009	2751	1820	2491	1662	2274	1502	2055	1371	1876
	8	1986	2711	1799	2454	1642	2240	1484	2023	1355	1847
	9	1962	2669	1777	2415	1622	2204	1465	1990	1337	1816
	10	1937	2625	1753	2375	1600	2166	1445	1955	1319	1784
	11	1911	2579	1729	2332	1578	2126	1425	1919	1300	1750
	12	1884	2531	1704	2288	1554	2085	1403	1881	1280	1715
	13	1856	2482	1678	2242	1530	2042	1381	1842	1259	1678
	14	1827	2430	1651	2194	1505	1998	1358	1801	1238	1641
	15	1797	2377	1623	2145	1479	1952	1334	1759	1216	1601
	16	1766	2322	1594	2094	1452	1905	1309	1715	1193	1561
	17	1733	2265	1565	2041	1425	1856	1284	1670	1169	1519
	18	1701	2206	1534	1987	1396	1805	1258	1623	1145	1476
	19	1667	2146	1503	1931	1367	1753	1231	1575	1120	1431
	20	1632	2084	1471	1873	1337	1699	1203	1526	1095	1386
	22	1560	1955	1404	1753	1275	1588	1146	1423	1041	1290
	24	1484	1819	1333	1627	1209	1470	1085	1314	985	1190
	26	1404	1675	1260	1494	1141	1346	1022	1200	927	1084
	28	1321	1525	1183	1354	1069	1216	956	1079	866	972
	30	1235	1366	1102	1206	994	1078	887	952	801	855
	32	1144	1205	1018	1061	916	948	815	837	734	751
	34	1050	1067	930	940	834	839	739	742	664	665
	36	951	952	839	839	749	749	661	661	594	594
	38	854	854	753	753	672	672	594	594	533	533
	40	771	771	679	679	606	606	536	536	481	481

Column Tables

<table>
<tr><td colspan="2" rowspan="3">F_y = 36 ksi
F_y = 50 ksi</td><td colspan="13" align="center">COLUMNS
W shapes</td></tr>
<tr><td colspan="13" align="center">Allowable axial loads in kips</td></tr>
</table>

Desig-nation		W12											
Wt./ft.		210		190		170		152		136		120	
F_y		36	50	36	50	36	50	36	50	36	50	36	50
	0	1335	1854	1205	1674	1080	1500	966	1341	862	1197	762	1059
	6	1266	1736	1142	1566	1023	1402	914	1253	815	1117	721	987
	7	1251	1711	1129	1543	1011	1381	903	1233	805	1100	712	972
	8	1236	1684	1115	1518	998	1359	891	1213	795	1082	702	956
	9	1219	1655	1100	1492	984	1335	879	1192	784	1062	692	938
	10	1202	1625	1084	1465	970	1310	866	1169	772	1042	682	920
	11	1185	1594	1068	1437	956	1285	853	1146	760	1021	671	901
	12	1166	1562	1051	1407	940	1258	839	1122	747	999	660	881
	13	1147	1528	1034	1376	924	1230	825	1096	734	976	648	860
	14	1127	1493	1016	1344	908	1200	810	1070	721	952	636	839
	15	1107	1457	997	1311	891	1170	794	1042	707	927	624	817
	16	1086	1419	978	1276	873	1139	778	1014	693	901	611	794
	17	1064	1381	958	1241	855	1107	762	985	678	875	597	770
	18	1042	1341	937	1204	837	1074	745	955	662	848	584	746
	19	1019	1300	916	1167	817	1039	728	924	647	819	569	720
	20	995	1257	894	1128	798	1004	710	892	630	790	555	694
	22	946	1169	849	1047	757	931	673	825	597	730	525	640
	24	894	1076	802	962	714	853	633	754	561	666	493	583
	26	840	977	752	872	668	771	592	680	524	598	460	522
	28	783	874	700	776	621	684	549	601	485	527	425	457
	30	723	766	645	679	571	597	504	524	444	459	388	398
	32	661	673	588	597	519	525	457	461	402	403	349	350
	34	596	596	529	529	465	465	408	408	357	357	310	310
	36	532	532	472	472	415	415	364	364	319	319	277	277
	38	477	477	423	423	372	372	327	327	286	286	248	248
	40	431	431	382	382	336	336	295	295	258	258	224	224

Effective length in ft. KL with respect to least radius of gyration r_y

<table>
<tr><td></td><td></td><td colspan="2">COLUMNS
W shapes</td><td colspan="2">F_y = 36 ksi
F_y = 50 ksi</td></tr>
</table>

Allowable axial loads in kips

Desig-nation		W12											
Wt./ft.		106		96		87		79		72		65	
F_y		36	50	36	50	36	50	36	50	36	50	36	50[†]
	0	674	936	609	846	553	768	501	696	456	633	413	573
	6	637	872	575	788	522	715	473	647	430	589	389	533
	7	629	858	568	775	515	703	467	637	424	579	384	524
	8	620	844	560	762	508	691	460	626	418	569	378	514
	9	611	828	552	748	501	678	453	614	412	558	373	504
	10	602	812	544	733	493	665	446	601	406	547	367	494
	11	593	795	535	718	485	650	439	588	399	535	361	483
	12	583	777	526	701	477	636	431	575	392	522	354	472
	13	572	759	516	685	468	620	423	561	385	509	348	460
	14	561	740	506	667	459	604	415	546	377	496	341	448
	15	550	720	496	649	450	588	407	531	369	482	334	435
	16	539	699	486	630	440	570	398	515	361	468	326	422
	17	527	678	475	611	430	553	389	499	353	453	319	408
	18	514	656	464	591	420	534	379	482	344	438	311	394
	19	502	634	452	570	409	515	370	465	336	422	303	380
	20	489	611	440	549	398	496	360	447	326	406	294	365
	22	462	562	416	505	376	455	339	410	308	372	277	334
	24	433	511	390	458	352	412	317	371	288	336	259	301
	26	404	457	362	408	327	367	294	329	267	297	240	266
	28	372	399	334	356	301	319	270	285	245	258	220	230
	30	340	348	304	310	273	278	245	249	222	225	199	201
	32	305	306	272	273	244	244	219	219	197	197	176	176
	34	271	271	242	242	216	216	194	194	175	175	156	156
	36	241	241	215	215	193	193	173	173	156	156	139	139
	38	217	217	193	193	173	173	155	155	140	140	125	125
	40	196	196	175	175	156	156	140	140	126	126	113	113

Effective length in ft. KL with respect to least radius of gyration r_y

Column Tables

<table>
<tr><td colspan="2">$F_y = 36$ ksi
$F_y = 50$ ksi</td><td colspan="9" align="center"><h2>COLUMNS
W shapes</h2></td><td></td></tr>
</table>

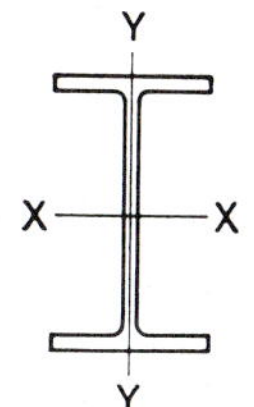

Allowable axial loads in kips

Desig- nation		W12									
Wt./ft.		58		53		50		45		40	
F_y		36	50	36	50	36	50	36	50	36	50‡
Effective length in ft. KL with respect to least radius of gyration r_y	0	367	510	337	468	318	441	285	396	255	354
	6	341	464	312	425	286	386	256	346	229	309
	7	335	454	307	416	279	374	250	335	223	299
	8	329	443	301	406	271	360	243	322	217	288
	9	322	432	295	395	263	346	235	309	210	276
	10	315	420	288	384	254	331	228	296	203	264
	11	308	407	282	372	246	315	220	281	196	251
	12	301	394	275	360	236	298	211	266	188	237
	13	293	380	268	347	226	281	202	250	180	222
	14	285	365	260	333	216	262	193	233	172	207
	15	276	351	252	319	206	243	183	216	163	191
	16	268	335	244	305	195	223	173	197	154	175
	18	249	302	227	274	171	181	152	159	135	141
	20	230	267	209	241	146	146	129	129	114	114
	22	209	229	189	206	121	121	106	106	94	94
	24	187	193	169	173	102	102	89	89	79	79
	26	164	164	147	147	87	87	76	76	67	67
	28	142	142	127	127	75	75	66	66	58	58
	30	123	123	111	111	65	65	57	57	51	51
	32	108	108	97	97	57	57	50	50	45	45
	34	96	96	86	86						
	38	77	77	69	69						
	41	66	66	59	59						

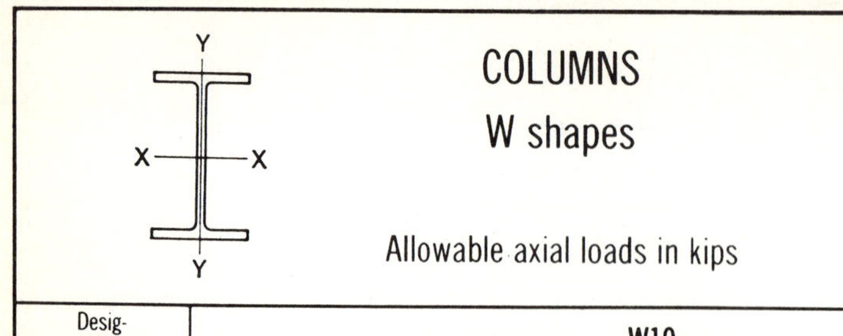

COLUMNS
W shapes

$F_y = 36$ ksi
$F_y = 50$ ksi

Allowable axial loads in kips

Desig-nation		W10								
Wt./ft.		112		100		88		77		68	
F_y		36	50	36	50	36	50	36	50	36	50
	0	711	987	635	882	559	777	488	678	432	600
	6	663	906	592	808	521	712	454	620	402	548
	7	653	888	583	792	513	697	447	607	395	537
	8	642	869	573	775	504	682	439	593	388	525
	9	631	848	562	756	495	665	431	579	381	512
	10	619	827	551	737	485	648	422	564	373	498
	11	606	805	540	717	475	630	413	548	365	484
	12	593	782	528	696	464	611	404	531	357	469
	13	579	757	516	674	453	591	394	513	348	454
	14	565	732	503	651	442	571	384	495	339	437
	15	550	706	489	627	430	550	373	476	330	421
	16	535	679	476	602	417	528	362	457	320	403
	17	519	651	461	577	405	505	351	437	310	385
	18	503	622	446	550	392	481	339	416	299	366
	19	486	591	431	523	378	457	327	394	289	347
	20	469	560	416	494	364	432	315	371	278	327
	22	433	495	383	435	335	379	289	324	255	285
	24	395	425	348	372	304	323	261	275	230	242
	26	355	362	312	317	271	275	232	234	204	206
	28	313	313	273	273	237	237	202	202	177	177
	30	272	272	238	238	206	206	176	176	155	155
	32	239	239	209	209	181	181	155	155	136	136
	34	212	212	185	185	161	161	137	137	120	120
	36	189	189	165	165	143	143	122	122	107	107
	38	170	170	148	148	129	129	110	110	96	96
	40	153	153	134	134	116	116	99	99	87	87

Effective length in ft. KL with respect to least radius of gyration r_y

COLUMNS
W shapes

Allowable axial loads in kips

Desig-nation		W10											
Wt./ft.		60		54		49		45		39		33	
F_y		36	50	36	50	36	50	36	50	36	50	36	50
Effective length in ft. KL with respect to least radius of gyration r_y	0	380	528	341	474	311	432	287	399	248	345	210	291
	6	353	482	317	433	289	394	260	351	224	303	189	255
	7	348	472	312	423	284	385	253	340	218	293	184	246
	8	341	461	306	414	279	376	247	328	213	283	179	237
	9	335	450	300	403	273	367	240	316	206	272	173	228
	10	328	437	294	392	268	357	232	303	200	260	167	217
	11	321	425	288	381	262	346	224	289	193	248	161	207
	12	313	412	281	369	256	335	216	274	186	235	155	196
	13	306	398	274	356	249	324	208	259	178	221	149	184
	14	297	383	267	343	242	312	199	243	170	207	142	171
	15	289	368	259	330	235	299	190	227	162	193	135	159
	16	280	353	251	316	228	286	180	209	154	177	127	145
	17	271	337	243	301	221	273	170	191	145	161	120	131
	18	262	320	235	286	213	259	160	172	136	144	112	117
	19	253	303	226	271	205	245	149	154	126	130	103	105
	20	243	285	217	255	197	230	138	139	116	117	95	95
	22	222	248	199	221	180	198	115	115	97	97	78	78
	24	201	209	179	186	161	167	97	97	81	81	66	66
	26	177	178	158	159	142	143	82	82	69	69	56	56
	28	154	154	137	137	123	123	71	71	60	60	48	48
	30	134	134	119	119	107	107	62	62	52	52	42	42
	32	118	118	105	105	94	94	54	54	46	46	37	37
	33	111	111	99	99	88	88	51	51	43	43		
	34	104	104	93	93	83	83						
	36	93	93	83	83	74	74						

Column Tables

<table>
<tr><td colspan="2"></td><td colspan="2" align="center">COLUMNS
W shapes</td><td align="center">$F_y = 36$ ksi
$F_y = 50$ ksi</td></tr>
</table>

Allowable axial loads in kips

Desig-nation		W8											
Wt./ft.		67		58		48		40		35		31	
F_y		36	50	36	50	36	50	36	50	36	50	36	50
	0	426	591	369	513	305	423	253	351	222	309	197	274
	6	387	525	336	455	276	375	229	310	201	272	178	241
	7	379	510	328	442	270	363	223	300	197	264	174	234
	8	370	494	320	428	263	352	218	290	191	255	170	226
	9	360	477	312	413	256	339	212	279	186	246	165	217
	10	350	459	303	397	249	326	205	268	180	236	160	208
	11	339	440	293	380	241	312	199	256	174	225	154	199
	12	328	420	283	363	233	297	192	244	168	214	149	189
	13	316	399	273	344	224	282	184	231	162	202	143	179
	14	304	378	263	325	215	266	177	217	155	190	137	168
	15	292	355	251	305	206	249	169	203	148	177	131	156
	16	279	331	240	284	196	232	160	188	141	164	124	145
	17	265	307	228	263	186	214	152	172	133	150	117	132
	18	251	281	216	240	176	195	143	156	125	136	110	119
	19	236	254	203	217	165	175	134	140	117	122	103	107
	20	221	230	190	196	154	158	124	126	109	110	95	97
	22	190	190	162	162	131	131	104	104	91	91	80	80
	24	159	159	136	136	110	110	88	88	76	76	67	67
	26	136	136	116	116	94	94	75	75	65	65	57	57
	28	117	117	100	100	81	81	64	64	56	56	49	49
	30	102	102	87	87	70	70	56	56	49	49	43	43
	32	90	90	76	76	62	62	49	49	43	43	38	38
	33	84	84	72	72	58	58	46	46	40	40	35	35
	34	79	79	68	68	55	55	44	44				
	35	75	75	64	64								

Effective length in ft. KL with respect to least radius of gyration r_y

Column Tables

COLUMNS
W shapes

Allowable axial loads in kips

Desig-nation		W8			W6					
Wt./ft.		28		24		25		20		15
F_y		36	50	36	50	36	50	36	50	36†	50†
	0	178	248	153	212	159	220	127	176	96	133
	6	155	208	133	178	136	182	109	145	81	108
	7	150	198	129	170	131	173	105	137	78	102
	8	144	188	124	161	126	163	100	129	75	96
	9	138	178	118	152	120	152	95	121	71	89
	10	132	166	113	142	114	141	90	112	67	82
	11	125	154	107	132	107	129	85	102	62	74
	12	118	142	101	121	100	117	79	92	58	66
	13	111	128	95	109	93	103	73	81	53	57
	14	103	114	88	97	85	90	67	70	48	49
	15	95	100	81	85	77	78	60	61	43	43
	16	87	88	74	74	69	69	54	54	38	38
	17	78	78	66	66	61	61	47	47	33	33
	18	69	69	59	59	54	54	42	42	30	30
	19	62	62	53	53	49	49	38	38	27	27
	20	56	56	48	48	44	44	34	34	24	24
	22	46	46	39	39	36	36	28	28	20	20
	24	39	39	33	33	31	31	24	24	17	17
	25	36	36	30	30	28	28	22	22		
	26	33	33	28	28						
	27	31	31								

Effective length in ft. KL with respect to least radius of gyration r_y

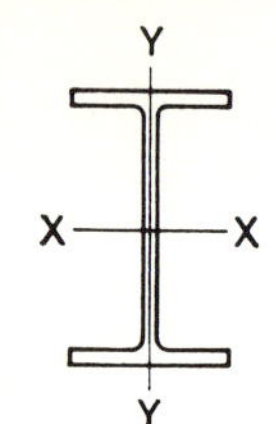

COLUMNS
W shapes

F_y = 36 ksi
F_y = 50 ksi

Allowable axial loads in kips

Designation		W6					W5				W4	
Wt./ft.		16		12		9		19		16		13
F_y		36	50	36	50	36	50	36	50	36	50	36	50
0		102	142	77	107	58	80	120	166	101	140	83	115
2		96	132	72	98	54	74	115	158	97	133	78	107
3		92	124	68	92	51	69	111	152	94	128	75	101
4		87	116	64	85	48	64	107	145	91	122	71	94
5		82	106	60	77	45	58	103	138	87	116	67	87
6		76	95	55	69	41	51	99	129	83	109	62	79
7		69	83	50	59	37	44	93	120	79	101	57	70
8		62	70	44	48	33	36	88	111	74	93	52	60
9		54	57	38	38	28	28	82	100	69	84	46	49
10		46	46	31	31	23	23	76	89	64	75	39	40
11		38	38	26	26	19	19	70	77	58	64	33	33
12		32	32	22	22	16	16	63	65	52	54	28	28
13		27	27	18	18	13	13	55	56	46	46	24	24
14		23	23	16	16	12	12	48	48	40	40	20	20
15		20	20	14	14	10	10	42	42	35	35	18	18
16		18	18					37	37	31	31	16	16
17								33	33	27	27		
18								29	29	24	24		
19								26	26	22	22		
20								24	24	20	20		
21								21	21	18	18		

Effective length in ft. KL with respect to least radius of gyration r_y

Column Tables

<table>
<tr><td rowspan="2"></td><td>$F_y = $ 36 ksi
$F_y = $ 50 ksi</td></tr>
</table>

COLUMNS
M shapes

Allowable axial loads in kips

Desig-nation		M6		M5		M4	
Wt./ft.		20		18.9		13	
F_y		36	50	36	50	36	50
	0	127	177	120	167	82	114
	2	122	168	114	157	77	105
	3	119	163	111	151	74	99
	4	116	157	106	143	70	92
	5	112	150	102	135	65	84
	6	107	142	96	126	60	75
	7	103	134	91	116	54	65
	8	98	125	85	105	48	54
	9	92	115	78	93	42	43
	10	87	105	71	81	35	35
	11	81	94	64	67	29	29
	12	74	83	56	57	24	24
	13	68	71	48	48	21	21
	14	61	61	42	42	18	18
	15	53	53	36	36	15	15
	16	47	47	32	32		
	17	41	41	28	28		
	18	37	37	25	25		
	19	33	33	23	23		
	20	30	30				
	21	27	27				
	22	25	25				
	23	23	23				

Effective length in ft. KL with respect to least radius of gyration r_y

$F_y = 36$ ksi

$F_y = 50$ ksi

Allowable axial loads in kips

Desig-nation		S6				S5				S4				S3			
Wt./ft.		17.25		12.5		14.75		10		9.5		7.7		7.5		5.7	
F_y		36	50	36	50	36	50	36	50	36	50	36	50	36	50	36	50
0		110	152	79	110	94	130	64	88	60	84	49	68	48	66	36	50
2		99	134	72	98	84	113	57	77	53	71	43	58	41	55	31	42
3		92	121	67	89	76	100	52	69	48	62	39	51	36	46	28	35
4		83	105	61	78	68	85	47	59	41	51	34	42	31	36	23	28
5		73	87	54	66	58	67	41	48	34	37	28	32	24	24	18	19
6		61	67	47	52	47	48	34	35	26	26	22	22	17	17	13	13
7		49	49	38	39	35	35	26	26	19	19	16	16	12	12	10	10
8		37	37	30	30	27	27	20	20	15	15	12	12	10	10	7	7
9		30	30	23	23	21	21	16	16	12	12	10	10				
10		24	24	19	19	17	17	13	13								
11		20	20	16	16												

Effective length in ft. KL with respect to least radius of gyration r_y

Column Tables

Properties of Pipe*

*Reprinted with permission from Granet, I., *Fluid Mechanics for Engineering Technology*. Englewood Cliffs, NJ: Prentice-Hall, 1981.

TABLE D1   Properties of Pipe
Schedules, Wall Thicknesses, and Weights

Nominal Pipe Size (in.)	Outside Diameter D (in.)	Wall Thickness t (in.)	Inside Diameter d (in.)	Inside Diameter d^2 (in.^{2})	Inside Diameter d^5 (in.^{5})	Area of Metal (in.^{2})	Internal Cross-Sectional Area (in.^{2})	Internal Cross-Sectional Area (ft^2)*	External Surface (ft^2)	Moment of Inertia (in.^{4})	Weight of Pipe (lb/ft)	Weight of Water per Foot of Pipe (lb/ft)
					Schedule 10							
14 O.D.	14.0	0.250	13.50	182.25	448 403	10.80	143.14	0.994	3.665	255.3	36.71	62.03
16 O.D.	16.0	0.250	15.50	240.25	894 660	12.37	188.69	1.310	4.189	383.7	42.05	81.74
18 O.D.	18.0	0.250	17.50	306.25	1 641 309	13.94	240.53	1.670	4.712	549.1	47.39	104.21
20 O.D.	20.0	0.250	19.50	380.25	2 819 505	15.51	298.65	2.074	5.236	756.4	52.73	129.42
24 O.D.	24.0	0.250	23.50	552.25	7 167 030	18.65	433.74	3.012	6.283	1315.0	63.41	187.95
30 O.D.	30.0	0.312	29.376	862.95	21 875 768	29.10	677.76	4.707	7.854	3206.0	98.93	293.72
					Schedule 20							
8	8.625	0.250	8.125	66.02	35 409	6.57	51.85	0.3601	2.258	57.72	22.36	22.47
10	10.75	0.250	10.25	105.06	113 141	8.24	82.52	0.5731	2.814	113.7	28.04	35.76
12	12.75	0.250	12.25	150.06	275 855	9.82	117.86	0.8185	3.338	191.8	33.38	51.07
14 O.D.	14.0	0.312	13.376	178.92	428 185	13.42	140.52	0.975	3.665	314.4	45.68	60.89
16 O.D.	16.0	0.312	15.376	236.42	859 442	15.38	185.69	1.290	4.189	473.2	52.36	80.50
18 O.D.	18.0	0.312	17.376	301.92	1 583 978	17.34	237.13	1.647	4.712	678.2	59.03	102.77
20 O.D.	20.0	0.375	19.25	370.56	2 643 344	23.12	291.04	2.021	5.236	1113	78.60	125.67
24 O.D.	24.0	0.375	23.25	540.56	6 793 832	27.83	424.56	2.948	6.283	1942	94.62	183.95
30 O.D.	30.0	0.500	29.00	841.0	20 511 149	46.34	660.52	4.587	7.854	5042	157.53	286.23
					Schedule 30							
8	8.625	0.277	8.071	65.14	34 248	7.26	51.16	0.3553	2.258	63.35	24.70	22.17
10	10.75	0.307	10.136	102.74	106 987	10.07	80.69	0.5603	2.814	137.4	34.24	34.96
12	12.75	0.330	12.09	146.17	258 304	12.87	114.80	0.7972	3.338	248.4	43.77	49.74
14 O.D.	14.0	0.375	13.25	175.56	408 394	16.05	137.88	0.9575	3.665	372.8	54.57	59.75
16 O.D.	16.0	0.375	15.25	232.56	824 801	18.41	182.65	1.268	4.189	562.1	62.58	79.12
18 O.D.	18.0	0.437	17.126	293.30	1 473 261	24.11	230.36	1.600	4.712	930.3	82.06	99.84
20 O.D.	20.0	0.500	19.0	361.00	2 476 099	30.63	283.53	1.969	5.236	1457	104.13	122.87
24 O.D.	24.0	0.562	22.876	523.31	6 264 703	41.39	411.00	2.854	6.283	2843	140.80	178.09
30 O.D.	30.0	0.625	28.75	826.56	19 642 160	57.68	649.18	4.508	7.854	6224	196.08	281.30

TABLE D1 (continued)

Nominal Pipe Size (in.)	Outside Diameter D (in.)	Wall Thickness t (in.)	Inside Diameter d (in.)	Inside Diameter d^2 (in.^{2})	Inside Diameter d^5 (in.^{5})	Area of Metal (in.^{2})	Internal Cross-Sectional Area (in.^{2})	(ft^2)*	External Surface (ft^2)	Moment of Inertia (in.^{4})	Weight of Pipe (lb/ft)	Weight of Water per Foot of Pipe (lb/ft)
					Schedule 40							
⅛	0.405	0.068	0.269	0.0724	0.00141	0.072	0.0569	0.00040	0.106	0.001064	0.24	0.0250
¼	0.540	0.088	0.364	0.1325	0.00639	0.125	0.1041	0.00072	0.141	0.003312	0.42	0.0449
⅜	0.675	0.091	0.493	0.2430	0.02912	0.167	0.1909	0.00133	0.177	0.007291	0.57	0.0830
½	0.840	0.109	0.622	0.3369	0.09310	0.250	0.3039	0.00211	0.220	0.01709	0.85	0.1317
¾	1.050	0.113	0.824	0.679	0.3739	0.333	0.5333	0.00371	0.275	0.03704	1.13	0.2315
1	1.315	0.133	1.049	1.130	1.270	0.494	0.8639	0.00600	0.344	0.08734	1.68	0.3744
1¼	1.660	0.140	1.380	1.904	5.005	0.669	1.495	0.01040	0.435	0.1947	2.27	0.6490
1½	1.900	0.145	1.610	2.592	10.82	0.799	2.036	0.01414	0.497	0.3099	2.72	0.8823
2	2.375	0.154	2.067	4.272	37.72	1.075	3.356	0.02330	0.622	0.666	3.65	1.454
2½	2.875	0.203	2.469	6.096	91.75	1.704	4.788	0.03322	0.753	1.530	5.79	2.073
3	3.5	0.216	3.068	9.413	271.8	2.228	7.393	0.05130	0.916	3.017	7.58	3.201
3½	4.0	0.226	3.548	12.59	562.2	2.680	9.888	0.06870	1.047	4.788	9.11	4.287
4	4.5	0.237	4.026	16.21	1 058	3.173	12.73	0.08840	1.178	7.233	10.79	5.516
5	5.563	0.258	5.047	25.47	3 275	4.304	20.01	0.1390	1.456	15.16	14.62	8.674
6	6.625	0.280	6.065	36.78	8 206	5.584	28.89	0.2006	1.734	28.14	18.97	12.52
8	8.625	0.322	7.981	63.70	32 380	8.396	50.03	0.3474	2.258	72.49	25.55	21.68
10	10.75	0.365	10.02	100.4	101 000	11.90	78.85	0.5475	2.814	160.7	40.48	34.16
12	12.75	0.406	11.938	142.5	242 470	15.77	111.93	0.7773	3.338	300.3	53.53	48.50
14 O.D.	14.0	0.437	13.126	172.3	389 638	18.61	135.32	0.9397	3.665	429.1	63.37	58.64
16 O.D.	16.0	0.500	15.000	225.0	759 375	24.35	176.72	1.2272	4.189	731.9	82.77	76.58
18 O.D.	18.0	0.562	16.876	284.8	1 368 820	30.79	223.68	1.5533	4.712	1172	104.75	96.93
20 O.D.	20.0	0.593	18.814	354.0	2 357 244	36.15	278.00	1.9305	5.236	1703	122.91	120.46
24 O.D.	24.0	0.687	22.626	511.9	5 929 784	50.31	402.07	2.7921	6.283	3424	171.17	174.23
					Schedule 60							
8	8.625	0.406	7.813	61.04	29 113	10.48	47.94	0.3329	2.258	88.73	35.64	20.77
10	10.75	0.500	9.75	95.06	88 110	16.10	74.66	0.5185	2.814	212.0	54.74	32.35
12	12.75	0.562	11.626	135.16	212 399	21.52	106.16	0.7372	3.338	400.4	73.16	46.00
14 O.D.	14.0	0.593	12.814	164.20	345 480	24.98	128.96	0.8956	3.665	562.3	84.91	55.86
16 O.D.	16.0	0.656	14.688	215.74	683 618	31.62	169.44	1.1766	4.189	932.4	107.50	73.42

TABLE D1   (continued)

Nominal Pipe Size (in.)	Outside Diameter D (in.)	Wall Thickness t (in.)	Inside Diameter d (in.)	Inside Diameter d^2 (in.^{2})	Inside Diameter d^5 (in.^{5})	Area of Metal (in.^{2})	Internal Cross-Sectional Area (in.^{2})	(ft^2)*	External Surface (ft^2)	Moment of Inertia (in.^{4})	Weight of Pipe (lb/ft)	Weight of Water per Foot of Pipe (lb/ft)
18 O.D. 18.0	18.0	0.750	16.500	272.25	1 222 981	40.64	213.83	1.4849	4.712	1515	138.17	92.80
20 O.D. 20.0	20.0	0.812	18.376	337.68	2 095 342	48.95	265.21	1.8417	5.236	2257	166.40	114.92
24 O.D. 24.0	24.0	0.968	22.064	486.82	5 229 029	70.04	382.35	2.6552	6.283	4654	238.11	165.94
					Schedule 80							
⅛	0.405	0.095	0.215	0.0462	0.000459	0.093	0.0363	0.00025	0.106	0.001216	0.31	0.0157
¼	0.540	0.119	0.302	0.0912	0.002513	0.157	0.0716	0.00050	0.141	0.003766	0.54	0.031
⅜	0.675	0.126	0.423	0.1789	0.01354	0.217	0.1405	0.00098	0.177	0.008619	0.74	0.0609
½	0.840	0.147	0.546	0.2981	0.04852	0.320	0.2341	0.00163	0.220	0.02008	1.09	0.1013
¾	1.050	0.154	0.742	0.5506	0.2249	0.433	0.4324	0.00300	0.275	0.04479	1.47	0.1875
1	1.315	0.179	0.957	0.9158	0.8027	0.639	0.7193	0.00499	0.344	0.1056	2.17	0.3112
1¼	1.660	0.191	1.278	1.633	3.409	0.881	1.283	0.00891	0.435	0.2418	3.00	0.5553
1½	1.900	0.200	1.500	2.250	7.594	1.068	1.767	0.01225	0.498	0.3912	3.63	0.7648
2	2.375	0.218	1.939	3.760	27.41	1.477	2.953	0.02050	0.622	0.8679	5.02	1.279
2½	2.875	0.276	2.323	5.396	67.64	2.254	4.238	0.02942	0.753	1.924	7.66	1.834
3	3.5	0.300	2.900	8.410	205.1	3.016	6.605	0.04587	0.917	3.894	10.25	2.859
3½	4.0	0.318	3.364	11.32	430.8	3.678	8.891	0.06170	1.047	6.280	12.51	3.847
4	4.5	0.337	3.826	14.64	819.8	4.407	11.50	0.07986	1.178	9.610	14.98	4.976
5	5.563	0.375	4.813	23.16	2 583	6.112	18.19	0.1263	1.456	20.67	20.78	7.875
6	6.625	0.432	5.761	33.19	6 346	8.405	26.07	0.1810	1.734	40.49	28.57	11.29
8	8.625	0.500	7.625	58.14	25 775	12.76	45.66	0.3171	2.257	105.7	43.39	19.79
10	10.75	0.593	9.564	91.47	80 020	18.92	71.84	0.4989	2.817	244.8	64.33	31.13
12	12.75	0.687	11.376	129.41	190 523	26.03	101.64	0.7958	3.338	475.1	88.51	44.04
14 O.D. 14.0	14.0	0.750	12.500	156.25	305 176	31.22	122.72	0.8522	3.665	687.3	106.13	53.18
16 O.D. 16.0	16.0	0.843	14.314	204.89	600 904	40.14	160.92	1.1175	4.189	1156	136.46	69.73
18 O.D. 18.0	18.0	0.937	16.125	260.05	1 090 518	50.23	204.24	1.4183	4.712	1833	170.75	88.50
20 O.D. 20.0	20.0	1.031	17.938	321.77	1 857 248	61.44	252.72	1.7550	5.236	2772	208.87	109.51
24 O.D. 24.0	24.0	1.218	21.564	465.01	4 662 798	87.17	365.22	2.5362	6.283	5672	296.36	158.26

Nominal Pipe Size (in.)	Outside Diameter D (in.)	Wall Thickness t (in.)	Inside Diameter d (in.)	Inside Diameter d^2 (in.²)	Inside Diameter d^5 (in.⁵)	Area of Metal (in.²)	Internal Cross-Sectional Area (in.²)	Internal Cross-Sectional Area (ft²)*	External Surface (ft²)	Moment of Inertia (in.⁴)	Weight of Pipe (lb/ft)	Weight of Water per Foot of Pipe (lb/ft)
					Schedule 100							
8	8.625	0.593	7.439	55.34	22 781	14.96	43.46	0.3018	2.258	121.3	50.87	18.83
10	10.75	0.718	9.314	86.75	69 357	22.63	68.13	0.4732	2.814	286.1	76.93	29.53
12	12.75	0.843	11.064	122.41	165 791	31.53	96.14	0.6677	3.338	561.6	107.20	41.66
14 O.D.	14.0	0.937	12.126	147.04	262 173	38.45	115.49	0.8020	3.665	824.4	130.73	50.04
16 O.D.	16.0	1.031	13.938	194.27	526 020	48.48	152.58	1.0596	4.189	1364	164.83	66.12
18 O.D.	18.0	1.156	15.688	246.11	950 250	61.17	193.30	1.3423	4.712	2180	207.96	83.76
20 O.D.	20.0	1.281	17.438	304.08	1 612 393	75.34	238.82	1.6585	5.236	3316	256.10	103.65
24 O.D.	24.0	1.531	20.938	438.40	4 024 179	108.07	344.32	2.3911	6.283	6853	367.40	149.43
					Schedule 120							
4	4.5	0.438	3.625	13.15	626.3	5.578	10.33	0.0717	1.178	11.65	19.01	4.47
5	5.563	0.500	4.563	20.82	1 978	7.953	16.35	0.1136	1.456	25.73	27.04	7.09
6	6.625	0.562	5.501	30.26	5 037	10.705	23.77	0.1650	1.734	49.61	36.39	10.30
8	8.625	0.718	7.189	51.68	19 202	17.84	40.59	0.2819	2.257	140.5	60.63	17.59
10	10.75	0.843	9.064	82.16	61 179	26.24	64.53	0.4481	2.817	324.2	89.20	27.96
12	12.75	1.000	10.750	115.56	143 563	36.91	90.76	0.6303	3.338	641.6	125.49	39.33
14 O.D.	14.0	1.093	11.814	139.57	230 134	44.32	109.62	0.7612	3.665	929.8	150.67	47.57
16 O.D.	16.0	1.218	13.564	183.98	459 133	56.56	144.50	1.0035	4.189	1555	192.29	62.62
18 O.D.	18.0	1.375	15.250	232.56	824 783	71.82	182.65	1.2684	4.712	2499	244.14	79.27
20 O.D.	20.0	1.500	17.000	289.00	1 419 857	87.18	226.98	1.5762	5.236	3754	296.37	98.35
24 O.D.	24.0	1.812	20.376	415.18	3 512 301	126.31	326.08	2.2644	6.283	7827	429.39	141.52
					Schedule 140							
8	8.625	0.812	7.001	49.0	16 819	19.93	38.50	0.2673	2.257	153.7	67.76	16.68
10	10.75	1.000	8.750	76.56	51 291	30.63	60.13	0.4176	2.817	367.8	104.13	26.06
12	12.75	1.125	10.500	110.25	127 628	41.08	86.59	0.6013	3.338	700.5	139.68	37.52
14 O.D.	14.0	1.250	11.500	132.25	201 136	50.07	103.87	0.7213	3.665	1027	170.22	45.01
16 O.D.	16.0	1.438	13.125	172.29	389 670	65.74	135.32	0.9397	4.189	1760	223.50	58.64
18 O.D.	18.0	1.562	14.876	221.30	728 502	80.66	173.80	1.2070	4.712	2749	274.23	75.32

Properties of Pipe

TABLE D1 (continued)

Nominal Pipe Size (in.)	Outside Diameter D (in.)	Wall Thickness t (in.)	Inside Diameter d (in.)	Inside Diameter d^2 (in.^{2})	Inside Diameter d^5 (in.^{5})	Area of Metal (in.^{2})	Internal Cross-Sectional Area (in.^{2})	(ft^2)*	External Surface (ft^2)	Moment of Inertia (in.^{4})	Weight of Pipe (lb/ft)	Weight of Water per Foot of Pipe (lb/ft)
20 O.D.	20.0	1.750	16.500	272.25	1 222 981	100.33	213.82	1.4849	5.236	4216	341.10	92.66
24 O.D.	24.0	2.062	19.876	395.09	3 102 022	142.11	310.28	2.1547	6.283	8625	483.13	134.45
Schedule 160												
½	0.840	0.187	0.466	0.2172	0.002197	0.3836	0.1706	0.00118	0.220	0.02212	1.30	0.074
¾	1.050	0.218	0.614	0.3770	0.08726	0.5698	0.2961	0.00206	0.275	0.05269	1.94	0.130
1	1.315	0.250	0.815	0.6642	0.3596	0.8365	0.5217	0.00362	0.344	0.1251	2.84	0.230
1¼	1.660	0.250	1.160	1.346	2.100	1.107	1.057	0.00734	0.435	0.2839	3.76	0.46
1½	1.900	0.281	1.338	1.790	4.288	1.429	1.406	0.00976	0.498	0.4824	4.86	0.61
2	2.375	0.343	1.689	2.853	13.74	2.190	2.241	0.01556	0.622	1.162	7.44	0.97
2½	2.875	0.375	2.125	4.516	43.33	2.945	3.546	0.02463	0.753	2.353	10.01	1.54
3	3.5	0.438	2.625	6.896	124.9	4.205	5.416	0.03761	0.917	5.032	14.32	2.35
3½	4.0	—	—	—	—	—	—	—	—	—	—	—
4	4.5	0.531	3.438	11.82	480.3	6.621	9.283	0.06447	1.178	13.27	22.51	4.02
5	5.563	0.625	4.313	18.60	1 492	9.696	14.61	0.1015	1.456	30.03	32.96	6.33
6	6.625	0.718	5.189	26.93	3 762	13.32	21.15	0.1469	1.734	58.97	45.30	9.16
8	8.625	0.906	6.813	46.42	14 679	21.97	36.46	0.2532	2.257	165.9	74.69	15.80
10	10.75	1.125	8.500	72.25	44 371	34.02	56.75	0.3941	2.817	399.3	115.65	24.59
12	12.75	1.312	10.126	102.54	106 461	47.14	80.53	0.5592	3.338	781.1	160.27	34.89
14 O.D.	14.0	1.406	11.188	125.17	175 292	55.63	98.31	0.6827	3.665	1117	189.12	42.60
16 O.D.	16.0	1.593	12.814	164.20	345 486	72.10	128.96	0.8955	4.189	1894	245.11	55.97
18 O.D.	18.0	1.781	14.438	208.46	627 412	90.75	163.72	1.1369	4.712	3021	308.51	71.05
20 O.D.	20.0	1.968	16.064	258.05	1 069 699	111.49	202.67	1.4074	5.236	4586	379.01	87.96
24 O.D.	24.0	2.343	19.314	373.03	2 687 570	159.41	292.98	2.0345	6.283	9458	541.94	127.15

*This column also represents the contents in cubic feet per foot of length.

Miscellaneous Tables

COEFFICIENTS OF EXPANSION

The coefficient of linear expansion (α) is the change in length, per unit of length, for a change of 1 degree of temperature. The coefficient of surface expansion is approximately two times the linear coefficient, and the coefficient of volume expansion, for solids, is approximately three times the linear coefficient.

A bar, free to move, will increase in length with an increase in temperature and will decrease in length with a decrease in temperature. The change in length will be $\alpha t l$, where α is the coefficient of linear expansion, t the change in temperature, and l the length. If the ends of a bar are fixed, a change in temperature t will cause a change in the unit stress of $E\alpha t$, and in the total stress of $AE\alpha t$, where A is the cross-sectional area of the bar and E the modulus of elasticity.

Table E1 gives the coefficient of linear expansion for 100°, or 100 times the value indicated above.

Example: A piece of medium steel is exactly 40 ft long at 60 °F. Find the length at 90 °F, assuming the ends are free to move.

$$\text{change in length} = \alpha t l = \frac{0.00065 \times 30 \times 40}{100} = 0.0078 \text{ ft}$$

The length at 90 °F is 40.0078 ft.

Example: A piece of medium steel is exactly 40 ft long, and the ends are fixed. If the temperature increases 30 °F, what is the resulting change in the unit stress?

$$\text{change in unit stress} = E\alpha t = \frac{29\ 000\ 000 \times 0.000\ 65 \times 30}{100} = 5655 \text{ lb/in.}^2$$

TABLE E1   Coefficients of Expansion for 100° = 100α*

Materials	Linear Expansion (°C)	(°F)	Materials	Linear Expansion (°C)	(°F)
Metals and Alloys:			Stone and Masonry:		
Aluminum, wrought	0.002 31	0.001 28	Ashlar masonry	0.000 63	0.000 35
Brass	0.001 88	0.001 04	Brick masonry	0.000 61	0.000 34
Bronze	0.001 81	0.001 01	Cement, portland	0.001 26	0.000 70
Copper	0.001 68	0.000 93	Concrete	0.000 99	0.000 55
Iron, cast, gray	0.001 06	0.000 59	Granite	0.000 80	0.000 44
Iron, wrought	0.001 20	0.000 67	Limestone	0.000 76	0.000 42
Iron, wire	0.001 24	0.000 69	Marble	0.000 81	0.000 45
Lead	0.002 86	0.001 59	Plaster	0.001 66	0.000 92
Magnesium, various alloys	0.0029	0.0016	Rubble masonry	0.000 63	0.000 35
Nickel	0.001 26	0.000 70	Sandstone	0.000 97	0.000 54
Steel, mild	0.001 17	0.000 65	Slate	0.000 80	0.000 44
Steel, stainless, 18-8	0.001 78	0.000 99			
Zinc, rolled	0.003 11	0.001 73			
Timber:			Timber:		
Fir parallel to fiber	0.000 37	0.000 21	Fir perpendicular to fiber	0.0058	0.0032
Maple parallel to fiber	0.000 64	0.000 36	Maple perpendicular to fiber	0.0048	0.0027
Oak parallel to fiber	0.000 49	0.000 27	Oak perpendicular to fiber	0.0054	0.0030
Pine parallel to fiber	0.000 54	0.000 30	Pine perpendicular to fiber	0.0034	0.0019

Expansion of Water
(Maximum density = 1)

°C	Volume	°C	Volume
0	1.000 126	50	1.011 877
4	1.000 000	60	1.016 954
10	1.000 257	70	1.022 384
20	1.001 732	80	1.029 003
30	1.004 234	90	1.035 829
40	1.007 627	100	1.043 116

*Reprinted with permission from AISC, *Manual of Steel Construction,* 8th ed. New York: American Institute of Steel Construction, 1980.

Miscellaneous Tables

 Wire and Sheet Metal Gages*
(in decimals of an inch)

Gage No.	U.S. Standard Gage for Uncoated Hot & Cold Rolled Sheets[b]	Galvanized Sheet Gage for Hot-Dipped Zinc Coated Sheets[b]	USA Steel Wire Gage	Gage No.	U.S. Standard Gage for Uncoated Hot & Cold Rolled Sheets[b]	Galvanized Sheet Gage for Hot-Dipped Zinc Coated Sheets[b]	USA Steel Wire Gage
7/0	—	—	.490	13	.0897	.0934	.092[a]
6/0	—	—	.462[a]	14	.0747	.0785	.080
5/0	—	—	.430[a]	15	.0673	.0710	.072
4/0	—	—	.394[a]	16	.0598	.0635	.062[a]
3/0	—	—	.362[a]	17	.0538	.0575	.054
2/0	—	—	.331	18	.0478	.0516	.048[a]
1/0	—	—	.306	19	.0418	.0456	.041
1	—	—	.283	20	.0359	.0396	.035[a]
2	—	—	.262[a]	21	.0329	.0366	—
3	.2391	—	.244[a]	22	.0299	.0336	—
4	.2242	—	.225[a]	23	.0269	.0306	—
5	.2092	—	.207	24	.0239	.0276	—
6	.1943	—	.192	25	.0209	.0247	—
7	.1793	—	.177	26	.0179	.0217	—
8	.1644	.1681	.162	27	.0164	.0202	—
9	.1495	.1532	.148[a]	28	.0149	.0187	—
10	.1345	.1382	.135	29	—	.0172	—
11	.1196	.1233	.120[a]	30	—	.0157	—
12	.1046	.1084	.106[a]				

[a] Rounded value. The steel wire gage has been taken from ASTM A510 "General Requirements for Wire Rods and Coarse Round Wire, Carbon Steel" Sizes originally quoted to 4 decimal equivalent places have been rounded to 3 decimal places in accordance with rounding procedures of ASTM "Recommended Practice" E29.

[b] The equivalent thicknesses are for information only. The product is commonly specified to decimal thickness, not to gage number.

AISI STANDARD NOMENCLATURE FOR FLAT ROLLED CARBON STEEL

Thickness (Inches)	Width (Inches)					
	To 3½ incl.	Over 3½ To 6	Over 6 To 8	Over 8 To 12	Over 12 To 48	Over 48
0.2300 & thicker	Bar	Bar	Bar	Plate	Plate	Plate
0.2299 to 0.2031	Bar	Bar	Strip	Strip	Sheet	Plate
0.2030 to 0.1800	Strip	Strip	Strip	Strip	Sheet	Plate
0.1799 to 0.0449	Strip	Strip	Strip	Strip	Sheet	Sheet
0.0448 to 0.0344	Strip	Strip				
0.0343 to 0.0255	Strip	Hot rolled sheet and strip not generally produced in these widths and thicknesses				
0.0254 & thinner						

*Reprinted with permission from AISC, *Manual of Steel Construction*, 8th ed. New York: American Institute of Steel Construction, 1980.

Miscellaneous Tables

Index

Roller, 69
Rolling friction, 125
Rolling resistance, coefficient of, 126
Rope, 68
Rotating beam test, 183
Rough surface, 69
Rupture, 182

Safety factor, 384
Scalar, 11
Scleroscope, 184
Screws, 130
SI system, 1
S-N curve (*see* Rotating beam test)
Second moment of area (*see* Moment
    of inertia)
Secondary creep, 182
Section modulus, 312
Section, transformed, 313
Sections, methods of, 101
Shafts
    angle of twist, 229
    combined stresses in, 410
    couplings, 231
    torsion, 224
Shear
    in beams, 273, 320
    horizontal, 323
    longitudinal, 322
    modulus, 146
    in rivets, 242
    in shafts, 224
    strain, 144
    stress, 142, 320, 400
    vertical, 278
Shear diagrams, 281
Shore scleroscope, 184
Slenderness ratio, 380
Slope in beams, 339
Smooth pin, 69
Smooth surface, 68
Space forces, 55
Sphere, 156
Square-threaded screws, 130
Static friction, 114
Statically indeterminate beams, 361
Statically indeterminate stresses, 149

Straight-line formula, 385
Strain
    generalized, 137
    shear, 143
Strength
    breaking, 180
    buckling, 381
    fatigue, 183
    fracture, 180
    of rivets, 242
    rupture, 180
    tensile, 176
    ultimate, 179
    yield, 179
Stress
    allowable, 242, 243, 260
    bearing, 141
    combined, 399
    in cylinders, 156
    engineering, 174
    fatigue, 183
    fiber, 308
    normal, 139
    principal, 402
    in rivets, 242
    shear, 142, 402
    in spheres, 156
    temperature, 151
    tension, 169
    in welds, 239, 257
Stress-strain curve
    in tension, 174, 176
    true, 174
Structures, 93
Structural loads, 137
Superposition method, 352

Tearing, 245
Temperature stress, 151
Tensile strength, effect of temperature
    on, 182
Tensile test, 169
Tension, 94, 138
Tertiary creep, 182
Testing
    charpy, 185
    compression, 181